Materials Forming, Machining and Tribology

Series Editor

J. Paulo Davim, Department of Mechanical Engineering, University of Aveiro, Aveiro, Portugal

This series fosters information exchange and discussion on all aspects of materials forming, machining and tribology. This series focuses on materials forming and machining processes, namely, metal casting, rolling, forging, extrusion, drawing, sheet metal forming, microforming, hydroforming, thermoforming, incremental forming, joining, powder metallurgy and ceramics processing, shaping processes for plastics/composites, traditional machining (turning, drilling, miling, broaching, etc.), non-traditional machining (EDM, ECM, USM, LAM, etc.), grinding and others abrasive processes, hard part machining, high speed machining, high efficiency machining, micro and nanomachining, among others. The formability and machinability of all materials will be considered, including metals, polymers, ceramics, composites, biomaterials, nanomaterials, special materials, etc. The series covers the full range of tribological aspects such as surface integrity, friction and wear, lubrication and multiscale tribology including biomedical systems and manufacturing processes. It also covers modelling and optimization techniques applied in materials forming, machining and tribology. Contributions to this book series are welcome on all subjects of "green" materials forming, machining and tribology. To submit a proposal or request further information, please contact Dr. Mayra Castro, Publishing Editor Applied Sciences, via mayra.castro@springer.com or Professor J. Paulo Davim, Book Series Editor, via pdavim@ua.pt

Yongxian Huang · Yuming Xie · Xiangchen Meng

Friction Stir Welding and Processing

哈爾濱工業大學出版社
HARBIN INSTITUTE OF TECHNOLOGY PRESS

Springer

Yongxian Huang
School of Materials Science
and Engineering
Harbin Institute of Technology
Harbin, China

Yuming Xie
School of Materials Science
and Engineering
Harbin Institute of Technology
Harbin, China

Xiangchen Meng
School of Materials Science
and Engineering
Harbin Institute of Technology
Harbin, China

ISSN 2195-0911 ISSN 2195-092X (electronic)
Materials Forming, Machining and Tribology
ISBN 978-981-99-8687-3 ISBN 978-981-99-8688-0 (eBook)
https://doi.org/10.1007/978-981-99-8688-0

Jointly published with Harbin Institute of Technology Press
The print edition is not for sale in China (Mainland). Customers from China (Mainland) please order the print book from: Harbin Institute of Technology Press.
ISBN of the Co-Publisher's edition: 978-7-5767-0505-8

This Springer imprint is published by the registered company Springer Nature Singapore Pte Ltd.
The registered company address is: 152 Beach Road, #21-01/04 Gateway East, Singapore 189721, Singapore

Preface

The field of friction stir welding and processing is over 30 years, and the idea of a book that could serve as a textbook has been around for several years. As the field grows and the number of engineers, workers, students, and researchers increases, a need has been felt for a book that could provide theoretical and technical guides. That is our motivation for putting this book together, which is written to serve as an initial book for students at senior undergraduate level or graduate level. Engineers and workers getting into the field of friction stir welding and processing may find the book useful as a resource for continuing education. We have purposely kept the book concisely and avoided the urge to make it a lengthy review. This book is certainly not aimed at involved researchers in this field or experts in this area, although they may find it helpful as they teach the subject to others. We do hope that such experts will find the book worth of recommendation. In our desire to finish the book in a timely manner, we have not been able to finish all chapters to the level we had initially wished. We hope and request readers to provide their comments and feedback for future editions.

Chapter 1 briefly introduced the principles and features of friction stir welding and processing, as well as our recently proposed method named deformation-driven metallurgy. The inherent issues of friction stir welding and processing were illustrated. This chapter was edited by Prof. Yongxian Huang, Ph.D.

Chapter 2 introduced the recent progress in the field of self-supported friction stir welding upon the inherent issue of rigid fixture necessity. The related technical developments, microstructures, and mechanical properties were discussed. This chapter was edited by Xiangchen Meng, Ph.D. and Prof. Long Wan, Ph.D.

Chapter 3 introduced the recent progress of non-weld-thinning friction stir welding upon the inherent issue of weld thinning. The related technical developments, microstructures, and mechanical properties were discussed. This chapter was edited by Xiangchen Meng, Ph.D. and Xiaotian Ma.

Chapter 4 introduced the recent progress of friction stir-based remanufacturing in vision of repairing defects. The elimination of inevitable keyholes and other defects in friction stir welding was discussed with emphasis. This chapter was edited by Xiangchen Meng, Ph.D. and Zhiwei Qin.

Chapter 5 introduced principles and basic methodology of numerical simulation in friction stir welding. High depth-to-width ratio friction stir welding was utilized as an exemplary method. Numerical design, joint formation mechanism, and grain growth behavior were discussed. This chapter was edited by Yuming Xie, Ph.D. and Prof. Yongxian Huang, Ph.D.

Chapter 6 introduced entire-process prediction method of microstructure and performance responses in friction stir welding. A combined model based on experimental validation, including computational fluid dynamics model, precipitation evolution model, dynamic recrystallization and recovery model, computation solid mechanics model, and neural network method, was adopted for the prediction of tensile strength. This chapter was edited by Yuming Xie, Ph.D. and Prof. Yongxian Huang, Ph.D.

Chapter 7 introduced recent progress of friction stir processing upon surface modification. Surface composite by direct friction stir processing, cryogenic surface-grinding assisted friction stir processing, and arc surface nitriding assisted friction stir welding were discussed. This chapter was edited by Prof. Yongxian Huang, Ph.D. Wei Wang, and Prof. Long Wan, Ph.D.

Chapter 8 introduced recent progress of friction stir processing upon bulk materials. Mg-based alloys and composites were discussed with emphasis in terms of microstructural evolution and mechanical properties. This chapter was edited by Prof. Yongxian Huang, Ph.D. Huizi Chen, and Prof. Long Wan, Ph.D.

Chapter 9 introduced deformation-driven metallurgy of graphene nanoplatelet-reinforced aluminum matrix composites upon strengthening-toughening demands. Feasibility verification, strength-ductility efficiency amelioration, and grain refinement mechanism were discussed. This chapter was edited by Yuming Xie, Ph.D. and Xiangchen Meng, Ph.D.

Chapter 10 introduced deformation-driven metallurgy of graphene nanoplatelet-reinforced aluminum matrix composites upon anti-corrosion demands. Corrosion-suppression effect and associated heteroatom modification were discussed. This chapter was edited by Yuming Xie, Ph.D. and Xiangchen Meng, Ph.D.

Chapter 11 introduced deformation-driven metallurgy of silicon carbide-reinforced aluminum matrix composites. Strength-ductility synergy was discussed based on optimization of particle sizes and mass fractions. This chapter was edited by Prof. Yongxian Huang, Ph.D. and Dongxin Mao.

Although we have not listed all the students by name, their tireless efforts are very much behind the body of this work. Many of their figures have been cited sometimes without full acknowledgment. Apologies to all those whose work we could not properly cite.

We are very thankful to Harbin Institute of Technology Press for providing opportunity to put this textbook together and pushing us to the finish line.

Finally, none of this would be possible without the support of our families and their sacrifices. So, we close with a great thank you to family members and friends.

Harbin, China
April 2022

Yongxian Huang

Contents

Abbreviations

ABS	Acrylonitrile butadiene styrene
AC-BM	As-cast base metal
AF	Active filling
AGG	Abnormal grain growth
AMC	Aluminum matrix composite
A-PFFSR	Active-passive filling friction stir repairing
AS	Advancing side
BM	Base material
BSE	Back scattered electron
BT-FSW	Bobbin tool friction stir welding
CAE	Computer-aided engineering
CDRX	Continuous dynamic recrystallization
CEL	Coupled Eulerian-Lagrangian
CFD	Computational fluid dynamics
CNT	Carbon nanotube
CPA	Corrosion-promotion activity
CPE	Constant phase element
CSA	Corrosion-suppression activity
CSM	Computational solid mechanics
CT	Computed tomography
DBT-FSW	Dual-rotation bobbin tool friction stir welding
DDM	Deformation-driven metallurgy
DDRX	Discontinuous dynamic recrystallization
DFFSR	Drilling-filling friction stir repairing
DFSP	Direct friction stir processing
DFT	Density functional theory
DRX	Dynamic recrystallization
DSC	Differential scanning calorimetry
EADS	European Aeronautic Defence and Space Company
EBSD	Electron backscattered diffraction
ECAP	Equal-channel angular pressing

EDS	Energy dispersive spectroscopy
EDX	Energy dispersive X-Ray
EIS	Electrochemical impedance spectroscopy
El.	Elongation
EPMA	Electron probe microanalyzer
FFPW	Filling friction plug welding
FFSJ	Friction-based filling stacking joining
FFSW	Filling friction stir welding
FFT	Fast Fourier transformation
F-GNP	Fluorinated graphene nanoplatelets
FIB	Focused ion beam
FPW	Friction plug welding
FSI	Fluid-solid interaction
FSLW	Friction stir lap welding
FSP	Friction stir processing
FSR	Friction stir repairing
FSW	Friction stir welding
FTPW	Friction taper plug welding
FZ	Filling zone
GDRX	Geometric dynamic recrystallization
GGA	Generalized gradient approximation
GNP	Graphene nanoplatelet
GP	Guinier-Preston
HAADF	High-angle annular dark field
HAGB	High-angle grain boundary
HAZ	Heat affected zone
HDP	Heat-dominated phase
HDPE	High-density polyethylene
HEBM	High-energy ball milling
HPT	High-pressure torsion
HR-TEM	High-resolution transmission electron microscope
HSTP	Heat-shear transition phase
HYB	Hybrid metal extrusion and bonding
IF	Interstitial-free
IFFT	Inverse fast Fourier transform
IMC	Intermetallic compound
IPF	Inverse pole figure
JLR	Joint line remnant
KAM	Kernel averaged misorientation
KWN	Wagner Kampmann
LAGB	Low-angle grain boundary
LAMMPS	Large-scale atomic/molecular massively parallel simulator
LCFSW	Level compensation friction stir welding
LPSO	Long-period stacking ordered
LSAZ	Lower shoulder affected zone

LSDZ	Lower shoulder dominated zone
LWNZ	Lower weld nugget zone
MDPE	Medium-density polyethylene
MEAM	Modified embedded atom method
MMC	Metal matrix composite
MPD-FSW	Micro plunge depth friction stir welding
ND	Normal direction
NN	Neural network
NRSA-FSW	Non-rotating shoulder assisted friction stir welding
NWT-FSW	Non-weld-thinning friction stir welding
NZ	Nugget zone
OM	Optical microscope
PD	Process direction
PDP	Potentiodynamic polarization
PE	Polyethylene
PEBsol	Perdew-Burke-Ernzerhof revised for solid
PF	Passive filling
PFSW	Penetrating friction stir welding
PFZ	Precipitation free zone
PID	Proportional-integral-derivative
PL	Pin length
PP	Polypropylene
PRD	Pin root diameter
PRZ	Plug recrystallized zone
PSN	Particle-simulated nucleation
PSZ	Pin stirred zone
PW	Plow width
PWHT	Post-weld heat treatment
PZ	Processed zone
QNZ	Quasi-nugget zone
RE	Rare earth
RS	Retreating side
SAED	Selected area electron diffraction
SAZ	Shoulder affected zone
SCE	Saturated calomel electrode
SCF/PEEK	Short carbon fiber-reinforced polyether ether ketone
SCFSW	Surface compensation friction stir welding
SD	Shoulder diameter
SDP	Shear dominated phase
SDZ	Shoulder-dominated zone
SEM	Scanning electron microscope
SiC/AMC	SiC particles reinforced aluminum matrix composite
SiC_{np}/AMC	Nano-SiC reinforced aluminum matrix composite
$SiC_{\mu p}$/AMC	Micro-SiC reinforced aluminum matrix composite
SMAT	Surface mechanical attrition treatment

SMGT	Surface mechanical grinding treatment
SMM	Sliding mesh model
SPD	Severe plastic deformation
SRAR	Single-roll angular rolling
$SRFSW_1$	Self-reacting friction stir welding
$SRFSW_2$	Self-refilling friction stir welding
SSAZ	Shoulder affected zone
SS-BM	Solid solution base metal
$SSFSW_1$	Self-support friction stir welding
$SSFSW_2$	Stationary shoulder friction stir welding
SS_LBT-FSW	Stationary lower shoulder bobbin tool friction stir welding
SSS	Supersaturated solid solution
SS_UBT-FSW	Stationary upper shoulder bobbin tool friction stir welding
STEM	Scanning transmission electron microscope
SZ	Stir zone
TD	Transverse direction
TDRX	Twin dynamic recrystallization
TEM	Transmission electron microscope
TERS	Tip-enhanced Raman spectroscopy
TIG	Tungsten inert gas
TMAZ	Thermo-mechanically affected zone
TWI	The Welding Institute
UFG	Ultrafine-grained
UGZ	Ultrafine grained zone
UHMW-PE	Ultrahigh molecular weight PE
USAZ	Upper shoulder affected zone
USDZ	Upper shoulder dominated zone
UTS	Ultimate tensile strength
UWNZ	Upper weld nugget zone
VCFSW	Vertical compensation friction plug welding
WBT-FSW	Water cooling bobbin tool friction stir welding
WNZ	Weld nugget zone
XPS	X-ray photoelectron spectroscopy
XRD	X-ray diffraction
YS	Yield strength

Chapter 1
Friction Stir Welding and Processing

1.1 Friction Stir Welding

Friction stir welding (FSW), a mature solid-state welding technique, involves temperature, mechanics, metallurgy and interactions. FSW has been widely applied in many fields, including aerospace, defence, railway, renewable energy and automobile, as shown in Fig. 1.1 [1]. Invented by The Welding Institute (TWI) of the UK in 1991, FSW is thought to be the most revolutionary welding technique due to its advantages of low processing time, low machine/tool consumable costs, low peak temperature, severe plastic deformation and high-quality joints [2–4]. Within the past 31 years of invention and rapid development, FSW has been successfully used to join different series of Al alloys in the automobile and aerospace sectors (2xxx [5–7], 5xxx [8–10], 6xxx [11, 12] and 7xxx [13–16]) and Mg alloys [17–19] that are hard to be welded by conventional fusion welding techniques [20]. FSW has also achieved partial successes in the joining of other metallic materials (Ti alloys [21, 22], Cu alloys [23–25] and steels [26–29]) and non-metallic thermoplastic polymers [30–32]. Moreover, FSW can improve the mechanical interlocking of the welded materials due to severe plastic deformation. Hence, FSW has been used to join many kinds of dissimilar materials according to interfacial reaction systems, as listed in Table 1.1.

FSW has achieved satisfactory results for joining similar and dissimilar materials. With further intensive research and applications, the inherent issues of FSW, which mainly consist of back support, weld thinning and keyhole defects, have attracted substantial attention. These issues are detrimental to the formation and integrity of joints. Currently, corresponding control strategies for the three inherent issues have been proposed and developed. There is a strong need for an insightful review of all these aspects to advance FSW from laboratory research into industrial manufacturing fields. Herein, all aspects of the inherent issues and control strategies, including technical characteristics and development, microstructures, mechanical properties, limitations, and prospects, are reviewed. Moreover, future aims for scientific research and engineering applications are pointed out according to the presented information.

Y. Huang et al., *Friction Stir Welding and Processing*, Materials Forming, Machining and Tribology, https://doi.org/10.1007/978-981-99-8688-0_1

Fig. 1.1 Typical applications of FSW in industrial manufacturing fields

Table 1.1. Classifications of dissimilar materials according to interfacial reaction systems

Interfacial reaction system	Example
Severe interfacial reaction	Al–Mg [33, 34]
Non-interfacial reaction	Mg-steel [35]
Medium-interfacial reaction	Al-steel [36, 37], Al-Ti [38, 39]
New dissimilar materials	Metal-polymer [40, 41]

A schematic diagram of conventional FSW is shown in Fig. 1.2. During conventional FSW, the workpieces to be welded are placed on a backing plate made of harder materials that can hold enough forging force. A rotating tool with a shoulder and a pin plunges to a certain depth in the top surface of the welding workpieces, causing frictional heat and material flow. With the rotation and advancement of the rotating tool, the welding workpieces behind the rotating tool are joined together. Upon completing FSW, the rotating tool is retracted, and then a keyhole defect is left at the end of the weld.

Based on the process characteristics, three inherent issues occur during conventional FSW.

(1) The necessity of strict assembly tools, rigid support and omnidirectional fixation results in welding difficulties and reduces mechanical properties. Moreover, the lack of root penetration easily occurs at the bottom of the joint.

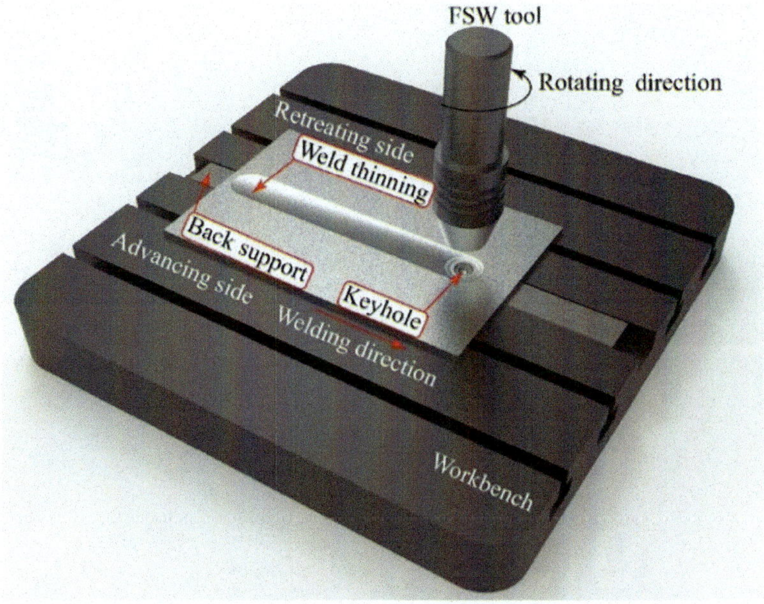

Fig. 1.2 Inherent issues in conventional FSW

(2) Weld thinning induced by a plunge depth of the shoulder is detrimental to joint integrity, which easily results in stress concentrations and fatigue damage.
(3) Keyholes left by the retraction of the pin and other welding defects induced by improper selections of welding tools and parameters appear in the weld, which causes the "bucket effect" due to the small joining area, diminishing mechanical properties.

FSW has been increasingly employed in the joining of large aerospace structures, such as fuel tanks of launch vehicles, space shuttles and space ships [42], as shown in Fig. 1.3a. In 2013, NASA established large-scale welding assembly tools for the rocket fuel tanks of its space launch system, which consisted of a circumferential dome welding tool, a gore welding tool and an enhanced robotic welding tool [43]. The full-scale assembly tools are exceedingly complex. Moreover, the structural design is very large and the manufacturing cost is remarkably high, as shown in Fig. 1.3b–d. Additionally, these welded structural parts always experience complex internal/external pressure and structural torque during their service life, which require highly reliable joints [42]. Conceivably, each of the residual inherent issues may lead to severe accidents during a space launch.

Therefore, there are three key issues that must be solved during conventional FSW: ① reduce the complexity of the fixtures and avoid the lack of root penetration defect; ② increase the effective area of load bearing and eliminate stress concentrations; ③ repair the keyholes and other defects and retain approximately equal strength after remanufacturing. New control strategies have prompted considerable scientific and

Fig. 1.3 Rocket fuel tanks of the SLS at NASA: **a** space launch system on its mobile launcher [44], **b** a liquid oxygen tank for a new rocket at Marshall Space Flight Center [45], **c** testing new FSW techniques before shuttle external tank production began on flight hardware [46] and **d** testing manufacturing processes and tools for the Orion structural test [46]

technological interests due to the difficulties and challenges of ensuring the reliability and structural integrity during conventional FSW. According to causes and characteristics of the three inherent issues, novel FSW techniques have been developed and expanded, as shown in Fig. 1.4. The back support and lack of root penetration defect are solved by self-supported FSW concepts, comprising bobbin tool FSW (BT-FSW), self-support FSW ($SSFSW_1$) and penetrating FSW (PFSW). Weld thinning is eliminated by stationary shoulder FSW ($SSFSW_2$), micro-plunge-depth FSW (MPD-FSW) and additive FSW. Keyholes and other defects are addressed by friction stir-based remanufacturing technologies, such as friction plug welding (FPW), filling FSW (FFSW) and progressive FSW. In the next discussion, the main control strategies for solving three inherent issues are introduced and reviewed in terms of the corresponding concepts, technical developments, key factors, microstructures, mechanical properties, current limitations and prospects.

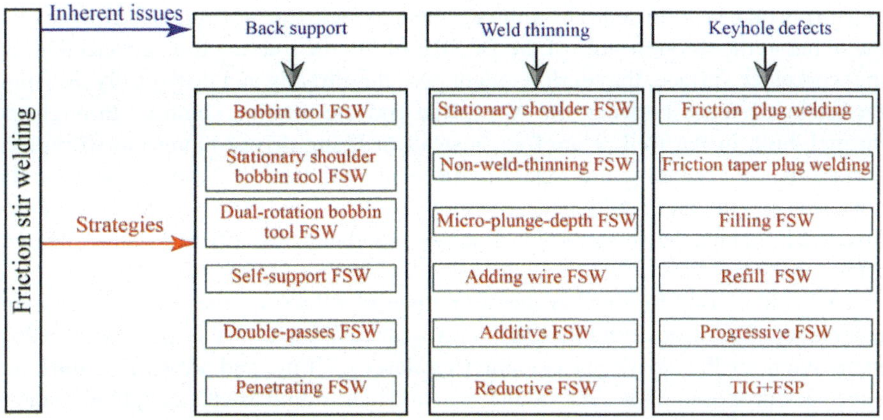

Fig. 1.4 Classifications of inherent issues and their control strategies in the FSW

1.2 Friction Stir Processing

Friction stir processing (FSP), a novel severe plastic deformation technique for materials solid state processing, is originated from FSW [4, 47]. During the FSP process, a rotating FSP tool is inserted into the base metal to modify the local microstructures for the specific and desired properties. The cast alloys, metal matrix composites, and surface composite on the base metal are usually modified or fabricated via FSP. Significant friction heating and intense plastic deformation are generated, which induce the dynamic recrystallization in the stirred zone. The performance of processed materials could be thus promoted with a structure of fine-equiaxed recrystallized grains and homogenously dispersed reinforcement phases. Furthermore, except for the refined microstructure, the processing of composite is carried out at temperature under the melting point of the matrix. The interface reaction between reinforcements and metal matrices is limited, which also does a great contribution to property enhancement.

For different application requirements, FSP technology has been continuously improved and developed. For the preparation of bulk composites or surface composites, the uniform distribution of reinforcements is considered as an important factor. Reinforcements needed to be placed on the base metal before FSP [47]. A groove is usually preset to fill the reinforcements [48–50]. To prevent the reinforcements from being displaced out of the groove during FSP, surface "repair" is accomplished with a modified FSP tool that only had a shoulder and no pin. The particles are confined in the groove after the surface "repair" process [51]. However, these preplaced particles are prone to be driven along the groove by the advancing tool. The remaining particles on the base metal are stirred into the substrate in a gathered way. Therefore, they had difficulty in dispersing well and tend to assemble especially when the size of reinforcement particles is under a certain value. To solve the above problem, a hollow direct friction stir processing (DFSP) tool consisting of a concave shoulder and no pin on it is proposed. The reinforcement particles are not preplaced on the base metal but in the through-hole of the DFSP tool. As the rotating tool advanced along the workpiece surface, the reinforcement particles directly and dispersedly flow into the enclosed space between the workpiece and the rotating shoulder through the through-hole in the tool. Therefore, instead of being driven along the workpiece surface by the advancing tool, these particles are confined in the space between the concave shoulder and workpiece. And they are stirred and pressed into the substrate dispersedly. Thus, they dispersed in a scattered and uniform way in the stir zone (SZ) after only one pass.

FSP is also used to improve the strengthening effect of other surface treatment techniques. To improve the hardness and wear resistance of the lightweight alloy, many methods like spark plasma sintering, laser melting, and arc surface nitriding have been proposed. However, taking arc surface nitriding as an example, the segregation of the nitride forms on the surface of the matrix reduced the hardness and wear properties of the coating [52, 53]. With the subsequent FSP, the segregation of the reinforcement phase could be inhibited and tend to distribute homogenously. The performance of the fabricated coating is thus promoted effectively.

The modified microstructure with refined grains could improve the strength of the materials. However, ductility and toughness frequently drop sharply with the grain size decrease. Combining grains with different sizes is recognized as an effective way to maintain the strengthening advantage of refined grains and reduce the loss of ductility. Based on the FSP tool, an adopted in-situ cryogenic grinding is proposed to get surficial ultra-fine grains, which could form a grained structure and achieve an extraordinary synergy of high strength and uniform tensile ductility.

1.3 Deformation-Driven Metallurgy

The development of light-weight structural applications is essential to fabricate novel materials with high specific strength that can withstand the complex environments encountered in aerospace and automotive industries. Metal matrix composites (MMCs) exhibit great promise and can in principle provide solutions to a wide range of these industrial applications [54–57]. For example, graphene nanoplatelets (GNPs) reinforced MMCs are believed to be state-of-the-art structural materials due to the extraordinary-high strength, large specific surface area and low density of the GNPs [58]. It is highly appreciated that comprehensive mechanical properties but not just one single property can be achieved without dramatic increased costs. However, most traditional and newly developed technologies are frequently accompanied by inherent issues due to limitations such as high price and time costs, reinforcement agglomeration, poor interfacial compatibility or processing difficulties [57, 59, 60].

In light of combining friction stir-based severe plastic deformation techniques and powder metallurgy, we recently proposed a novel composite preparation method named deformation-driven metallurgy (DDM) [61]. During the DDM process (Fig. 1.5), composite powders or green compacts are placed in to a constrained

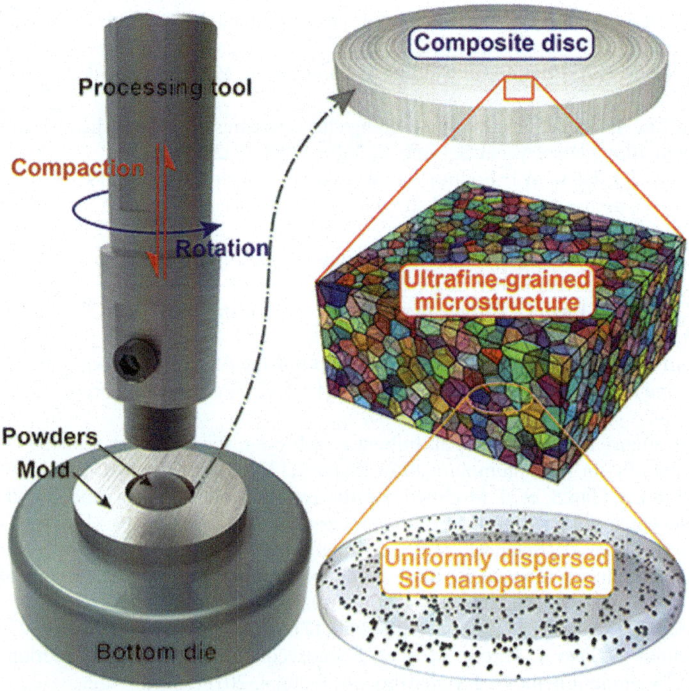

Fig. 1.5 Schematic of the DDM method

mold with a circular hole, and then a cylinder tool is plunged into the hole with high rotational velocity. Severe plastic deformation and corresponding frictional/deformation heat was applied to the composites to form a densified bulk with ultra-fine-grained microstructures and homogeneously dispersed reinforcements. Graphene nanoplatelets and silicon carbides were used as the reinforcements. Heteroatom modification, size selection, and mass fraction of those reinforcements were investigated. We found that the DDM composites showed great potential in the structural and functional applications, including high-strength-ductility materials and anti-corrosion materials. Our results indicate that DDM can be used to achieve the high-performance metal matrix composites and to overcome the disadvantages of other technologies.

References

1. Magalhães VM, Leitão C, Rodrigues DM. Friction stir welding industrialisation and research status. Sci Technol Weld Join. 2018;23(5):400–9.
2. Thomas WM, Nicholas ED, Needham JC, et al. Friction stir butt welding. Frict Weld. 1991:1–19.

3. Ma ZY. Friction stir processing technology: a review. Metall Mater Trans A. 2008;39 A(3):642–58.
4. Mishra RS, Ma ZY. Friction stir welding and processing. Mater Sci Eng R Reports. 2005;50(1–2):1–78.
5. Chen Y, Liu H, Feng J. Friction stir welding characteristics of different heat-treated-state 2219 aluminum alloy plates. Mater Sci Eng A. 2006;420(1–2):21–5.
6. Liu HJ, Li JQ, Duan WJ. Friction stir welding characteristics of 2219–T6 aluminum alloy assisted by external non-rotational shoulder. Int J Adv Manuf Technol. 2013;64(9–12):1685–94.
7. Zhang ZH, Li WY, Feng Y, et al. Global anisotropic response of friction stir welded 2024 aluminum sheets. Acta Mater. 2015;92:117–25.
8. Zhou C, Yang X, Luan G. Fatigue properties of friction stir welds in Al 5083 alloy. Scr Mater. 2005;53(10):1187–91.
9. Lombard H, Hattingh DG, Steuwer A, et al. Optimising FSW process parameters to minimise defects and maximise fatigue life in 5083–H321 aluminium alloy. Eng Fract Mech. 2008;75(3–4):341–54.
10. Hirata T, Oguri T, Hagino H, et al. Influence of friction stir welding parameters on grain size and formability in 5083 aluminum alloy. Mater Sci Eng A. 2007;456(1–2):344–9.
11. Shi L, Wu CS, Gao S, et al. Modified constitutive equation for use in modeling the ultrasonic vibration enhanced friction stir welding process. Scr Mater. 2016;119:21–6.
12. Maggiolini E, Tovo R, Susmel L, et al. Crack path and fracture analysis in FSW of small diameter 6082–T6 aluminium tubes under tension–torsion loading. Int J Fatigue. 2016;92:478–87.
13. Ma ZY, Mishra RS, Mahoney MW. Superplastic deformation behaviour of friction stir processed 7075Al alloy. Acta Mater. 2002;50(17):4419–30.
14. Kawashima T, Sano T, Hirose A, et al. Femtosecond laser peening of friction stir welded 7075–T73 aluminum alloys. J Mater Process Technol. 2018;262:111–22.
15. Bayazid SM, Farhangi H, Asgharzadeh H, et al. Effect of cyclic solution treatment on microstructure and mechanical properties of friction stir welded 7075 Al alloy. Mater Sci Eng A. 2016;649:293–300.
16. Ji SD, Jin YY, Yue YM, et al. Effect of Temperature on material transfer behavior at different stages of friction stir welded 7075–T6 aluminum alloy. J Mater Sci Technol. 2013;29(10):955–60.
17. Wang K, Shen Y, Yang X, et al. Evaluation of microstructure and mechanical property of FSW welded MB3 magnesium alloy. J Iron Steel Res Int. 2006;13(4):75–8.
18. Singh K, Singh G, Singh H. Review on friction stir welding of magnesium alloys. J Magnes Alloy. 2018;6:399–416.
19. Pan F, Xu A, Deng D, et al. Effects of friction stir welding on microstructure and mechanical properties of magnesium alloy Mg-5Al-3Sn. Mater Des. 2016;110:266–74.
20. Oliveira JP, Miranda RM, Braz Fernandes FM. Welding and joining of NiTi shape memory alloys: a review. Prog Mater Sci. 2017;88:412–66.
21. Gangwar K, Ramulu M. Friction stir welding of titanium alloys: a review. Mater Des. 2018;141:230–55.
22. Mironov S, Sato YS, Kokawa H. Friction-stir welding and processing of Ti-6Al-4V titanium alloy: a review. J Mater Sci Technol. 2018;34(1):58–72.
23. Shinde G, Gajghate S, Dabeer PS, et al. Low cost friction stir welding: a review. Mater Today Proc. 2017;4(8):8901–10.
24. Guan W, Shen Y, Yan Y, et al. Fabrication of ultra-thin copper foil pressure welding using FSW equipment. J Mater Process Technol. 2017;2018(251):343–9.
25. Heidarzadeh A, Laleh HM, Gerami H, et al. The origin of different microstructural and strengthening mechanisms of copper and brass in their dissimilar friction stir welded joint. Mater Sci Eng A. 2018;735(August):336–42.
26. Liu FC, Hovanski Y, Miles MP, et al. A review of friction stir welding of steels: tool, material flow, microstructure, and properties. J Mater Sci Technol. 2017;34(1):39–57.

27. El-Batahgy AM, Miura T, Ueji R, et al. Investigation into feasibility of FSW process for welding 1600MPa quenched and tempered steel. Mater Sci Eng A. 2016;651:904–13.
28. Zhang H, Wang D, Xue P, et al. Achieving ultra-high strength friction stir welded joints of high nitrogen stainless steel by forced water cooling. J Mater Sci Technol. 2018;34(11):2183–8.
29. Lee SJ, Sun Y, Fujii H. Stacking-fault energy, mechanical twinning and strain hardening of Fe-18Mn-0.6C-(0, 1.5)Al twinning-induced plasticity steels during friction stir welding. Acta Mater. 2018;148:235–48.
30. Banjare PN, Sahlot P, Arora A. An assisted heating tool design for FSW of thermoplastics. J Mater Process Technol. 2017;239:83–91.
31. Ratanathavorn W. Hybrid joining of aluminum to thermoplastics with friction stir welding engineering. [S.l.]:[s.n.] 2012.
32. Pramanik A, Basak AK, Dong Y, et al. Joining of carbon fibre reinforced polymer (CFRP) composites and aluminium alloys—A review. Compos Part A Appl Sci Manuf. 2017;101:1–29.
33. Shi H, Chen K, Liang Z, et al. Intermetallic compounds in the banded structure and their effect on mechanical properties of Al/Mg dissimilar friction stir welding joints. J Mater Sci Technol. 2017;33(4):359–66.
34. Liu L, Ren D, Liu F. A review of dissimilar welding techniques for magnesium alloys to aluminum alloys. Materials (Basel). 2014;7(5):3735–57.
35. Kasai H, Morisada Y, Fujii H. Dissimilar FSW of immiscible materials: steel/magnesium. Mater Sci Eng A. 2015;624:250–5.
36. Zheng Q, Feng X, Shen Y, et al. Dissimilar friction stir welding of 6061 Al to 316 stainless steel using Zn as a filler metal. J Alloys Compd. 2016;686:693–701.
37. Thomä M, Wagner G, Strass B, et al. Ultrasound enhanced friction stir welding of aluminum and steel: process and properties of EN AW 6061/DC04-joints. J Mater Sci Technol. 2018;34(1):163–72.
38. Huang Y, Lv Z, Wan L, et al. A new method of hybrid friction stir welding assisted by friction surfacing for joining dissimilar Ti/Al alloy. Mater Lett. 2017;207:172–5.
39. Choi JW, Liu H, Fujii H. Dissimilar friction stir welding of pure Ti and pure Al. Mater Sci Eng A. 2018;730(May):168–76.
40. Eslami S, Tavares PJ, Moreira PMGP. Friction stir welding tooling for polymers: review and prospects. Int J Adv Manuf Technol. 2017;89(5–8):1677–90.
41. Wu LH, Nagatsuka K, Nakata K. Achieving superior mechanical properties in friction lap joints of copper to carbon-fiber-reinforced plastic by tool offsetting. J Mater Sci Technol. 2018;34:1628–37.
42. Wang G, Zhao Y, Hao Y. Friction stir welding of high-strength aerospace aluminum alloy and application in rocket tank manufacturing. J Mater Sci Technol. 2018;34(1):73–91.
43. Space launch system: Tooling up to build the world's largest rocket. [2017-08-07]. https://www.nasa.gov/centers/marshall/news/news/releases/2013/13-080.html.
44. Schorr AA, Creech SD, Ogles M, et al. Space launch system spacecraft and payload elements: Making progress toward first launch. AIAA Sp 2016:1–14.
45. Burkey MA. (much) closer look at how we build SLS. [2016-06-02]. https://blogs.nasa.gov/Rocketology/2016/06/02/a-much-closer-look-a-how-we-build-sls/
46. Slotwinski JA, Additive manufacturing: the current state of the art and future potential. [2021-03-04]. https://secwww.jhuapl.edu/techdigest/Content/techdigest/pdf/V35-N04/35-04-Slotwinski.pdf
47. Mishra R, Ma ZY, Charit I. Friction stir processing: a novel technique for fabrication of surface composite. Mater Sci Eng A. 2003;341(1–2):307–10.
48. Morisada Y, Fujii H, Nagaoka T, et al. Effect of friction stir processing with SiC particles on microstructure and hardness of AZ31. Mater Sci Eng A. 2006;433(1–2):50–4.
49. Morisada Y, Fujii H, Nagaoka T, et al. MWCNTs/AZ31 surface composites fabricated by friction stir processing. Mater Sci Eng A. 2006;419(1–2):344–8.
50. Sharifitabar M, Sarani A, Khorshahian S, et al. Fabrication of 5052Al/Al_2O_3 nanoceramic particle reinforced composite via friction stir processing route. Mater Des. 2011;32(8–9):4164–72.

51. Lee C, Huang J, Hsieh P. Mg based nano-composites fabricated by friction stir processing. Scr Mater. 2006;54(7):1415–20.
52. Yu S, Li D, Sun H, et al. Microanalysis of single-phase AlN nanocrystals and AlN-Al nanocomposites prepared by DC arc-discharge. J Cryst Growth. 1998;183(3):284–8.
53. Huashun Y, Kim JD, Kang SB. The formation of AlN and TiN particles during nitrogen bearing gas injection into Al-Mg-Ti melt. Mater Sci Eng A. 2004;386(1–2):318–25.
54. Papageorgiou DG, Kinloch IA, Young RJ. Mechanical properties of graphene and graphene-based nanocomposites. Prog Mater Sci. 2017;90:75–127.
55. Suryanarayana C, Al-Aqeeli N. Mechanically alloyed nanocomposites. Prog Mater Sci. 2013;58(4):383–502.
56. Olakanmi EO, Cochrane RF, Dalgarno KW. A review on selective laser sintering/melting (SLS/SLM) of aluminium alloy powders: processing, microstructure, and properties. Prog Mater Sci. 2015;74:401–77.
57. Azarniya A, Azarniya A, Sovizi S, et al. Physicomechanical properties of spark plasma sintered carbon nanotube-reinforced metal matrix nanocomposites. Prog Mater Sci. 2017;90:276–324.
58. Yu Z, Yang W, Zhou C, et al. Effect of ball milling time on graphene nanosheets reinforced Al6063 composite fabricated by pressure infiltration method. Carbon N Y. 2019;141:25–39.
59. Guo B, Zhang X, Cen X, et al. Enhanced mechanical properties of aluminum based composites reinforced by chemically oxidized carbon nanotubes. Carbon N Y. 2018;139:459–71.
60. Fan G, Jiang Y, Tan Z, et al. Enhanced interfacial bonding and mechanical properties in CNT/Al composites fabricated by flake powder metallurgy. Carbon N Y. 2018;130:333–9.
61. Xie Y, Meng X, Huang Y, et al. Deformation-driven metallurgy of graphene nanoplatelets reinforced aluminum composite for the balance between strength and ductility. Compos Part B Eng. 2019;177:107413.

Chapter 2
Self-Supported Friction Stir Welding

2.1 Introduction

With the rapid development of industrial manufacturing fields, friction stir welding (FSW) has achieved satisfactory results and has been extensively applied in many domains, including aerospace and railway, wherein back support is provided by rigid backing plates. Gantry-type FSW equipment is used for hollow extrusions, tubes and enclosed structural parts, in which the welding workpieces need to be fixed and rigidly supported, increasing the manufacturing cost and welding difficulties. Moreover, the insufficient plunge depth of the rotating pin during conventional FSW results in the lack of root penetration which subsequently diminished mechanical properties. As such, self-supported FSW techniques, including bobbin tool FSW (BT-FSW) and self-support FSW ($SSFSW_1$), were invented to solve these problems, which can decrease the structural rigidity and reduce the difficulties in design and manufacturing. Self-supported FSW can decrease the flexibility of a welding assembly and realize three-dimensional welding of single-curvature or double-curvature structural parts. Based on previous publications, the related technical developments, microstructures and mechanical properties are reviewed in the following chapters.

2.2 Bobbin Tool Friction Stir Welding

2.2.1 Principle

BT-FSW, also referred to as self-reacting FSW ($SRFSW_1$), is a welding technique derived from conventional FSW that consists of an upper shoulder, a lower shoulder and a rotating pin, which has been widely used to join Al alloys [1, 2], Mg alloys [3] and few ferrous alloys [4]. Figure 2.1 shows a schematic of BT-FSW. The BT-FSW tool plunges into the preset hole or traverses along the butt interface between the

Y. Huang et al., *Friction Stir Welding and Processing*, Materials Forming, Machining and Tribology, https://doi.org/10.1007/978-981-99-8688-0_2

Fig. 2.1 Schematic of the BT-FSW technique [5]

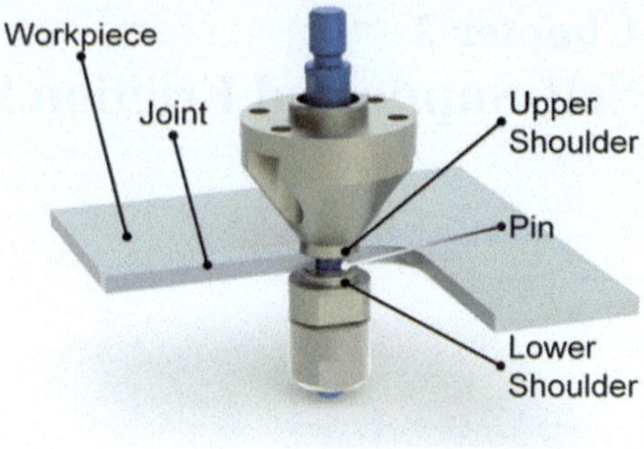

welding workpieces. The upper and lower shoulders closely contact with the upper and lower surfaces of the welding workpieces, while the rotating pin stirs and blends the welding materials, generating frictional heat and material flow. Additionally, the lower shoulder can act as a backing plate, providing forging effects on the plasticized materials at the bottom of the joints. With the rotation and advancement of the welding tool, the plasticized materials flow in both welding workpieces under the action of the friction and extrusion induced by the shoulder and pin, realizing high-quality metallurgical bonding without the support of a backing plate.

Compared with conventional FSW, BT-FSW has many advantages, as detailed hereafter.

(1) The load along the vertical direction of the FSW equipment is reduced, and the rigidity requirements are also decreased.
(2) Rigid back support is avoided, which reduces the complexity of the welding fixtures.
(3) The adaptability of the working condition is broadened, which can realize a variety of welds, such as welds with a tailored thickness and three-dimensional welds.
(4) The lack of root penetration defect is eliminated.
(5) The heat input along the thickness direction is relatively uniform, decreasing the welding distortion and improving the mechanical property homogeneity.

2.2.2 *Technical Development*

BT-FSW technique can be mainly divided into fixed-gap and adjustable-gap or adaptive-gap BT-FSW. With the improvement of the welding equipment and the increase in control accuracy, other novel BT-FSW technologies have been proposed and optimized [3, 6]. The fixed-gap welding tool is the most extensively used during BT-FSW process, for which the welding equipment requirements are relatively low. The design of the welding tool is simply integrated with the ability to adjust the lower shoulder. The main process parameters during BT-FSW are rotational velocity and welding speed, though other variables, such as dwelling time, tool gap, support/clamp settings and plate conditions, may also influence the joint strength. The process conditions also depend upon variations in several factors, such as tool geometry, features, and other process settings. The recommended parameters for BT-FSW of Al alloys are rotational velocities of 450–600 r/min and welding speeds of 75–100 mm/min for thin sheets (4–8 mm), while 170–300 r/min and 100–500 mm/min for thick plates (approximately 25 mm). Table 2.1 lists the functions and effects of different topological parameters of the BT-FSW tools, which can be adjusted according to the different welding requirements, such as the thickness and types of materials to be welded.

Sued et al. [8] stated that the best joint for thin Al alloy plates was produced by tool pins with four facets followed by threaded tool pins with three facets, as displayed in Fig. 2.2. Pirizadeh et al. [9] determined that a convex pin with 7% greater contact area with the welding workpieces resulted in a higher tensile strength efficiency (60.6%) than a simple pin (45.6%) due to higher frictional heat and better mixing of the materials during BT-FSW of thermoplastics. Additionally, the materials of the BT-FSW tools, particularly focusing on refractory alloys, have been further optimized. Thomas et al. [4] designed a bobbin tool consisting of a pin manufactured from refractory alloys with a high tungsten content and shoulders made of refractory alloys with a low tungsten content to weld 8-mm-thick 12% Cr alloyed steel, through which they produced excellent BT-FSW joints.

Table 2.1 Functions and effects of different topological parameters of the BT-FSW tools [7]

Topological parameter	Function	Dimensional effect
Shoulder morphology	Mixing and stirring of materials in the surface of joints	The more suitably complex the shoulder morphology, the better the material flow
Shoulder diameter	Heat generation, material transfer and plastic deformation	The larger the shoulder diameter, the higher the heat generation
Shoulder angle	Entrapment of materials	A larger concave angle improves material accumulation, whereas a larger convex angle results in material overflow
Pin morphology	Mixing and transferring of materials around the pin	The more suitably complex the pin morphology, the better the heat generation and material flow
Pin angle	Movement of materials along the thickness direction	A large angle increases material flow
Pin diameter	Heat generation and plastic flow	A large diameter increases heat generation and decreases material flow
Pin length	Material forging and stirring	The longer the pin length, the stronger the material forging and stirring effects, and the better the defect elimination

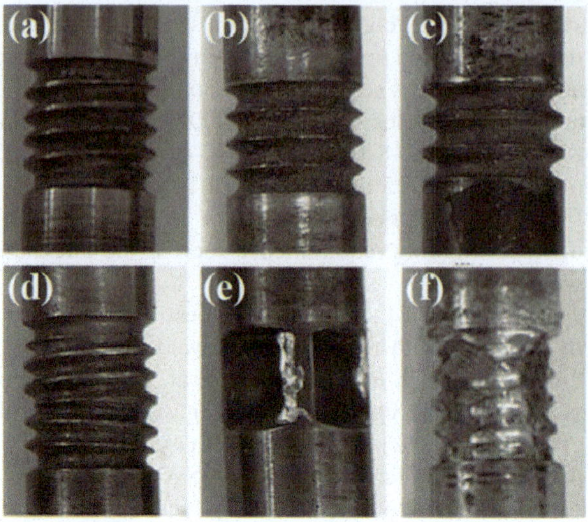

Fig. 2.2 BT-FSW tools: **a** threaded cylindrical pin, **b** cylindrical pin with 1.5-mm-pitch threads, **c** cylindrical pin with 2-mm-pitch threads, **d** cylindrical pin with left-hand and right-hand threads, **e** cylindrical pin with four facets and **f** threaded cylindrical pin with three facets [8]

Fixed-gap BT-FSW has some shortcomings [3, 10]. During BT-FSW, the frictional heat and advancing resistance easily lead to the distortion of the welding workpieces, causing the contact conditions between the welding workpieces and the welding tool cannot be adjusted. As a result, welding defects form at the welds or even lead to the fracture of the rotating pin. To achieve satisfactory results with fixed-gap BT-FSW, many preliminary experiments need to be performed to determine a suitable gap between the upper and lower shoulders.

TWI invented a novel welding mode called floating BT-FSW, in which the gap between the upper and lower shoulders was adjusted via displacement control or pressure control, as shown in Fig. 2.3 [11]. Floating BT-FSW has been successfully used to weld Al alloys [11] and steel [4]. There are few reports on floating BT-FSW due to the relatively complex welding control. Adjustable-gap BT-FSW is usually conducted under displacement control or pressure control modes. This technique can momentarily adjust the gap between the upper and lower shoulders to eliminate the thickness variations induced by thermal expansion. Hence, this control technique can realize BT-FSW of plates with unequal thickness. Figure 2.4 shows a schematic of adjustable-gap BT-FSW, which was proposed by Marshall Space Flight Center and has been used to manufacture airplanes [12].

Counterrotating-shoulder BT-FSW was also proposed by NASA as an alternative to the mechanism and fixtures used in conventional BT-FSW [13], as shown in Fig. 2.5. The mechanism internally induces major or all the forces and torques exerted on the workpieces, simplifying massive external fixtures. By reducing or eliminating (relative to the use of a "self-reacting" tool) the torque that must be externally induced, the counterrotating-shoulder BT-FSW reduces the tendency towards distortion or slippage of the welded workpieces. In addition, dos Santos and Hilgert [14] invented stationary shoulder BT-FSW (SSBT-FSW), for which the aim was to complete the welding process at the lowest peak temperature via selective stationary

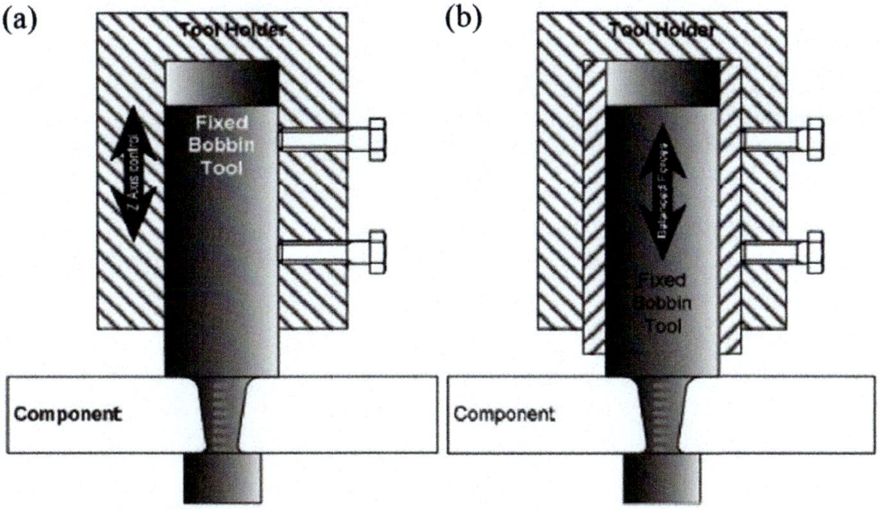

Fig. 2.3 Different kinds of bobbin tools: **a** fixed and **b** floating welding tools [11]

upper and lower shoulders. The stationary conditions for the shoulder in SSBT-FSW mainly consist of stationary upper shoulder BT-FSW (SS_UBT-FSW) [15], stationary lower shoulder BT-FSW (SS_LBT-FSW) or synchronous stationary upper and lower shoulders BT-FSW. Goebel et al. [16] produced a high-quality surface finish on the stationary side and an ultimate tensile strength efficiency of 82% by SS_UBT-FSW of AA2198-T851.

The primary control methods for BT-FSW are displacement control and pressure control. For fixed-gap and adaptive-gap BT-FSW, the control modes present significant differences. Displacement control is primarily employed during fixed-gap BT-FSW. When the coincidence between the welding route and the rotating tool is confirmed, the fixed-gap BT-FSW can be started, which requires few welding parameters. Moreover, this welding procedure is relatively simple. Floating BT-FSW further simplifies the welding control, which can be realized by using the sensor responsible for balancing the axial mechanics of the rotating tool. This sensor is adapted for displacement control and pressure control modes. Both welding control modes require complex control devices, and the feedback information in the upper and lower shoulders as well as the rotating pin need to be precisely collected by the sensor system. Both of adaptive-gap BT-FSW and floating BT-FSW mainly depend on the configuration of the sensor system and its related selection. According to previously published papers, both of welding control modes have certain advantages, whereas deficiencies still exist in terms of welding adaptability. BT-FSW of plates greater than 4 mm in thickness has been widely reported. For BT-FSW of thin-walled structures, especially for sheets thinner than 2 mm, many problems must still be addressed, such as welding distortion that results in an incomplete weld. Therefore, further optimizing the welding control methods plays a significant role in promoting

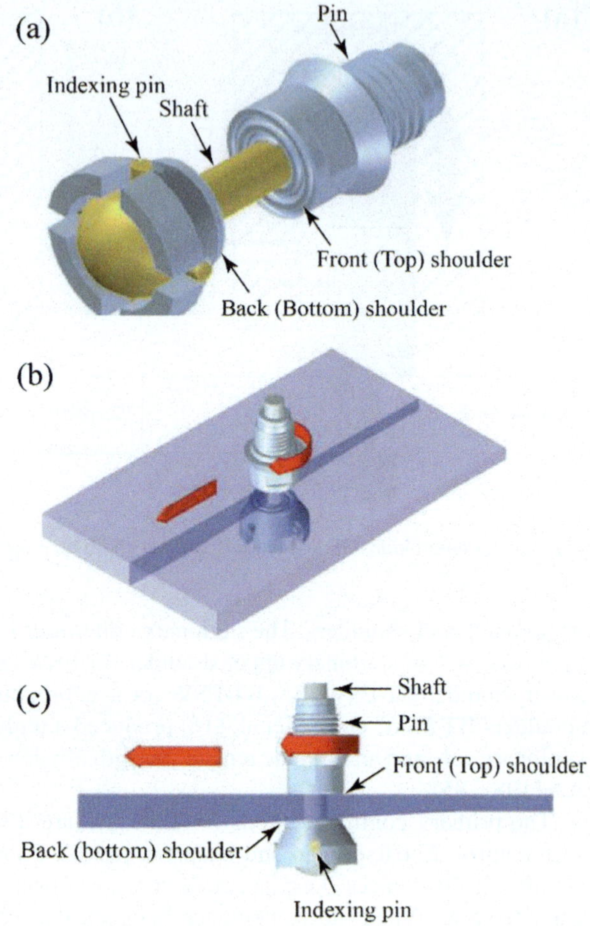

Fig. 2.4 Schematic view of adjustable-gap or adaptive-gap BT-FSW: **a** tool including a gimballed bottom shoulder, **b** tool in use on a tapered-thickness workpiece and **c** enlarged side view of tool in use on a tapered-thickness workpiece [12]

the welding stability of thin-walled structures. Moreover, there are still two challenges in BT-FSW that differ from conventional FSW [17]. One is the considerably increased load on the BT-FSW pin. In particular, the strength and lifetime of the pin is a major limiting factor for this process. The welding stability, especially in the BT-FSW of thin-walled sheets, is the other critical point. The transient temperature and material flow behaviour can lead to instabilities that cause defective welds and tool fracture, as shown in Fig. 2.6a. These instabilities are not observed in conventional FSW, as the transient temperature and material flow are more constrained due to the presence of a backing plate. The unstable material flow easily results in welding defects, as shown in Fig. 2.6b. Moreover, there is notable tool wear, which results from two factors. One source for excessive wear is debris from a fractured upper shoulder ceramic inlay, in which the surface of the pin at the interface with the upper shoulder is considerably damaged (Fig. 2.6c). This phenomenon needs to be considered when using ceramic inlays. The torque transmitter for the lower shoulder

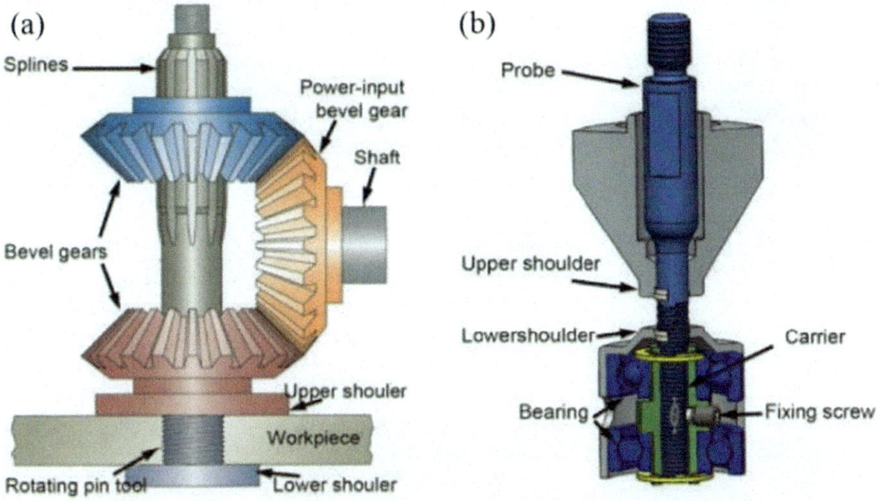

Fig. 2.5 Schematics of **a** counterrotating-shoulder BT-FSW [13] and **b** SSBT-FSW [15, 16]

also shows signs of wear and plastic deformation due to the overloading after a large number of welding experiments, as shown in Fig. 2.6d. Therefore, understanding the joint formation mechanism is very important for high-quality BT-FSW welds.

2.2.3 Formation Mechanism

The joint formation mechanism is the beginning of theoretical research for welding techniques. The intrinsic aspect of BT-FSW is that the original welding interfaces are changed into a new joining interface via plastic flow and transfer of the plasticized materials induced by thermo-mechanical behaviours, which are based on the friction between the rotating tool and the welding workpieces. The BT-FSW process involves complex heat generation and material flow behaviour, which are primarily governed by welding tool geometries, process parameters and properties (thermal and mechanical properties) of base material (BM). Determining the heat generation and material flow mechanism can provide valuable references for welding tool design, parameter selection and technical innovation. The heat generation and material flow behaviour are mainly investigated through experiments and numerical simulations.

Welding heat generation directly influences the interfacial material flow, microstructural evolution, welding defects and so on, thereby affecting the global mechanical properties and residual stress distribution [18]. The heat generation during BT-FSW is primarily influenced by the contact conditions between the rotating

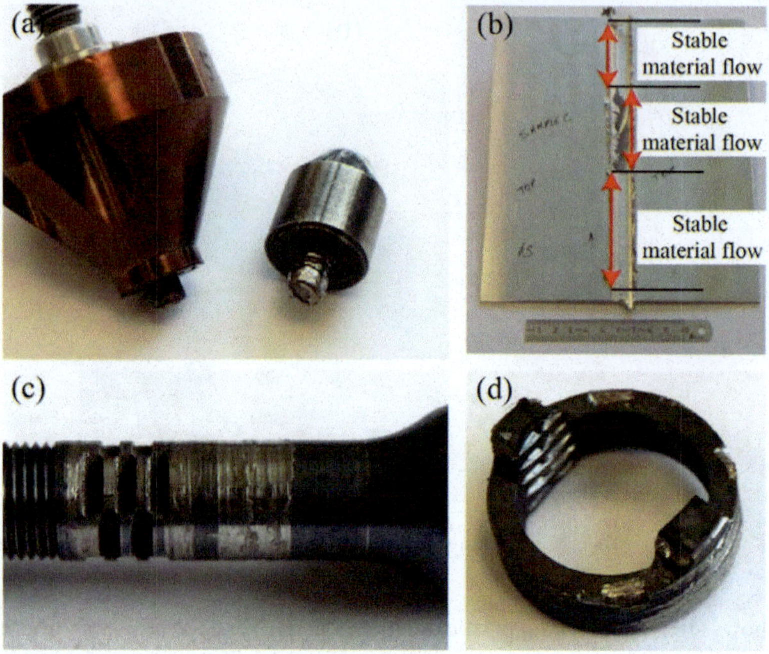

Fig. 2.6 Challenges in BT-FSW: **a** pin fracture, **b** spontaneous flow instability, **c** pin wear due to a fractured inlay and **d** wear of a lower shoulder holder [17]

tool and the welding workpieces [19]. Trueba et al. [1] measured the spatial distribution of welding temperatures adjacent to the rotating tool. They found that the temperatures at the retreating side (RS) tended to be greater than those at the advancing side (AS), as shown in Fig. 2.7. The welding temperatures were also the highest at the beginning and ending of the weld due to the initial dwelling time and the exit from the workpieces, respectively. Chen et al. [6] developed a temperature measurement and control system, containing a closed-loop control system and the Smith predictive proportional-integral-derivative (PID) control method. In their study, the variations at the interface temperature within one rotation of the rotating tool were experimentally measured during BT-FSW for the first time. The characteristic parameters of the BT-FSW process were as followed: at 350 r/min, the steady-state temperature in the nugget zone (NZ) was maintained at 495 °C (the temperature was set to 480 °C). The temperature control error was approximately 15 °C, and the torque was maintained in the range of 48–50 N·m, which further improved the uniformity and stability of the BT-FSW process. The numerical simulation of heat generation during BT-FSW was based on conventional FSW, which simplified the boundary conditions to obtain a model of heat generation based on the constitutive relation of material mechanics.

In an experimental study of conventional FSW, the material tracer technique was used to investigate the material flow behaviour [20, 21]. Guerra et al. [22]

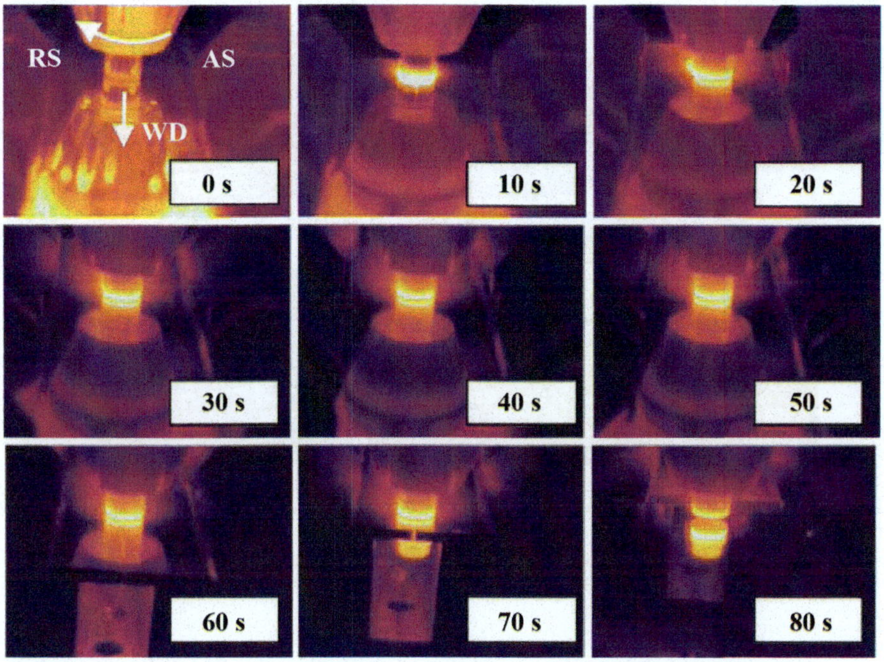

Fig. 2.7 Thermal infrared images during a trial weld [1]. *Note* WD, Welding direction

inserted a copper foil at the faying surface/joint line of butt-welded workpieces—overlapping both dissimilar materials—and examined the vertical flow pattern during welding. The transfer process of plasticized materials can be inferred by observing and analysing the spatial distribution of the tracing materials [20]. Schmidt et al. [23] employed conventional metallography, X-ray and computed tomography (CT) to investigate material flow by introducing a thin copper strip in the workpieces as a marker material, and then estimated the flow velocity for the first time. Tamadon et al. [24] used stacked layers of multicolour plasticine as the tracer materials to investigate the material flow behaviour and formation mechanisms of the defects in the entry and exit, as shown in Fig. 2.8.

For the entry zone, as the bobbin tool enters the workpiece from the free edge of the butt line, the material flow from the RS to the AS is disrupted. The plasticized mass flow makes a spray zone outside of the workpiece at the RS along with a discontinuity line at the AS of the welding line. The multi-layered plasticine analogue model provided visualization of the plastic flow of the materials. This model accurately replicated the localized disruption for the positions of the severe plastic deformation. Although the material tracer technique has helped researchers comprehend the material flow, some limitations remain. Material transfer occurs in the NZ, which cannot be directly detected by common devices. The spatial distribution of the tracer materials during or after FSW is difficult to be quantized point by point to analyse the actual material flow around the rotating pin. There are differences between the

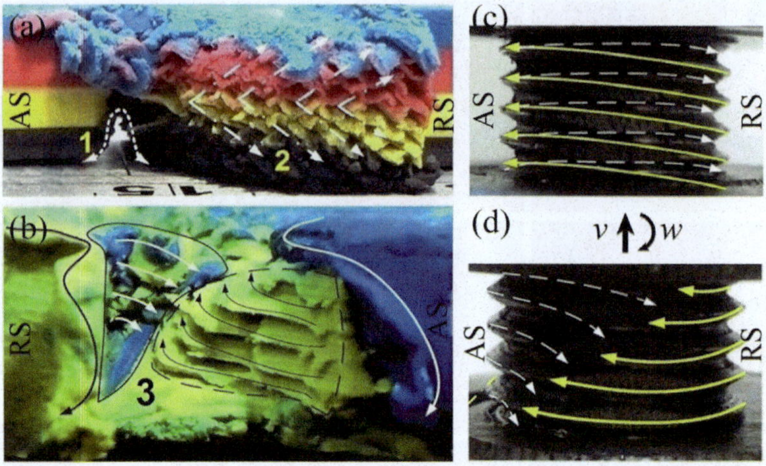

Fig. 2.8 Material flow behaviours: effect of threads **a** in the entry zone and **b** the exit zone replicated in the analogue model; tunnel void, schematics of the shear bands in the entry zone and **d** exit zone; schematics of the shear flow around the pin for the **c** entry zone and exit zone, respectively [24] 1—tunnel void; 2—entry zone; 3—exit zone

tracer materials and the welding workpieces, including the BMs, dimensions and spatial distribution, which all inevitably result in differences between the tracer and actual experiments. Currently, there is still a lack of experimental studies on the joint formation mechanism of BT-FSW; related studies mainly focus on numerical simulations from the viewpoints of heat generation and material flow.

Schmidt and Hattel [25] proposed a new thermal pseudo-mechanical model in which the yield stress of the weld was the driver for heat generation. Hilgert et al. [26] presented a thermal 3D model that used a thermal pseudo-mechanical heat source and included tool rotation, an analytical shear layer model and ambient heat sinks, such as the machine and the atmosphere. The predictions of all models were in excellent agreement with each other and the experiment. Liu et al. [27] showed that the temperature field in the cross-section was symmetric about the mid-thickness of the welded workpieces. The high-temperature zones at the top and bottom surfaces near the shoulders were higher than those at the mid-thickness.

Compared with the numerical simulation of heat generation, the simulation of material flow behaviour is more mature. Hilgert et al. [5, 28] proposed a computational fluid dynamics (CFD) model that treated the materials in the NZ as the highly viscous non-Newtonian shear thinning liquid. A customized parametric solver was used to solve the highly nonlinear Navier–Stokes equations. The contacts between the rotating tool and the welding workpieces were determined by coupling the torque within the CFD model to a thermal pseudo-mechanical model. An existing analytical shear layer model was calibrated using artificial neural networks trained with the predictions of the CFD model, as shown in Fig. 2.9. They also emphasized that the mixed approach was designed for fast predictions only. A deeper insight into the physics of the problem only could be gained through a 3D coupled thermomechanical model. Chen et al. [29] reported that a local "suction-filling" cycle appeared in front of the rotating pin ranging the shoulder radius. The flow field distribution presented a "butterfly" shape in the perpendicular section, while the distance between the butterfly region and surface was approximately 10–15% of the thickness between the upper and lower surfaces of the welding workpieces. Despite the process development, there are still unknown variables in the characterization of the process parameters that can cause uncontrolled welding defects. The entry zone and exit zone contain two discontinuity defects, and removing them is one of the current challenges for improving weld quality. Tamadon et al. [24] stated that numerical models showed that the entry zone and exit zone in BT-FSW were affected by the inhomogeneity of the material flow regime, which caused the ejection or disruption of the plastic flow in the gap between the bobbin shoulders. Fraser et al. [30] used smoothed particle hydrodynamics to simulate the BT-FSW process. This method allowed for easy tracking of the free surface for the welded workpieces, which permitted the visualization of welding defects. This method also showed the dispersion of the mixed materials in both workpieces. Predictions of the defects at the entry and exit regions were obtained via the numerical model (Fig. 2.10).

Based on a coupled Eulerian–Lagrangian model, Wen et al. [31] studied the formation mechanism of the band pattern zone during BT-FSW of AA2219-T87, where welding defects easily formed [32, 33]. As shown in Figs. 2.11 and 2.12, the dynamic

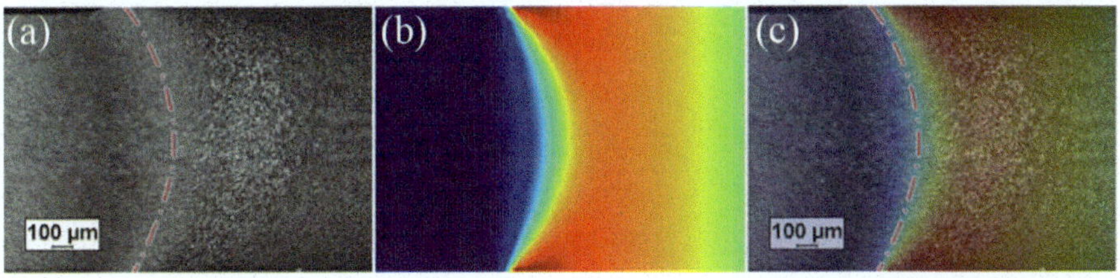

Fig. 2.9 Comparison between the predicted shear layer (red colour indicates high velocity) and microstructure: **a** microstructure at the RS, **b** velocity profile and **c** combination [5]

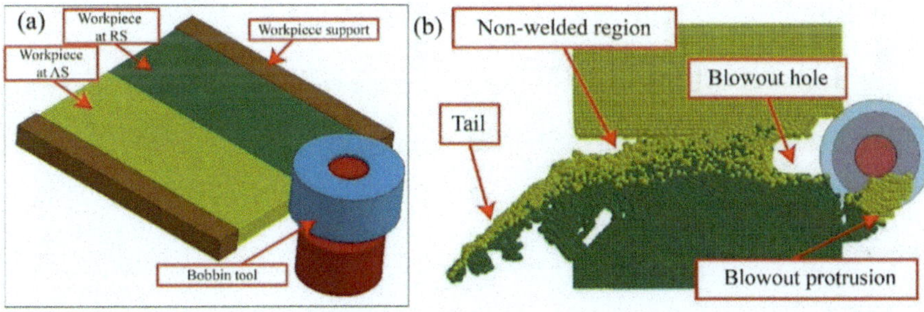

Fig. 2.10 Numerical simulation using smoothed particle hydrodynamics method: **a** model and **b** defect predictions at the entry and exit regions in BT-FSW [30]

forming processes of the band pattern zone in the vertical and horizontal directions were calculated. First, the materials were heated and softened rapidly when the tool was in contact with the workpieces, and then the materials were pushed from the AS to the RS along the thickness direction. Second, two streams of material flow induced by the upper and lower shoulders were formed and mixed with each other during the movement of the rotating tool. Third, the band pattern zone was formed when the two material streams were accumulated. The shoulder function was the main factor affecting this process. In the horizontal direction, plasticized materials adhering to the tool moved from the RS to the AS of the joint, as shown by the arrow, which had higher and lower velocities around the rotating tool, as illustrated in Fig. 2.12. Subsequently, the plasticized materials continued to move towards the AS and were released from the rotating tool to join the materials at the AS.

Numerical simulations of the heat generation, material flow, axial force and torque in BT-FSW are still in progress. Exploring an effective numerical simulation method is a hotspot for BT-FSW research, in which studies focus on the choice of numerical

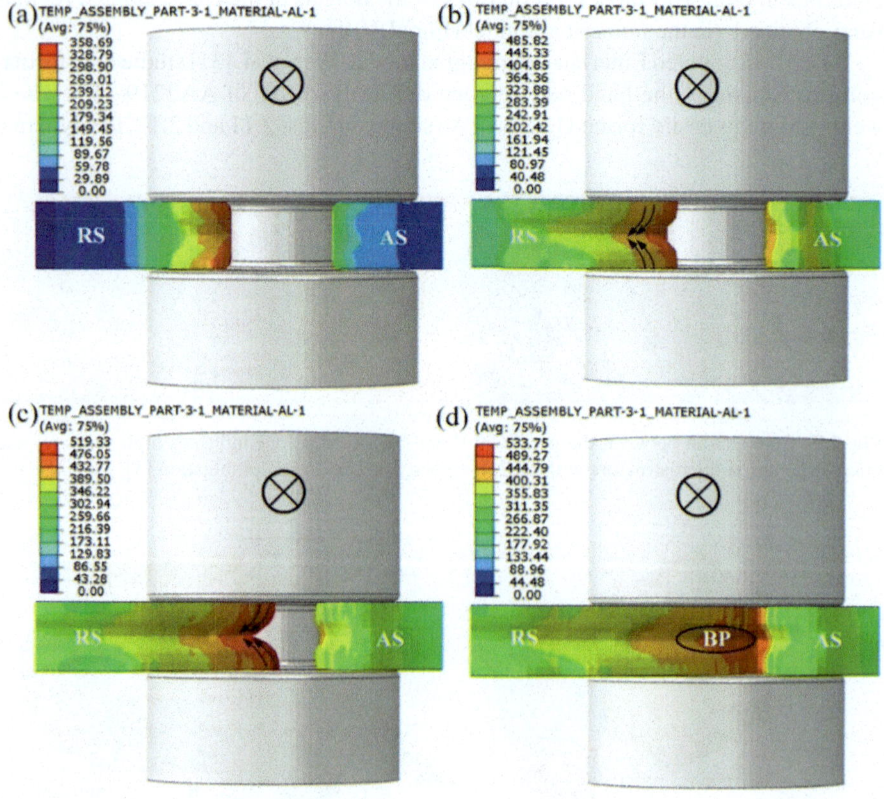

Fig. 2.11 Band pattern formations in the vertical direction at different times: **a** the tool only contacts the workpieces, **b** the tool plunges into the workpieces, **c**)the tool moves away from the workpieces and (d) BT-FSW is completed [31]

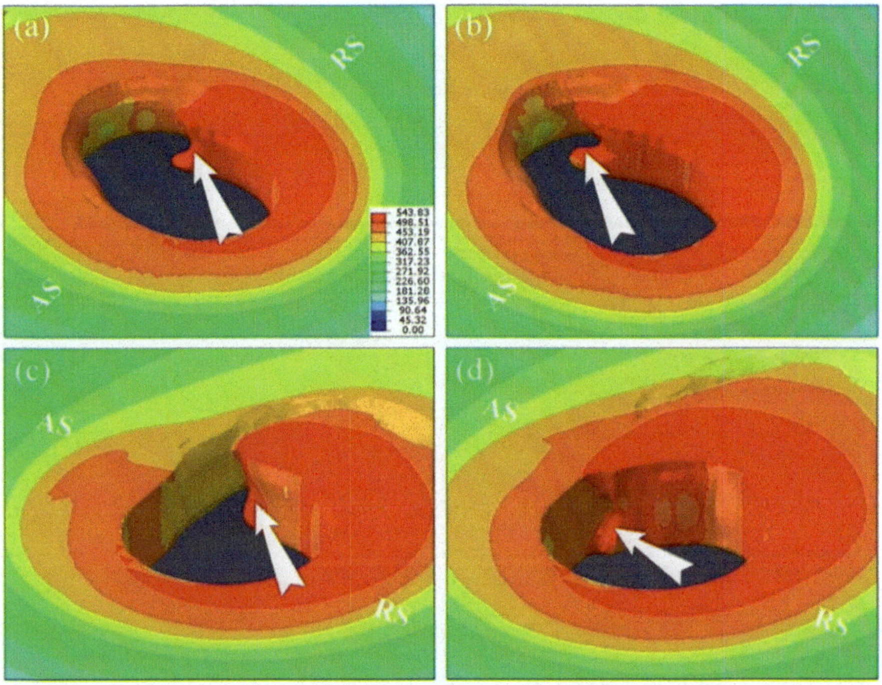

Fig. 2.12 Band pattern formations in the horizontal direction at different times: **a** the tool only contacts the workpieces, **b** the tool plunges into the workpieces, **c** the tool moves away from the workpieces and **d** BT-FSW is completed [31]

methods, the judgement of contact conditions, the selections of material constitutive relations and thermal conditions. This understanding is helpful for parameter optimization, avoiding defect formation and tool breakage.

2.2.4 Microstructural Characteristics

During BT-FSW, severe thermo-mechanical behaviours are attributed to the microstructural evolution of the joint, which can be divided into four regions: NZ, thermo-mechanically affected zone (TMAZ), heat affected zone (HAZ) and BM, as shown in Fig. 2.13. The macrostructure in the cross-section presents a symmetrical deformed dumbbell shape, which is associated with thermo-mechanical effects. In contrast to conventional FSW, the central NZ of a BT-FSW joint presents a “finger” pattern resulting from the stacking of multiple “onion rings” induced by the complex metal flow [1].

Under a large strain rate and high peak temperature, dynamic recrystallization (DRX) occurs, which is attributed to fine and equiaxed grains [34, 35]. The grain

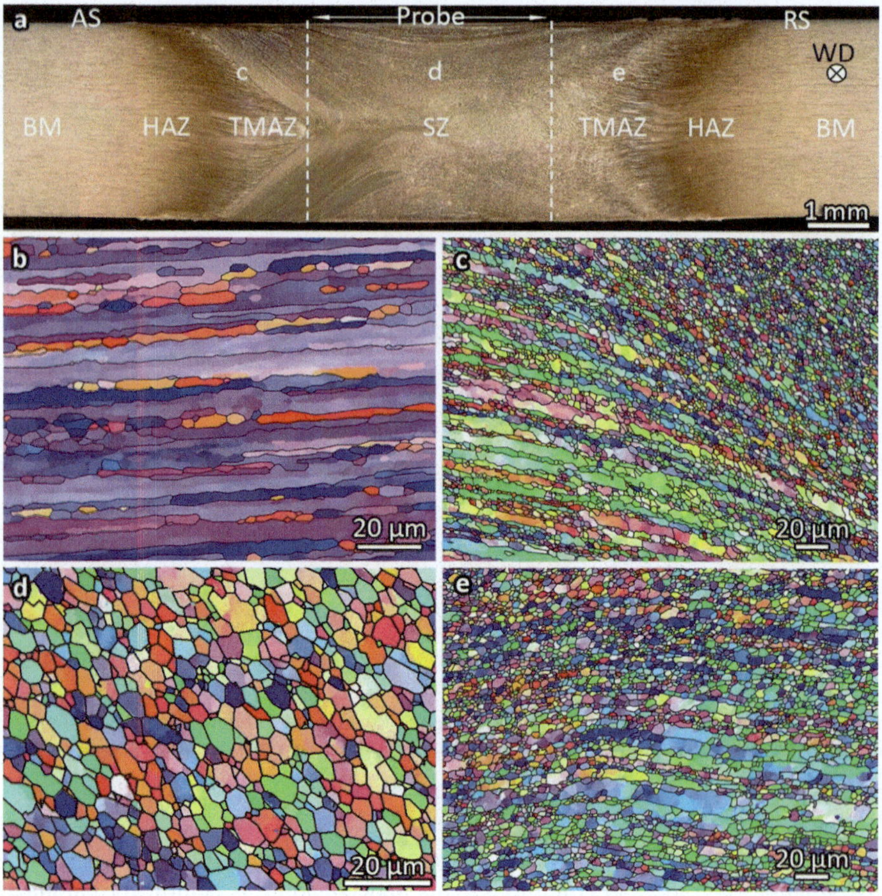

Fig. 2.13 Microstructures in a BT-FSW joint: **a** macrograph showing a typical BT-FSW joint; electron back-scattered diffraction micrographs showing the grain morphologies of **b** BM, **c** TMAZ at the AS, **d** NZ and **e** TMAZ at the RS [39]

sizes in the upper shoulder affected zone (USAZ) and lower shoulder affected zone (LSAZ) are larger than those in the centre of the NZ. The reason for the coarser grains is that the heat generation near the shoulder affected zone (SAZ) is higher than that in the centre of the NZ. Padmanaban and Sarin Sundar [36] found that the grains in the NZ near the shoulders were coarser than the others in FSW joints of AZ31B Mg alloys. Moreover, it is difficult to dissipate heat from the welding tool because the lower shoulder also produces a lot of frictional heat and no heat sink is available during BT-FSW, similar to the backing plate in conventional FSW [37]. The grains in the TMAZ are deformed and elongated because of the plastic deformation caused by the interaction between the rotating tool and the welding workpieces, and the recrystallization of these grains generally does not occur for BT-FSW of Al alloys or conventional FSW of Al and Mg alloys [38].

DRX more easily occurs in Mg alloys than in Al alloys, which tends to equiaxed grains in the TMAZ of BT-FSWed Mg alloys [40–42]. Compared with conventional FSW, BT-FSW more easily generates DRX outside of the NZ due to the additional heat generation provided by the lower shoulder [3]. The HAZ is located between the TMAZ and the BM, wherein the materials only suffer from thermal effects without mechanical stirring, which results in grain coarsening. Esmaily et al. [43] indicated that the faster corrosion process in these regions could be linked to the progression of crevice corrosion. Threadgill et al. [44] found that a high-level substructure appeared in the TMAZ and a low-level substructure was observed in the NZ through electron back-scattered diffraction analysis of AA6082-T6 joints produced by fixed-gap BT-FSW. Xu et al. [45] reported that more inhomogeneous refined grains, high-angle grain boundaries (HAGBs) in the NZ along the thickness direction and a larger percentage of low-angle grain boundaries (LAGBs) were achieved in conventional FSW joints than in BT-FSW joints of AA7085-T7452 due to the frictional heat and strain rate of plastic deformation combined with a certain degree of discontinuous DRX.

For precipitation-strengthened Al alloys, the dissolution, coarsening and reprecipitation of the precipitation phases occur during BT-FSW, which affect the mechanical properties. Figure 2.14 shows bright-field transmission electron microscopy (TEM) micrographs of the BM and NZ. High-density fine needle-shaped precipitates and low-density coarse block-shaped precipitates in the BM were composed of β'' and β' phases, which provided good mechanical properties. First, all metastable phases β'' and β' were diminished in the NZ, but some block-shaped equilibrium phases were randomly distributed in the NZ irrespective of the rotational velocity. This implied that the peak temperature in the NZ had surpassed the dissolution of β' (the dissolution of β' is higher than that of β''). Second, the dislocation densities increased with increasing rotational velocity [37]. The HAZ was characterized by the coarsening and transformation of the β' precipitates. Liu et al. [33] stated that the β'' and β' phases were diminished in the TMAZ of AA6061-T6 BT-FSW joints, which is different from the results reported by Sato et al. [46], who found that β' in the TMAZ region could remain in the matrix because the peak temperature was lower than the solvus temperature of β'. It is postulated that the thermal cycle in the BT-FSW process is more serious than that in conventional FSW. The combined effects were attributed to joint softening and reduced mechanical properties. Welding parameters also play significant role in the microstructural evolution and subsequent corrosion resistance. A fast BT-FSW process could improve the presence of the HAZ and lead to smaller intermetallic compounds (IMCs) particles in the NZ, effectively avoiding more pronounced intergranular corrosion attack caused by the formation of large Cu-rich IMCs in a slow BT-FSW process.

One of the most important characteristics in BT-FSW joints is joint line remnants (JLRs). Esmaily et al. [43, 47] reported that JLRs resulted from the presence of oxide particles, which formed semi-continuous bands from the top to the bottom of the NZ, which were also known as kissing bonds or lazy S, as shown in Fig. 2.15. Kainuma et al. [48] and Zhou et al. [49] reported that the existence of kissing bonds would seriously deteriorate the fatigue properties of joints in both low-cycle and high-cycle

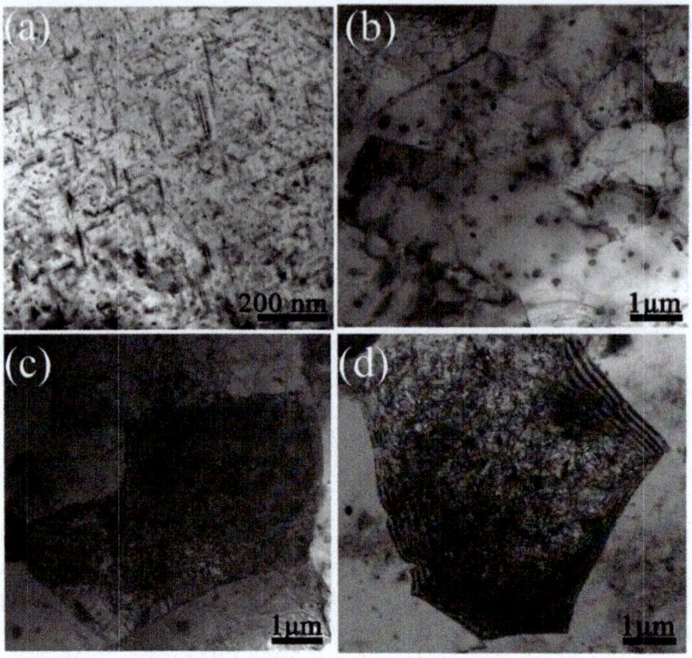

Fig. 2.14 Bright-field TEM micrographs: **a** BM; the NZ at rotational velocities of **b** 400 r/min, **c** 500 r/min and **d** 600 r/min [37]

tests. Moreover, a severe corrosion attack at the location of JLRs was formed in the earlier stages of atmospheric corrosion (≤200 h), and it was suggested that the faster corrosion could be related to the progression of crevice corrosion at the oxide particles/Al matrix and/or the presence of extrusion-induced contamination in these regions [43]. In fact, the formation of the JLRs is related to the local material flow induced by the shoulder-dominated zone (SDZ) and pin-stirred zone (PSZ) [39]. Increasing rotational velocity could evidently enhance the local material flow in the SDZ for thin sheets. However, the rotational velocity has a limited effect on the local material flow in the PSZ for thick sheets. In particular, a cylindrical pin with a small diameter and fewer features was less functional. To eliminate JLRs defects, a reasonable design of welding tools that can improve the local material flow in the NZ is essential.

Wang et al. [50] stated that another typical feature of the BT-FSW joint was void defects formed at the triple junction of the TMAZ, upper shoulder dominated zone (USDZ) and lower shoulder dominated zone (LSDZ) of the AS, which reduced mechanical properties, as shown in Fig. 2.16. The formation of these voids was linked to the accumulation of the symmetrical material flow generated by the upper and lower shoulders at the AS, as shown in Figs. 2.16 (a) and (b). During BT-FSW, the layered structures of the USDZ and LSDZ had the same λ value, and the crests and troughs collided synchronously in the triple junction region (the periodic distance

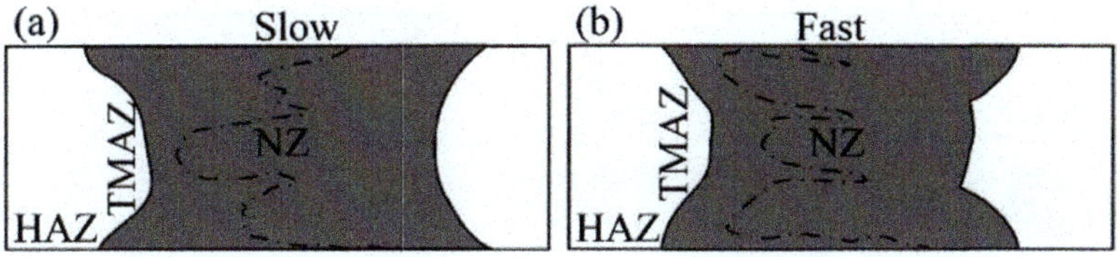

Fig. 2.15 Illustration showing the fraction of JLRs with respect to the area fraction of the NZs: **a** slow BT-FSW and **b** fast BT-FSW [43]

(λ) between each layer corresponded to the ratio of v/ω [51]). Apparently, some large voids were left between the troughs of the USDZ and LSDZ that were hard to be refilled, resulting in the void defects. They developed dual-rotation BT-FSW (DBT-FSW), in which the upper and lower shoulders rotated at certain rotational velocities, to avoid void defects by implementing a staggered layered structure and an unbalanced force between the upper and lower shoulders [50]. In DBT-FSW, the λ value was different for the USDZ and LSDZ. The LSDZ had a smaller λ value than that of the USDZ due to the higher rotational velocity, which indicated that the crests and troughs were staggered between the USDZ and LSDZ, consequently avoiding void defects.

2.2.5 Mechanical Properties

Table 2.2 lists the mechanical properties of BT-FSW joints composed of 2xxx and 6xxx Al alloys, partial Mg alloys and a few polymers. The mechanical properties of the BT-FSW joints are related to joint softening and welding defects. According to published studies, mechanical properties of BT-FSW joints are all lower than those of conventional FSW joints due to severe joint softening induced by both of shoulders. For Al alloys and Mg alloys, the thermo-mechanical behaviors affect the microstructural evolutions, comprising dissolution, coarsening and reprecipitation of the precipitation phases as well as coarsening of the grains, which are attributed to joint softening and result in joint fracture. Li et al. [3] stated that a high welding speed was beneficial to grain refinement, which was attributed to a higher restriction on dislocation motion and localized plastic deformation due to the large numbers of grain boundaries. Yang et al. [52] stated that the slow cooling rate in the HAZ induced by the higher peak temperature during BT-FSW provided sufficient temperature and time for precipitation evolution. Hou et al. [37] reported that the dissolution and coarsen of precipitation phases in the HAZ occurred at higher rotational velocities, thereby decreasing mechanical properties. To reduce the severe joint deterioration induced by the higher welding heat input, Zhao et al. [53] proposed water cooling BT-FSW (WBT-FSW) in which water was sprayed on the surface of the workpieces. Using this approach, both the peak temperature and the high-temperature exposure time were significantly decreased. Hence, this process only induced the coarsening

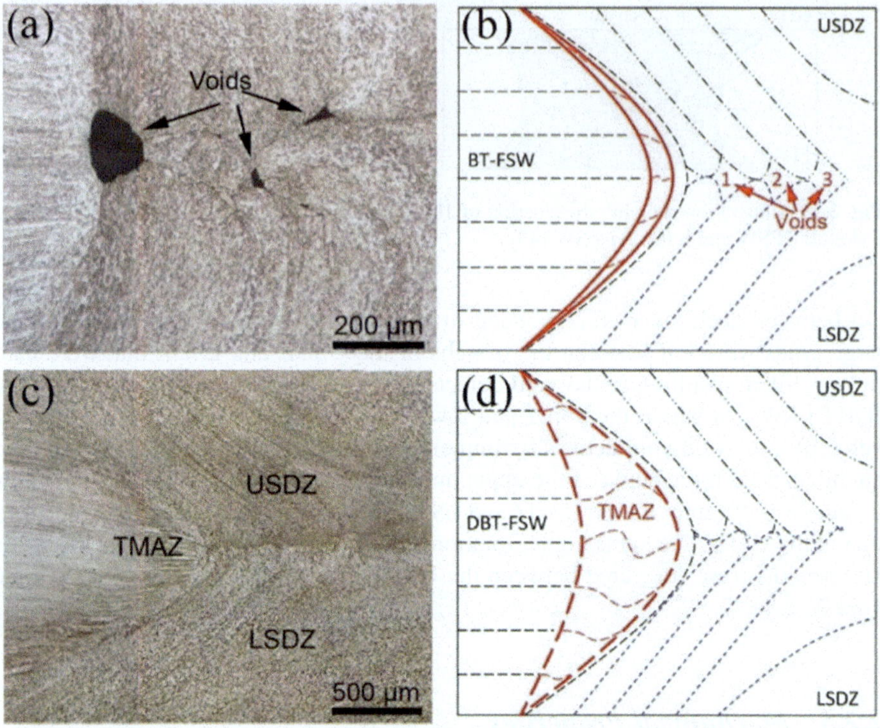

Fig. 2.16 Typical microstructures on the AS and illustrations for material flow: **a**, **b** BT-FSW and **c**, **d** DBT-FSW [50]

of the β'' precipitates but not the dissolution and transformation of the β'' precipitates. Consequently, the tensile strength of the WBT-FSW joint reached 178 MPa, which was 11.4% higher than that of the conventional BT-FSW joint. Moreover, some cooling mediums, such as CO_2 and argon gas, can be explored to improve the properties of BT-FSW joints. Post-weld heat treatment (PWHT) is another effective way to enhance mechanical properties of BT-FSW joints. Dalder et al. [54] heat treated AA2219-T62 BT-FSW joints under the naturally aged T4 condition, which increased the total elongation from 10.5 to 13% and tensile strength from 280 to 305 MPa. For workpieces in the O condition, the welding joints were heat treated to the T62 condition, and ductility increased to 6.6%.

In addition to joint softening, welding defects, the sites of stress concentration, easily become the nucleus of tensile cracks, subsequently decreasing mechanical properties. The common defects in BT-FSW joints are voids, tunnels and lazy S defects, which are generally found close to the centre of the TMAZ or the joining interface. Warsinski et al. [55] verified that the concentration of oxygen along the fracture surface of lazy S was roughly twice that on the fracture surface of the NZ. There are three mechanisms for defect formation: ① volume deficiency, ② inadequate material flow and mixing induced by low heat input [56], and ③ accumulation of

Table 2.2 Mechanical properties of BT-FSW and FSW joints of various materials

Material	Thickness/mm	Tensile strength /MPa			Elongation/%			References
		BM	FSW	BT-FSW	BM	FSW	BT-FSW	
6061-T6	4	284		196	14		11.1	[33]
2195	8	577*	427	468	6.4*		12	[57]
2198-T851	3.2	474		379	12.5		7.5	[39]
6056-T78	4	322	268	257	12	2.1	1.1	[58]
2219-T87		459	321	331	13	7.8	7.1	[2]
2024-T3	4	469		375	18.7		3.58	[59]
6061-T651	8	290		243	10		9.2	[60]
6061-T4	6.35	246	189	229	10.5*			[52]
6082-T6	6	331*		232	12*			[61]
6061-T6	4	284		187	14		10.3	[37]
6082-T6	4	290		198	6		16	[8]
2A12-T6	6	460		345	12		5.1	[32]
6061-T6	5	300		224	13		8.9	[10]
AZ31B-O	2	345		231.9	18		2	[3]
AZ61	5	267		213	8.3		4.8	[35]
6061-T6	5	300		221	13		9.4	[62]
6063-T3	4	260		170	12		10.1	[53]
ABS	5	34.14		20.7	50*			[9]

Note The numbers with "*" have been added by our research team from other literatures. Among them, 577* and 6.4* are from results of Zhang et al. [63]; 10.5* is referenced via the literature [64]; 331* and 12* are from Scialpi et al. [65]; 50* is from the literature [66]

symmetrical material flow induced by the upper and lower shoulders in the centre of the TMAZ [50]. Hou et al. [37] and Pirizadeh et al. [9] stated that void formation due to the overflow of materials at certain high rotational velocities deteriorated the joint strength. Moreover, a high welding speed easily caused insufficient material flow, which further resulted in deficient pin-driven material flow, leading to void formation.

During universal mechanical tests, the integral mechanical properties of the joints were considered in greater detail. The characterization of the local mechanical properties is always neglected. Only Wang et al. [67] investigated the local mechanical properties of the NZ, TMAZ and HAZ via digital image correlation. They reported that the local elongation in the HAZ and TMAZ (13.1%) was higher than the global elongation in the whole joint (5.1%). Moreover, the NZ exhibited a local elongation of 13.8%, whereas the BM exhibited a local elongation of 13.5%, indicating that the BM exhibited better ductility, which was attributed to the heterogeneous deformation induced by the highly inhomogeneous microstructure in the whole joint. During conventional FSW, the tensile properties and the hardness have good correspondence.

For third-generation Al-Li alloys with complex strengthening phases or Mg alloys, in which the texture is strongly dominant, the failure mechanism of the joint is not clear. Moreover, the relationships between the microstructures and mechanical properties need to be investigated systematically.

2.2.6 Prospects

Although BT-FSW has achieved many satisfactory results in laboratory or manufacturing fields, some technical problems still exist.

(1) Severe stress in the welding tool: Because the diameter of the pin is far lower than those of the shoulders, the welding pin easily fractures due to the advancing resistance and the large torque induced by the lower shoulder. The design of welding tools according to different welding requirements needs to be studied.
(2) Higher requirements of welding control. Compared to conventional FSW, BT-FSW has more complex contact conditions and a higher welding difficulty, especially for thin-walled structures or materials with low thermal conductivity.
(3) Preset hole at the beginning and ending stages during the BT-FSW process. Before or after welding, the preset holes need to be fabricated to facilitate the plunge and retraction of the BT-FSW pin. Especially for the enclosed structure, the installation and disassembly of the lower shoulder are more difficult.
(4) High welding heat input. The existence of a lower shoulder causes the total heat input of BT-FSW to be far higher than that of conventional FSW, which results in severe joint softening, thereby deteriorating mechanical properties. Moreover, aiming at thin-walled structures, the higher heat input is prone to cause serious buckling distortion, resulting in bad formation and joint integrity.
(5) Narrow welding process window. The tilt angle in the BT-FSW process is zero; hence, this process cannot provide sufficient forging force, which results in the narrow welding process window and high requirements of welding equipment fixtures.

2.3 Penetrating Friction Stir Welding

To improve surface finishing and eliminate the kissing bond in the root of the weld, Liu et al. [68] proposed penetrating friction stir welding (PFSW), which used stationary-shoulder support and a tilting pin. This method can guarantee a sufficient forging force, which improved joint formation and mechanical properties by increasing the back supported force.

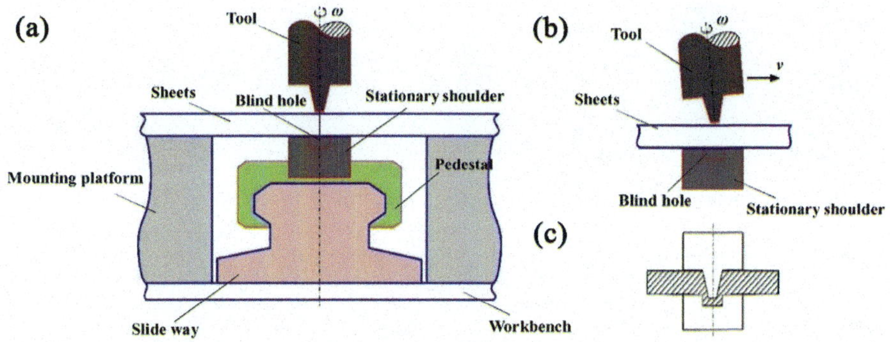

Fig. 2.17 Schematics of the PFSW process: **a** equipment and relative positions, **b** lateral view parallel to the welding direction and **c** local cross-sectional view [68]

2.3.1 Principle

Figure 2.17 shows schematics of the PFSW process, which consists of a slide, a pedestal, a stationary shoulder and a mounting platform including two separate segments. The slide realizes relative motion between the welding workpieces and the stationary shoulder. The plasticized materials under the rotating shoulder are supported by the stationary shoulder. To guarantee the complete penetration of the rotating pin, a blind hole is made in the centre of the stationary shoulder. The two separate parts of the mounting platform are placed beside the slide with the purpose of fixing the sheets. Before the PFSW process, the slide and the mounting platform need to be fastened to the workbench. The relative positions of these components are shown in Fig. 2.17a. The tilt angle of the rotating tool is kept constant, as illustrated in Fig. 2.17b. Under a pseudo-steady state, the cross-sectional view of softened materials surrounding the rotating pin tool is depicted in Fig. 2.17c [68].

2.3.2 Joint Formation

Figure 2.18 depicts the upper and lower surface appearances of a typical PFSW joint, in which no grooves or other visible defects are observed. The upper surface of the PFSW joint is similar to that in conventional FSW, and the width is equal to the diameter of the rotating shoulder. The lower surface presents good surface finishing due to the relative motion between the welding workpieces and the stationary shoulder, which indicates that the plasticized materials in the NZ are successfully supported by the stationary shoulder.

The macrostructure in the cross-section of the PFSW joint differs from that in conventional FSW joint, presenting an hourglass shape that is obviously asymmetric along the thickness direction, as shown in Fig. 2.19. The upper and lower materials in the NZ are subjected to different deformation and heat exposure during the PFSW

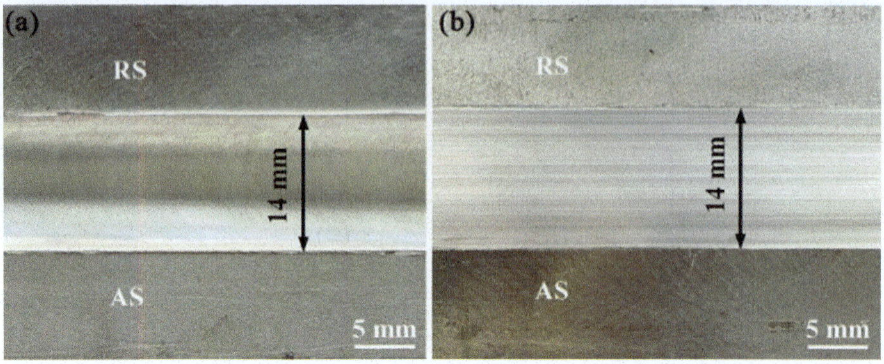

Fig. 2.18 Surface appearances of a typical PFSW joint: **a** upper surface and **b** lower surface [68]

process, resulting in the formation of the asymmetric hourglass shape. Furthermore, the NZ is divided into the SAZ, PAZ (pin affected zone) and stationary shoulder affected zone (SSAZ). The SSAZ is a typical feature of a PFSW joint. Moreover, no kissing bond defect is detected in the root, which indicates that the continuous oxide film in a conventional FSW joint can be successfully destroyed when the welding workpieces are completely penetrated by the rotating pin during the PFSW process.

2.3.3 Mechanical Properties

Figure 2.20 illustrates the mechanical properties of the PFSW joints. The tensile strength and elongation of the PFSW joint reached 354 MPa and 7.5%, respectively, which were approximately 79.2% and 58.6% of those in the BM (447 MPa and 12.8%) [69]. By comparison, the tensile strength and elongation of the conventional FSW joint (CFSW) at the same welding speed of 300 mm/min were 317 MPa and 3.1%, respectively, which were lower than those of the PFSW joint. It is concluded that the PFSW effectively realizes back support from the stationary shoulder and then eliminates the kissing bond, improving mechanical properties.

2.3.4 Prospects

PFSW has been used to successfully join 2219-T6 Al alloys, wherein back support was realized and a sound joint with a smooth surface was obtained. This method can reduce the softening degree and eliminate weld thinning of the joint, thereby improving mechanical properties. PFSW has the potential to join fuel tanks in rockets. The structural complexity of the welding tool restricts its broader adoption, including applications in hollow or enclosed structures with narrow spaces.

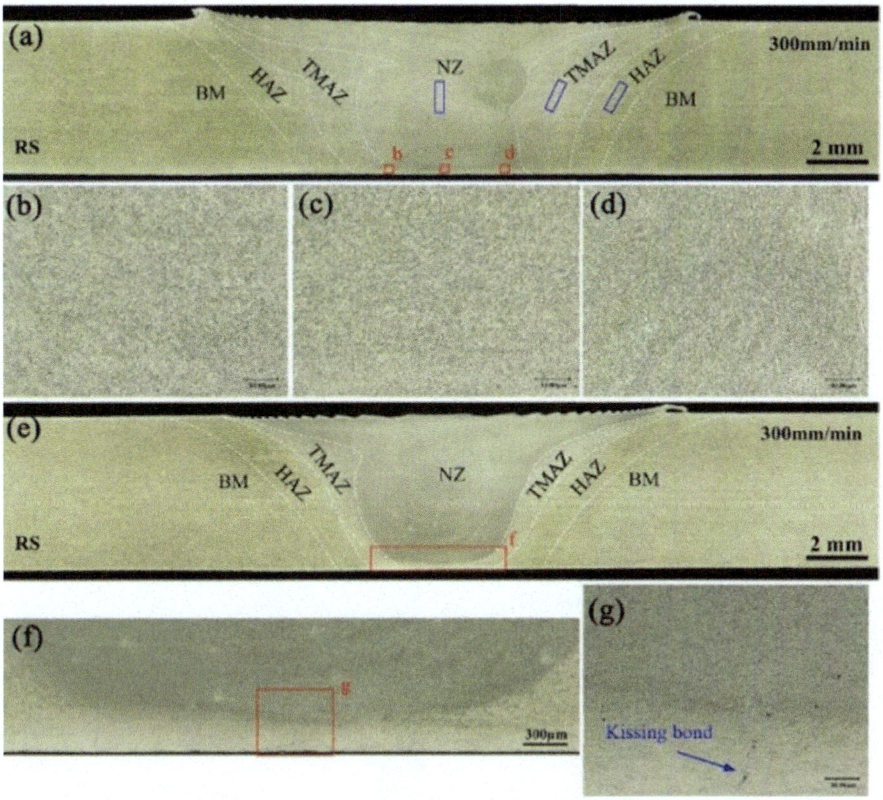

Fig. 2.19 Macrostructure and microstructures of 2219-T6 Al alloys in the root: **a–d** a PFSW joint and **e–g** a conventional FSW joint [69]

2.4 Self-Support Friction Stir Welding

2.4.1 Principle

To overcome the disadvantages of severe joint softening and preset holes in the BT-FSW process, Huang et al. [70] proposed $SSFSW_1$ technique, which overcome the inherent issue of a single shoulder in conventional FSW. The welding tool used for $SSFSW_1$ is composed of a concave upper shoulder with a large diameter, a convex lower shoulder with a small diameter and a rotating pin. The high-quality joining of hollow or enclosed structures can be realized by adjusting the tilt angle and the asymmetric shoulders, as shown in Figs. 2.21 and 2.22. In contrast to BT-FSW, the convex lower shoulder in $SSFSW_1$ can realize the adjustment of the tilt angle, which is beneficial to broadening welding process windows. The thickness adaptation of $SSFSW_1$ is achieved without a tilt angle (Fig. 2.21). During $SSFSW_1$, the preset holes during the beginning and ending of BT-FSW are avoided, simplifying welding procedures.

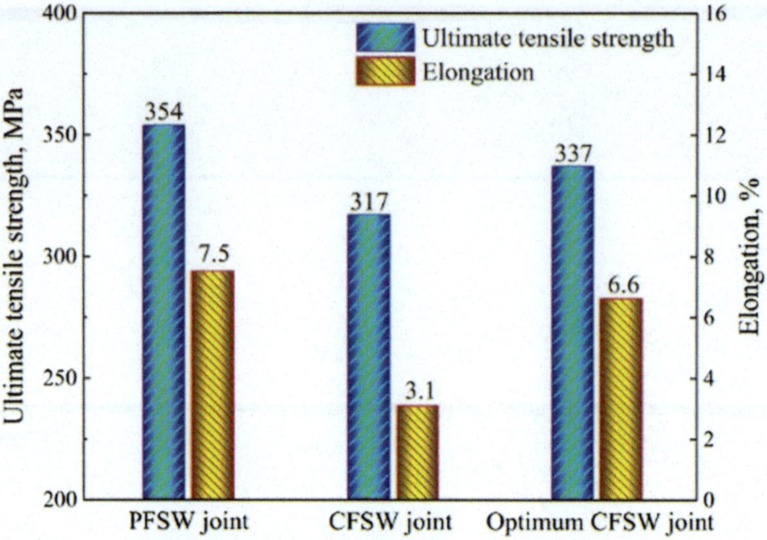

Fig. 2.20 Comparison of mechanical properties between PFSW and conventional FSW joints of 5-mm-thick 2219-T6 Al alloy plates [69]

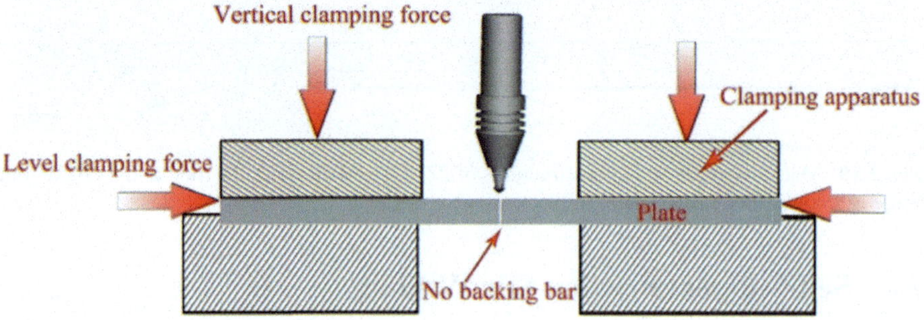

Fig. 2.21 Schematic of SSFSW1 [71]

Due to the complete plunge of the rotating pin along the thickness direction, the lack of root penetration can be avoided. Additionally, aiming at special structures with a narrower welding area at the back of a workpiece, the smaller convex shoulder enables this method to be more effective than BT-FSW.

2.4.2 Design Criterion of Welding Tools

The $SSFSW_1$ tool can realize the thickness adaptation of the workpieces due to the convex lower shoulder, as shown in Fig. 2.23. During $SSFSW_1$, the tilt angle can be

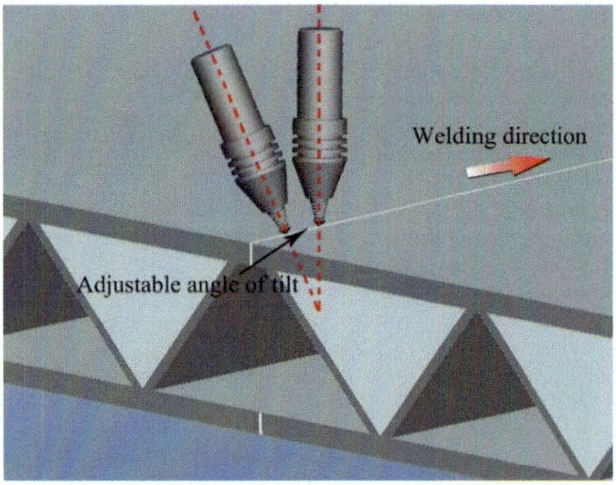

Fig. 2.22 Adjustable tilt angle of the $SSFSW_1$ welding tool [71]

adjusted from 1° to 4°. The distance between the concave upper shoulder and the root of the tapered pin is H, whereas h represents the distance between the root of the tapered pin and the convex lower shoulder. The angle between the centre axis and the lower shoulder is β. Moreover, α represents the tilt angle that the $SSFSW_1$ tool can be adjusted.

The thickness of the workpieces can be adjusted from d to $d + d_I$. From Fig. 2.24, the following equations can be drawn as

$$H = R \tan\alpha + d/\cos\alpha - r_1 \tan\alpha. \tag{2.1}$$

$$H = R \tan\alpha + d/\cos\alpha - r_1 \tan\alpha. \tag{2.2}$$

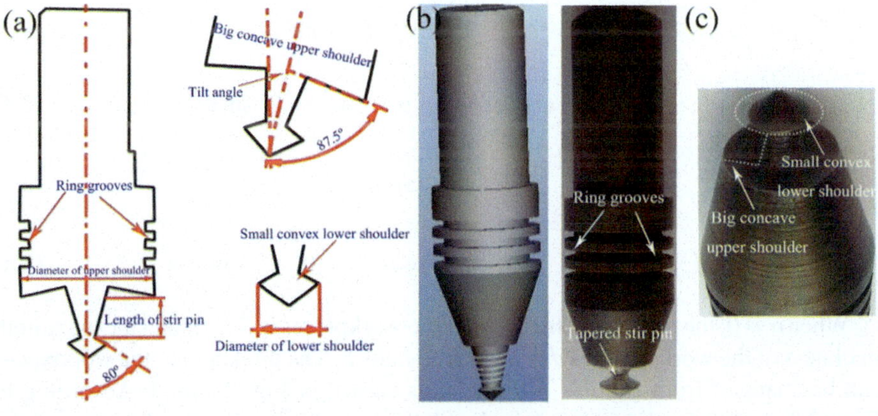

Fig. 2.23 Welding tools of $SSFSW_1$: **a** design view, **b** model and **c** physical picture [72, 73]

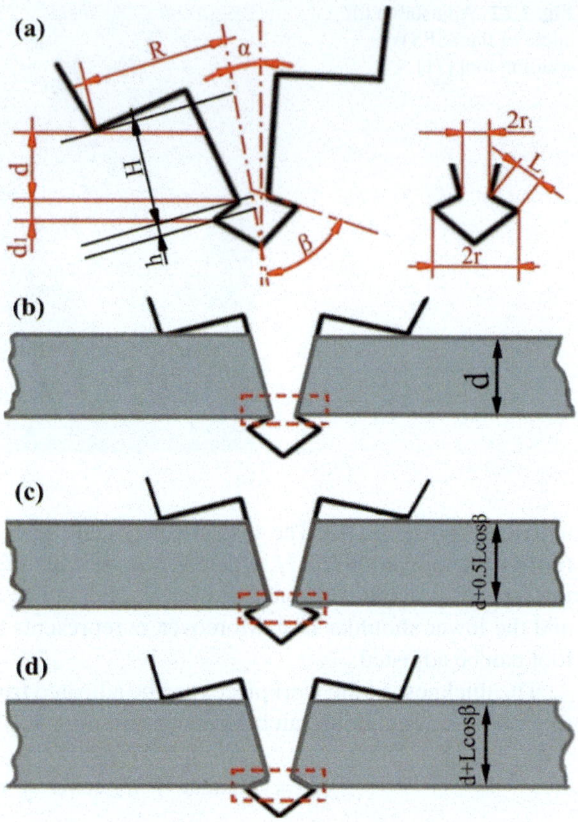

Fig. 2.24 Thickness adaptability of the $SSFSW_1$ tool: **a** key parameters related to the adjustability of the workpiece thickness when α is 0, **b** minimum workpiece thickness, **c** workpiece thickness changing from d to $d + 0.5L\cos\beta$ and **d** maximum workpiece thickness [70]

where R is the radius of the upper shoulder, r is the radius of the lower shoulder; r_1 is the radius of the root tapered pin.

$$d = H\cos\alpha + (r_1 - R)\sin\alpha \tag{2.3}$$

$$d + d_1 = (H + h)\cos\alpha + (r - R)\sin\alpha \tag{2.4}$$

$$h = L\cos\beta \tag{2.5}$$

$$d + d_1 = (H + L\cos\beta)\cos\alpha + (r - R)\sin\alpha \tag{2.6}$$

When α is 0, the minimum thickness of the workpieces is $d = H$, and the maximum thickness of the workpieces is $d + d_1 = H + L\cos\beta$. The thickness of the workpieces can be changed from H to $H + L\cos\beta$, as shown in Fig. 2.24b–d. According to Eqs. (2.3) and (2.6), the variation in the tilt angle α is related to the thickness of the

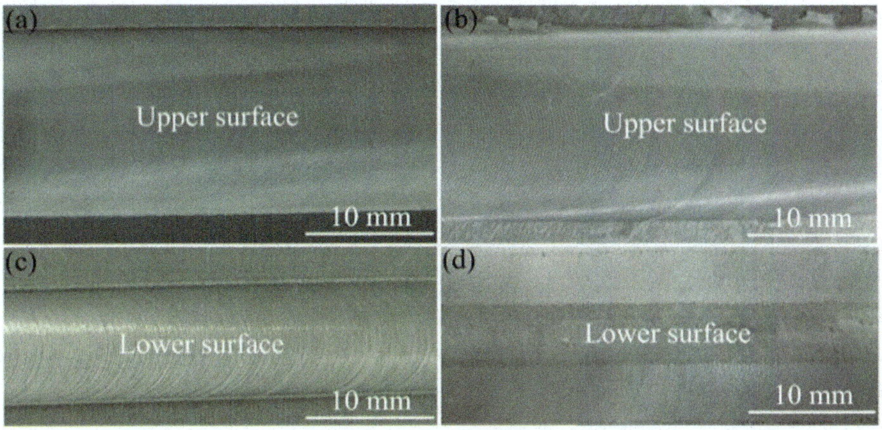

Fig. 2.25 Surface appearances of the typical $SSFSW_1$ and conventional FSW joints: **a** upper surface of a $SSFSW_1$ joint; **b** upper surface of a conventional FSW joint; **c** lower surfaces of a $SSFSW_1$ joint; **d** lower surface of a conventional FSW joint

workpieces to which the $SSFSW_1$ tool can adapt. When other parameters are fixed, the range of thickness of the workpieces is reduced as the tilt angle α increases [70]. As shown in Fig. 2.25, the upper and lower surface appearances produced by $SSFSW_1$ are free-defects, which avoid the indentation induced by the shoulder plunge.

The design of the welding tool is closely correlated with the thermal and mechanical properties [74]. The mechanical properties of the welded workpieces, such as yield strength, ductility and hardness, significantly influence plastic deformation and heat generation. The thermal properties, such as thermal conductivity and specific heat, affect the heat transfer in the welding workpieces. Focusing on different materials, rational selections of the materials and topologies of the welding tools are essential. In published studies, 6xxx Al alloys with low yield strength and good flowability, such as 6061-T6 and 6005-T6 Al alloys, have been successfully joined via the $SSFSW_1$ tool containing a threaded pin and a lower shoulder with a small diameter. Compared with 6xxx Al alloys, 2xxx and 7xxx Al alloys have higher yield strength, lower ductility and worse flowability. A $SSFSW_1$ welding tool with good ductility and improved material flow effect is urgent.

Figure 2.26 shows the design of the $SSFSW_1$ tools, which are all made of H13 steel due to the better balance between strength and ductility at high temperature; note that W6 steel was not used due to the high loss in ductility. The circular pin and the three-facet pin do not easily fracture, whereas they have difficulties in producing sufficient material flow and deteriorate joint formation. The threaded pin can realize the material flow along the thickness direction, obtaining good joint integrity. For threaded pins with three facets, the material flow can be promoted, and the fracture risk of the welding tool is reduced during conventional FSW [75, 76]. The partial stress concentrations and reduction of load bearing easily improve the fracture risk during $SSFSW_1$ of 7xxx Al alloys due to the high torque. In addition to the rotating

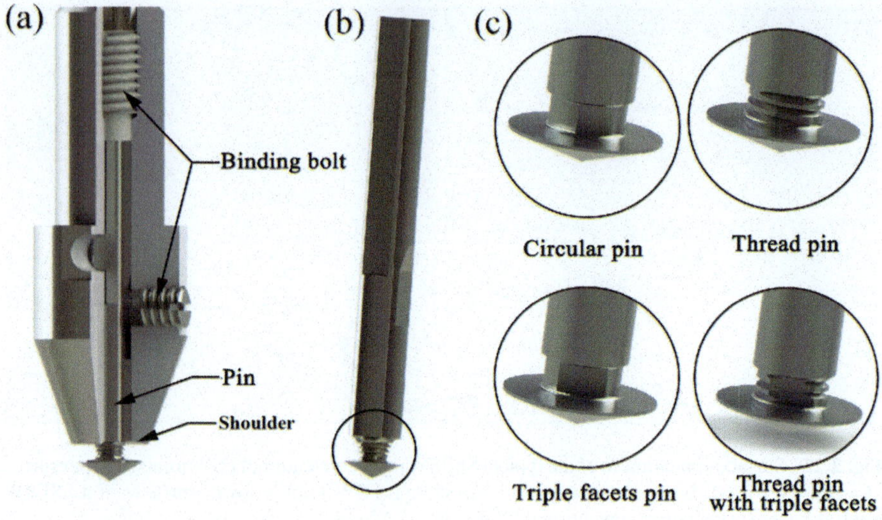

Fig. 2.26 Schematics of **a** detachable welding tool, **b** macroscopic rotating pin and **c** different rotating pin morphologies

pin, which can improve the material flow, the diameter of the lower shoulder is another important parameter that provides sufficient forging force and frictional heat, thereby reducing the welding torque and avoiding pin fracture risks. For 3-mm-thick 2xxx or 7xxx Al alloy sheets, the diameter of the rotating shoulder should be twice as high as the thickness of the workpieces to be welded.

2.4.3 Formation Mechanism

During the $SSFSW_1$ process, the material flow in the TMAZ is demonstrated schematically in Fig. 2.27. When the $SSFSW_1$ pin is plunged into the welding workpieces, a cavity remains besides the rotating pin, and the shape of the cavity is determined by the profile of the rotating pin. The volume of the plasticized materials below the upper shoulder and upon the lower shoulder depends on the rotational velocity, welding speed and tilt angle. When the $SSFSW_1$ tool goes forward, the materials in the RS are dragged by the shoulder over the weld to the AS and flow into the pin-affected zone. The materials in the pin-affected zone flow in layers around the pin in depth, and the layers are stacked on the welding line. Once the restriction of material flow from the leading edge to the trailing edge increases, flashes should occur on the RS. This results in insufficient material filling in the AS, leading to the formation of tunnel or pore defects. The upward pressure, which prevents the collapse of the NZ, depends on the diameter of the lower shoulder and the heat input. When

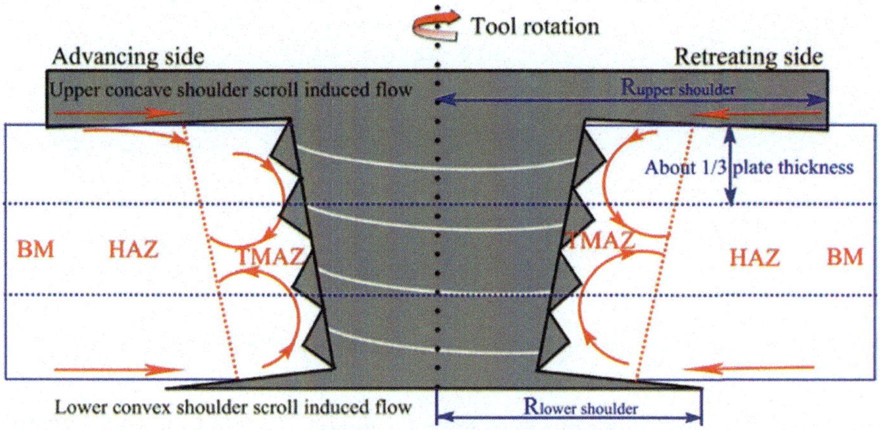

Fig. 2.27 Material plastic flow model during the $SSFSW_1$ process [77]

the diameter of the lower shoulder is small, the upward pressure is not sufficient to prevent the collapse of the NZ, and the defect or converged point shifts downward.

2.4.4 Microstructural Characteristics

The macroscopic aspect of the typical joint is shown in Fig. 2.28, the weld shape differs from the conventional FSW weld shape, and is slightly hourglass shaped. This is because the upper and lower surface experience extreme deformation and frictional heating during $SSFSW_1$, thereby resulting in generation of waist-shaped NZ. The $SSFSW_1$ tool moves downwards and spins counter clockwise, thus the AS is on the left and the RS is on the right. The macrostructure of the $SSFSW_1$ joint is more symmetric compared with conventional FSW joint. From the macroscopic appearance, root flaws existed in the conventional FSW welded joint have been eliminated completely. Single onion ring pattern observed in the welding nugget in the conventional FSW welds is replaced by series of onion rings stacked vertically through the thickness on the AS. Onion rings have alternate bright and dark rings. At the same time, a layer resulting from the outflow of plastic material is formed on the RS at the lower surface.

Based on macrostructural characterization shown in Fig. 2.28, four distinct zones, HAZ, TMAZ, upper weld NZ (UWNZ) and lower weld NZ (LWNZ), have been identified. The cross-section typical feature of the $SSFSW_1$ joint is shown in Fig. 2.29. There is a smooth narrow interface between weld NZ (WNZ) and TMAZ on the AS, which can also be found in the conventional FSW joints. This is because grains or deformation bands in the TMAZ have flowed only a little into the region behind the pin on the AS, however, on the RS, such flow is greater due to the rotating tool which shears metal from before it to that side and additionally sweeps it into

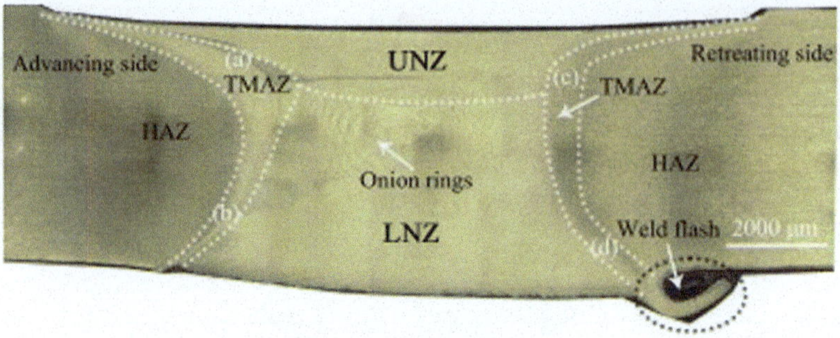

Fig. 2.28 Macrostructure of a $SSFSW_1$ joint of 5-mm-thick 6082-T6 Al alloys [73]

the region behind resulting in a very irregular interface between elongated TMAZ grains and the WNZ. The UWNZ is directly influenced by the stir pin and the concave upper shoulder, the LWNZ is directly influenced by the stir pin and the convex lower shoulder. The WNZ comprises the material strongly affected by the tool rotation. It is subjected to a high level of plastic deformation and frictional heating. The nugget size is typically slightly greater than the pin diameter. The TMAZ is produced by friction between the tool shoulder and the plate surface, as well as by plastic deformation of the material in contact with the stir pin. The TMAZ comes in for the plastic shear stress around the plastic flow of material and the grains are elongated along the direction of maximum shear stress. The HAZ just experiences a thermal cycle and does not undergo appreciable plastic deformation. Therefore, the microstructure in the HAZ is partially slack quenched and the grains are coarsened.

Figure 2.30 shows the optical microstructures of $SSFSW_1$ joints produced at the same welding parameter. As can be seen, the smooth narrow interface among UWNZ, LWNZ and TMAZ on the AS and four distinct zones were existed. This was different from the conventional FSW joint. The optical microstructures of $SSFSW_1$ joints produced at different welding speeds on the AS were shown in Fig. 2.31. What appear to be band pattern is discernible in the WNZ from the AS to the WNZ center [also can be seen in Fig. 2.29a], and the band geometry is different for different welding speed. This is consistent with the result reported by Horton who observed the similar band pattern formed in the WNZ of the 2014/2219 dissimilar alloy BT-FSW joints. The overaging effect and the coarser second-phase particles can also be found in the UWNZ.

Low magnification optical macrographs of cross section of $SSFSW_1$ joints at the welding speed varied from 10 to 200 mm/min are shown in Fig. 2.32. With respect to conventional FSW, all the joints exhibit four distinctive zones, i.e. HAZ, TMAZ, UWNZ and LWNZ are depicted in Fig. 2.32a, d. In WNZ, the structure of band pattern is discernable. The UWNZ means the upper part of the structure of band pattern and is mainly influenced by the upper shoulder. The LWNZ means the lower part of the structure of band pattern and is mainly influenced by the lower shoulder. The weld shape differs from the weld made from conventional FSW, and is slightly hourglass shaped. Such a shape is mainly determined by the material flowing during

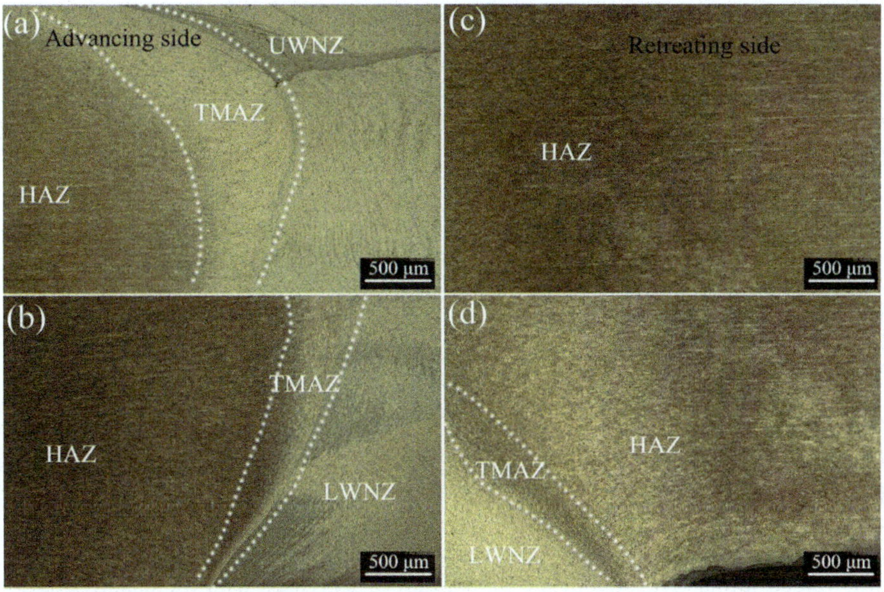

Fig. 2.29 Photo of microstructure of joint in different regions taken from top and bottom along thickness direction of Fig. 2.28 [73]

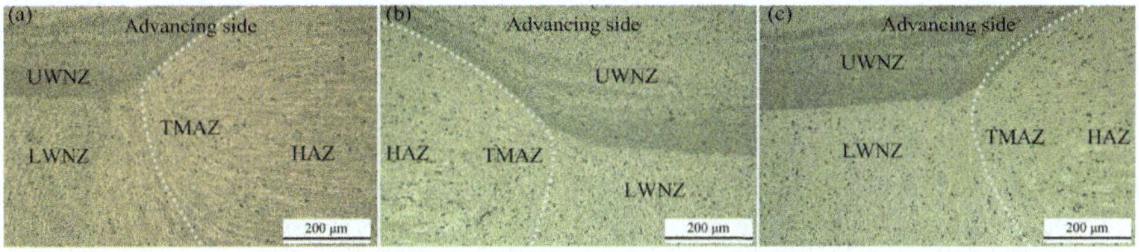

Fig. 2.30 Optical microstructures of $SSFSW_1$ joints produced at the same welding parameters on the AS [73]

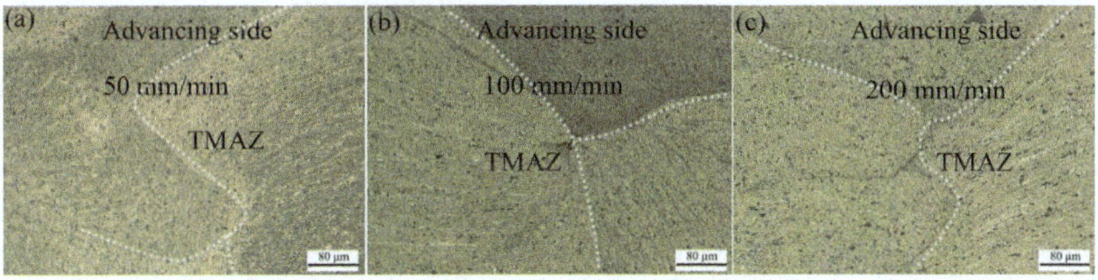

Fig. 2.31 Optical microstructures of $SSFSW_1$ joints produced at different welding speeds. **a** 50 mm/min, **b** 100 mm/min and **c** 200 mm/min [73]

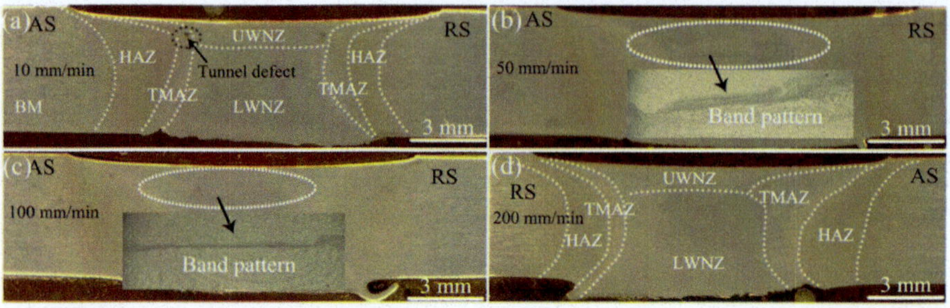

Fig. 2.32 The cross-sections of $SSFSW_1$ joints welded for different welding speeds: **a** 10 mm/min, **b** 50 mm/min, **c** 100 mm/min, and **d** 200 mm/min [72]

the $SSFSW_1$. The upper and lower surface experience extreme deformation and frictional heating by the contact with the upper and lower shoulder during $SSFSW_1$ process, thereby resulting in such NZ. The interface between the recrystallized NZ and the parent metal is relatively diffuse on the RS of the $SSFSW_1$ tool, but quite sharp on the AS of the tool. The tunnel defect forms at the AS of the self-support friction stir welded joints. The size of the tunnel defect decreases with increasing welding speed. The formation of the weld defects at the top of the AS is closely related to the metal flow driven by shoulder extrusion and pin stirring. This observation also shows that the metal around the defects exhibits vortex-like flow features, which is often considered abnormal and should be avoided in FSW process, that cause disordered flow.

Single onion ring pattern observed in the welding nugget in the conventional FSW welds is replaced by series of onion rings stacked vertically through the thickness on the AS, as shown in Fig. 2.33a–c. By employing a threaded tool the plastic material is forced from the plate down into the weld and may travel several times depending on the rotational and welding speed. Onion rings have alternate bright and dark rings. The gap between two consecutive rings was more at center than those at the outer edge. Onion rings found in the welded zone is a direct evidence of characteristic material transport phenomena occurring during $SSFSW_1$. The reduced stirring due to the higher speed or lower rotation speed will result in the partial breaking of natural Al_2O_3 oxide layer and low heat input retards flowability of plastic material. This weak flow causes localized presence of broken oxide particle in the form of dark wavy zigzag line or kissing bond defect, as shown in Fig. 2.33d. It is imperative that proper selection of welding parameters can effectively eliminate the formation of zigzag line, and which even can change failure location from WNZ to TMAZ or HAZ thus improved mechanical performance of friction stir welded joints.

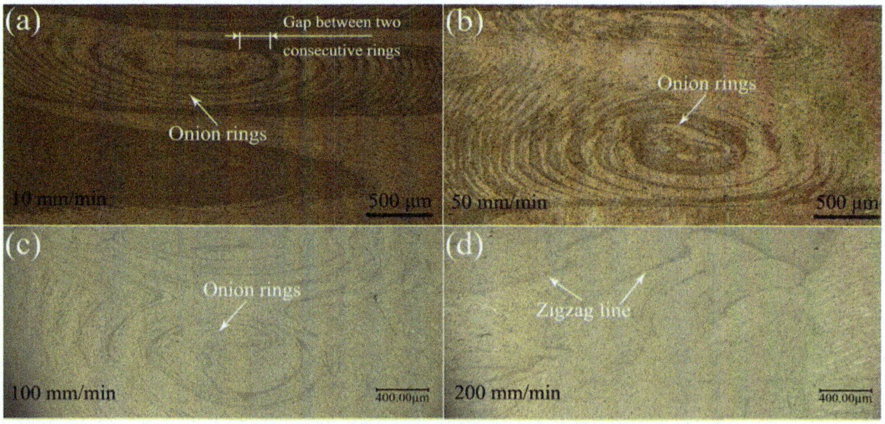

Fig. 2.33 Effect of welding parameters on the formation of onion ring and zigzag line: **a** 10 mm/min, **b** 50 mm/min, **c** 100 mm/min, and **d** 200 mm/min [72]

2.4.5 Second-Phase Particles and Grain Morphology

The WNZ comprises the material strongly affected by the tool rotation. It is subjected to a high level of plastic deformation and frictional heating. The UWNZ is directly influenced by the stir pin and the concave upper shoulder, however, the LWNZ is directly influenced by the stir pin and the convex lower shoulder. The highest strain and strain rates take place in WNZ. Intense plastic deformation and frictional heating during FSW result in generation of a recrystallized fine-grained microstructure within stirred zone. Figure 2.34 shows the typical cross-sectional second-phase particles distribution of the WNZ of the $SSFSW_1$ joint. It can be seen that a considerable amount of constituent particle are randomly distributed in the BM. However, there were far fewer large precipitates in the NZ than in the BM. The second-phase particles in the WNZ are more uniform and intensive. There existed much more uniform and finer dispersoids (AlSiMnFe). The WNZ is surrounded by the TMAZ, having highly deformed and bent inhomogeneous coarser aluminum grains due to stirring by tool. The TMAZ grain structure appears to be bent towards the tool shoulders, as a result of material flow induced during the FSW. However, in the region of TMAZ plastic deformation and recrystallization is somewhat lesser than weld NZ. The TMAZ consists of larger recrystallized grains than that in the WNZ on the AS, which is shown in Fig. 2.35. The regions bounded by dashed lines in Fig. 2.35 are a part of TMAZ. It has been suggested that the differences in microstructural observations may be resolved by considering a mechanism of continuous dynamic recrystallisation in the TMAZ. In the HAZ, remote from the center of the weld, there is no obvious change to the grain structure compared with the BM, however, the grain size decreases with increasing distance of the weld centerline, which are shown in Fig. 2.36.

In any FSW process, the properties and performance of the weld are dictated by the microstructure, which in turn is determined by the thermal cycle of the welding

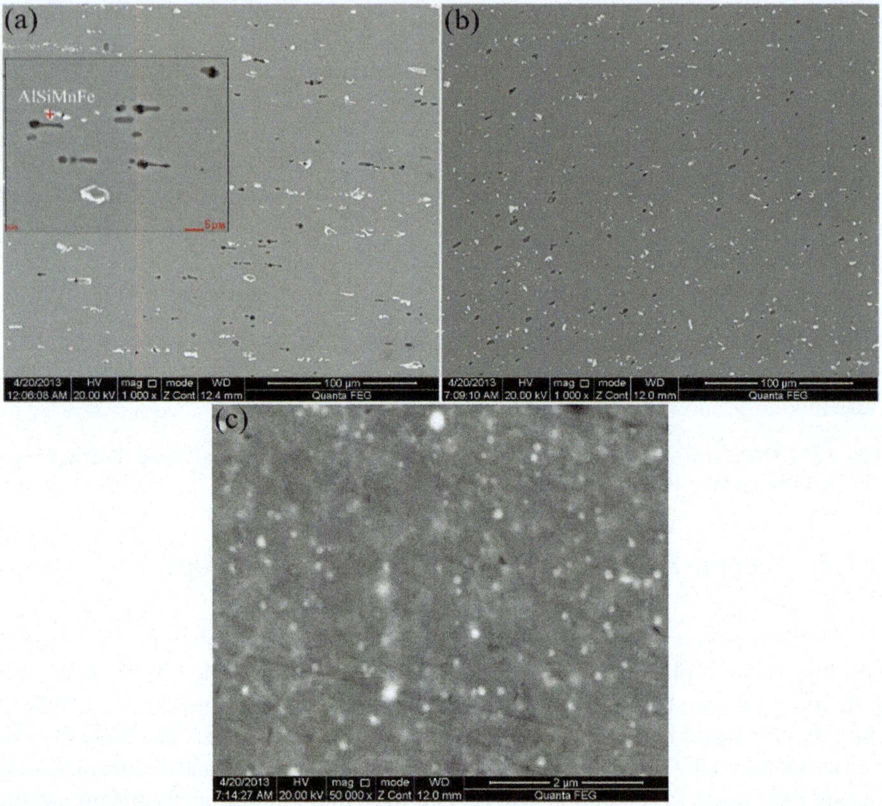

Fig. 2.34 The comparison of precipitates distribution: a BM, **b** WNZ and **c** higher magnification of WNZ [72]

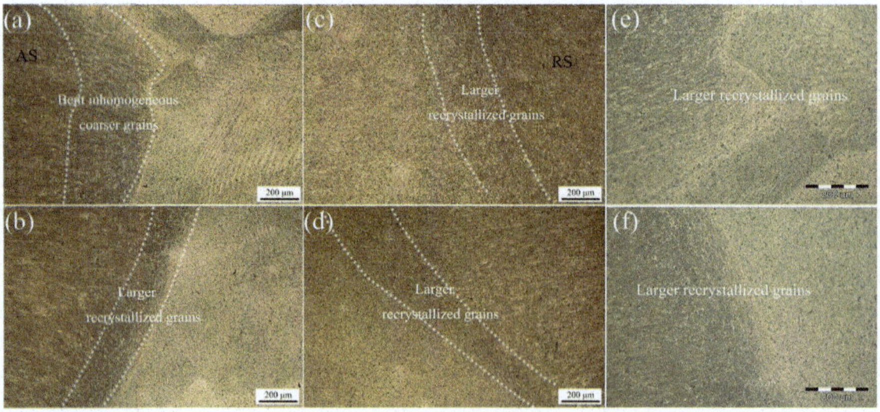

Fig. 2.35 The variation of microstructure of the $SSFSW_1$ joint in different zones with the welding speed 50 mm/min [72]

Fig. 2.36 Grain size and morphology with increasing distance to the weld centerline [72]

process. From previous research, FSW results in the temperature increase up to 400–550 °C within the NZ due to friction between tool and workpieces and plastic deformation around rotating pin. The precipitates in aluminum alloys can coarsen or dissolve into aluminum matrix depending on alloy type and maximum temperature produced in the welding process. Figure 2.37 shows the microstructure of the UWNZ, TMAZ, HAZ and unaffected BM. The precipitate evolution was noticeable in the TMAZ and HAZ. The dissolution of strength precipitates was observed in the TMAZ, as shown in Fig. 2.37a. The extent of dissolution, of course, depends on the thermal cycle experienced by TMAZ. The unaffected BM consists of needle-shaped precipitates, rod-shaped precipitates and α-Al matrix, and the main hardening precipitates in 6082-T6 aluminum alloy are Mg_2Si. The homogenously distributed precipitates are generally smaller in the NZ and there were far fewer large precipitates in the NZ than in the BM, as shown in Figs. 2.37b. A considerable amount of coarser second-phase particles are randomly distributed in the UWNZ, as shown in Fig. 2.38a. This was the reason why the microhardness in this region was the lowest. However, a large amount of finer and uniform second-phase particles are distributed in the TMAZ, as shown in Fig. 2.38b. The microhardness in this region is relatively high due to the good solid solution hardening and the higher density of second-phase particles, resulting in more obstacles to the movement of dislocations. At the same time, the coarsening of second-phase particles and the dissolution of the rod-shaped precipitate and needle-shaped precipitate were also observed in the HAZ. Both coarsening and dissolution of the precipitates lead to a drop in the microhardness, which is consistent with the experimental results.

Recrystallization occurred in the TMAZ due to the plastic deformation and the function of thermal cycle. The equiaxed grains were observed in the TMAZ, as shown in Figs. 2.39a and 2.40a. The TMAZ consists of larger recrystallized grains than that in the WNZ on the AS. It has been suggested that the differences in microstructural observations may be resolved by considering a mechanism of continuous dynamic recrystallisation in the TMAZ. The deformation process associated with welding introduces a large quantity of dislocations, while at the same time grain growth occurs as the temperature rises. A similar observation that dissolution of strength precipitates in the TMAZ and HAZ has also been found in the lower part of the joint on the AS, as shown in Fig. 2.39b. In the HAZ, remote from the center of the weld, there is no obvious change to the grain structure compared with the BM, however,

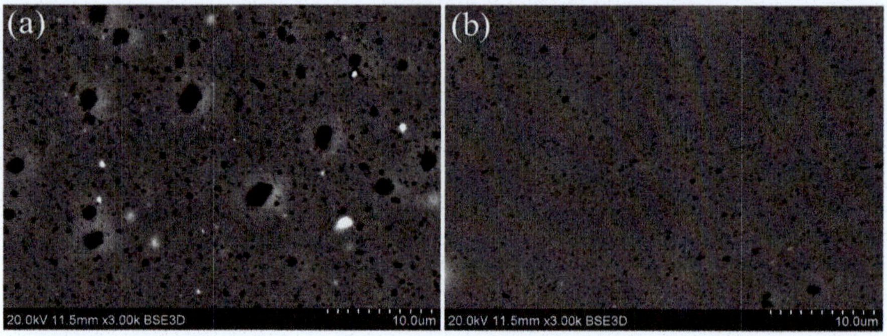

Fig. 2.37 The comparison of the microstructure of the UWNZ, TMAZ, HAZ and unaffected BM

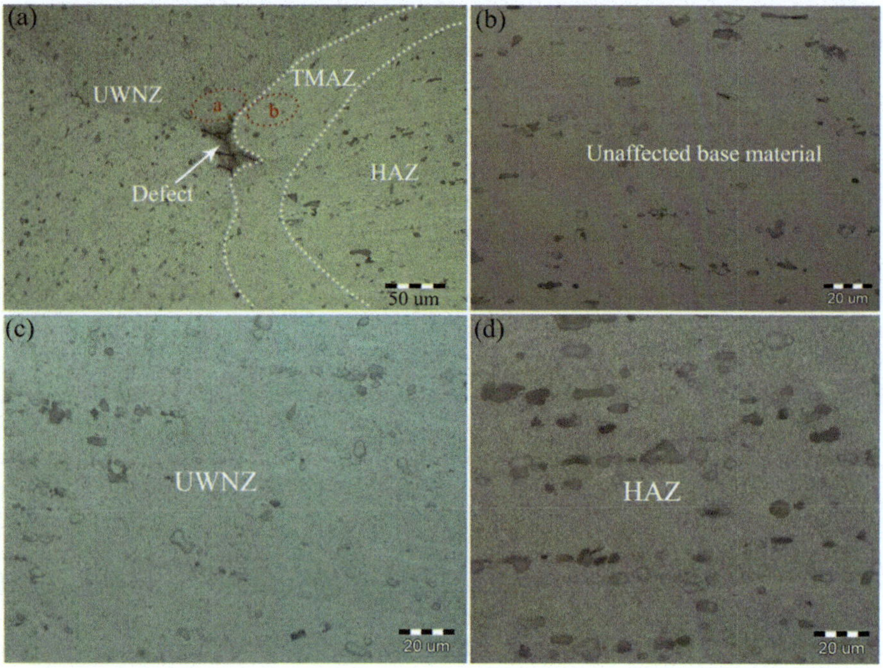

Fig. 2.38 SEM micrographs of second-phase particles distribution in the WNZ and TMAZ [73]

the grain size decreases with increasing distance of the weld centerline, as shown in Fig. 2.40b, c.

In all the cases, the average microhardness in the WNZ (UWNZ and LWNZ) in proximity of the TMAZ was found to be significantly lower because of an overaging effect and the coarser second-phase particles. The drop in microhardness in WNZ with respect to the unaffected BM is due to the decomposition of needle-shaped β” precipitates which is the main source of hardening and strengthening, the coarsening

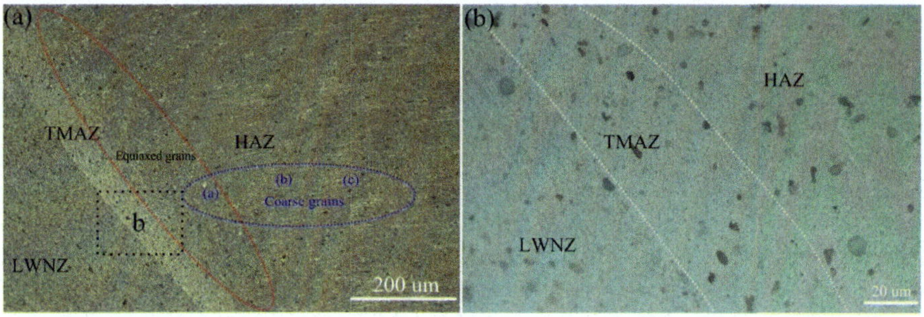

Fig. 2.39 The features of the microstructure in the lower part of the $SSFSW_1$ joint on the AS [73]

Fig. 2.40 The variation of grain size and morphology with increasing distance of the weld centerline [73]

of precipitates into semi- and non-coherent rod-shaped β′ precipitates (overaged precipitate structure) and the low dislocation density associated with the dynamically recrystallized structure. Another probable reason is that intense deformation causes fragmentation of second-phase particles and precipitates leading them to be redistributed in new locations at the interior of grains. The results achieved from microhardness test show that the slightly higher microhardness was found in the HAZ. This is due to the fact that the HAZ has been deformed very slightly, and has different thermo-mechanical behavior with respect to the TMAZ. Some scientists demonstrated such effect by analyzing the reinforcing particles' behavior in terms of dimension and aspect ratio in the FSW 6056 alloy by employing a deep transmission electron microscopy analysis. The low microhardness was explained to be due to the disappearance of Guinier–Preston (GP) zones and formation of overaged precipitates.

The mechanical properties of Al–Mg-Si alloys are strongly dependent on the metastable precursors of the equilibrium β (Mg_2Si) phase. The precipitation behavior of Al–Mg-Si alloys is generally reported to be supersaturated solid solution (SSS) $\rightarrow$ cluster $\rightarrow$ GP zones $\rightarrow \beta'' \rightarrow \beta' \rightarrow \beta$ [78, 79]. The dissolution of the precipitates is observed in the TMAZ. The homogenously distributed precipitates are generally smaller in the NZ, and there are far fewer large precipitates in the NZ than in the BM. A considerable number of precipitates are randomly distributed in the UNZ. Dong et al. [80] found that the materials in the NZ experienced the highest peak temperature and the precipitates β'' dissolved into the α-Al matrix during FSW of AA6005A-T6. Moreover, coarsening of second-phase particles and dissolution of the rod-shaped precipitates and needle-shaped precipitates are also observed in the HAZ. This easily results in severe joint softening in the HAZ.

PWHT is performed to change the microstructural morphologies and improve the mechanical properties. The nanoparticles become increasingly denser. The particles include the precipitates along grain boundaries, which are broken by the rotating pin. The heat input coarsens the small metastable precipitates. The maximum temperature of 723 K cannot induce the dissolution of the stable Mg_2Si phase particles but benefits atomic diffusion. The diffusive flows of Mg and Si atoms from nanoprecipitates and the α-Al matrix strongly contribute to the growth of the grain boundaries (Fig. 2.41). As a result, the concentrations of Mg and Si at the interfaces of grain boundaries and matrix decrease [81]. With the increase in holding temperature, the microhardness increases and tends to uniform due to the complete solution of the overaged strengthening particles and new precipitation of fine needle-shaped precipitates.

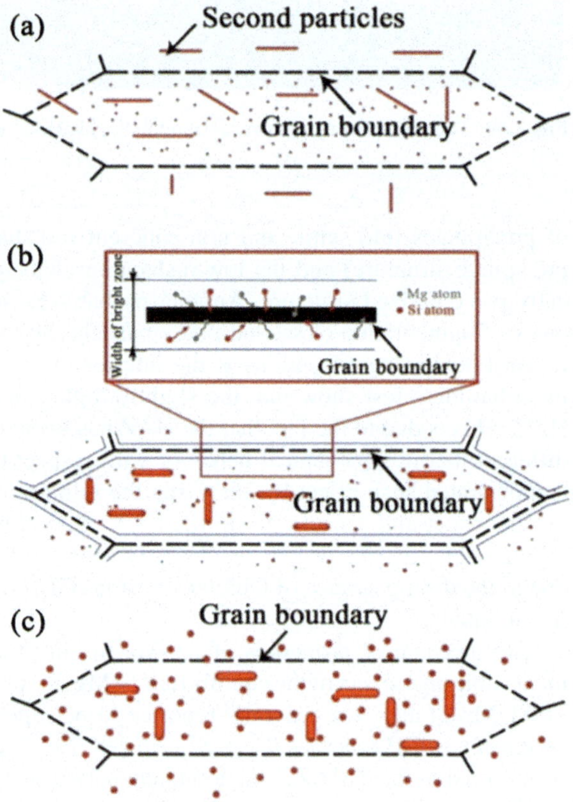

Fig. 2.41 Schematic maps of microstructural evolution with PWHT: **a** BM, **b** precipitate-free zone and the precipitates and **c** coarsening of precipitates [81]

2.5 Plastic Deformation Analysis of $SSFSW_1$ Joint

As shown in Fig. 2.42, the TMAZ experiences both temperature and deformation during $SSFSW_1$ and is characterized by a highly deformed structure. In the outer part of the TMAZ, the original grains remain identifiable in the deformed structure with the formation of subgrain structures. The parent metal elongated grains were deformed in an upward and downward flowing pattern around the NZ. In the conventional FSW process, the bottom of WNZ is in contact with the tool pin and the backing plate. However, the LWNZ is influenced by the tool pin and the lower shoulder. The degree of plastic deformation in the lower part of the $SSFSW_1$ joint is higher than that of the conventional FSW joint. The TMAZ grain structure appears to be bent towards the tool shoulders, as a result of material flow induced during the FSW. The HAZ experiences a thermal cycle, but does not undergo any plastic deformation. The HAZ retains the same grain structure with the parent material and is detected generally by a change in etching response. The transition appears sharp on the AS of the weld, particularly at the bottom of the weld. An abrupt change in structure is observed at the edge of TMAZ defined the TMAZ/HAZ boundary, separating the comparatively large grains from the fine grains associated with HAZ. It is obvious that the degree of plastic deformation in the black region (in Fig. 2.42) is the largest and the HAZ almost has no deformation. The degree of plastic deformation in the lower part of the joint is lower than that in the upper part of the joint. This is also another reason why there existed the variation of the microhardness in different regions of the TMAZ and the microhardness of the TMAZ is relatively higher than that in HAZ.

For 6082-T6 aluminum alloy, the microhardness profiles obtained from different regions of the upper part on the AS are shown in Fig. 2.43. The microhardness of the LWNZ near the TMAZ was relatively lower, especially in the red region, as shown in Fig. 2.43a. The values of microhardness of the TMAZ were relatively high which can reach 89.4 HV and 84.7 HV, respectively, as shown in the black and yellow region in Fig. 2.43a. The microhardness variation from UWNZ to HAZ was significant. The

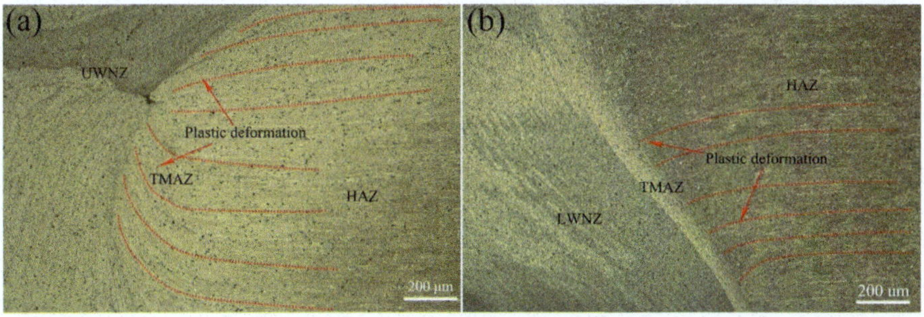

Fig. 2.42 The degree of plastic deformation of $SSFSW_1$ in different zones: **a** upper part of the joint on the AS and **b** lower part of the joint on the AS [73]

minimum microhardness recorded in all the regions of the joint was average 64.9 HV, however, it was evident that the maximum microhardness was average 89.4 HV.

Worth mentioning is that the similar results are also obtained from different zones of the lower part of the $SSFSW_1$ joint on the AS, which is shown in Fig. 2.44. The microhardness of the UWNZ near the TMAZ was the lowest [average 60.9 HV, as shown in Fig. 2.44b], but the average microhardness value of the TMAZ was the highest [average 77.1 HV, as shown in Fig. 2.44b].

The microstructural evolution of the $SSFSW_1$ joint markedly affects the mechanical properties associated with joint softening, which is dominated by the welding parameters, such as rotational velocity, welding speed and shoulder diameter. The microhardness distribution in the cross-section presents a "W" shape, in which the lowest microhardness value is located at the interface between the TMAZ and HAZ due to severe joint softening induced by the coarsening and dissolution of the precipitates. With the increase in welding speed, the softening degree is reduced because

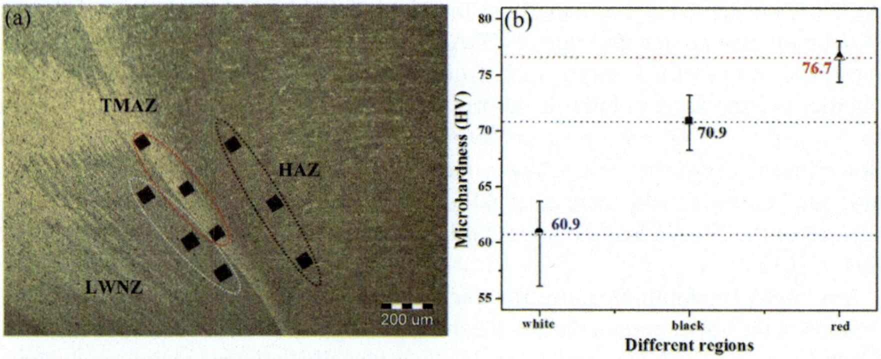

Fig. 2.43 The comparison of microhardness values obtained from different regions of the upper part of the $SSFSW_1$ joint on the AS [73]

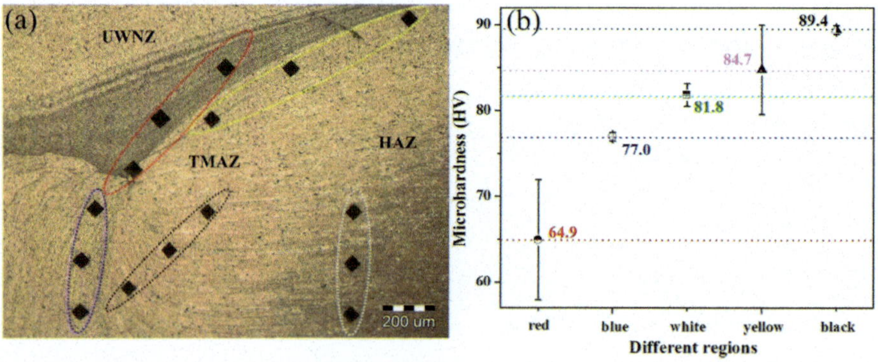

Fig. 2.44 The comparison of microhardness values obtained from different regions of the lower part of the $SSFSW_1$ joint on the AS [73]

of the short exposure time at high temperature. Table 2.3 lists the tensile properties of $SSFSW_1$ joints. For 5-mm-thick 6082-T6 Al alloy sheets, the maximum tensile strength with a corresponding joint efficiency of 69% was achieved at a welding speed of 200 mm/min, a rotational velocity of 800 r/min, a tilt angle of 4° and a plunge depth of 0.1 mm [72]. Moreover, the tensile fracture all located at the AS rather than at the RS due to the more severe joint softening in the AS induced by the high welding peak temperature. By increasing the lower shoulder diameter to a certain extent, the overflow of the plasticized materials is reduced, contributing to an increase of mechanical properties. Only the influences of welding speed and shoulder diameter are investigated in a narrow process range. The other welding parameters, including rotational velocity, tilt angle and plunge depth, still need to be studied to enable high-speed FSW with high quality to satisfy the requirements in the manufacturing fields.

The tensile properties of the joints as a function of welding speed are shown in Fig. 2.45. The tensile strength increases with the increase of welding speed from 50 to 200 mm·min^{-1}. The maximum tensile strength is reached at the welding speed of 200 mm·min^{-1}, and the corresponding joint efficiency is 69%. The joint ductility obeys the opponent tendency to the tensile strength, and its maximum value is 12.58%.

Table 2.3 Tensile properties for $SSFSW_1$ joints of different Al alloys

Material	Thickness/mm	Tensile strength / MPa		Elongation/%		References
		BM	$SSFSW_1$	BM	$SSFSW_1$	
6005A-T6	5	270	190	16.5	6.9	[70]
6082-T6	5	320	220	14	12.8	[72]

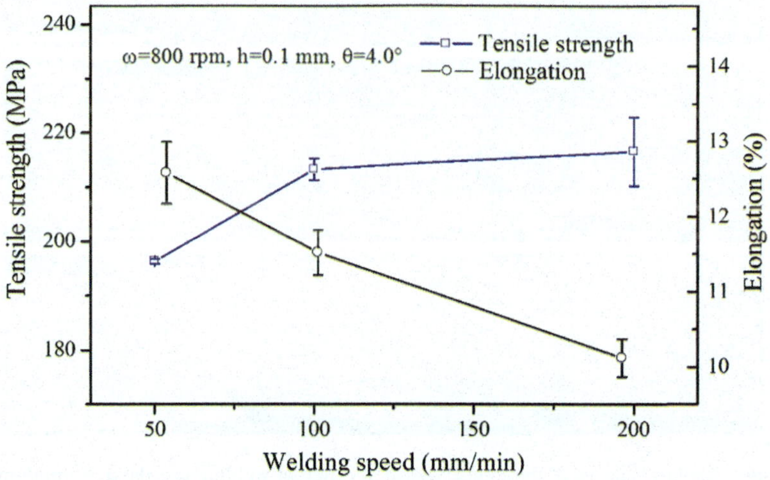

Fig. 2.45 Tensile properties of the joints for different welding speeds [72]

The fracture features of the joints welded at different welding speeds are shown in Fig. 2.46. The macro-fracture surface of tensile tested specimen presents 45° angle shear fracture along the tensile axis and the appearance indicates a possible failed type of ductile fracture. In the present study, the fracture occurs in the HAZ adjacent to the TMAZ on the AS, which is corresponding to the lowest hardness of the joint, as shown in Fig. 2.44. The fracture surfaces of tensile specimens are characterized using SEM to understand the failure patterns, as shown in Fig. 2.47. From the results, we can see that they invariably consist of dimples, which indicates the specimens fail in ductile manner under the action of the tensile loading. Because fine dimples are the characteristic feature of ductile fracture, this feature validates the high ductility of the $SSFSW_1$ joint during the tensile testing.

The above-mentioned results indicate that the welding speed has a significant effect on the mechanical properties of the $SSFSW_1$ joints of 5-mm-thick 6082-T6 aluminum alloy. The welding speed influences the heat input per unit length of weld which controls the degree of softening and flowability of plasticized material. On the one hand, at lower welding speed the amount of heat supplied to the deforming material in weld area is greater and therefore wider is the softened area around the stirring tool leading to more improved metal flow and more effective bonding in the weld. At high welding speed $SSFSW_1$ tool results in lower heat input per unit length of weld which in turn reduces stirring of material. On the other hand, increasing welding speed can decrease the heat input into the joints, and thus reducing the coarsening and transformation of the precipitates. In this case, the precipitations

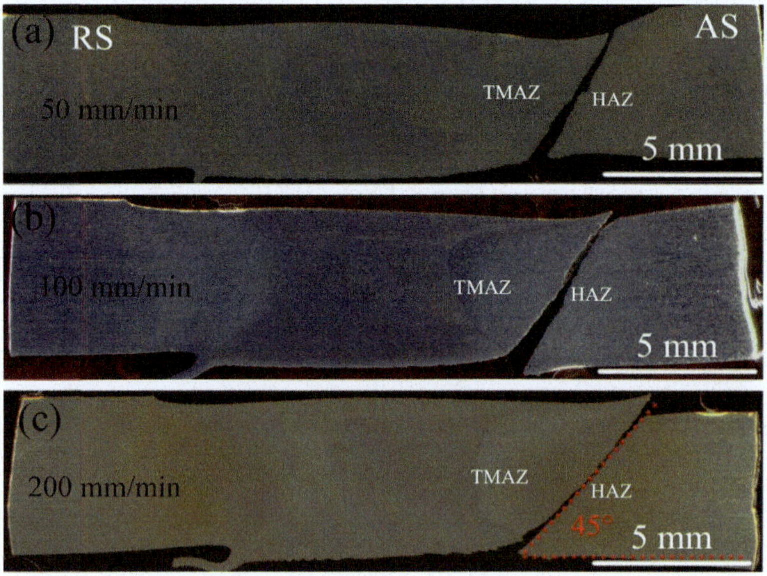

Fig. 2.46 Fracture features of the joints welded at different welding speeds: **a** 10 mm/min, **b** 50 mm/min, and **c** 200 mm/min [72]

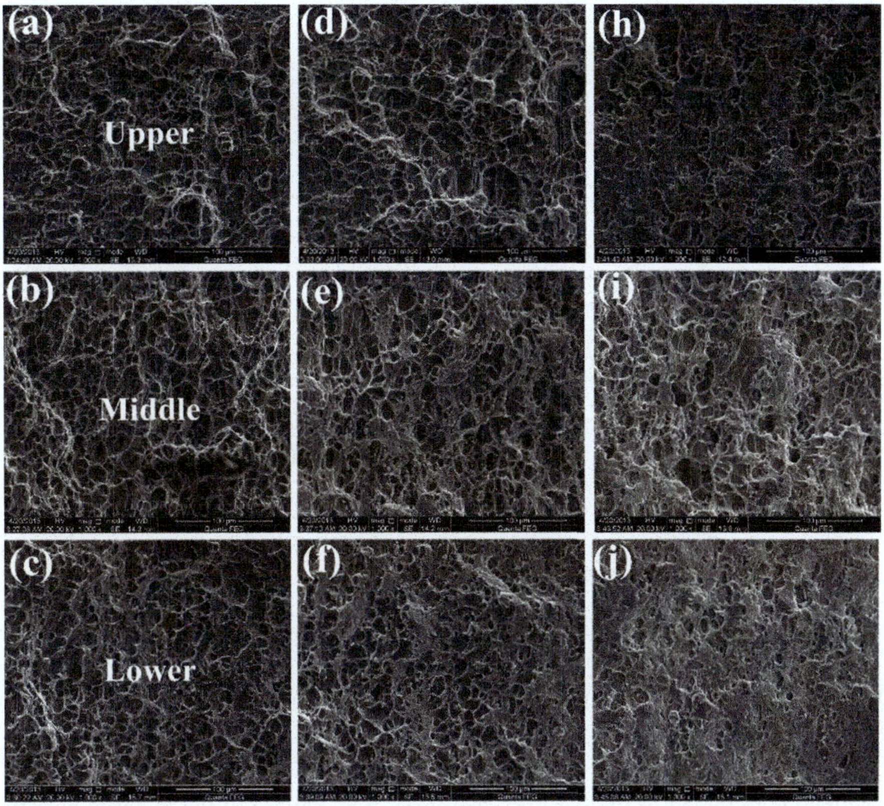

Fig. 2.47 Effect of welding speeds on fracture morphology: **a–c** 10 mm/min, **d–f** 50 mm/min, and **h**–j 200 mm/min [72]

deterioration is weakened in the HAZ corresponding to failure location of the tensile specimen.

As can be seen from Fig. 2.48, the structure of band pattern is discernable in the UWNZ from the AS to the WNZ center. The similar band pattern formed in the WNZ of the 2014/2219 dissimilar alloy BT-FSW joints. The unsteady flow of some sort must be the genesis of the cyclically varying microstructure observed in FSW. It is clear that there is a smooth narrow interface between WNZ and TMAZ on the AS compared with that on the RS. The plastic material in the TMAZ have flowed only a little into the region behind the pin on the AS, however, on the RS, such flow is greater due to the rotating tool shears meta before it to that side and additionally sweeps it into the region behind resulting in a very irregular interface.

The upper and lower surface are in contact with the tool shoulders therefore experiences more frictional heating and plastic flow. It has been reported that flow of the material near top surface layers of weld nuggets (up to a depth of about one third of weld thickness) is caused by FSW tool shoulder while that in sub-surface region

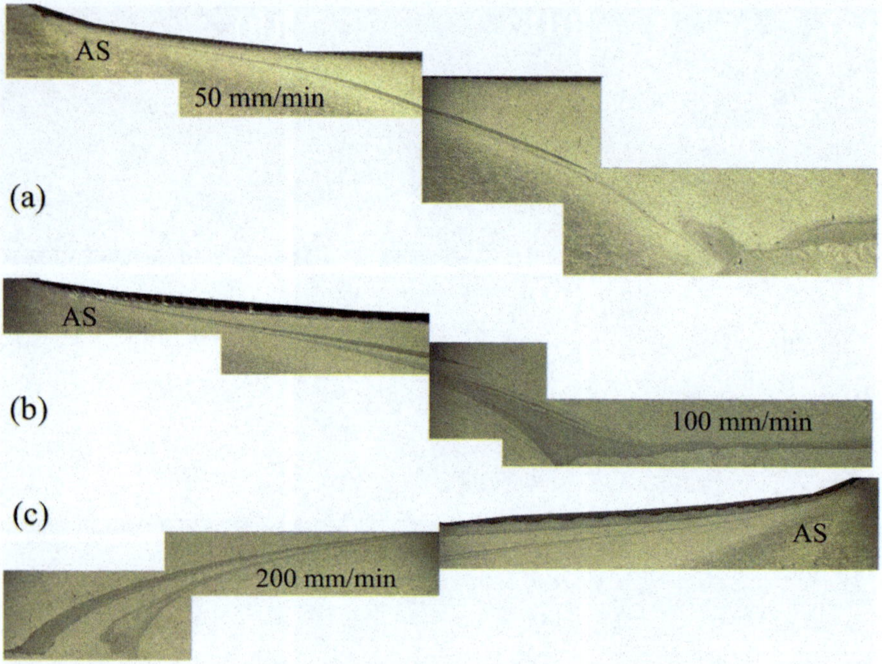

Fig. 2.48 The structure of band pattern at different welding speeds: **a** 50 mm/min, **b** 100 mm/min, and **c** 200 mm/min [72]

occurs due to thread of the stir pin. Therefore, flow pattern of the material near the top and the bottom of the weld surface becomes different from the middle part of the WNZ. As illustrated in Fig. 2.49, the weld zones of $SSFSW_1$ joints are divided into three parts. The first is zone I, where the softened metal driven by the big concave upper shoulder occurs via a horizontal flow around the axle of the $SSFSW_1$ tool. The second is zone II, where the softened metal driven by the screw thread on the pin is positioned upward in the spiral flow. The third is zone III, where the softened metal driven by the small convex lower shoulder. The regions bounded by dashed lines in Fig. 2.49b–d are the locations of onion rings or zigzag line for those micrographs given in Fig. 2.49b–d. The stacked onion ring features found in the WNZ are an evidence of the characteristic material transport occurring during $SSFSW_1$ process. It means that the material transfer in the WNZ takes place layer by layer under the function of the $SSFSW_1$ tool. In zone I, aside from the horizontal flow of the metal driven by the pressure force of the upper shoulder, an additional flow runs from upper surface to a depth of about one forth of weld thickness driven by the upper shoulder. In zone II, by employing a threaded tool the material is forced from the plate down into the weld and may travel several times depending on the rotational and welding speed.

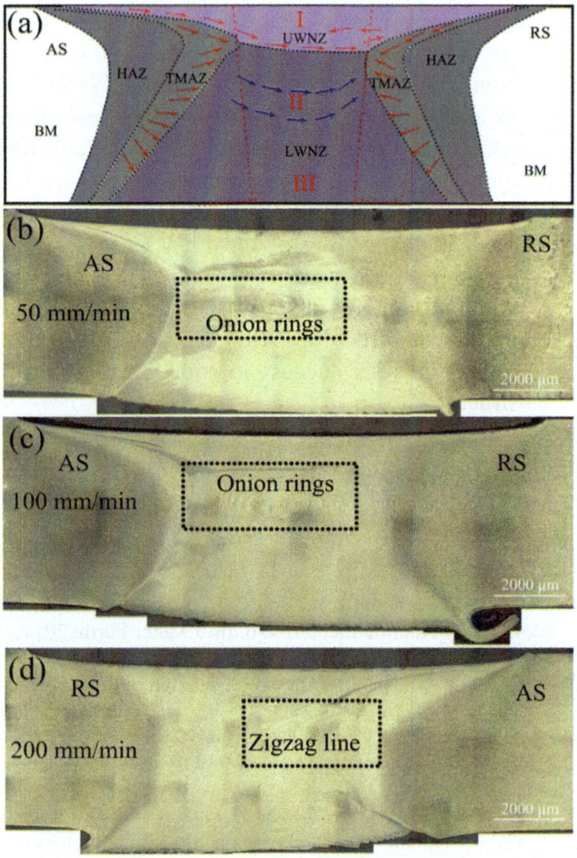

Fig. 2.49 Material flow around the $SSFSW_1$ tool: **a** Material plastic flow model for $SSFSW_1$, **b** 50 mm/min, **c** 100 mm/min and **d** 200 mm/min [72]

2.6 Prospects

The formation characteristics, material flow processes and mechanical properties of $SSFSW_1$ joints of 6xxx Al alloys have been extensively reported by our research group. However, 2xxx and 7xxx high-strength Al alloys have not been joined via $SSFSW_1$. Potential topics include but are not limited to the following.

(1) Design criterion of the welding tool. Tool materials and design of the tool shape are the key factors influencing the service life of the welding tool and joint quality. For 2xxx and 7xxx high-strength Al alloys, high-quality welding tools with high strength, good ductility and special tool shape need to be designed to guarantee the service life of the welding tool and the joint integrity.
(2) Static and fatigue performances. The tensile performance has been extensively studied, whereas the other static performances, such as shear tests, impact tests and corrosion tests, need to be further investigated to evaluate the overall

static performances. Fatigue performance, which affects the service life of the structural parts, is immature and should be emphasized to ensure stability.

(3) Non-weld-thinning $SSFSW_1$. During $SSFSW_1$, due to the plunge of the rotating shoulder and the small lower shoulder, weld thinning easily appears in the joint, which decreases joint strength. Hence, adjusting the structures of the welding tools and workpieces is extremely necessary.

(4) Commercial applications of $SSFSW_1$. With further research efforts and thorough understanding of the FSW process, an increasing number of applications have been realized for conventional FSW in the fabrication of Al alloys with high quality in aerospace and transportation fields. Although many challenges and difficulties still exist, $SSFSW_1$ provides very attractive possibilities for commercial success in some special parts, such as hollow and enclosed structures.

References

1. Trueba L, Torres MA, Johannes LB, et al. Process optimization in the self-reacting friction stir welding of aluminum 6061–T6. Int J Mater Form. 2017;11(4):559–70.
2. Marie F, Alléhaux D, Esmiller B. Development of the bobbin tool technique on various aluminum alloys. In: Annotated 5th international symposium friction stir welding, Metz, France: [s.n.], 2004. p. 39–52.
3. Li WY, Fu T, Hütsch L, et al. Effects of tool rotational and welding speed on microstructure and mechanical properties of bobbin-tool friction-stir welded Mg AZ31. Mater Des. 2014;64:714–20.
4. Thomas WM, Wiesner CS, Marks DJ, et al. Conventional and bobbin friction stir welding of 12% chromium alloy steel using composite refractory tool materials. Sci Technol Weld Join. 2009;14(3):247–53.
5. Hilgert J, Santos JF, Huber N, et al. Shear layer modelling for bobbin tool friction stir welding. Sci Technol Weld Join. 2012;17:454–9.
6. Chen S, Li H, Lu S, et al. Temperature measurement and control of bobbin tool friction stir welding. Int J Adv Manuf Technol. 2016;86(1–4):337–46.
7. Wahid MA, Khan ZA, Siddiquee AN. Review on underwater friction stir welding: a variant of friction stir welding with great potential of improving joint properties. Trans Nonferrous Met Soc China. 2018;28(2):193–219.
8. Sued MK, Pons D, Lavroff J, et al. Design features for bobbin friction stir welding tools: development of a conceptual model linking the underlying physics to the production process. Mater Des. 2014;54:632–43.
9. Pirizadeh M, Azdast T, Ahmadi SR, et al. Friction stir welding of thermoplastics using a newly designed tool. Mater Des. 2014;54:342–7.
10. Zhou L, Li GH, Liu CL, et al. Effect of rotation speed on microstructure and mechanical properties of self-reacting friction stir welded Al-Mg-Si alloy. Int J Adv Manuf Technol. 2017;89(9–12):3509–16.
11. Wang GQ, Zhao YH, Tang YY. Research progress of bobbin tool friction stir welding of aluminum alloys: a review. Acta Metall Sinica (Eng Lett). 2020;33:13–29.
12. Carter R, Kirby L. Gimballed shoulders for friction stir welding. NASA Tech Briefs 2008(January):27–28.
13. Nunes AC. Counterrotating-shoulder mechanism for friction stir welding. Huntsville: NASA Marshall Space Flight Center; 2007.

14. Dos Santos JF, Hilgert J. Apparatus for sriction stir welding: 中国. US20120248174A1. 2012-10-4.
15. Scupin P. Semi-stationary shoulder bobbin tool (S3BT): a new approach in high speed friction stir welding; 2015.
16. Goebel J, Martin R, Norman A, et al. Semi-stationary shoulder bobbin tool friction stir welding of AA2198-T851. J Mater Process Technol. 2017;245:37–45.
17. Hilgert J. Knowledge based process development of bobbin tool friction stir welding. Hamburg: Technische Universitat Hamburg; 2012.
18. Tongne A, Desrayaud C, Jahazi M, et al. On material flow in friction stir welded Al alloys. J Mater Process Technol. 2017;239:284–96.
19. Nandan R, Debroy T, Bhadeshia HKDH. Recent advances in friction-stir welding—process, weldment structure and properties. Prog Mater Sci. 2008;53(6):980–1023.
20. Huang Y, Wang Y, Wan L, et al. Material-flow behavior during friction-stir welding of 6082–T6 aluminum alloy. Int J Adv Manuf Technol. 2016;87(1–4):1115–23.
21. Colligan K. Material flow behavior during friction welding of aluminum. Weld J. 1999;75(7):229–37.
22. Guerra M, Schmidt C, McClure JC, et al. Flow patterns during friction stir welding. Mater Charact. 2002;49(2):95–101.
23. Schmidt HNB, Dickerson TL, Hattel JH. Material flow in butt friction stir welds in AA2024-T3. Acta Mater. 2006;54(4):1199–209.
24. Tamadon A, Pons D, Sued K, et al. Formation mechanisms for entry and exit defects in bobbin friction stir welding. Metals (Basel). 2018;8(1):33.
25. Schmidt HB, Hattel JH. Thermal modelling of friction stir welding. Scr Mater. 2008;58(5):332–7.
26. Hilgert J, Schmidt HNB, dos Santos JF, et al. Thermal models for bobbin tool friction stir welding. J Mater Process Technol. 2011;211(2):197–204.
27. Liu XM, Yao JS, Cai Y, et al. Simulation on the temperature field of bobbin tool friction stir welding of AA 2014 aluminium alloy. Appl Mech Mater. 2013;2013(433–435):2091–5.
28. Hilgert J, Dos Santos JF, Huber N. Investigation of the material shear layer in bobbin tool friction stir welding. In: Annotated Friction Stir Welding Process, IV. [S.l.]: [s.n.], 2011. p. 187–193.
29. Chen S, Lu A, Yang D, et al. Analysis on flow pattern of bobbin tool friction stir welding for 6082 aluminum. In: Annotated proceedings of 1st international joint symposium joint welding; 2013. p. 353–358.
30. Fraser KA, St-Georges L, Kiss LI. Numerical simulation of bobbin tool friction stir welding. In: 10th Friction stir welding symposium, Beijing: [s.n.]; 2014. p. 1–15.
31. Wen Q, Li WY, Gao YJ, et al. Numerical simulation and experimental investigation of band patterns in bobbin tool friction stir welding of aluminum alloy. Int J Adv Manuf Technol. 2019;100(9/12):2679–87.
32. Zhang H, Wang M, Zhang X, et al. Microstructural characteristics and mechanical properties of bobbin tool friction stir welded 2A14-T6 aluminum alloy. Mater Des. 2015;65:559–66.
33. Liu HJ, Hou JC, Guo H. Effect of welding speed on microstructure and mechanical properties of self-reacting friction stir welded 6061–T6 aluminum alloy. Mater Des. 2013;50:872–8.
34. Ji S, Huang R, Meng X, et al. Enhancing friction stir weldability of 6061–T6 Al and AZ31B Mg alloys assisted by external non-rotational shoulder. J Mater Eng Perform. 2017;26(5):2359–67.
35. Zhou L, Li GH, Zha GD, et al. Effect of rotation speed on microstructure and mechanical properties of bobbin tool friction stir welded AZ61 magnesium alloy. Sci Technol Weld Join. 2018;23(7):596–605.
36. Padmanaban G, Balasubramanian V, Sarin Sundar JK. Influences of welding processes on microstructure, hardness, and tensile properties of AZ31B magnesium alloy. J Mater Eng Perform. 2010;19:155–65.
37. Hou JC, Liu HJ, Zhao YQ. Influences of rotation speed on microstructures and mechanical properties of 6061–T6 aluminum alloy joints fabricated by self-reacting friction stir welding tool. Int J Adv Manuf Technol. 2014;73(5–8):1073–9.

38. Mishra RS, Ma ZY. Friction stir welding and processing. Mater Sci Eng R Reports. 2005;50(1–2):1–78.
39. Wang FF, Li WY, Shen J, et al. Effect of tool rotational speed on the microstructure and mechanical properties of bobbin tool friction stir welding of Al-Li alloy. Mater Des. 2015;86:933–40.
40. Ion SE, Humphreys FJ, Whi SH. Dynamic recrystallisation and the development of microstructure during the high temperature deformation of magnesium. Acta Metall. 1982;30(10):1909–19.
41. Singh K, Singh G, Singh H. Review on friction stir welding of magnesium alloys. J Magnes Alloy. 2018;6(4):399–416.
42. Park SHC, Sato YS, Kokawa H. Effect of micro-texture on fracture location in friction stir weld of Mg alloy AZ61 during tensile test. Scr Mater. 2003;49:161–6.
43. Esmaily M, Mortazavi N, Osikowicz W, et al. Corrosion behaviour of friction stir-welded AA6005-T6 using a bobbin tool. Corros Sci. 2016;111:98–109.
44. Threadgill PL, Ahmed MMZ, Martin JP, et al. The use of bobbin tools for friction stir welding of aluminium alloys. Mater Sci Forum. 2010;638–642:1179–84.
45. Xu WF, Luo YX, Fu MW. Microstructure evolution in the conventional single side and bobbin tool friction stir welding of thick rolled 7085–T7452 aluminum alloy. Mater Charact. 2018;138:48–55.
46. Sato YS, Kokawa H, Enomoto M, et al. Microstructural evolution of 6063 aluminum during friction-stir welding. Metall Mater Trans A. 1999;30(9):2429–37.
47. Esmaily M, Mortazavi N, Osikowicz W, et al. Bobbin and conventional friction stir welding of thick extruded AA6005-T6 profiles. Mater Des. 2016;108:114–25.
48. Kainuma S, Katsuki H, Iwai I, et al. Evaluation of fatigue strength of friction stir butt-welded aluminum alloy joints inclined to applied cyclic stress. Int J Fatigue. 2008;30(5):870–6.
49. Zhou C, Yang X, Luan G. Effect of kissing bond on fatigue behavior of friction stir welds on Al 5083 alloy. J Mater Sci. 2006;41(10):2771–7.
50. Wang FF, Li WY, Shen J, et al. Improving weld formability by a novel dual-rotation bobbin tool friction stir welding. J Mater Sci Technol. 2018;34(1):135–9.
51. Chen ZW, Pasang T, Qi Y. Shear flow and formation of Nugget zone during friction stir welding of aluminium alloy 5083-O. Mater Sci Eng A. 2008;474(1–2):312–6.
52. Yang C, Ni DR, Xue P, et al. A comparative research on bobbin tool and conventional friction stir welding of Al-Mg-Si alloy plates. Mater Charact. 2018;145(August):20–8.
53. Zhao Y, Wang C, Dong C. Microstructural characteristics and mechanical properties of water cooling bobbin-tool friction stir welded 6063–T6 aluminum alloy. MATEC Web Conf. 2018;206(2):03002.
54. Dalder ENC, Pastrnak JW, Engel J, et al. Bobbin-tool friction-stir welding of thick-walled aluminum alloy pressure vessels. Weld J. 2007:1–17.
55. Warsinski K, West M, Freeman J, et al. Investigation of lazy S feature in self-reacting tool friction stir welds. In: Annotated friction stir welding process, VI. [S.l.]: [s.n.]; 2011. p. 171–176.
56. Cao X, Jahazi M. Effect of welding speed on the quality of friction stir welded butt joints of a magnesium alloy. Mater Des. 2009;30(6):2033–42.
57. Skinner M, Edwards RL. Improvements to the FSW process using the self-reacting technology. Mater Sci Forum. 2003;426–432:2849–54.
58. Oonne CD, Biallas G. Microstructure and mechanical properties of the aluminium alloy 6056 welded by friction stir welding techniques. Weld World. 2006;50(11–12):98–106.
59. Neumann T, Zettler R, Vilaça P, et al. Analysis of self-reacting friction stir welds in a 2024-T351 alloy. In: Annotated friction stir welding process, IV. Orlando: [s.n.]; 2007.
60. Adriana M, Obregon T. Effect of process parameters on temperature distribution, microstructure, and mechanical properties of self-reacting friction stir welded aluminum alloy 6061–T651. El Paso: The University of Texas; 2011.
61. Okamoto K, Sato A, Park SHC, et al. Microstructure and mechanical properties of FSWed aluminum extrusion with bobbin Tools. Mater Sci Forum. 2012;706–709:990–5.

62. Zhou L, Li GH, Liu CL, et al. Microstructural characteristics and mechanical properties of Al–Mg–Si alloy self-reacting friction stir welded joints. Sci Technol Weld Join. 2017;22(5):438–45.
63. Zhang J, Feng XS, Gao JS, et al. Effects of welding parameters and post-heat treatment on mechanical properties of friction stir welded AA2195-T8 Al-Li alloy. J Mater Sci Technol. 2018;34(1):219–27.
64. Shen Z, Yang X, Zhang Z, et al. Mechanical properties and failure mechanisms of friction stir spot welds of AA 6061–T4 sheets. Mater Des. 2013;49:181–91.
65. Scialpi A, de Filippis LAC, Cavaliere P. Influence of shoulder geometry on microstructure and mechanical properties of friction stir welded 6082 aluminium alloy. Mater Des. 2007;28(4):1124–9.
66. Mendes N, Loureiro A, Martins C, et al. Effect of friction stir welding parameters on morphology and strength of acrylonitrile butadiene styrene plate welds. Mater Des. 2014;58:457–64.
67. Wang FF, Li WY, Shen J, et al. Global and local mechanical properties and microstructure of bobbin tool friction-stir-welded Al-Li alloy. Sci Technol Weld Join. 2016;21(6):479–83.
68. Liu H, Hu Y, Wang H, et al. Stationary shoulder supporting and tilting pin penetrating friction stir welding. J Mater Process Technol. 2018;255:596–604.
69. Hu Y, Liu H, Li S, et al. Improving mechanical properties of a joint through tilt probe penetrating friction stir welding. Mater Sci Eng A. 2018;731:107–18.
70. Huang YX, Wan L, Lv SX, et al. Novel design of tool for joining hollow extrusion by friction stir welding. Sci Technol Weld Join. 2013;18(3):239–46.
71. Wan L, Huang Y, Wang Y, et al. Friction stir welding of aluminium hollow extrusion: weld formation and mechanical properties. Mater Sci Technol. 2015;31(12):1433–42.
72. Wan L, Huang Y, Guo W, et al. Mechanical properties and microstructure of 6082–T6 aluminum alloy joints by self-support friction stir welding. J Mater Sci Technol. 2014;30(12):1243–50.
73. Wan L, Huang Y, Lv Z, et al. Effect of self-support friction stir welding on microstructure and microhardness of 6082–T6 aluminum alloy joint. Mater Des. 2014;55:197–203.
74. Balasubramanian V. Relationship between base metal properties and friction stir welding process parameters. Mater Sci Eng A. 2008;480(1–2):397–403.
75. Huang Y, Xie Y, Meng X, et al. Numerical design of high depth-to-width ratio friction stir welding. J Mater Process Technol. 2018;252:233–41.
76. Huang Y, Meng X, Wang Y, et al. Joining of aluminum alloy and polymer via friction stir lap welding. J Mater Process Technol. 2018;257(March):148–54.
77. Huang Y, Wan L, Huang T, et al. The weld formation of self-support friction stir welds for aluminum hollow extrusion. Int J Adv Manuf Technol. 2016;87(1–4):1067–75.
78. Edwards GA, Stiller K, Dunlop GL, et al. The precipitation sequence in Al-Mg-Si alloys. Acta Mater. 1998;46(11):3893–904.
79. Simar A, Bréchet Y, de Meester B, et al. Integrated modeling of friction stir welding of 6xxx series Al alloys: process, microstructure and properties. Prog Mater Sci. 2012;57(1):95–183.
80. Dong P, Li H, Sun D, et al. Effects of welding speed on the microstructure and hardness in friction stir welding joints of 6005A–T6 aluminum alloy. Mater Des. 2013;45:524–31.
81. Huang YX, Wan L, Lv ZL, et al. Microstructure and microhardness of aluminium alloy friction stir welds with heat treatment. Sci Technol Weld Join. 2016;21(8):638–44.

Chapter 3
Non-weld-Thinning Friction Stir Welding

Regardless of conventional fiction stir welding (FSW) [1] or its derived welding technologies [2, 3], weld thinning always forms at the joints and easily results in the strength loss. This is because that a rotating shoulder plunges into the welding workpieces to a certain thickness ($\geq$ 0.05 mm, which is determined by the thickness of the workpieces) to produce intense frictional heat and result in the overflow of the plasticized materials. The plunge of the rotating shoulder cannot be neglected, and weld thinning cannot be avoided, as shown in Fig. 3.1. With wide applications of conventional FSW in manufacturing fields, such as aerospace, railway and automobile industries, the negative problems caused by weld thinning have been progressively visualized [4]. First, the formation of weld thinning always produces flash defects and deteriorates the surface finishing and integrity. Second, weld thinning is prone to decrease the area of effective load bearing [5]. Kumar et al. [6] stated that when the plunge of the rotating shoulder increased to a certain extent, weld thinning markedly increased, which easily caused joint fracture in the weak region of the weld, drastically deteriorating the mechanical properties. Third, the acute edge induced by weld thinning usually forms between the base material (BM) and the nugget zone (NZ). Although mechanical properties of the joint with weld thinning can satisfy the service requirements, the acute edge still becomes a potential safety hazard during service. Under the action of load, the crack source is prone to form at the acute edge, which further reduces the anti-fatigue properties and service life. Especially in some fields that need a long service period, the acute edge of the workpieces must be avoided. Even if the acute edge can be burnished and transformed into a smooth transition, the potential safety hazard cannot be eliminated completely. Last, accurate welding and conformal welding concepts are widely mentioned and focused, which mean strict requirements for the appearance dimension of the structural parts without weld thinning. Hence, weld thinning is a substantial obstacle to achieving high-quality FSW joints.

Y. Huang et al., *Friction Stir Welding and Processing*, Materials Forming, Machining and Tribology, https://doi.org/10.1007/978-981-99-8688-0_3

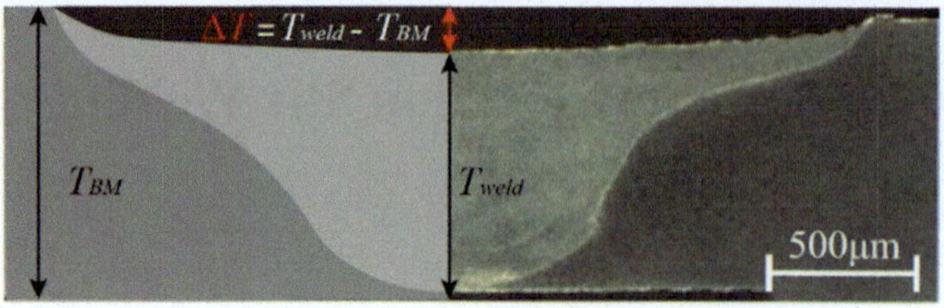

Fig. 3.1 Weld thinning of a typical FSW joint [the value of weld thinning (ΔT) is defined by the difference between the weld thickness (T_{weld}) and the BM thickness (T_{BM})]

Especially in military manufacturing fields, mechanical properties of the joints are not the sole criterion to evaluate the FSW joint quality. With the increasing requirements of joint quality, the FSW joint not only needs excellent mechanical properties but also requires superior surface integrity and outstanding service reliability. How to reduce or eliminate weld thinning and then achieve high-quality FSW joints is a key to expand FSW technological applications. To reduce or eliminate the weld thinning of the FSW joint and improve joint quality, researchers have performed extensive investigations from the viewpoints of the design in both of welding tool and workpiece structures as well as the assisted processes. The used welding technologies to reduce or eliminate weld thinning mainly consist of three categories. The first technology category seeks to prevent the overflow of plasticized materials, which comprises stationary shoulder FSW ($SSFSW_2$) or non-rotating shoulder assisted FSW (NRSA-FSW). The second technology category seeks to avoid the loss of plasticized materials, which is characterized by micro plunge depth FSW (MPD-FSW) or zero-plunge-depth FSW. The last technology category seeks to increase the additional materials to compensate for the loss of plasticized materials under the premise of adequate frictional heat and material flow, which comprises additive FSW (surface compensation FSW, T-plate assisted FSW and in situ additive FSW) and reductive FSW.

3.1 Stationary Shoulder Friction Stir Welding

According to the previously presented information, weld thinning is attributed to the plunge of the rotating shoulder that provides a sufficient forging force on the materials of the NZ. The frictional heat generated by the rotating shoulder always results in weld overheating, which deteriorates mechanical performance. Moreover, the generated frictional heat is useless for materials with low thermal conductivity, such as Ti alloys [7, 8] and polymers [9–11]. $SSFSW_2$ or NRSA-FSW, a solid-state joining technology derived from conventional FSW by TWI, has been extensively studied in the scientific community and applied in industry, which has achieved satisfactory results for butt-type, lap-type and T-type joints [12]. $SSFSW_2$ is deemed to be the next generation FSW technology by the European aeronautic defence and space company (EADS) [13]. Compared with conventional FSW, $SSFSW_2$ has many advantages, which are detailed hereafter:

(1) A smooth joint surface is beneficial to reducing surface machining after FSW and improving fatigue properties.
(2) Low heat input decreases the width of the HAZ, and the temperature distribution along the thickness direction is uniform.
(3) Stationary shoulder can transform the tensile stress in conventional joints into compressive stress, reducing the residual stress and welding distortion.
(4) Quasi-non-weld-thinning increases the effective area of load bearing.
(5) $SSFSW_2$ has the advantage of joining materials with low thermal conductivity and welding T joints or corner joints.
(6) $SSFSW_2$ can realize the filling of materials and provide a technical method for exploring additive manufacturing FSW.
(7) $SSFSW_2$ reduces the torque and advancing resistance, which is helpful to the development of robotic FSW.

In this chapter, welding tools, process parameters, material compatibilities, technical development, microstructural evolution and mechanical properties of $SSFSW_2$ joints are comprehensively reviewed. Additionally, the advantages, limitations and industrial applications of $SSFSW_2$ are neutrally discussed, laying the theoretical foundation for expanding the industrial manufacturing fields of $SSFSW_2$.

3.1.1 Principle

$SSFSW_2$ is a novel solid-state welding method, whose welding tool consists of a stationary shoulder and a rotating pin, as shown in Fig. 3.2 [12]. During $SSFSW_2$, the rotating pin plunges into the workpieces, while the stationary shoulder tightly contacts the top surface of the workpieces without a plunge depth. The welding workpieces can be heated and stirred by the thermo-mechanical effects induced by the rotating pin alone, which is different from that of conventional FSW. Under the enclosed chamber, which is composed of the stationary shoulder, rotating pin, surrounding hard materials and backing plate, the overflow of the plasticized materials is prevented, thereby achieving a sound non-weld-thinning joint.

The heat generation mechanism during $SSFSW_2$ is very different from that during conventional FSW, which further influences joint surface integrity and microstructural characteristics. During conventional FSW, approximately 70 ~ 80% of the main heat generation is produced by the friction between the rotating shoulder and the welding workpieces, whereas 20 ~ 30% of the heat generation is from the frictional heat induced by the rotating pin and the plastic deformation heat. Due to the relatively high heat input on the surface of the workpieces, the materials are easier to be softened and plasticized, and then overflow from the weld, causing weld thinning. For $SSFSW_2$, the primary heat input is produced by the rotating pin with

Fig. 3.2 Illustration of $SSFSW_2$ [12]

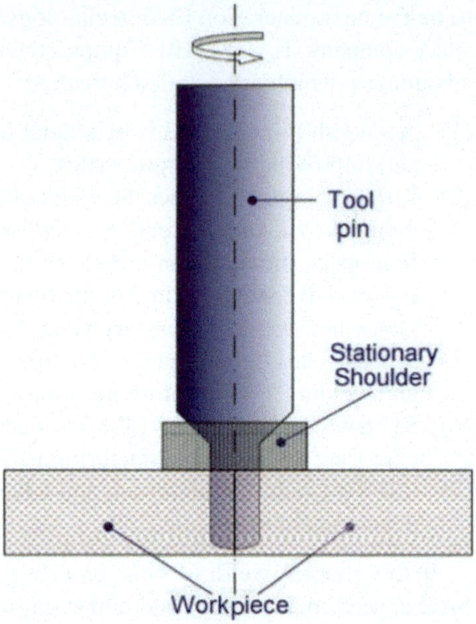

either no or low heat generation provided by the rotating shoulder. $SSFSW_2$ effectively reduces the softening degree of the materials on the surface of the workpieces and is beneficial to reducing or eliminating weld thinning. Moreover, for FSW of thick plates, when the heat generation and material flow induced by the rotating pin are optimized to guarantee structural integrity during conventional FSW, surface overheating forms and deteriorates the joint performance. $SSFSW_2$ can produce a uniform temperature distribution along the thickness direction and attribute to homogeneous microstructures, thereby improving mechanical properties. During conventional FSW, the translation of the rotating shoulder generally produces some flashes and regular semi-circular shoulder marks on the surface, as shown in Fig. 3.3a. The smoother surface results from the sliding effect of a stationary shoulder across the weld surface (Fig. 3.3b). The roughness in the $SSFSW_2$ joint is markedly reduced by over an order of magnitude compared with that in the conventional joint (Fig. 3.3c), whereas the weld thinning in the $SSFSW_2$ joint was also substantially reduced from about 1 mm to less than 0.2 mm (Fig. 3.3d) [14].

3.1.2 Welding Tools

During $SSFSW_2$, the rotating pin is the key factor of frictional heat and material transfer, which is primarily responsible for achieving high-quality joints. First, the rotating pin squeezes the plasticized materials and is subjected to bending force and torque, which requires the tool materials to have high strength, stability, ductility and

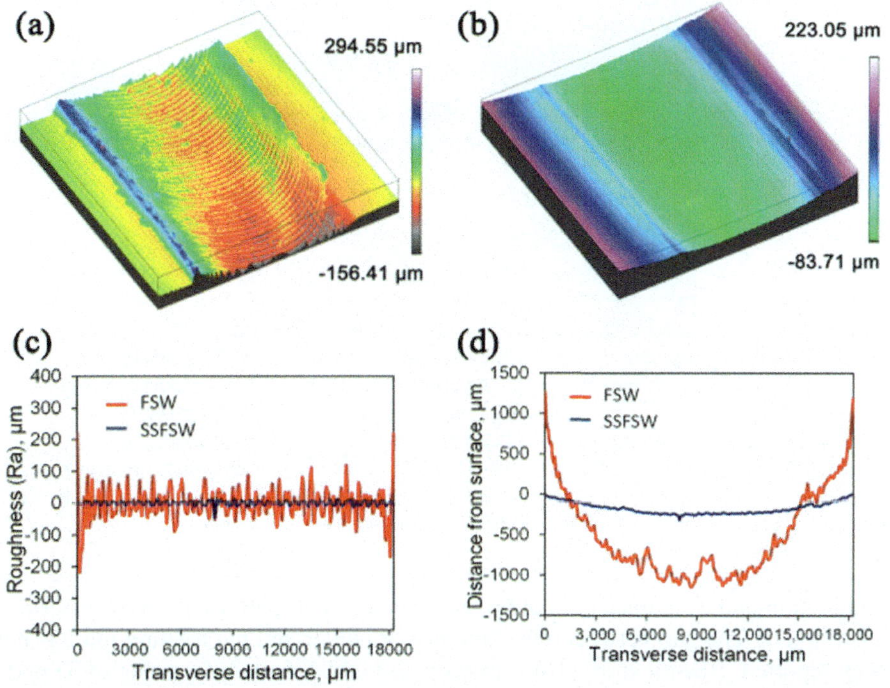

Fig. 3.3 Comparison of the surface quality with both processes: surface displacement maps of **a** conventional FSW and **b** $SSFSW_2$; **c** surface roughness (Ra) and **d** local displacement relative to the original sheet surface [14]

anti-fatigue properties at high temperature. Second, the wear resistance is another important factor to guarantee welding stability due to the severe friction between the rotating pin and the workpieces. Third, to improve the material flow around the surface of the rotating pin, the tool materials demand good machinability so that the surface morphologies of the rotating pin can be carved, such as threads and facets. Last, an important aspect is the gap between the stationary shoulder and the rotating pin. During $SSFSW_2$, the viscoplastic materials under the forging effects are easily squeezed into the gap, which result in weld thinning induced by the loss of plasticized materials or even the fracture of the rotating pin for long-term work conditions. This problem is the main difficulty facing $SSFSW_2$, which should be solved urgently in the industrial manufacturing fields.

TWI first designed an $SSFSW_2$ system for Ti alloys, as shown in Fig. 3.4, which mainly consisted of a stationary shoulder (10), a rotating pin (11), an inert gas protection system (13–15) and water cooling jackets (4) [15]. Water cooling jackets can protect the safety of welding equipment due to high heat generation induced by the high strength of Ti alloys, whereas inert gas protection can avoid oxides and nitrides. These tools significantly increase the complexity of the $SSFSW_2$ system. To reduce or eliminate tool wear and increase the homogeneity of microstructures, a tapered

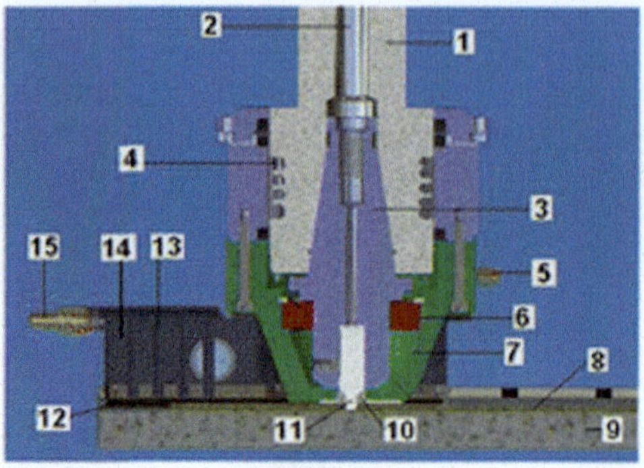

(1) Rotating spindle | (2) Draw bar | (3) Tool holder
(4) Water cooling jackets | (5) Argon input | (6) Support bearing
(7) Tool head | (8) Ti workpiece | (9) Backing plate
(10) Stationary shoulder | (11) Rotating probe | (12) Sliding seal
(13) Argon supply | (14) Gas chamber | (15) Inert gas input

Fig. 3.4 A cross-sectional view of the $SSFSW_2$ assembly [18]

rotating pin and a stationary shoulder are made of W–Re alloys and WC materials, respectively. The gap between the stationary shoulder and the rotating pin has not been reported. Russell et al. [16] also proposed additive manufacturing of Ti parts by multiple $SSFSW_2$ processes using both 1D and 2D welding paths, as displayed in Fig. 3.5. Brassington and Colegrove [17] also stated that the new development of $SSFSW_2$ tools was considered as a step change in the process evolution, in which the low heat input and smooth surface profile could have the potential to join Ti propellant tanks. EADS developed a robotic $SSFSW_2$ approach for Al alloys called Delta-N [13]. The welding system was significantly simplified without the gas protection or cooling device used for Al alloy butt joints, which only contained a stationary shoulder and a rotating tool, as shown in Fig. 3.6. The robotic $SSFSW_2$ approach successfully manufactured a 3D space structure.

Liu et al. [19] and Ji et al. [20] used $SSFSW_2$ to join AA2219-T8 and AA6005A-T6, respectively, wherein the rotating tools consisted of a small rotating shoulder and a rotating pin, as shown in Fig. 3.7. Here, the stationary shoulder only assisted the surface formation of joints. Due to the small rotating shoulder, this method can only be used for butt or lap welds, and it is difficult to join corner or complex structural parts.

The heat generation during FSW depends on the strength and hardness of the workpieces. The higher the strength and hardness, the higher the heat generation, the better the material flow. Due to the higher strength and hardness of metals, such as Ti alloys and Al alloys, relatively higher heat generation can be produced. Thermoplastic polymers or polymer matrix composites have relatively low strength and hardness, making it difficult to generate sufficient frictional heat and material flow. Moreover, the fluidity of the thermoplastic polymer is good, which easily overflows from the NZ, causing groove defects or weld thinning [22]. A modified $SSFSW_2$ called stationary

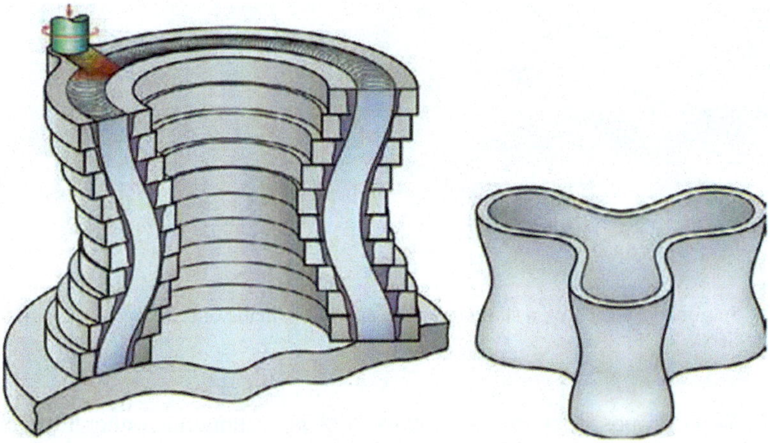

Fig. 3.5 Additive manufacturing by $SSFSW_2$ [16]

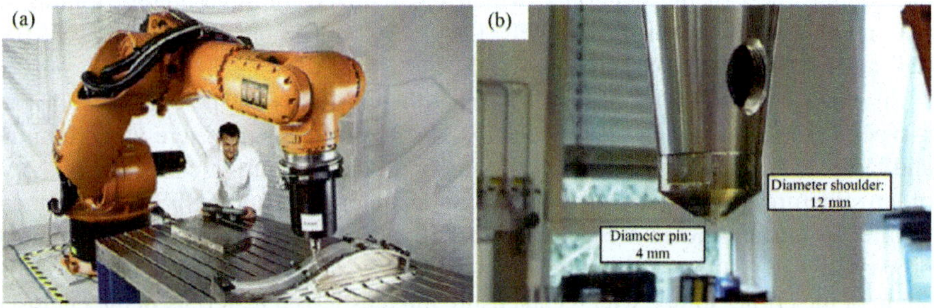

Fig. 3.6 Robotic $SSFSW_2$ by EADS: **a** welding equipment and **b** stationary shoulder [13]

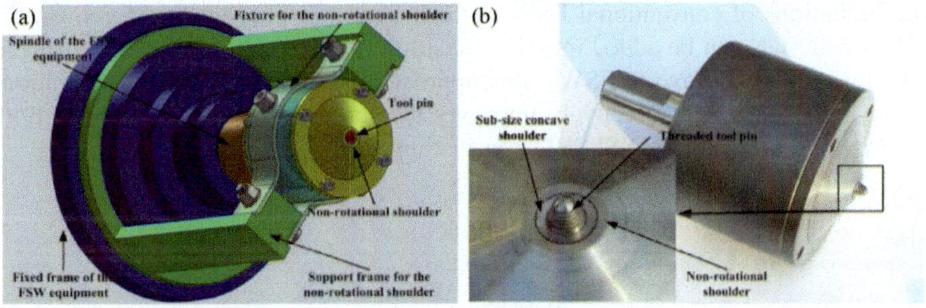

Fig. 3.7 Design of the $SSFSW_2$ tool: **a** model and **b** photograph [21]

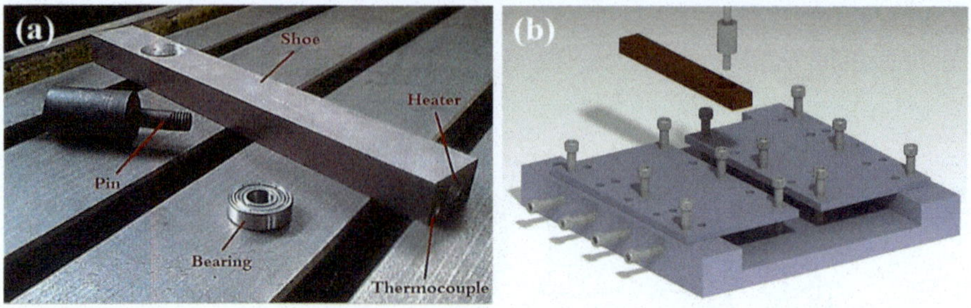

Fig. 3.8 Stationary shoe FSW: **a** stationary shoe and **b** schematic of the designed fixture for heat-assisted $SSFSW_2$ [25]

shoe FSW is proposed, whose stationary tool is rectangular rather than circular to guarantee joint integrity, as shown in Fig. 3.8. A closed-loop thermo-controller was responsible for supplying additional heating to the workpieces and slowing the cooling rates of the materials, which has realized the high-quality joining of polymers, such as polypropylene (PP) [23], acrylonitrile butadiene styrene (ABS) [24] and high-density polyethylene (HDPE) [10].

In addition to butt joints, $SSFSW_2$ also has the feasibility and potential to fabricate T-type or corner joints [26–29], which avoids the problems of severe weld thinning and lap defects, such as cold lap or kissing bond defects during conventional FSW of lap structures in T joints [30–32]. Schematic diagrams of the $SSFSW_2$ equipment for the T joints are shown in Figs. 3.9 and 3.10, containing non-additive and additive processes. The stationary shoulder is machined into a corner structure similar to the angle of the corner joints, which can ensure the plasticized materials in the enclosed chamber and then avoid the overflow of materials. The weld thinning in the skin and the lap defects between the skin and stringer are completely eliminated, overcoming the limitations of conventional FSW in the T-type or corner joints. Moreover, the filling materials can be added into the transition between the skin and the stringer, which is called additive $SSFSW_2$; this approach reduces the stress concentrations at the corner and then effectively improves the static and fatigue properties. Both additive and non-additive $SSFSW_2$ joints exhibit satisfactory joint coefficients.

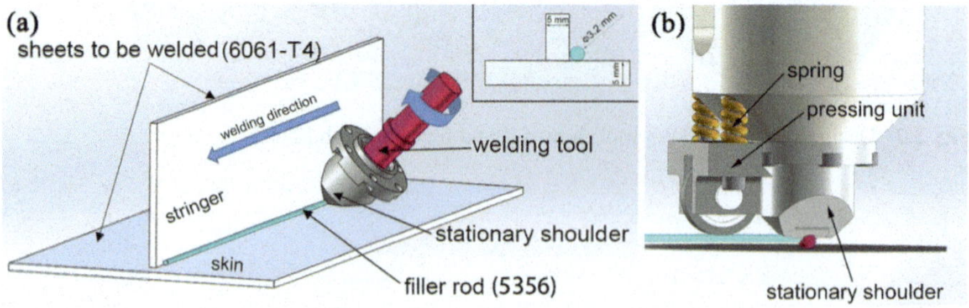

Fig. 3.9 Schematics: **a** welding process and **b** tool package for an additive T joint [26]

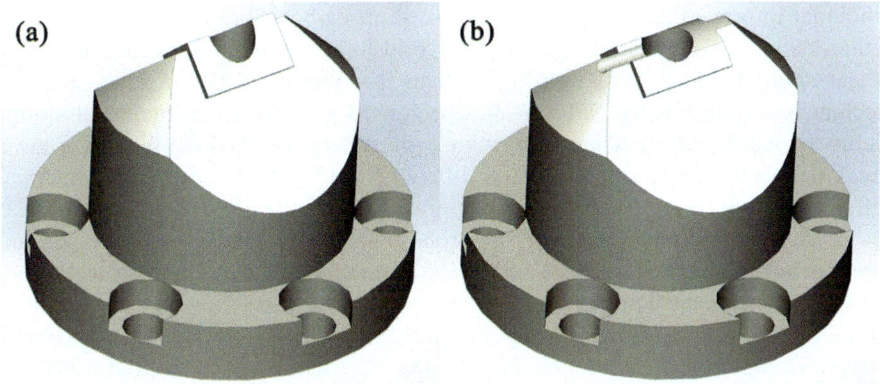

Fig. 3.10 Illustrations of stationary shoulders: **a** non-additive and **b** additive $SSFSW_2$ [26]

3.1.3 Material Compatibilities

Table 3.1 highlights the recent progress of $SSFSW_2$ in chronological order with a summary of research findings, in which one can note that the publication rate in this area has steadily increased since 2004 [27]. $SSFSW_2$, which is characterized by several factors, including uniform temperature distribution, non-weld-thinning, and smooth surface, has been successfully used to join Ti alloys, steels [28], coppers, Al alloys [especially high-strength Al alloys (2xxx [19] and 7xxx [33]) that cannot be welded by conventional fusion welding] and thermoplastic polymers [24]. Moreover, the other advantages induced by the stationary shoulder are remarkable and have also been used to join dissimilar materials in butt and lap joints, such as Al/Mg alloys [34, 35] and Al/polymers [36]. Aiming at materials with low thermal conductivity, $SSFSW_2$ produces a more uniform temperature distribution along the thickness direction and is firstly developed to join Ti alloys [18]. Al alloys, including the 2xxx, 6xxx and 7xxx series, have been most commonly used and extensively studied from the viewpoints of welding parameters and design of welding tools [19, 20, 37]. This is hardly surprising, considering that these materials are widely used in the industrial sectors because of their excellent properties [38]. Additionally, with the increasing demands of light weight, thermoplastic polymers and polymer matrix composites, $SSFSW_2$ has also attracted extensive attention for these materials. Due to the good flowability, the weld thinning phenomenon easily appears due to the overflow of the plasticized materials during conventional FSW [22]. Moreover, the extremely low thermal conductivity of the polymer (compared with that of Al alloys) always results in a non-uniform temperature distribution and overheating of materials at the top surface of the joint. Strand [39] firstly employed stationary shoe FSW, named $SSFSW_2$, to join PP and achieved remarkable results. Subsequently, Mostafapour and Taghizad Asad [25] proposed a stationary shoe with a heating system to preheat the materials in front of the rotating tool, further improving the surface quality and mechanical properties of the joints. Huang et al. [36] also reported that the stationary shoulder improved the surface finishing of short carbon fibre-reinforced polyether ether ketone (SCF/PEEK)/AA2060-T8 hybrid joint and that the maximum tensile shear strength of 33 MPa was higher than that produced by the state-of-the-art techniques. Additionally, $SSFSW_2$ has been used to create high-quality dissimilar joints of Al/Mg alloys. Compared with conventional FSW, $SSFSW_2$ has a lower peak temperature and simpler material flow driven by the rotating pin, which can effectively reduce the intermetallic compounds (IMCs) and the cold lap defect. Ji et al. [40, 41] first noted that $SSFSW_2$ could not only reduce the peak temperature and pin adhesion but also improve the joint formation and material mixing, thereby enhancing mechanical properties. They also combined $SSFSW_2$ with ultrasonication to join dissimilar Al/Mg alloys and eliminated pin adhesion, producing a maximum tensile strength of 152 MPa [35, 42, 43]. During friction stir lap welding (FSLW), the material flow behaviour is dominated by two modes: rotating pin affecting the hook and rotating shoulder controlling the cold lap [44, 45]. The application of $SSFSW_2$ in FSLW of similar or dissimilar materials significantly reduces lap defects and improves mechanical properties [46–48].

3.1.4 Microstructural Characteristics

Because the shoulder is static, the material flow and heat generation are significantly reduced, thereby decreasing the inhomogeneity degree of microstructures in the through-thickness direction. The primary frictional heat is generated around the rotating pin, further affecting the microstructural evolution, which presents significant differences relative to conventional FSW. The NZ in the conventional FSW joint presents an irregular shape, and the thermo-mechanically affected zone (TMAZ) in the conventional FSW joint exhibits asymmetry. These findings are attributed to the differences in shear stress and material flow on the advancing side (AS) and retreating side (RS) [62]. The NZ in the $SSFSW_2$ joint is similar to the shape of the rotating pin because the shoulder is static, and the width of the shoulder affected zone (SAZ) in this joint is far lower than that in a conventional FSW joint. In fact, the microstructural evolution is closely connected with the welding thermal cycle that depends on the thermal and mechanical properties of the welding workpieces. In this chapter, three kinds of typical materials are emphasized: Ti alloys with low thermal conductivity and high strength, Al alloys with high thermal conductivity and medium strength, and thermoplastic polymers with low thermal conductivity and low strength.

1. Ti alloys

Ti alloys, due to their high specific strength and good corrosion resistance, have been the first choice for the next generation of jet engines in the aerospace industry. FSW of Ti alloys is the primary focus of researchers. Due to low thermal conductivity, a large temperature gradient along the thickness direction easily results in inhomogeneous microstructures or tearing defects in the NZ, reducing mechanical properties [12, 63].

Table 3.1 $SSFSW_2$ in chronological order with focus/findings

Authors (References)	Year	Materials	Focus/findings
Russell et al. [15]	2007	Ti-6Al-4V	$SSFSW_2$ was first developed by TWI primarily to improve FSW of Ti-based alloys
Russell et al. [16]	2008	Ti-6Al-4V	Additive manufacturing of Ti parts by $SSFSW_2$ processes performed using both 1D and 2D welding
Ahmed et al. [49]	2011	AA6082-T6	Plastic deformation in the through thickness direction was dominated by the pin in $SSFSW_2$
Li and Liu [21]	2013	AA2219-T6	$SSFSW_2$ prevented the plasticized materials from escaping from the NZ and eliminated flash defects and weld thinning, achieving a tensile strength of 307 MPa
Liu et al. [19]	2013	AA2219-T6	Microstructure and weld formation were dominated by the rotating pin and small concave shoulder
Li et al. [50]	2014	AA6061-T6	AA6061-T6 $SSFSW_2$ joint had narrower TMAZ and HAZ, the microstructures were more symmetrical and homogeneous, and the maximum tensile strength was 77.3% of that of the BM
Ji et al. [20]	2014	AA6005-T6	The width of SAZ was small. The tensile strength and elongation of the joint reached 82% and 65% of those of the BM
Wu et al. [14]	2015	AA7050-T7651	The cold surface temperature dramatically decreased the normal workpiece thickness and can lead to "speed cracking" similar to that found in extrusion under hotter welding conditions
Ji et al. [41]	2016	AA6061-T6/ AZ31B	Excellent material mixing happened and IMCs were reduced due to heat sink from stationary shoulder
Sun et al. [51]	2017	AA7010-T7651	A considerably smaller thermal field was developed with $SSFSW_2$, which reduced the size of the region of inhomogeneous compression created by the local temperature rise ahead of the tool
Sun et al. [52]	2018	AA7050	The longitudinal residual stress distributions from all three techniques showed good agreement within the uncertainty limits. A high tensile residual stress field (~ 170 MPa) was found in the NZ
Barbini et al. [53]	2018	AA2024/AA7050	$SSFSW_2$ decreased both the welding area and the diffusion at the interface of the two alloys

(continued)

Table 3.1 (continued)

Authors (References)	Year	Materials	Focus/findings
Meng et al. [43]	2018	AA6061-T6/ AZ31B	Stationary shoulder coupled with ultrasonic improved the surface formation of Al/Mg joint and broke the IMCs into pieces, thereby increasing mechanical properties
Regensburg et al. [54]	2018	AW1050 H14/ CW024A R240	$SSFSW_2$ was applied to produce 2-mm-thick EN AW1050/CW024A lap joints with minimized intermixing at the interface, lap defects and significant IMCs
Huang et al. [36]	2018	SCF/PEEK and AA2060-T8	Stationary shoulder enhanced the surface finishing, and the maximum tensile shear strength of 33 MPa was higher than that achieved by the-state-of-the-art technology
Strand [39]	2004	PP	Stationary shoe FSW was first proposed to join thermoplastic PP
Mostafapour [55]	2012	HDPE	HDPE could be welded at a high shoulder temperature and rotational velocity and a low welding speed. Tool coating improved mechanical properties and surface qualities
Kiss and Czigany [56]	2012	PETG	A stationary shoe without a heating system was used to join amorphous PETG, which was difficult to be welded using conventional tools. The rotating pin could break and randomly distribute spherulite
Bagheri et al. [57]	2013	ABS	A high rotational velocity and shoe temperature and a low welding speed improved welding quality
Rahbarpour et al. [58]	2014	Wood/plastic	FSW of wood/polymer composites by a stationary shoe with a heating system was performed to eliminate voids and poor mixing. The maximum tensile strength of the joint was 92.95% of that of the BM
Mendes et al. [59]	2014	ABS	A high-quality joint was obtained for a tool temperature of 115 °C, an axial force greater than 1.5 kN, a rotational velocity higher than 1250 r/min and a low welding speed of 50 ~ 100 mm/min
Eslami et al. [11]	2015	PP/PC	One of the main challenges was to prevent material overflow inside the shoulder, which caused shoulder failure, especially in long joints. A Teflon polymeric stationary shoe was the best option
Czigány and Kiss [60]	2015	Glass fibre-reinforced PP	$SSFSW_2$ realized the homogenization and interlocking of fibres. A screwed pin with eight teeth reduced the length of fibres, thereby decreasing mechanical properties

(continued)

Table 3.1 (continued)

Authors (References)	Year	Materials	Focus/findings
Mostafapour and Taghizad Asad [25]	2016	Nylon 6	Nylon 6 was joined via heat-assisted $SSFSW_2$ with an external heat source. A high-quality joint with over 98% of the base strength was obtained
Eslami et al. [61]	2018	HMHDPE	The vertical positioning of the tool was responsible for the axial force, thereby providing both the forging pressure and frictional heat

A back-heat-assisted FSW method was proposed to improve the uniform temperature distribution and eliminate tearing defects [8, 63], which always required heating equipment. $SSFSW_2$ arises at a historic moment to improve the homogeneous distribution of microstructures and phases. Based on the characteristics of Ti alloys, the FSW joint is mainly divided into three distinct zones: NZ, TMAZ and HAZ, as shown in Fig. 3.11. Some researchers have stated that no TMAZ occurred in the $SSFSW_2$ joints of Ti alloys [64], whereas scholars claimed that a narrow TMAZ existed when observing an obvious microstructural evolution due to the limited plastic deformation and heat input [65, 66]. The microstructural evolution in the TMAZ and heat affected zone (HAZ) is primarily affected by the microstructures of the BM. The microstructure in the NZ is a result of processing conditions, tool geometry, cooling rates and material compositions.

The microstructural evolution of Ti alloys involves transformation, nucleation, and growth of the α and β phases in the BM. Ti-6Al-4V, a kind of Ti alloy, is the most extensively studied Ti alloy due to its extensive industrial applications. Figure 3.12 shows a schematic of the microstructural evolution of Ti-6Al-4V during FSW, which is divided into the welding stage and cooling stage. When the peak temperature in the NZ is above the β-transus temperature, the BM is destroyed and dynamic recrystallization (DRX) occurs in the welding stage. Then, fine, equiaxed and transformed β phases are achieved at the end of the welding stage. In the cooling stage, the materials in both the upper and lower surfaces of the NZ suffer from fast cooling, and small grain boundary α phases are formed. At the end, smaller lamellar

Fig. 3.11 Macrostructure of a Ti-6Al-4V $SSFSW_2$ joint [12]. *Note ND* normal direction; *TD* transverse direction

$\alpha + \beta$ phases within the prior β grains are obtained. Due to low thermal conductivity, the materials in the middle region of NZ undergo longer high-temperature exposure times. The slower cooling process is attributed to the larger lamellar $\alpha + \beta$ phases within the prior β grains in the end. In the HAZ, the temperature is below the β-transus temperature. However, the heat cycle still causes grain coarsening, leading to an $\alpha \rightarrow \beta$ transus. In the cooling stage, $\beta \rightarrow \alpha + \beta$ occurs; hence, bimodal microstructures, which is characterized by primary coarser α grains and lamellar $\alpha + \beta$ phases within prior β grains, can be observed. In the welding stage, after being broken by the rotating tool, a higher temperature but below the β-transus can be obtained at the NZ top. Hence, some $\alpha \rightarrow \beta$ occurs. Upon cooling, $\beta \rightarrow$ lamellar $\alpha + \beta$ occurs; hence, small lamellar structures can be obtained. In the middle and bottom regions of NZ, the temperatures are far below the β-transus temperature. The stirring action plays a more important role, and only fine and equiaxed grains can be observed. Reducing the temperature gradient can be beneficial to achieving uniformly distributed microstructures by $SSFSW_2$.

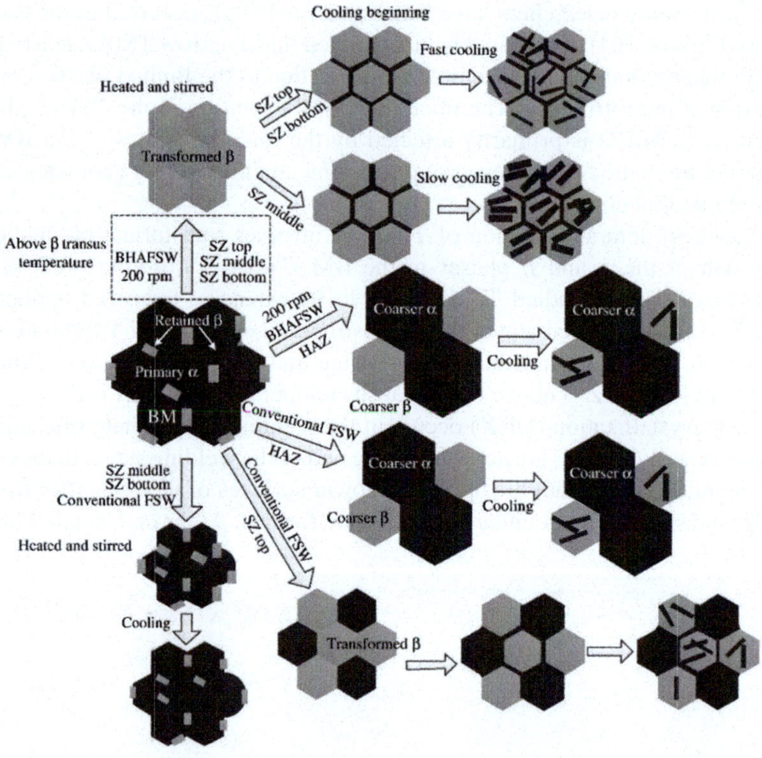

Fig. 3.12 Diagrammatic sketch of microstructural evolution during FSW of Ti-6Al-4V alloys [8]

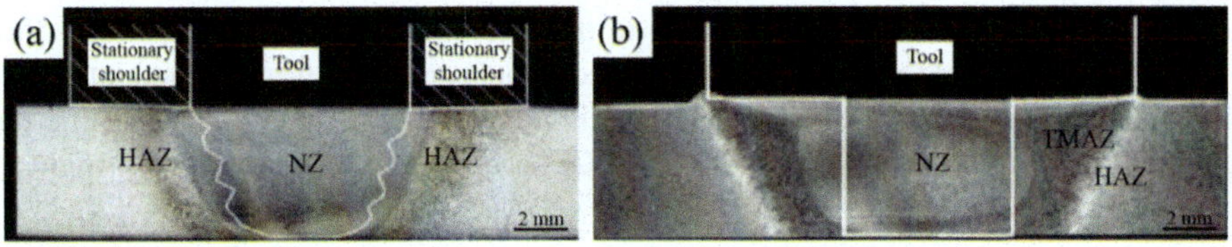

Fig. 3.13 Macrostructures: **a** $SSFSW_2$ and **b** conventional FSW joints [50]

2. Al alloys

Al alloys are classified into precipitation-strengthened alloys and solid-solution-strengthened alloys. Precipitation-strengthened Al alloys, such as 2xxx, 6xxx and 7xxx Al alloys, are the most investigated materials by $SSFSW_2$ due to their increasing demands and extensive applications in railway and aerospace fields [67]. It is known that the peak temperature during FSW can approach about 0.8 T_m or even 0.95 T_m, which is higher than the solution temperature of many precipitation-strengthened Al alloys [68]. The welding heat cycle and mechanical stirring influence the evolution of the grain size, dislocations and especially precipitates, which further influence the subsequent mechanical properties. The thermo-mechanical behaviours simultaneously affect the joint formation. As shown in Fig. 3.13, the SAZ is significantly reduced or even eliminated due to the stationary shoulder, and the widths of the TMAZ and HAZ are decreased due to the reduced heat input on the top surface of the joint compared with that in the conventional FSW joint. For the sound $SSFSW_2$ joint of the precipitation-strengthened Al alloys, the fracture location of the joints easily occurs at the HAZ with the severest softening degree. Therefore, the microstructural evolution in the HAZ directly determines the global mechanical properties of the $SSFSW_2$ joints.

For the typical age-hardened 2xxx Al–Cu–(Mg) Al alloys, the evolution sequence of precipitates is as follows: SSS $\rightarrow$ GP(I) $\rightarrow$ GP(II) $\rightarrow$ metastable phase θ'' $\rightarrow$ metastable phase θ' $\rightarrow$ stable phase θ (Al_2Cu). In 6xxx Al–Mg–Si Al alloys, Mg and Si are the major alloying elements, contributing to precipitation strengthening. The precipitation sequence is as follows: SSS $\rightarrow$ clusters $\rightarrow$ initial β'' $\rightarrow$ (pre-β''/needle-shaped β'' precipitate) $\rightarrow$ (rod-shaped precipitates β'/lath-shaped β' precipitates/U1, U2) $\rightarrow$ (β-Mg_2Si/Si). In the 7xxx Al–Zn–Mg–(Cu) Al alloys, the precipitation sequence is as follows: SSS $\rightarrow$ metastable phase η' $\rightarrow$ stable phase η [69–71]. The BM consists of a large number of uniformly distributed fine precipitates (approximately 5 nm in size) and some larger dispersoids. Li et al. [33] reported that the distribution and quantity of dispersoids in the HAZ are similar to those in the BM. However, the precipitates in the HAZ are larger than those in the BM. In the NZ, fine precipitates are absent, and the dispersoids are much larger than those in the BM. Liu et al. [19, 21, 72] examined the microstructural evolution of SSFSW of 2219-T6 Al alloys. The thermal cycle drives metastable phases (GP zones and θ'') to transform to θ' and θ phases and then inhibits the growth of recrystallized grains in the NZ. The microhardness is increased by the formation of θ' phases, but the coarsening of

θ' phases explicitly causes a reduction in the microhardness of Al–Cu alloys. Sun et al. [73] also stated that the needle-like precipitates grew and transformed into short rod-like precipitates as a result of the welding thermal cycle in the HAZ, while the NZ was the only one that showed a large GP zone dissolution peak and a high β'' precipitation peak.

3. Thermoplastic polymers

Thermoplastic polymers were firstly joined via $SSFSW_2$ by Strand [39], whose cross-sectional macrostructure is shown in Fig. 3.14. The $SSFSW_2$ joint is separated into BM, NZ, TMAZ and bottom disturbance. Due to the extremely low thermal conductivity, no HAZ appears in the $SSFSW_2$ joints of polymers. The NZ is characterized by onion rings and a relatively narrow TMAZ. A typical defect occurs near the bottom disturbance due to insufficient stirring and low thermal conductivity and becomes the weakest region during the tensile test, which can be eliminated by double-pass FSW [74, 75]. Kiss and Czigany [56] stated that stirring action of the rotating pin could break and randomly distribute spherulite. Additionally, for fibre-reinforced thermoplastic composites, the main difficulty is the lack of continuity between the fibres in the BM and NZ. Moreover, Czigány and Kiss [60] performed $SSFSW_2$ of glass fibre-reinforced composites and reported that this method could realize the homogenization and interlocking of the fibres, as indicated in Fig. 3.15. The fracture surface showed that fibres are extracted from the NZ, indicating interlocking with each other inside the NZ.

Gao et al. [77] performed differential scanning calorimetry (DSC) for different regions of the polyethylene (PE) joint and analysed the crystallization behaviour induced by the uneven temperature distribution with the following formula:

$$W_c = \frac{\Delta H_m}{\Delta H_m^0} \times 100\% \tag{3.1}$$

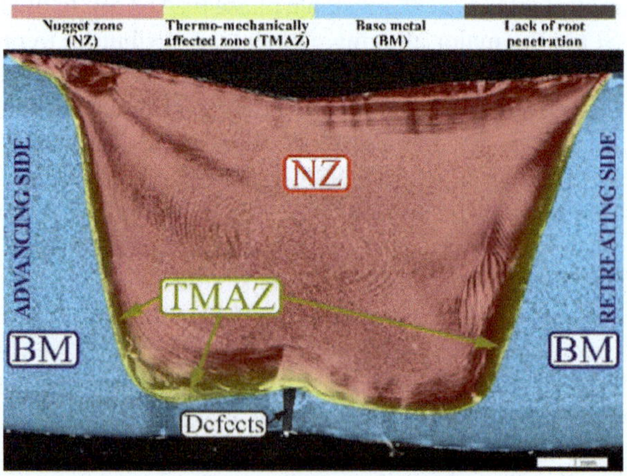

Fig. 3.14 Macrostructure of a typical $SSFSW_2$ joint of PP [76]

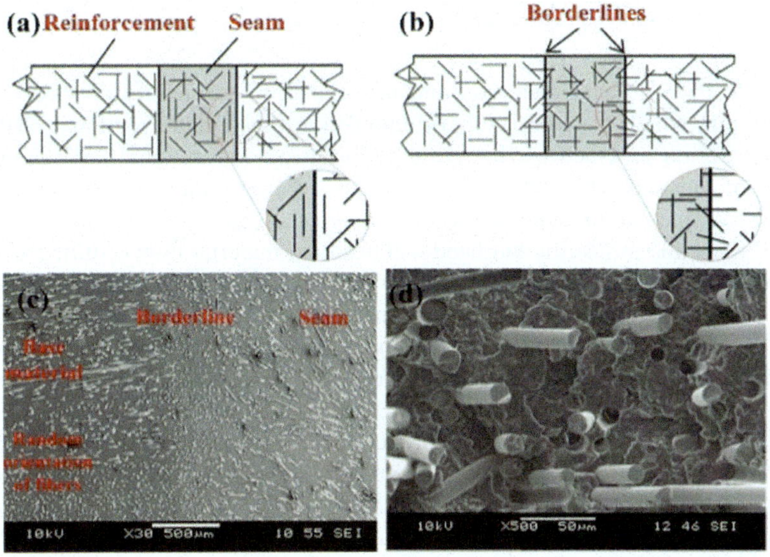

Fig. 3.15 Welded zone in fibre-reinforced composites with a thermoplastic matrix: **a** non-intermeshing fibres and **b** intermeshing fibres at the interface and **c** borderline between the NZ and the BM and **d** the fracture surface [60]

where ΔH_m is the melting enthalpy of the sample and ΔH_m^0 is the melting enthalpy of perfectly crystalline PE, for which the value is 293 J/g. The crystalline contents of the BM, HAZ, TMAZ and NZ were 54.5%, 54.0%, 51.1% and 48.2%, respectively (Fig. 3.16). There is no obvious HAZ due to the low thermal conductivity. The crystallinity of the TMAZ and NZ was significantly reduced due to the low cooling rates.

Fig. 3.16 DSC curves of a PE FSW joint [77]

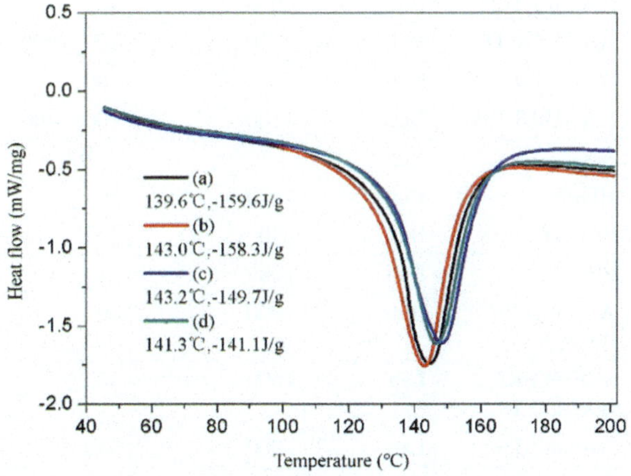

3.1.5 Process Development

During $SSFSW_2$, heat generation and material flow are mainly provided by the rotating pin, which are completely different from those in conventional FSW. The heat generation provided by the rotating pin during conventional FSW is only 20 ~ 30% of the total heat input. The optimum welding parameters during conventional FSW that can be beneficial to obtaining excellent joints may not be suitable for $SSFSW_2$ due to insufficient frictional heat and material flow resulting from the loss of frictional heat generated by the rotating shoulder. There are obvious differences in welding parameters between conventional FSW and $SSFSW_2$. Moreover, $SSFSW_2$ can also prevent the plasticized materials from overflowing from the weld at extremely high heat inputs during conventional FSW and broaden the welding parameter windows. According to the theories of heat generation during FSW [78–80], heat input is closely correlated with base properties of workpieces (yield strength, hardness, etc.), welding parameters (welding speed, rotational velocity, plunge depth, etc.) and equipment parameters (shoulder or pin diameter, pin morphology, etc.). Table 3.2 lists the optimum welding parameter windows for $SSFSW_2$ of different materials with different thermal and mechanical properties.

For materials (Ti, Ti alloys, etc.) with high yield strength and low thermal conductivity, the relatively low rotational velocity and welding speed are feasible. Rai et al. [18] stated that sound joints could be achieved at rotational velocities of 400 ~ 500 r/

Table 3.2 Optimum welding parameter windows for $SSFSW_2$ of different materials

Materials	Joint type	Rotational velocity (rpm)	Welding speed (mm/min)	Preheating temperature (°C)	References
T-i6Al-4V	Butt	400 ~ 500	60 ~ 80		[18]
Steel	Butt	200	120 ~ 140		[28]
AA2219-T6	Butt	600 ~ 900	100		[81]
AA6005A-T6	Butt	2000	300 ~ 600		[20]
AA6061-T6	Butt	750 ~ 1500	100		[50]
AA7050-T7651	Butt	1500	100 ~ 400		[14]
ABS	Butt	$\geq$ 1250	$\leq$ 100	115	[24]
HDPE	Butt	1400	25	100	[10]
Nylon 6	Butt	630	20	150	[25]
PP	Butt	1080	51	160	[39]
AA6061-T6/ AZ31B	Butt	1000	60		[35]
SCF/PEEK/ AA2060-T8	Lap	1600	30		[36]
AA6082-T6	Lap	2000	840		[27]
Pure copper	Butt	3000	300		[27]

min and welding speeds of 60 ~ 80 mm/min during $SSFSW_2$ of Ti-6Al-4V alloys. Maltin et al. [28] also stated that defect-free $SSFSW_2$ joints of steels can be produced at a rotational velocity of 200 r/min and a welding speed of 120 or 140 mm/min. Aiming at Cu or Cu alloys with high thermal conductivity [82], the frictional heat quickly transfers from the weld to the surrounding materials, and then results in welding defects, which always demand high heat input (high rotational velocity or low welding speed). Wei and Martin [27] stated that Cu sheets with a thickness of 0.85 mm were successfully joined using a rotational velocity of 3000 r/min, a welding speed of 300 mm/min and a Z-direction force of 6 kN, in which the specially high welding speed was used to avoid the distortion of the ultra-thin sheets. No remarkable wear was observed on the tool pin or stationary shoulder.

For Al alloys, Liu et al. [81] reported that sound joints were only obtained at a rotational velocity of 800 r/min in conventional FSW, whereas $SSFSW_2$ could produce defect-free joints in a wider range of rotational velocities of 600 ~ 900 r/min at a constant welding speed of 100 mm/min for AA2219-T6. Wu et al. [14] compared conventional FSW with $SSFSW_2$ of AA7050-T7651 based on the same size rotating pin and a welding speed of 400 mm/min. The optimum rotational velocities in conventional FSW and $SSFSW_2$ were 700 r/min and 1500 r/min, respectively. $SSFSW_2$ generated a about 30% lower heat input than conventional FSW, which achieved a narrow HAZ. Ji et al. [20] stated that a sound joint could be attained over a wide welding speed range, from 300 to 600 mm/min, based on a small rotating shoulder and a constant rotational velocity of 2000 r/min during $SSFSW_2$ of AA6005A-T6. Li et al. [50] indicated that tensile properties for $SSFSW_2$ joints of AA6061-T6 were nearly unaffected by rotational velocities of 750 ~ 1500 r/min at a constant welding speed of 100 mm/min, whereas increasing the welding speed from 100 to 300 mm/min markedly increased the tensile properties at a constant rotational velocity of 1500 r/min. The maximum joint efficiency of 77.3% (256 MPa) was achieved under welding parameters of 1500 r/min and 300 mm/min. To attain sound $SSFSW_2$ Al alloy butt joints with high strength, the rotational velocities of 500 ~ 2000 r/min and the welding speeds of 50 ~ 600 mm/min are appropriate.

For polymer materials (thermoplastic polymers or polymer matrix composites) with extremely low yield strength and thermal conductivity [83, 84], it is difficult to transfer heat input to preheat the materials at the front of the rotating tool. Higher heat input (high rotational velocity and low welding speed) or a secondary heat source is required to attain sound joints. Mendes et al. [24] stated that high strength efficiency was achieved only when a high rotational velocity and a high axial force were used during $SSFSW_2$ without a heating system in 6-mm-thick ABS sheets. Rotational velocity and axial force values above a certain threshold were required to produce defect-free joints. A high-quality joint was obtained for a tool shoulder temperature of 115 °C, an axial force higher than 1.5 kN, rotational velocities greater than 1250 r/min and low traverse speed ranging between 50 and 100 mm/min [59]. Azarsa and Mostafapour [10] performed $SSFSW_2$ with HDPE and stated that a combination of high rotational velocity and low welding speed increased flexural strength by reducing the sizes of defects. The optimum welding parameters to achieve maximum flexural strength were a rotational velocity of 1400 r/min, a welding speed of 25 mm/

min and a shoulder temperature of 100 °C. Aydin [85] adopted backing preheating at 50 °C to join a 4-mm-thick ultrahigh molecular weight PE (UHMW-PE) sheet and then improved joint formation due to a uniform temperature gradient through the thickness direction. Vijendra and Sharma [86] stated that induction heating enabled materials to be plasticized in a short time and to be easily stirred. The optimum conditions for the maximum joint strength were a tool pin temperature of 45 °C, a welding speed of 50 mm/min and a rotational velocity of 2000 r/min.

Additionally, aiming at the $SSFSW_2$ of dissimilar materials, the welding parameters depend on the materials or joining types (butt or lap joints). For dissimilar materials with severe interfacial reactions, mechanical interlocking and metallurgical bonding are attributed to the joint strength. Meng and his co-workers [35, 42, 43] stated that a rotational velocity of 1000 r/min could be responsible for the sufficient heat input, while a low welding speed of 60 mm/min could provide much more stirring time to improve the mechanical interlocking for $SSFSW_2$ of Al and Mg alloys. A high welding speed higher than 60 mm/min easily resulted in welding defects, which can be further increased to 80 mm/min assisted by ultrasonication [87]. When using a lap configuration of polymer at the upper and metal at the lower, the selection of welding parameters should be in accord with the joining of polymers, which needs relatively lower welding speeds lower than 50 mm/min without a secondary heating system. For lap joints of similar or dissimilar materials, the process variations are different from those of butt joints, whose rotational velocity is relatively lower [47, 88] and welding speed is slightly higher due to the original interface in butt joints disappearing in the upper sheet of lap welding.

3.1.6 Mechanical Properties

$SSFSW_2$ not only inherits the advantages including low peak temperature, small residual stress and distortion, simple process and non-pollution of conventional FSW, but also generates a concentrated heat input, reducing the joint softening degree and substantially improving static mechanical properties. Figure 3.17 shows the mechanical properties of $SSFSW_2$ joints. Wu et al. [14] compared the mechanical properties between $SSFSW_2$ and conventional FSW of AA7050-T7651. $SSFSW_2$ provided an approximately 30% lower heat input and 20 ~ 40 MPa higher yield and tensile strength than those of conventional FSW joints, which were attributed to the compatible deformation induced by uniform thickness properties. Huang et al. [36] established a radar chart to describe the relationships between weld thinning, microstructural characteristics and mechanical properties (Fig. 3.18). They stated that $SSFSW_2$ could control the microstructural evolution and subsequently affect mechanical properties, which were better than those of the state-of-the-art techniques. The microstructure, surface stress state and stress concentration were investigated at the fatigue crack initiation sites. The stress concentration determines the fracture location, and the surface microstructure and surface stress state influence the fatigue properties. $SSFSW_2$ can reduce the residual stress to a certain extent since the stationary shoulder exerts

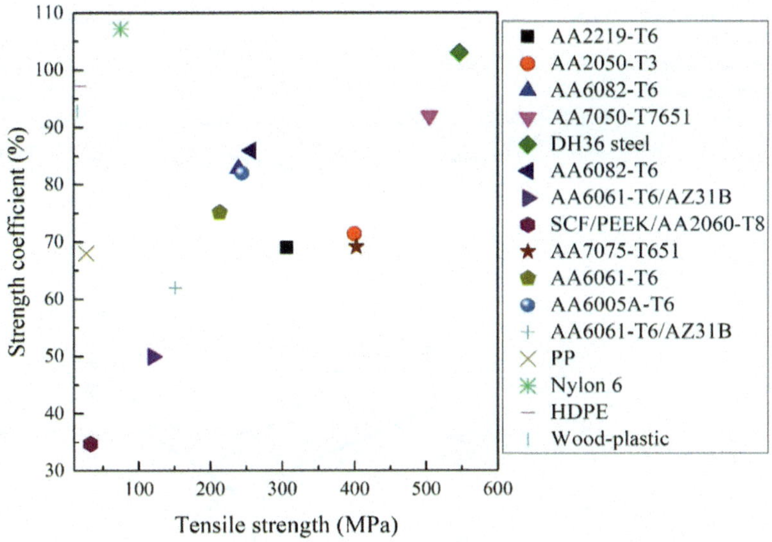

Fig. 3.17 Mechanical properties and strength coefficient (%) of $SSFSW_2$ in the current research and published open literature and current research

residual compressive stress on the materials on the top surface, which is beneficial to improving fatigue performances. Jayaraman et al. [89] indicated that FSW increased the high-cycle fatigue endurance of Al alloys by 80% due to a deep surface layer of compressive residual stress by low plasticity burnishing. Huang et al. [90–92] proposed in situ rolling FSW and fabricated ultrafine grains and a compressive residual stress layer on the surface of the joint, improving the corrosion resistance and reducing residual stress and distortion. The surface of the $SSFSW_2$ joint was characterized by ultrafine grains, which evidently enhanced the fatigue properties [26]. Li et al. [26] also stated that the stress concentration at the internal corner for additive $SSFSW_2$ joints was evidently lower than that for non-additive joints due to the additional materials filling into the corner, as shown in Fig. 3.19. EADS [13] also applied $SSFSW_2$ in the complex 3D structural window frame of an aircraft and stated that no failure occurred after 140,000 cycles, indicating excellent fatigue properties.

3.1.7 Prospects

$SSFSW_2$ has been used to join many kinds of materials with varied thicknesses, such as Al alloys, Ti alloys, steels and polymers. Many outstanding advantages have been applied in the manufacturing fields of aerospace and transportation, such as smooth surface, low residual stress and distortion, and high joint quality. However, some areas need to be addressed, which are detailed hereafter:

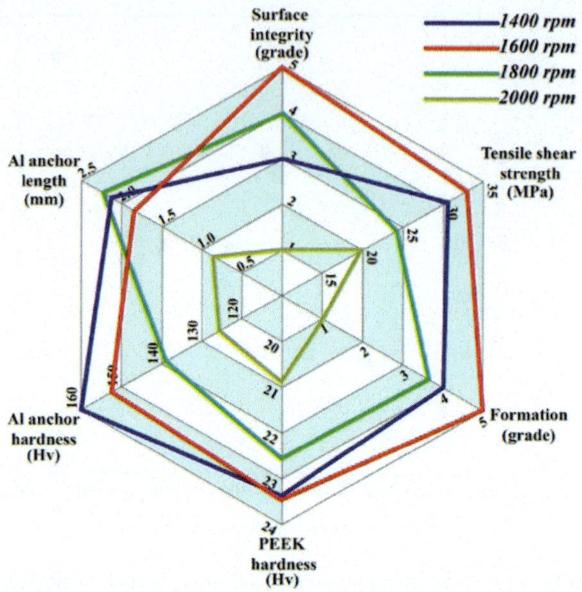

Fig. 3.18 Radar chart of the microstructural characteristics versus mechanical properties [36]

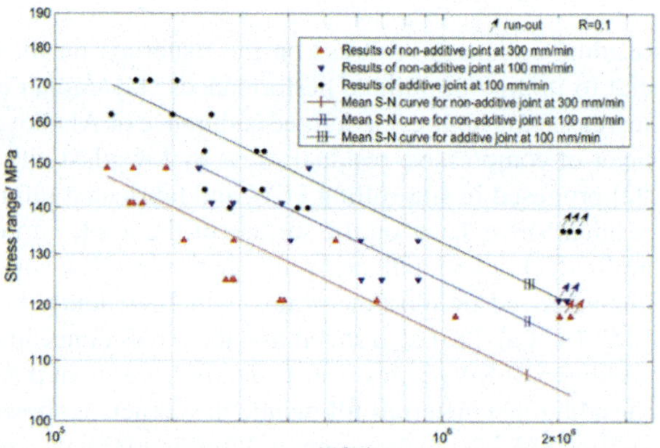

Fig. 3.19 S–N curves for additive and non-additive T joints produced by $SSFSW_2$ [26]

(1) Long service welding tool: During $SSFSW_2$, the plasticized materials easily overflow into the gap between the stationary shoulder and the rotating tool, which results in the overheating of the rotating pin and then acutely deteriorates the service life. A tool coating technology shall be developed to improve the heat insulation and wear resistance, increasing the service life of the rotating tool.

(2) $SSFSW_2$ of metal matrix composites (MMCs): Initial tool wear can also proceed without an optimized shape at high wear rates. The tool wear in FSW of MMCs is attributed to particle erosion and abrasion from the reinforcements during rotating and translational vortex flow. In addition to the excessive cost effect

of tool wear, the contamination of the weld by debris from the worn tool is of considerable concern. Premature failure due to the brittleness and high tool cost is the challenge encountered with these materials.

(3) $SSFSW_2$ of polymer matrix composites and between these composites and metals: $SSFSW_2$ has the remarkable advantage and feasibility to join polymer matrix composites with low thermal conductivity and strength due to the focused heat source and assisted heating. Moreover, $SSFSW_2$ has the potential to join dissimilar or monolithic materials.

(4) High corrosion and wear resistance of $SSFSW_2$ joints: Although $SSFSW_2$ can achieve high-quality joints, the corrosion and wear resistances are reduced due to the dissolution, coarsening and reprecipitation of the strengthening phases, especially in the 2xxx and 7xxx Al alloys. In addition to depositing particles, such as pure Al, Al_2O_3, SiC_p, multi-wall carbon nanotubes or graphene hybrids with surface friction stir processing (FSP) or cold spraying, may become further hotspots to improve the wear and corrosion resistance.

3.2 Additive Friction Stir Welding

In addition to the methods that prevent or avoid the overflow of the plasticized materials, under the condition of a sufficient forging force, additive FSW can solve these issues through the introduction of additional materials.

3.2.1 Friction Stir Welding

Ji et al. [93] proposed a hybrid metal extrusion and bonding (HYB) process by adding filler materials and introducing plastic deformation. During this process, an extruder head was clamped against the welding workpieces. The welding workpieces were separated from each other so that an I-groove formed between them. The diameter of the rotating pin was slightly larger than the width of the groove to ensure complete contact between the sidewalls of the groove and the pin, as indicated in Fig. 3.20. During the rotation and advancement of the pin, the materials along with the oxide layer on the groove sidewalls were stirred and mixed as they flowed downward into the groove and consolidated behind the pin. A sound joint with a thickness of NZ greater than the thickness of BM was obtained. Typically, the peak temperature in the groove between the welding workpieces is between 350 and 400 °C, which is below the peak temperature reported for FSW; this lower temperature is beneficial to reducing the joint softening and improving joint quality. Sandnes et al. and Li et al. [94, 95] employed cold spraying with FSW joints to improve corrosion resistance, in which weld thinning was eliminated by the addition of pure Al or Al_2O_3 particles. Partial weak bonding easily formed at the interface between the coating and the workpieces; meanwhile, cold spraying equipment is relatively expensive. Li et al.

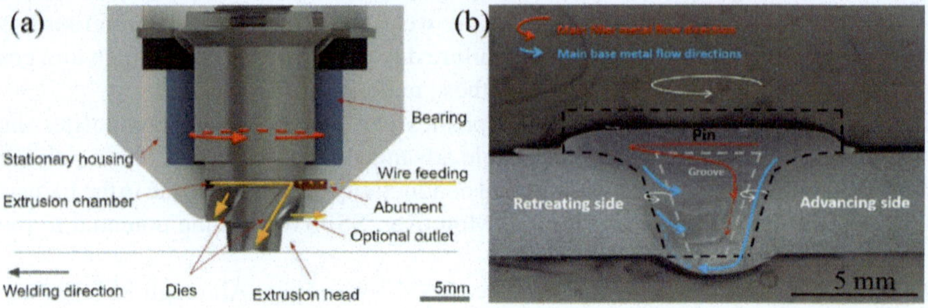

Fig. 3.20 Schematic and microstructure of the HYB process: **a** schematic and **b** macrostructure [93]

[96] applied for a patent called in situ additive FSW to solve the weld thinning problem, in which a banding wire was introduced into the surface of the welding workpieces by a wire feed system and a locating guide rail that made the banding wire tightly contact the welded workpieces. Under the thermo-mechanical effects, the banding wire and the welding workpieces could be joined, and weld thinning was avoided. Moreover, the type of banding wire could be changed according to the surface requirements, such as wear and corrosion resistances, thereby creating a stronger interface than that produced with cold spraying [97]. However, note that the joining between the banding wire and the workpieces involved a butt-lap interface, which easily led to lap welding defects, such as hook or cold lap defects [47, 98].

3.2.2 Compensation Friction Stir Welding

Russell et al. [16], Ogura et al. [99] and Meng et al. [100] introduced additional solid-state materials to the joint region along the welding line during FSW of butt workpieces, which were named surface compensation FSW (SCFSW) or level compensation FSW (LCFSW), as shown in Fig. 3.21a. The achieved joints had a larger cross-sectional thickness than conventional joints. After FSW, the materials in the upper surface higher than the planar surface of the base workpieces were machined, producing a non-weld-thinning joint. Although the weld thinning problem was solved, its disadvantages were also obvious. First, the addition and removal of materials before and after welding made the FSW process relatively complicate, reducing the welding efficiency. Second, many workpieces require three-dimensional contours. When FSW was utilized to join complex three-dimensional structures, it was difficult to add materials to the joint region. This method presented a relatively narrow application. Third, the upper part of the joint generally possessed more superior mechanical properties than the middle and lower parts of FSW joints, such as pure Al or 2219-O Al alloys [101, 102]. Even though weld thinning could be eliminated, the mechanical properties are more prone to deterioration due to the removal of materials from the upper part of the joints. The opposite reductive FSW was also proposed, which was similar to additive FSW, wherein the thickness of the whole workpieces was higher than the thickness of the present weld. After FSW, the materials higher than the top surface of the weld were removed, and the stress concentration caused by the weld thinning was avoided. Alternatively, before FSW, the abundant materials besides the weld were milled and formed a boss, which was further joined by FSW as in additive FSW. Although this method could eliminate weld thinning, the resource waste was very severe, and the operating procedure was more complex.

Korayem et al. [104] proposed a FSW method for eliminating weld thinning by an additional T-plate to simplify the procedures in SCFSW that required a complex extrusion process to fabricate a convex welding line. During welding, the T-plate was placed at the interface between the both workpieces, which made the surface of the weld higher than the surface of the workpieces. The lateral force ensured the T-plate was static. The frictional heat produced by the shoulder was attributed to the joining between the flange and the materials on the top surface of the welding workpieces, while the thermo-mechanical behaviours improved the welding of the web and the materials on the vertical surface of the workpieces, thereby eliminating weld thinning. Because the joining between the flange and the top surface of the welding workpieces only depended on the thermal conduction from the heat generation of the shoulder, the thickness of the flange was limited to less than 2 mm in thickness [105–107]. Moreover, the ultra-thin flange was easily deformed due to high frictional heat, resulting in poor joint integrity. Additionally, both of the joining interfaces between the flange and the web appeared at the butt joint [108, 109], which always required a higher heat input to ensure good joint formation. Otherwise, the joining difficulties

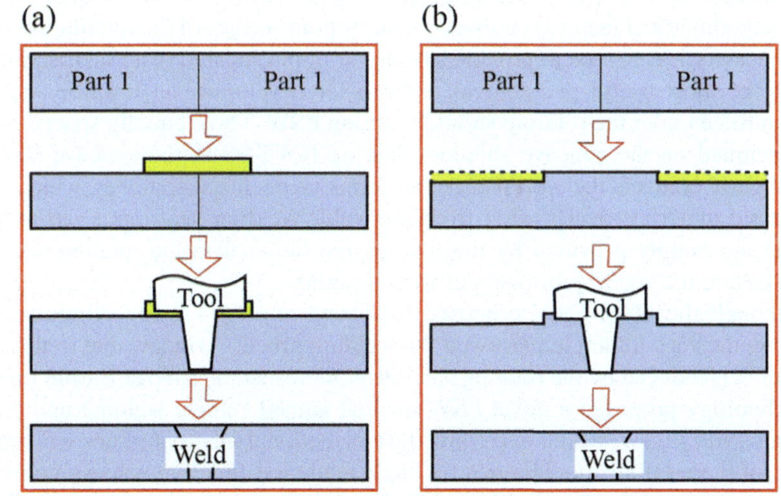

Fig. 3.21 Schematics of **a** additive FSW [103] and **b** reductive FSW

were greater due to the insufficient material flow induced by low heat input. Li and Shen [110] also stated that a sound joint could be achieved at a low welding speed and a high rotational velocity when an additional filler was introduced at the butt interface. The extremely high heat input was prone to severe joint softening, consequentially deteriorating the joint strength.

3.3 Zero-Plunge-Depth Friction Stir Welding

The welding tool is the foundation of FSW, playing significant roles in microstructures, defects and mechanical properties. During conventional FSW, the welding tool always contains a shoulder and a pin, which can realize high-quality welding under a tilt angle. The application of the shoulder plunge depth during FSW is the intrinsic reason for the occurrence of weld thinning. It can be inferred that weld thinning can be eliminated with a zero plunge depth, thereby avoiding the loss of plasticized materials. Based on this concept, a novel zero-plunge-depth FSW was proposed by Zhang et al. [111, 112] and our workgroup [113].

3.3.1 Principle

Zhang et al. [111, 112] developed the novel non-weld-thinning FSW (NWT-FSW), which could be performed at a zero shoulder plunge depth, as shown in Fig. 3.22. The weld thinning phenomenon is effectively eliminated without adding or removing any materials before or after welding, and without other assisted shoulder tools. A concave geometrical feature is utilized on the bottom surface of the rotating shoulder. The concavity is designed to provide a reservoir of pin sheared materials. Scrolls are among the most useful geometrical features for developing an inwardly directed traction force under the rotating shoulder during FSW. Three equally spaced scrolls are machined on the concave shoulder. During NWT-FSW, the concave shoulder only slightly contacts the workpieces and forms an enclosed chamber, which avoid plasticized materials overflowing from the weld. Welding heat input and material transfer are mainly provided by the rotating pin rather than the rotating shoulder, which ensure the sound joint without weld thinning.

Although the NWT-FSW proposed by Zhang et al. [111] can eliminate weld thinning, the zero tilt angle narrowed the welding process window due to the small forge effects induced by the rotating shoulder. Our research group at Harbin Institute of Technology proposed a novel FSW method named "single welding and double forming" MPD-FSW, which overcome the conventional inherent issues wherein the plunge of the rotating shoulder at a tilt angle made it difficult to achieve non-weld-thinning joints [113]. This welding tool was designed to be detachable, as shown in Fig. 3.23, which was composed of an outer rotating shoulder, an inner rotating

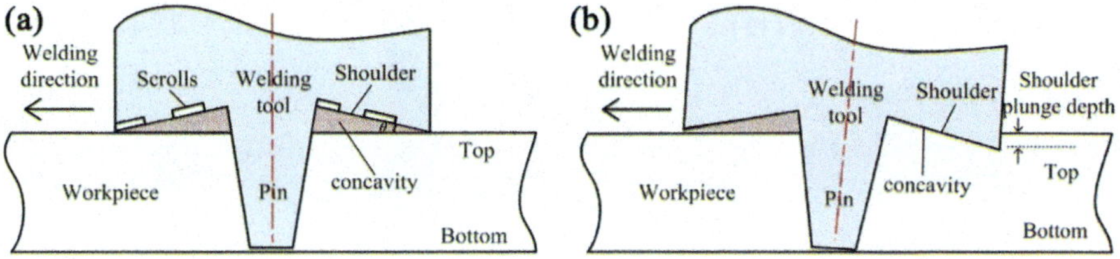

Fig. 3.22 Schematic views: **a** NWT-FSW and **b** conventional FSW [111, 114]

shoulder and a rotating pin. The plunge depths of the inner and outer rotating shoulders could be changed by controlling the relative height between the outer and inner shoulders. The concave shoulder morphology is specially designed to increase the forging and gathering effects on the plasticized materials, thereby improving the joint formation. During welding, the inner rotating shoulder and pin plunge into the workpieces to a certain depth, which are responsible for the "single welding", and then ensure the joint inner integrity. Without the plunge depth, the outer rotating shoulder only slightly contacts with the workpieces and was attributed to "double forming", which further rolls the plasticized materials on the surface and subsurface. The flash defects are eliminated, and weld thinning is avoided. Moreover, the rational choice of the rotating pin is essential to increase the stirring effect and heat generation due to the decrease in heat generation induced by the relatively smaller inner rotating shoulder. A tapered threaded pin with three facets was designed to improve the material flow behaviour.

3.3.2 *Formation Mechanism*

Figure 3.24 shows the surface formation of MPD-FSW joints. The surface appearance is divided into two affected zones: the outer rotating shoulder affected zone, and the inner rotating tool affected zone. The plunge depth of the inner rotating shoulder is larger than that of the outer rotating shoulder, which corresponds to the subzone.

The MPD-FSW process was mainly separated into "single welding" and "double forming". During welding, the inner rotating shoulder and pin plunged into the workpieces and generated frictional heat and deformation heat, which further attributed to the transfer of the plasticized materials, and then a metallurgical bond was achieved. The heat generation of the "single welding" was completely from the frictional heat between the inner rotating tool and the workpieces. Furthermore, the outer concave shoulder slightly contacted the workpieces and formed an enclosed chamber, which restricted the overflow of the plasticized materials that were squeezed from the inner rotating shoulder affected zone. With the rotation and advancement of the outer rotating shoulder, the plasticized materials from the inner rotating tool affected zone were refilled into the weld thinning location induced by the plunge of the inner rotating shoulder, thereby achieving non-weld-thinning joints (Fig. 3.25). The outer

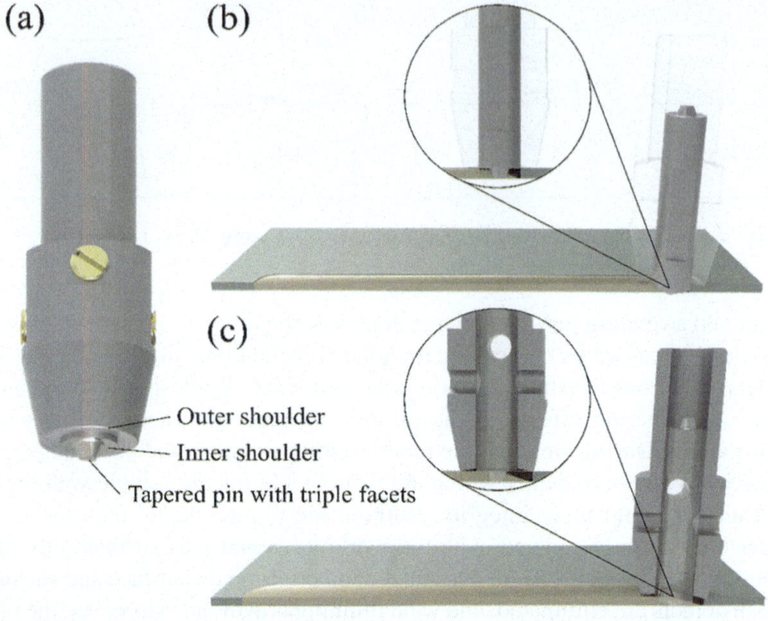

Fig. 3.23 **a** Profile of the tool; schematics of the "single welding, double forming" concept: **b** the first forming and **c** the second forming

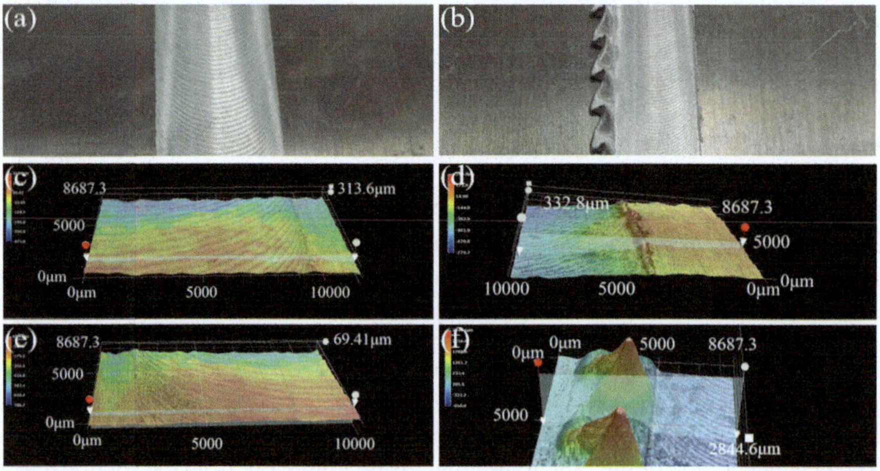

Fig. 3.24 Surface morphologies of MPD-FSW and conventional joints: schematics of surface appearances in **a** MPD-FSW and **b** conventional joints, 3D morphologies at the AS of **c** MPD-FSW and **d** conventional joints, and 3D morphologies at the RS of **e** MPD-FSW and **f** conventional joints

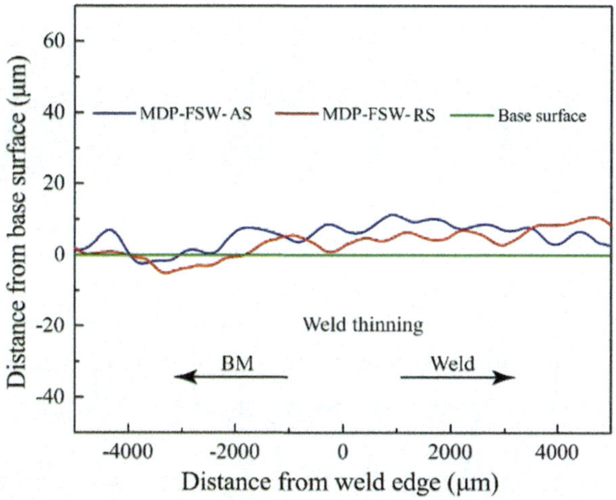

Fig. 3.25 Variations in the heights of the different joints

rotating shoulder only squeezed and forged the plasticized materials, which caused the total heat generation of the MPD-FSW to be lower than that of conventional FSW. No-weld-thinning appeared in the MPD-FSW joints.

In addition to heat generation, the material flow behaviour could be separated into three types. The first type was the layer flow induced by the welding speed, in which the plasticized materials around the rotating tool flowed backward as the tool advanced. The second type was the rotation flow induced by the rotating welding tool via friction, during which the flow rate gradually decreased as distance from the centre of the welding tool increased. The last prominent type was the flow induced by the rotating pin, which made the plasticized materials near the subsurface of the pin flow downward and then upward away from the pin. In particular, the special design of the pin morphologies, such as threads and facets, can significantly improve this flow.

An illustration of the material flow behaviour induced by the novel welding tool is displayed in Fig. 3.26 [115]. During the MPD-FSW process, the material flow rate was divided into v_{shear} and v_{normal} due to the presence of the three facets on the rotating pin. The v_{shear} mainly provided shear force to improve material transfer, whereas the v_{normal} significantly pushed the plasticized materials and further promoted the material flow rate, as shown in Fig. 3.26a, b. Under the relatively lower peak temperature induced by the inner small rotating shoulder, the improved material flow rate effectively avoided the macroscopic welding defects and further enhanced the material transfer rate through the thickness direction. The plasticized materials flowed downward to the bottom near the pin surface [marked by red arrows in Fig. 3.26b] and then flowed upward away from the pin surface [marked by blue arrows in Fig. 3.26b]. As shown in Fig. 3.26c, the partially plasticized materials easily flowed out of the NZ due to the small shoulder diameter when the outer rotating shoulder was not employed [marked by green arrows in Fig. 3.26c]. Moreover, the outer rotating

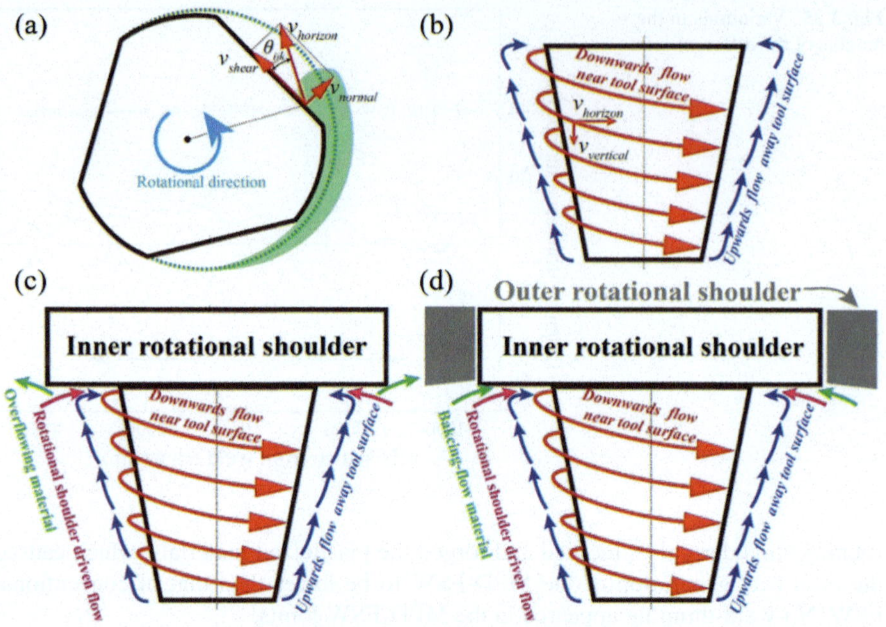

Fig. 3.26 Material flow and velocity distributions during FSW with controlled shape and performance: velocity distributions in the **a** horizontal section and **b** vertical section; material flow in **c** conventional FSW and **d** MPD-FSW [115]

shoulder contacted the rotating tool in the clearance fit, which can preheat the overflowed materials and provide a downward pressure. This pressure can prevent the plasticized materials from overflowing from the weld and refill these materials into the NZ, as shown in Fig. 3.26d. Hence, favourable surface integrity without weld thinning can be achieved.

3.3.3 Microstructural Characteristics

The microstructure of an MPD-FSW joint is shown in Fig. 3.27. The morphology is mainly divided into the outer SAZ, inner SAZ, TMAZ, HAZ and NZ. During welding, the plunge depth of the outer shoulder was only 0.05 mm and that of the inner shoulder was 0.15 mm, which caused the outer SAZ to be located only at the subsurface of the workpieces. In the top, the grains were the smallest in the whole joint, implying that this zone experienced severe plastic deformation at high strain rates. After being subjected to the rotating pin, the upper part of the weld underwent additional deformation by the rotating tool shoulder. Huang et al. [90, 91] also stated that a plunge depth of 0.05 mm of rolling balls could fabricate a gradient surface layer on Al alloys, thereby improving the microhardness on the subsurface.

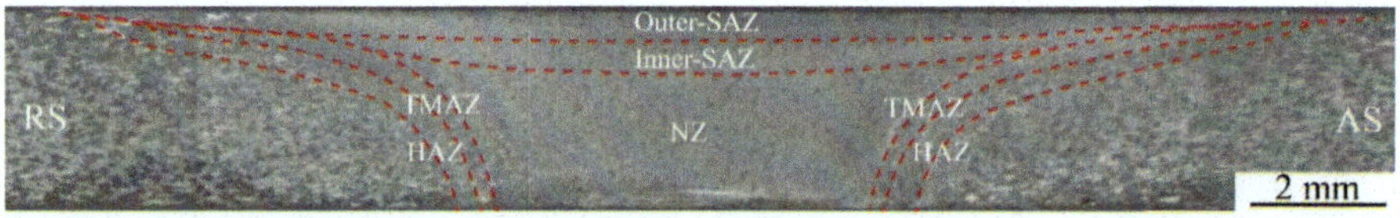

Fig. 3.27 Macrostructure of a typical MPD-FSW joint

Apparently, the grain structures for each NZ layer of the NWT-FSW joint exhibited different sizes from those of conventional FSW joints. Furthermore, the grains of both joints presented different distributed characteristics along the thickness direction, as shown in Fig. 3.28. First, the grain size of the NWT-FSW joint was smaller than that of the conventional FSW joint at each layer of the NZ. Second, the grain size of the conventional FSW joint gradually decreased from the upper to lower parts of the NZ, as commonly observed in the conventional FSW joints of other Al alloys [102, 116]. For the NWT-FSW joint, the grain size was nearly uniform along the thickness direction. Compared with conventional FSW, the thermal and mechanical effects of the rotating shoulder on NWT-FSW or MPD-FSW joints were weakened and became more uniform along the thickness direction.

Figure 3.29 shows the macro and microstructures in the cross-section of joints for 6 mm thick 2219-T6 Al alloys. The welds can be divided into four zones separated by different mechanical action and heat input: welding nugget zone (WNZ), TMAZ, HAZ, and BM. The WNZ shows a typical bowl shape with slight asymmetry. The interface between the two BMs shows a solid-state bonding and changes from an original single interface to a complex interface with interactive mixing (Fig. 3.29a). The fluctuating effects by the pin with a thread and milling facet feature have compression on the plasticized materials during the welding process, which further improve the flow of the plasticized materials. The severe plastic deformation promotes the nucleation of recrystallized grains [117, 118]. The grains in WNZ are significantly fine and equiaxed due to the occurrence of dynamic recrystallization. From the top to the bottom of the WNZ, the interface is well joining, and there are no defects such as cavities and cracks (Fig. 3.29b–d). Due to the difference in the corrosion characteristics of the two materials, the corrosion of the grain boundary at the interface is more severe. In addition, the black spots are corroded precipitates, and appear on the side of AA2219-T6 alloys. The defect-free bonding on the interfaces under high frictional heat and severe plastic deformation could enhance the strength of WNZ [119]. Generally, the material flow is weak at the bottom of the weld when the heat input is insufficient, such as the welding speed is faster and the rotational velocity is lower than the acceptable weld parameter window, leading to defect such as kissing bond [120, 121]. In this work, the kissing bond is suppressed at the root interface of the WNZ (Fig. 3.29d), which also can be demonstrated by fracture locations of the joints. The protruding inner shoulder with considerable plunge depth enhances the forging effect, which causes more plasticized materials to flow towards the bottom. The pin with triple milling facets increases the plastic deformation of materials, further facilitating the flow of the materials at the bottom.

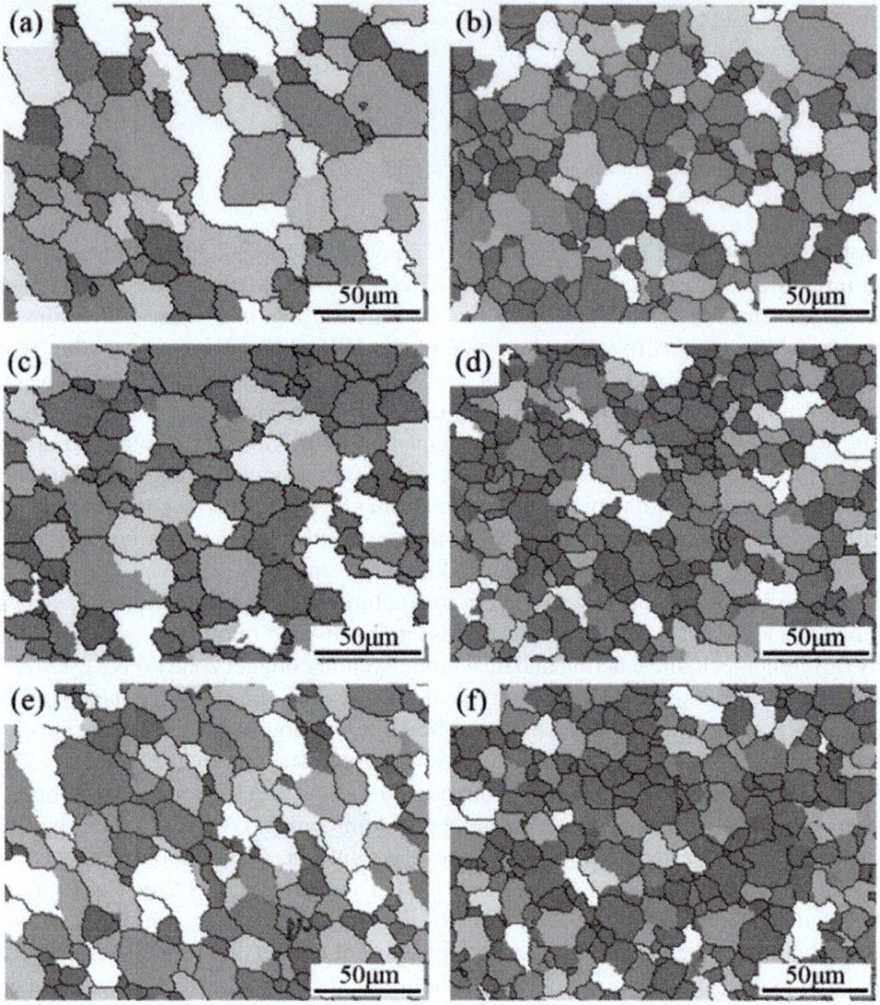

Fig. 3.28 Grain structures of the NZ: **a–c** upper, middle and lower parts of the conventional FSW joint and **d–f** upper, middle and lower parts of the NWT-FSW joint [102, 116]

Figure 3.30 shows the distribution of precipitates in different micro-zones at the side of AA2219-T6 alloy. The precipitates in HAZ are slightly coarsened compared to BM. The original larger precipitates are not completely dissolved during the thermal process, which become nucleation sites in the subsequent cooling process and gradually coarsening. The insignificant change in HAZ is due to that the protruding inner shoulder with a small diameter and the outer shoulder with a small plunge depth reduce the frictional heat, which effectively reduce the softening of the HAZ. The precipitates in the TMAZ are coarsened further than that in the HAZ (Fig. 3.30c). Some large precipitates are broken slightly under slight plastic deformation. The

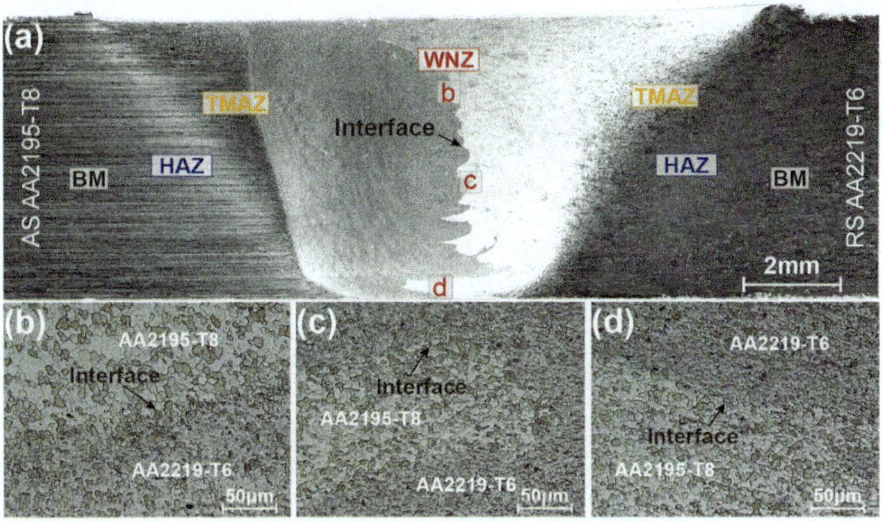

Fig. 3.29 Macro- and microstructures of the typical joint: **a** macrostructures, **b** top-WNZ, **c** middle-WNZ, and **d** bottom-WNZ

segregation of precipitates at the grain boundaries is also observed [red arrows in Fig. 3.30c], which can result in a wide precipitation free zone (PFZ). The coarsening of precipitates and wide PFZ can promote the propagation of cracks [122], which result in the TMAZ becoming the weakest area compared to the HAZ. With the increase of plastic deformation, the precipitates in WNZ are smaller with a uniform distribution (Fig. 3.30d). The precipitates are broken into finer particles by the large strain rate due to the pin with milling facets. The high dislocation density caused by severe plastic deformation contributes to the nucleation of new precipitates. This makes the distribution of precipitates more uniform during the process of solid solution and re-precipitating.

3.3.4 Mechanical Properties

The decreased heat input during zero-plunge-depth FSW (compared with that in conventional FSW) was obtained due to the loss of frictional heat generated by the rotating shoulder, which was the intrinsic reason for grain refinement and the reduction in the SAZ. These useful microstructural evolutions further narrowed the softening regions and degree, which provided the higher mechanical properties of the NWT-FSW joint. The maximum tensile strength of the optimized joint reached 218 MPa, whereas the maximum elongation of 24.3% was comparable with that of base 5052 Al alloys [111], as shown in Fig. 3.31a. Additionally, the effect of the

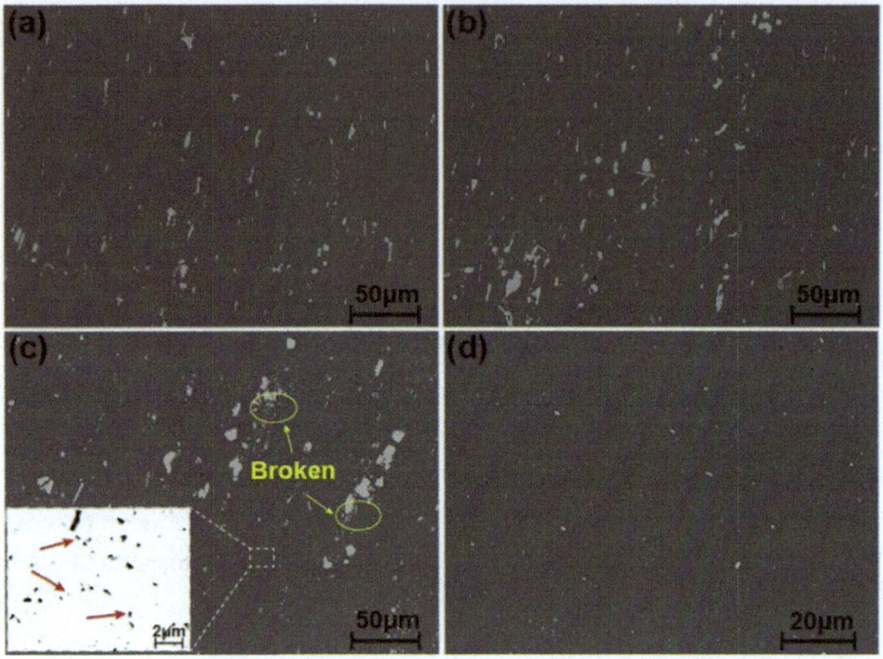

Fig. 3.30 Distribution of precipitates at different zones: **a** 2219-BM, **b** 2219-HAZ, **c** 2219-TMAZ, and **d** 2219-WNZ

outer rotating shoulder and welding parameters on the tensile properties of the MPD-FSW is shown in Fig. 3.32b. The addition of the outer rotating shoulder provided significantly higher tensile strength than that provided by only the inner shoulder. Moreover, when the rotational velocity was 600 rpm, the tensile properties first increased and then decreased as the welding speed varied from 100 to 600 mm/min. The maximum tensile strength of the AA6082-T4 joint reached 231 MPa, which was equivalent to 90.8% of the BM [113].

Figure 3.32 summarizes the influence of welding speeds on the tensile properties. The ultimate tensile strength firstly increases and then decreases with increasing welding speeds. The tensile properties are related to the complicated evolution of precipitates and grain size [123, 124]. The tensile strength of joints with welding speed lower than 100 mm/min remains at a low level. This is because the precipitates have undergone severe dissolution and coarsening under higher heat input at lower welding speeds. As the welding speeds increase to 200 ~ 300 mm/min, the decrease of heat input reduces the softening degree of materials. According to the established model [125], the lower the peak temperature, the less the dissolution and the greater the retained precipitates volume fraction. The ultimate tensile strength of the joint at a welding speed of 300 mm/min reaches (352 $\pm$ 6) MPa, which is equivalent to 81.7 $\pm$ 1.4% of base AA2219-T6 alloys. Although the performance of the joint depends on the coarsening and dissolution of precipitates, it is seriously affected by

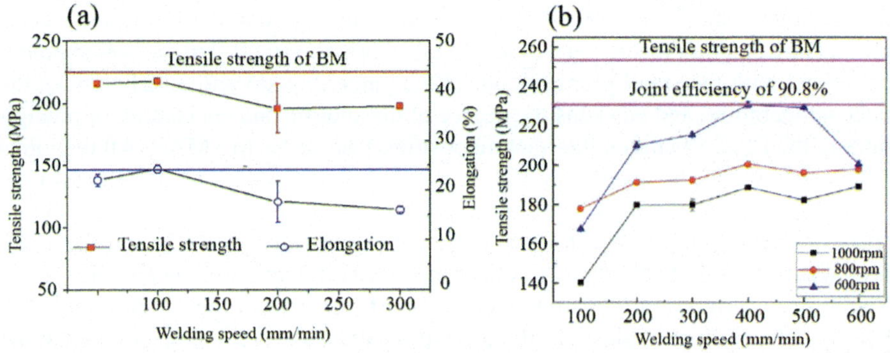

Fig. 3.31 Tensile properties of zero-plunge-depth FSW joints: **a** AA5052 [111] and **b** AA6082-T4

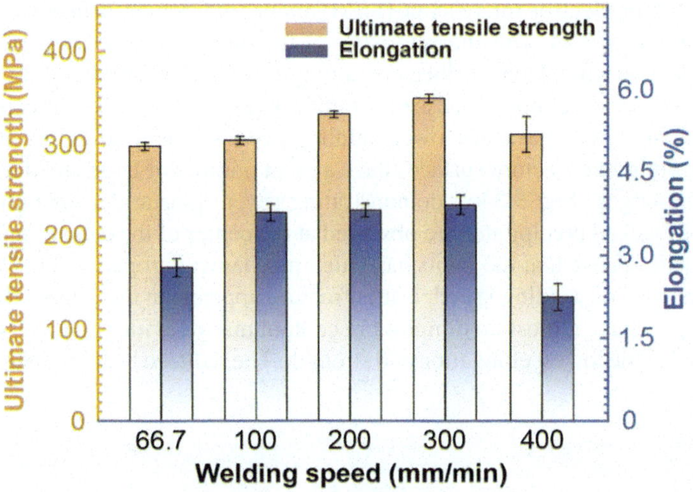

Fig. 3.32 Tensile properties of joints at different welding speeds for 6 mm thick 2219-T6 Al alloys

welding defects. The further reduction of heat input is insufficient for the material to flow adequately as the welding speed further increases. The formation of cavity defects and the kissing bonding at the interface cause the initiation and propagation of creaks, which deteriorate the tensile properties.

Figure 3.33 displays the fracture locations of the joints at different welding speeds. The joints are fractured at the TMAZ on the side of AA2219-T6 alloys when the welding speeds are lower than 300 mm/min (Fig. 3.33a–d). This indicates that the interface between the two BMs has a higher strength for the stepped-shoulder FSW joints, but the TMAZ is a relatively weak zone due to the transition of microstructures and the coarsening of precipitates. For the FSW of dissimilar aluminum alloys, the mechanical properties of the WNZ are enough for all the joints when the welds are defect-free and the materials are adequately mixed [126]. Ji et al. [20] reported that stationary shoulder FSW reduced the softening degree of the HAZ, and caused the fracture position located in TMAZ. In this work, the small size of inner shoulder combined with the small plunge depth of the outer shoulder effectively reduce the peak temperature, and weakens the degree of dissolution and coarsening of precipitates in the HAZ, changing the fracture position from HAZ to TMAZ. All the joints with welding speeds lower than 300 mm/min are fractured along a path at 45° to the tensile direction, which is a typical ductile fracture pattern. When the welding speed reaches 400 mm/min, the cavity defects appear at the WNZ, and the crack propagates irregularly along the interface and cavity defects, as shown in Fig. 3.33e. The lower heat input further strengthens the performances of the TMAZ and HAZ, but the reduced flow of plasticized materials weakens the solid-state bonding at the interface between the two BMs. The cavity also decreases load-bearing area, and promotes the initiation and propagation of cracks. These factors change the fracture positions from the TMAZ to the WNZ.

The scanning electron microscope (SEM) images of fracture surface morphologies are shown in Fig. 3.34. The characteristics of fracture surfaces change significantly with welding speeds. A flat surface containing few shallow and flat dimples can be observed when the welding speed is 66.7 mm/min (Fig. 3.34a), which is caused by the severe softening of material at a lower welding speed. As the welding speed increases to 300 mm/min, the fracture surfaces have a great number of large and deep dimples and tearing edges (Fig. 3.34c), demonstrating severe plastic deformation. A large amount of cracked precipitates are observed at the center of the deep dimples. These characteristics prove that the joints have high mechanical properties [127]. With the further increase of welding speed, brittle fracture appears on the joints. The fracture morphologies have almost no dimples but contain many cavities (Fig. 3.34d), which correspond to the lowest elongation and strength. The reduced heat input under higher

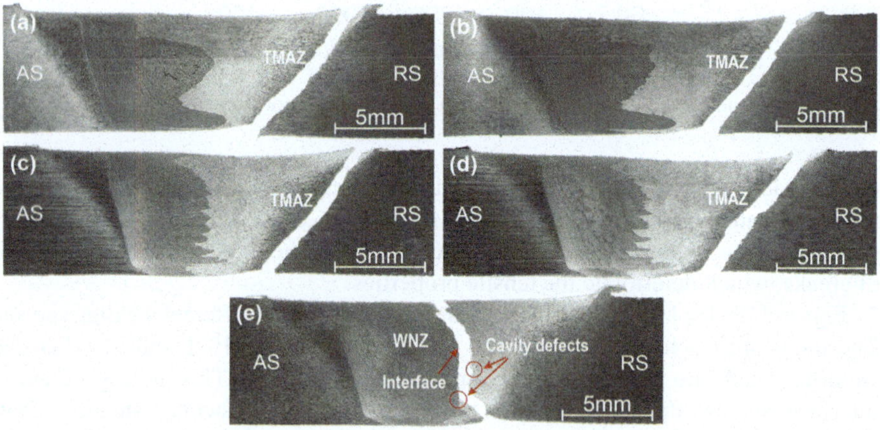

Fig. 3.33 Fracture locations of joints 6 mm thick 2219-T6 Al alloys: **a** 66.7 mm/min, **b** 100 mm/min, **c** 200 mm/min, **d** 300 mm/min, and **e** 400 mm/min

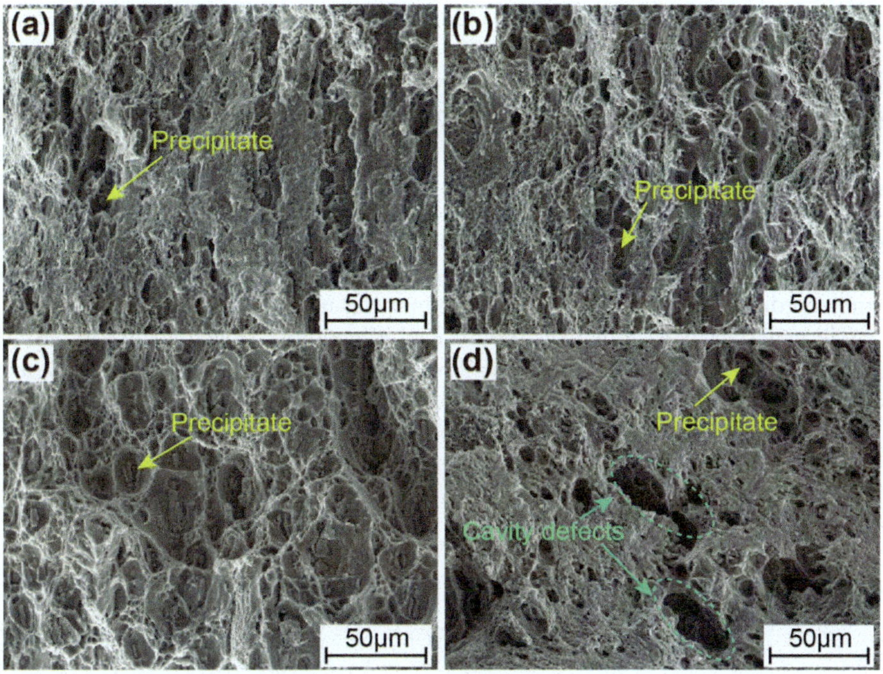

Fig. 3.34 Fracture morphologies of joints 6 mm thick 2219-T6 Al alloys: **a** 66.7 mm/min, **b** 100 mm/min, **c** 300 mm/min, and **d** 400 mm/min

welding speeds results in insufficient materials flow, which weakens the interface bonding and even produces welding defects.

3.3.5 Corrosion Behaviour

The welding of these two alloys leads to significant microstructural modifications, such as recrystallization, precipitate dissolution, and precipitate coarsening [128]. The major strengthening phases of AA2195-T8 alloy are composed of T_1 (Al_2CuLi), and θ''/θ' (Al_2Cu, the main strengthening phase of Al–Cu alloys) [129]. The T_1 phase has an equilibrium semi-coherent hexagonal close-packed structure, and the θ''/θ' phases are metastable phases with cubic structures which are respectively coherent and semi-coherent with the aluminum matrix [130]. It is this unique θ''/θ' phase that is thought to be responsible for the bulk of precipitation strengthening. These microstructural changes promote different anti-corrosion performances [131–133]. Therefore, the present work aims to examine the precipitation-driven corrosion behavior of heterogeneous joints between AA2195-T8 and AA2219-CS. Friction stir welding, based on the principle of severe plastic deformation and low heat input [134–137], was chosen as the welding techniques. The microstructural factors, including precipitation and matrix evolution, were revealed. Electrochemical and immersion measurements were taken to evaluate the corrosion resistance. The effect of microstructures at different zones on the corrosion behavior was discussed.

1. Computational fluid dynamics model

A computational fluid dynamics (CFD) model is served as a converter from primitive causative variables to field variables, including strain rate and temperature. FLUENT software was used for the calculation of the CFD model. The domain for calculation with coordinates of 60 mm × 90 mm × 6 mm was used.

The materials in the domain were described as incompressible fluid with a constant density [138]. The deformation and flow behaviors followed the Navier–Stokes equations and Boussinesq's approximation:

$$\nabla \cdot v = 0 \tag{3.2}$$

$$\rho \frac{dv}{dt} = -\nabla P + \nabla \cdot \sigma \tag{3.3}$$

$$\rho c_p \frac{dT}{dt} = -\nabla(k \nabla T) + (\sigma : \nabla v) - \phi \tag{3.4}$$

where ρ is the density, t is the step time, v is transient velocity, $\sigma = \mu(\nabla v + \nabla v^t)$ is the deviatoric stress tensor, μ is the non-Newtonian viscosity, P is the structural pressure, T is the absolute temperature, ϕ is the source term which represents the heat generation of volumetric deformation heat and surficial frictional heat, and the heat loss of convection and radiation, c_P is the specific heat capacity, and k is the thermal conductivity.

The equation between viscosity, equivalent strain rate, and flow stress is defined by Perzyna's viscoplasticity model [139]:

$$\mu = \frac{\sigma_e}{3\dot{\varepsilon_e}} \tag{3.5}$$

where σ_e is the flow stress or called yield stress which can be defined by the constitutive equation of aluminum alloys, $\dot{\varepsilon_e}$ is the equivalent strain rate:

$$\sigma_e = \sqrt{\frac{3}{2}\sigma_{ij}\sigma_{ij}} \quad i, j = 1, 2, 3 \tag{3.6}$$

$$\dot{\varepsilon_e} = \sqrt{\frac{2}{3}\dot{\varepsilon_{ij}}\,\dot{\varepsilon_{ij}}} \quad i, j = 1, 2, 3 \tag{3.7}$$

Here, σ_{ij} and $\dot{\varepsilon_{ij}}$ are the components of the second-order deviatoric stress tensor and the second-order strain rate tensor, respectively. The strain rate tensor can be defined by Helmholtz velocity decomposing theorem:

$$\dot{\varepsilon_{ij}} = \frac{1}{2}\left(v_{i,j} + v_{j,i}\right) \quad i, j = 1, 2, 3 \tag{3.8}$$

The flow stress can be calculated by Sellars–Tegart's constitutive equations based on strain rate and temperature [140]:

$$\sigma_e = \frac{1}{\alpha}\sinh^{-1}\left[\left(\frac{Z}{A}\right)^{\frac{1}{n}}\right] \tag{3.9}$$

Here, α, A and n are material constants. Z is the Zener–Hollomon parameter, which is defined as:

$$Z = \dot{\varepsilon}_e e^{\frac{Q_{zh}}{RT}} \tag{3.10}$$

where Q_{zh} is the activation energy, R is the gas constant.

However, there are limitations and drawbacks of Eq. (3.9) when the temperature rises near melting point [141]. Thus, a modified constitutive equation is proposed:

$$\sigma_e = \left[1 - \left(\frac{T}{T_s}\right)^{\gamma}\right]\frac{1}{\alpha}\sinh^{-1}\left[\left(\frac{Z}{A}\right)^{\frac{1}{n}}\right] \tag{3.11}$$

where T_s is the solidus temperature, and γ is the influencing factor.

There are two parts of heat source for the heat boundary conditions: frictional heat and deformation heat. Frictional heat is a surficial heat source applied to the interface between welding tool and workpiece, which is described as:

$$F = \beta_1 R\omega \cdot \left[\frac{\delta\sigma_e}{\sqrt{3}} + (1-\delta)\mu P \sin\xi\right] \cdot \cos\theta \tag{3.12}$$

The deformation heat is a volumetric heat source, which can be defined as:

$$\Phi = \beta_2\mu\left[2\left(\frac{\partial u}{\partial x}\right)^2 + 2\left(\frac{\partial v}{\partial y}\right)^2 + 2\left(\frac{\partial w}{\partial z}\right)^2 + \left(\frac{\partial u}{\partial y} + \frac{\partial v}{\partial x}\right)^2 + \left(\frac{\partial u}{\partial z} + \frac{\partial w}{\partial x}\right)^2 + \left(\frac{\partial w}{\partial y} + \frac{\partial v}{\partial w}\right)^2\right] \tag{3.13}$$

where β_1 and β_2 are the empirical coefficients, μ is the frictional coefficient, R is the radius of elements to axis, ω is the rotational velocity, δ is the slip rate, P is the axial pressure, ξ and θ represent the orientation of the element. The slip rate can be defined as [142]:

$$\delta = 1 - e^{-\frac{\omega R}{\delta_0 \omega_0 R_p}} \quad (3.14)$$

where δ_0 is an interfacial constant, ω_0 is a constant for nondimensionalization, and R_s is the radius of the pin. Besides, in consideration of heat dissipation by the welding tool, the interfacial thermal efficiency is defined by:

$$\lambda = \frac{J_D}{J_D + J_T} \quad (3.15)$$

$$\mathrm{J} = \sqrt{k\rho C_p} \quad (3.16)$$

where J_D and J_T represent the domain and the tool, respectively. k, ρ and C_p are the thermal conductivity, the density, and the heat capacity, respectively.

The boundary heat convection and radiation can be given as follows [78]:

$$-k\frac{\partial T}{\partial z} = h(T - T_\infty) + \sigma\varepsilon\left(T^4 - T_\infty^4\right) \quad (3.17)$$

where h is the heat transfer coefficient, σ is the Stefan–Boltzmann constant, and ε is the emissivity of the domain.

For the motion boundary conditions, the thermo-plasticized alloy, which is regarded as non-Newtonian fluid, is driven by the velocity equation [143]:

$$\begin{cases} v_{horizon} = \eta\omega R \\ v_{shear} = v_{horizon}\cos\xi\cos\theta \\ v_{normal} = v_{horizon}\cos\xi\sin\theta \\ v_{vertical} = v_{horizon}\sin\xi \end{cases} \quad (3.18)$$

Moreover, all the physical and empirical parameters can be found in Table 3.3 for calculation.

2. Precipitation evolution model

The evolution of precipitates has a great influence on the corrosion behavior in the heterogeneous joint. Many published papers consistently suggested that tailoring the distribution of precipitates could improve the corrosion resistance of the FSWed joint [118, 144–147]. A modified Wagner–Kampmann model was thus proposed to calculate the precipitation evolution. The input data of temperature and strain rate were obtained from the CFD simulation results and our previous literature [143]. This evolution algorithm is based on the discretization of time and particle size [117]. The overall calculation process mainly includes three parts: ① Calculating the growth rate of precipitation particles such as dissolving and coarsening; ② Calculating the nucleation rate of particles with different sizes at each time step; ③ Recording the continuum equation of each element between the matrix and precipitations. θ''/θ' phases were the only considered precipitates in AA2219. Since δ' is not the major

Table 3.3 Physical and empirical parameters used in CFD models

Parameter	Symbol	Value
Material constant #1 of AA2195-T8	α	1.1×10^{-8} Pa^{-1}
Material constant #2 of AA2195-T8	A	3.77×10^8
Material constant #3 of AA2195-T8	n	3.55
Influencing factor of AA2195-T8	γ	0.4
Activation energy of Sellars–Tegart constitutive equation of AA2195-T8	Q_{zh}	1.52×10^5 J/mol
Solidus temperature of AA2195-T8	Ts	807 K
Material constant #1 of AA2219-CS	α	1.3×10^{-8} Pa^{-1}
Material constant #2 of AA2219-CS	A	3.25×10^8
Material constant #3 of AA2219-CS	n	2.76
Influencing factor of AA2219-CS	γ	0.5
Activation energy of Sellars–Tegart constitutive equation of AA2219-CS	Q_{zh}	1.55×10^5 J/mol
Gas constant	R	8.314 J/(mol K)
Solidus temperature of AA2219-CS	Ts	807 K
Empirical coefficient #1	β_1	0.8
Empirical coefficient #2	β_2	0.7
Friction coefficient	μ	0.4
Axial pressure	P	65 MPa
Interfacial constant	δ_0	0.4
Thermal conductivity of alloys	$k_D(T)$	$92.5 + 0.17 \times$ T W/(m K)
Thermal conductivity of H13	k_T	29 W/(m K)
Density of alloys	ρ_D	2700 kg/m^3
Heat capacity of alloys	$c_{pD}(T)$	$703 + 0.373 * \text{T} + 0.00155 * \text{T}^2$ J/(kg K)
Heat capacity of H13	c_{pT}	470 J/(kg K)
Heat transfer coefficient of upper and side wall	h_w	15 W/(m^2 K)
Heat transfer coefficient of lower wall	h_{lw}	130 W/(m^2 K)
Stefan–Boltzmann constant	σ	5.67×10^{-8} W/(m^2 K^4)
Emissivity	ε	0.35
Diffusion factor of Cu element	$D_{0,Cu}$	6.5×10^{-5} m^2/s
Diffusion factor of Li element	$D_{0,Li}$	4.7×10^{-5} m^2/s
Diffusion activation energy of Cu element	$Q_{d,Cu}$	1.35×10^5 J/mol
Diffusion activation energy of Li element	$Q_{d,Li}$	1.03×10^5 J/mol
Interfacial energy of T_1	$\gamma_{int,T1}$	0.23 J/m^2
Interfacial energy of θ variants	$\gamma_{int,\theta}$	0.2 J/m^2

(continued)

Table 3.3 (continued)

Parameter	Symbol	Value
Molar volume of T_1	$V_{m,T1}$	3.15×10^{-5} m^3/mol
Molar volume of θ variants	$V_{m,\theta}$	2.85×10^{-5} m^3/mol
Concentration constant of T_1	$C_{e0,T1}$	10.35
Concentration constant of θ variants	$C_{e0,\theta}$	9.68
Dissolution enthalpy of T_1	$Q_{e,T1}$	4.87×10^4 J/mol
Dissolution enthalpy of θ variants	$Q_{e,\theta}$	4.62×10^4 J/mol
Nucleation constant #1 of T_1	$J_{0,T1}$	1.035×10^{35} (m^3 s)
Nucleation constant #1 of θ variants	$J_{0,\theta}$	9.67×10^{34} (m^3 s)
Nucleation constant #2 of T_1	$A_{0,T1}$	1.92×10^4 J/mol
Nucleation constant #2 of θ variants	$A_{0,\theta}$	1.63×10^4 J/mol
Geometric constant	K_s	11.0

precipitation for the alloys whose element ratio of Cu to Li is larger than 2.5, there are two kinds of major strengthening particles in AA2195 alloy [148]. Cu element and Li element, which constitute the T_1 and θ″/θ′ phases, are chosen to be the binary diffusion controlling elements. These two elements are involved in the calculation, and the element with a lower diffusion rate at the specific time shall prevail. Besides, the shapes of the two major phases are platelet rather than spherical. For the convenience of calculation, all the particles precipitating during the welding process were treated as globular. We considered the ratio of radius to the thickness of the precipitations as a specific value k_s, then the equivalent radii can be extracted. There is the relationship between thickness and equivalent radius of the platelet-like precipitations.

$$r = \sqrt[3]{\frac{3k_s^2\delta^3}{4}} \tag{3.19}$$

where r is the equivalent radius, δ is the thickness.

The first part is to calculate the growth rate of precipitation particles. It depends on the solute concentration at the interface between the particles with a certain radius and the matrix. The growth rate can be described as [149]:

$$v_i = \frac{\partial r_i}{\partial t} = \frac{C_{ss} - C_{int}}{C_p - C_{int}}\frac{D_f}{r_i} \tag{3.20}$$

where C_{ss} is the solute concentration of diffusion controlling element far away from the adjacent area of precipitations, C_{int} is the concentration at the interface between precipitations and matrix, which can be calculated from Gibbs–Thomson equation [150], C_p is the concentration of the diffusion controlling element in the precipitations, D_f is the diffusion coefficient, and r_i is the radius of the particle groups. Due to the severe plastic deformation essence of FSW, the dislocation density increases in an order of magnitude during the welding process, contributing to the presence of pipe diffusion [151]. An additional coefficient is introduced into the diffusion equation to reflect this characteristic, which can be written as:

$$D_f = D_0\sqrt{1+\dot{\varepsilon}}\exp\left(-\frac{Q_d}{RT}\right) \tag{3.21}$$

where D_0 is the diffusion factor; Q_d is the diffusion activation energy. The interfacial concentration is given by [124]:

$$C_{int} = C_e \exp\left(\frac{2\gamma_{int}V_m}{r_i RT}\right) \tag{3.22}$$

where C_e is the equilibrium concentration of solute element at the interface of matrix and precipitation, γ_{int} is the interfacial energy per unit area of particles and matrix, which is influenced by the strain rate during the welding process, and V_m is the molar volume of the precipitations. The equilibrium concentration C_e can be calculated by:

$$C_e = C_{e0}\exp\left(-\frac{Q_e}{RT}\right) \tag{3.23}$$

where C_{e0} is a constant and Q_e is the dissolution enthalpy of the precipitations. According to Eqs. (3.20)–(3.23), the critical radius between coarsening and dissolution can be given if $C_{ss} = C_{int}$:

$$r^* = 2\gamma_{int}V_m\left[RT\prod_{i=1}^{n}\ln\left(\frac{C_{iss}}{C_{ie0}}\right) + Q_e\right]^{-1} \tag{3.24}$$

The second part is to calculate the nucleation rate at each time step. Ignoring the incubation time required to generate nuclei for precipitations, the nucleation rate can be expressed as based on the effect of strain rate [152]:

$$J_{nucl} = \frac{J_0\sqrt{1+\dot{\varepsilon}}}{\sqrt{T}}\exp\left\{-\frac{A_0}{(RT)^3[\ln(C_{ss}/C_e)]^2} - \frac{Q_d}{RT}\right\} \tag{3.25}$$

$$J_0 = \frac{2\rho_{dis}V_a D_0 C_{at}\sqrt{\gamma_{int}}}{a^4 b\sqrt{k}} \tag{3.26}$$

$$A_0 = \frac{16\pi\varphi_i\gamma_{int}^3 N_A V_m^2}{3C_p^2} \tag{3.27}$$

where ρ_{dis} is the dislocation density which provides the position for nucleation, V_a is the volume of the matrix atom, k is the Boltzmann constant, C_{at} is the atom fraction of diffusion controlling element in the matrix, a is the lattice constant of the matrix, b

is the burger's vector, φ_i is the wetting constant at dislocation, and N_A is Avogadro's number.

The last part is the continuum equation. The core of particle grouping strategy is to discretize particles according to radii, which can simultaneously simulate nucleation, dissolution, and coarsening. Two kinds of precipitates were divided from small to large into a series of groups with a fixed radius. Each particle group can be defined as a control body. Therefore, the nucleation, dissolution, and coarsening can be written as methods of the group class, which symbolizes the source term, inflow, and outflow of the group instance. The material flow equation based on the material flow can be written as:

$$\frac{\partial \mathrm{N}}{\partial t} = -\frac{\partial F_n}{\partial r} + \phi_n \tag{3.28}$$

where N is the number density of the particles in each particle group, F_n is the material flow, and ϕ_n is the source term which refers to the nucleation rate at each time step. The material flow is described based on the equation of growth rate:

$$F_n = \Delta N v \tag{3.29}$$

Therefore, the first-order upwind material flow equation is further expressed as:

$$\frac{\partial \mathrm{N}_i}{\partial t} = \begin{cases} \frac{\partial F_{n,i-1}}{\partial r_{i-1}} - \frac{\partial F_{n,i+1}}{\partial r_{i+1}} + \phi_{n,i} & F_{n,i-1} > 0,\ F_{n,i+1} < 0 \\ -\frac{\partial F_{n,i+1}}{\partial r_{i+1}} + \phi_{n,i} & F_{n,i-1} \le 0,\ F_{n,i+1} < 0 \\ \frac{\partial F_{n,i-1}}{\partial r_{i-1}} + \phi_{n,i} & F_{n,i-1} > 0,\ F_{n,i+1} \ge 0 \\ \phi_{n,i} & F_{n,i-1} \le 0,\ F_{n,i+1} \ge 0 \end{cases} \tag{3.30}$$

The explicit continuum equation among all the particle groups and matrix can be given as:

$$C_{ss} = \frac{C_0 - C_p \sum f_v^i}{1 - \sum f_v^i} \tag{3.31}$$

$$f_v^i = \frac{4}{3}\pi r_i^3 N_i \tag{3.32}$$

where C_0 is the total concentration of the diffusion controlling element, f_v^i is the volume fraction of each particle group based on the globular assumption, and $\sum f_v^i$ is the volume fraction of the precipitation. Moreover, the involved parameters were given in Table 3.3.

Figure 3.35 shows the typical macrostructure of the heterogeneous FSW joint between AA2195-T8 and AA2219-CS. Several microzones, including BM of AA2195 (BM-2195), HAZ of AA2195 (HAZ-2195), thermo-mechanically affected zone of AA2195 (TMAZ-2195), WNZ of AA2195 (WNZ-2195), WNZ of AA2219 (WNZ-2219), TMAZ of AA2219 (TMAZ-2219), HAZ of AA2219 (HAZ-2219), and

BM of AA2219 (BM-2219), can be identified by their features. Equiaxed grains and elongated grains can be seen in respectively WNZ and TMAZ, while there are slightly grain coarsening existing in the HAZ, which are consistent with the common FSW process [128]. Due to the difficulty to the separation of TMAZ via waterproof tapes, only BM, HAZ, and WNZ were considered for convenience. Figure 3.36 shows the macroscopic corrosion morphologies of the heterogeneous joint. The obvious mixing zone of AA2219 and AA2195 can be found at all the macroscopic corrosion morphologies, implying that the corrosion degrees of the two dissimilar alloys are different. The corrosion pits at WNZ are distributed more densely and uniformly, indicating that the precipitates are dissolved and re-precipitated during the welding process. In contrast, the pits at BM are more sparse and uneven. The density and average size of corrosion pits are decreased with the increase of the welding speed at WNZ, while there are no obvious changes at BM. Figures 3.37 and 3.38 show the corrosion morphologies and extracted average corrosion pit depth from SEM images of each microzone at the welding speeds of 100 and 300 mm/min. WNZ-2219 exhibits the best corrosion resistance, followed by WNZ-2195, HAZ-2219, and HAZ-2195. The BMs show the most severe corrosion morphologies.

Heterogeneous strengthening phases, including precipitates [153, 154] and reinforcements [155–157], are often the origin of pitting corrosion. The electrochemical heterogeneity between those particles, such as T_1 phase [158] and θ''/θ' phase [159], and the matrix triggered the galvanic corrosion near the phase interface. The corrosion of each microzone starts in the form of pitting corrosion at different welding speeds. A higher dislocation density is generated by severe plastic deformation during the FSW process, which produces a large number of nucleation sites for precipitation [125, 160]. The precipitate tends to grow when its size is larger than a critical radius but solid-solute when its size is smaller than a critical radius (Eq. 3.24) [161]. It makes the precipitate distribution of WNZ-2219 more uniform with a low volume fraction during the process of solid soluting and re-precipitating. As the welding speed increases, the heat input decreases rapidly, which leads to an increase in the content of solid-soluted Cu in the matrix and fewer precipitates. The corrosion resistance of the heterogeneous joint at low heat input is then enhanced. The corrosion potentials versus saturated calomel electrode (SCE) of the θ''/θ' and T_1 are about − 0.70 V and

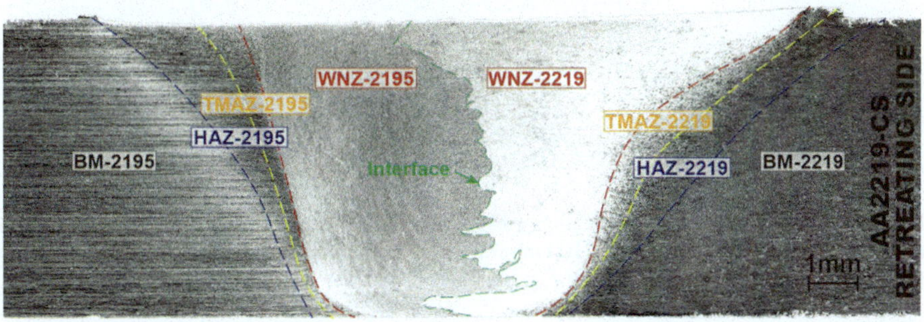

Fig. 3.35 Typical macrostructure of the heterogeneous joint

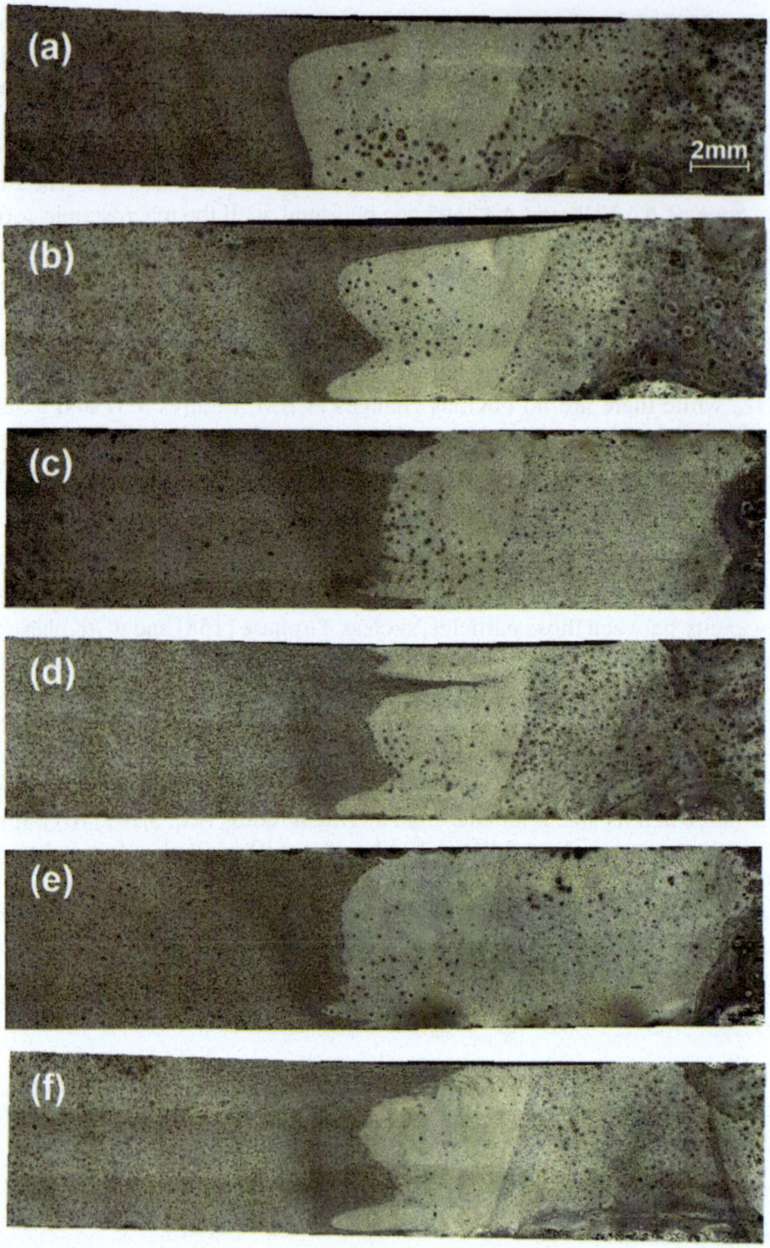

Fig. 3.36 Corroded morphologies after immersion corrosion tests at different welding speeds: **a** 33.3 mm/min, **b** 66.7 mm/min, **c** 100 mm/min, **d** 200 mm/min, **e** 300 mm/min, and **f** 600 mm/min

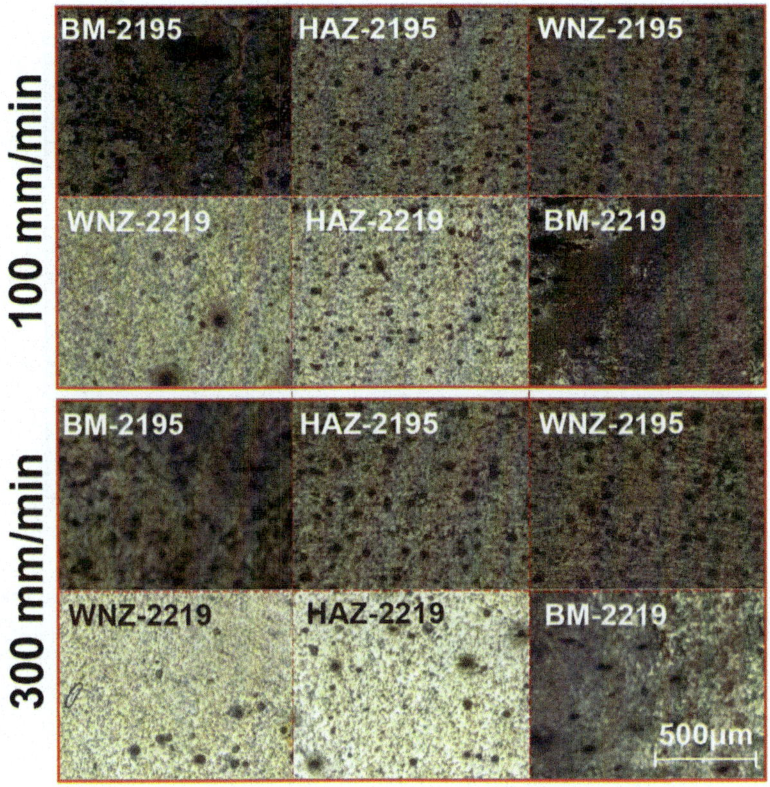

Fig. 3.37 Corrosion morphologies of each microzone at different welding speeds

− 1.09 V, respectively [162]. The lithium element with high reactional activity leads to a very high dissolution rate of T_1 phase, so the lithium-containing phases on the surface disappear in the corrosion solution. The corrosion of the aluminum matrix dominates the corrosion behavior. The main reactions occurring in the corroded surface in oxygen-containing neutral solution are as follows:

Anodic reaction:

$$Al = Al^{3+} + 3e^- \tag{3.33}$$

Cathodic reaction:

$$O_2 + 2H_2O + 4e^- = 4OH^- \tag{3.34}$$

In addition, due to the presence of Cl^- in the solution, great escalation of the corrosion activity is introduced, there promoting the dissolution of the Al matrix and speeding up the corrosion:

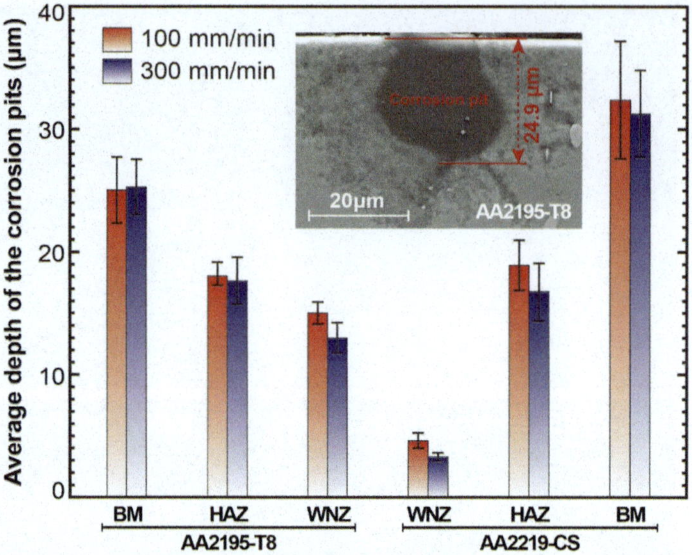

Fig. 3.38 Depth of the corrosion pits of each microzone at different welding speeds extracted from the cross-section of immersion corrosion specimens

$$Al^{3+} + Cl^- = AlCl^{2+} \tag{3.35}$$

$$AlCl^{2+} + Cl^- = AlCl_2^+ \tag{3.36}$$

$$AlCl_2^+ + Cl^- = AlCl_3 \tag{3.37}$$

The precipitation phases are at their peak content with a relatively large average size for BM-2195 and BM-2219. The solid-soluted Cu element in the matrix is in a low content, indicating that the corrosion potential difference between the matrix and the precipitates is high. Micro-batteries are formed near the larger precipitates, leading to the generation of pitting corrosion. In addition, the number density of θ''/θ' formed is higher than the T_1 and θ''/θ' phases in AA2195-T8 alloy due to the high Cu content in AA2219-CS alloy. The corrosion of BM-2219 is more serious than that of BM-2195. For both HAZs, partial precipitates are dissolved into the matrix during the heat stage, so that the corrosion potential of the matrix is raised. For WNZs, the peak temperature during the welding process is higher than the AA2219-CS alloy at the RS because the AA2195-T8 alloy at the AS has higher flow stress. More heat generation contributes to more precipitates. Also, due to the existence of the lithium element in AA2195-T8 alloy, the corrosion degree of WNZ-2195 is more serious than that of WNZ-2219.

3. Intergranular corrosion

Intergranular corrosion characterizes the corrosion resistance of heterogeneous joints in an oxidative corrosion environment [163]. Figures 3.39 and 3.40 depict the depth-wise expansion of corrosion morphologies at the welding speeds of 100 and 300 mm/min. BM-2195 has obvious corrosion structures along the grain boundaries in the depth direction. The corrosion at BM-2219 is more serious. Some grains at the edge of BM-2219 have been corroded off and developed into exfoliation corrosion. In addition, the corrosion degree of each microzone gradually increases in the order of WNZ-2219, WNZ-2195, HAZ-2195, and HAZ-2219. WNZ-2219 and WNZ-2195 show the morphology with almost no intergranular corrosion, which is related to the severe plastic deformation process of FSW. The more refined equiaxed grains and the suppression of the precipitation behavior on grain boundaries significantly reduce the heterogeneity in WNZ and weakens the degree of intergranular corrosion. The HAZs on both sides shows corrosion morphologies like their respective BMs, but the corrosion degree is weaker, which is consistent with the previous immersion tests.

Figure 3.41 shows the corrosion depth of each microzone after intergranular corrosion tests. As the welding speed increased, the corrosion depth shows a downward trend. WNZ-2219 has nearly zero corrosion depth at the welding speed of 300 mm/min, which means that this zone is protected as a cathode in intergranular corrosion.

4. Exfoliation corrosion

The static weight loss curve is a reliable and straightforward method to determine the corrosion rate [164]. The weight loss curve can be obtained by measuring the relative change of weight and fitting with the following formula:

$$\Delta m = M + Ae^{-\frac{T}{t}} = \frac{m_0 - m_T}{S} \tag{3.38}$$

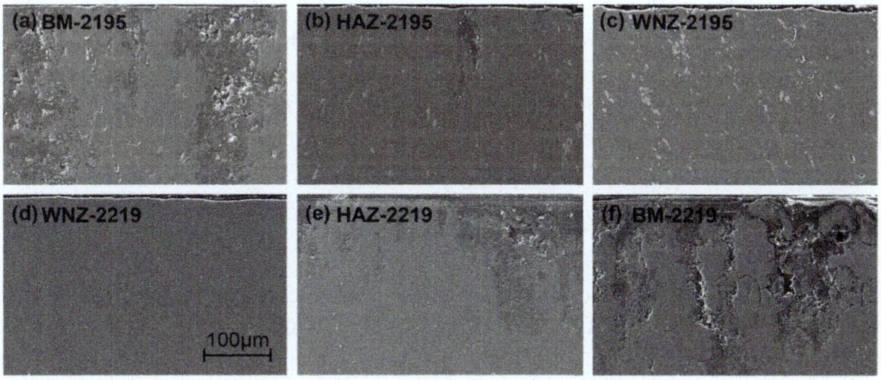

Fig. 3.39 Corrosion depth of each microzones after intergranular corrosion at the welding speed of 100 mm/min: **a** BM-2195, **b** HAZ-2195, **c** WNZ-2195, **d** WNZ-2219, **e** HAZ-2219, and **f** BM-2219

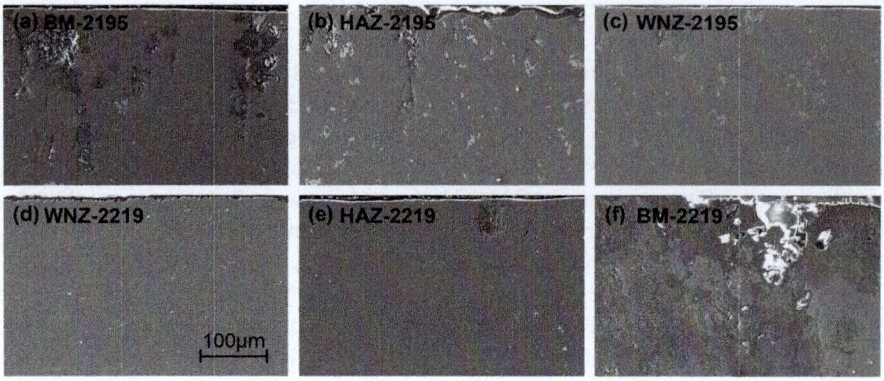

Fig. 3.40 Corrosion depth of each microzones after intergranular corrosion at the welding speed of 300 mm/min: **a** BM-2195, **b** HAZ-2195, **c** WNZ-2195, **d** WNZ-2219, **e** HAZ-2219, and **f** BM-2219

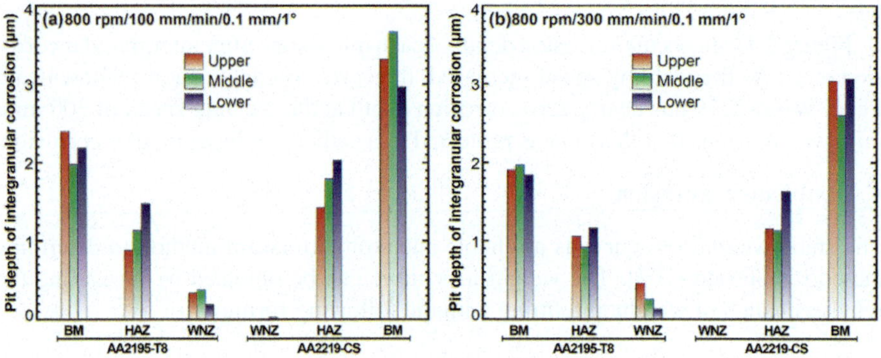

Fig. 3.41 Statistics of the corrosion depth of heterogeneous joints after intergranular corrosion tests at different welding speeds: **a** 100 mm/min, and **b** 300 mm/min

where M, A, and t are the corrosion constants, T is the corrosion time, m_0 is the initial mass before corrosion, m_T is the measured mass after corrosion, and S is the exposed area for corrosion. Based on the weight loss curve, we can obtain the corrosion rate through its derivative with respect to time:

$$v = -\frac{Ae^{-\frac{T}{t}}}{t} = \frac{m_0 - m_T}{ST} \tag{3.39}$$

Figure 3.42 shows the comparison of corrosion rate. The corrosion rate at high welding speed is lower, indicating that reducing heat input is beneficial to improve the corrosion resistance. Figures 3.43 and 3.44 show the corroded surfaces at different microzones with respectively 6 h and 48 h corrosion time. Many corrosion pits at the initial stage of corrosion are formed in BM-2219, showing a typical exfoliation corrosion morphology. The corrosion degree is deepened during the entire corrosion process, resulting in more corrosion products unremovable by concentrated nitric acid. The corrosion degree of BM-2195 is close to that of the BM-2219, but the corroded surface shows a preferred orientation of corrosion caused by the cold working of T8 heat treatment. The original morphology cannot be distinguished after 12 h. It is worth noting that the corrosion rate of the two BMs in the welds is faster than that of the homogeneous AA2195-T8 or AA2219-CS alloy plates, implying that the BM is preferentially corroded during its lower corrosion potential. HAZ-2219 and HAZ-2195 has a low initial corrosion rate which is gradually accelerated with the increase of corrosion time. The corrosion degree at HAZ-2219 is more serious. In the early stage of corrosion, HAZ and its adjacent BM form a macro galvanic couple. The HAZ becomes the protected zone, resulting in a low corrosion rate of the HAZ itself. However, after the BM generates more oxidation products to form a self-protection layer, the corrosion behavior of HAZ begins to be accelerated significantly. After 96 h, the corrosion morphologies of HAZ are close to their corresponding BM. WNZ-2219 is the minor corroded area, implying that a large amount of Cu element dissolved in the matrix increases the corrosion potential of the microzone. After 48 h, the exposed surface begins to deteriorate, and eventually the surface morphology of fine corrosion pits is formed. In addition, corrosion crevices can be observed in all areas during the later stage of corrosion, indicating that the corrosive medium has begun to invade the inside of the joints. Since the surface layer has formed dense oxidation corrosion products as a protective layer, the corrosion rate of the surface layers is suppressed. The weight loss behavior in the later stage of corrosion gradually turns to be dominated by the internal corrosion reaction of the corrosion crevices.

5. Corrosion by electrochemical methods

As shown in Fig. 3.45 and Table 3.4 the higher corrosion rate of BM-2219 than that of BM-2195 proved the more serious corrosion in BM-2219 during the corrosion process. The corrosion potentials of both BMs are lower than those of their corresponding HAZs and WNZs, indicating that all the welding-affected zones in the macroscopic galvanic couples are nobler and protected during the corrosion process. The BMs are preferentially corroded as an anode.

Due to microstructural differences among each microzone, there is uneven corrosion potential existing in the heterogeneous joint. This distribution of the precipitates will be subjected to the welding heat input and the initial heat treatment. Widener et al. [165] investigated the exfoliation corrosion of Al–Cu and Al–Cu–Li alloys, including AA2024, AA2219, and AA2198 alloys. They found that when the heat treatment was T3 or T4, the WNZ exhibited anode activity relative to BM. When the heat treatment was CS or T8, the WNZ was nobler relative to BM, and the corrosion of BM was accelerated. The WNZ and HAZ in our study play the role of a cathode in the galvanic couples, while the BM acts as the anode. This is because after solution treatment (cold working), and artificial aging treatment of both BMs in T6 or T8 states, large amounts of θ''/θ' and T_1 phases are precipitated. The matrix contains only a small amount of solid solution elements. By contrast, the WNZ is affected

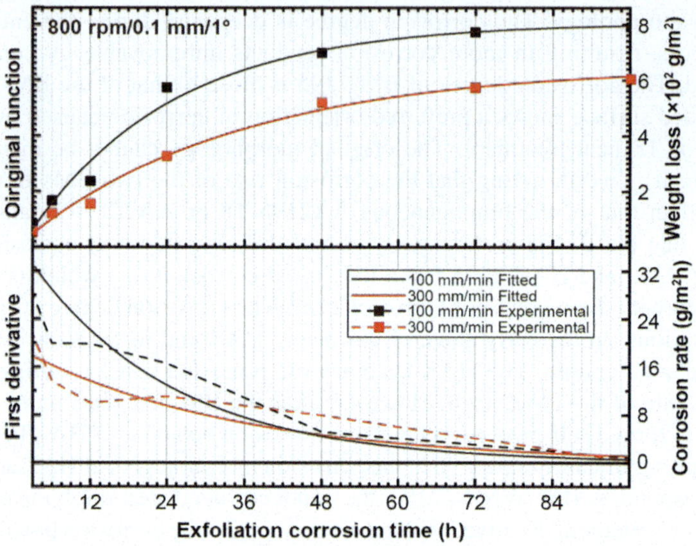

Fig. 3.42 Weight loss and corrosion rate calculated from exfoliation corrosion tests at different welding speeds

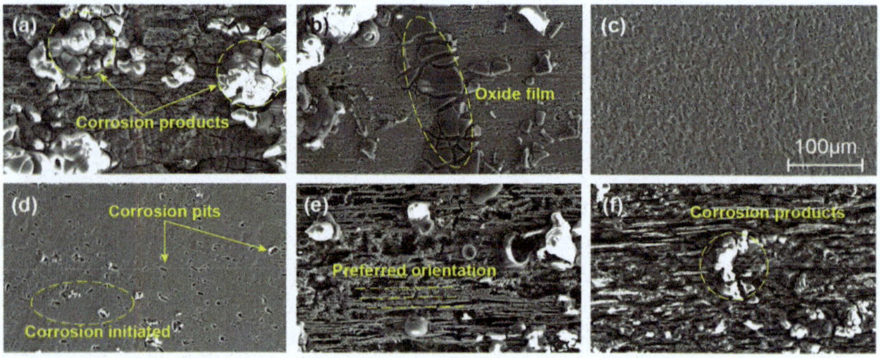

Fig. 3.43 Typical corroded surfaces at different microzones with 6 h corrosion time: **a** BM-2219, **b** HAZ-2219, **c** WNZ-2219, **d** WNZ-2195, **e** HAZ-2195, and **f** BM-2195

by the heat input and mechanical stirring induced by the FSW process. Precipitates dissolute at the heating stage, leading to an increase in the content of solid-soluted Cu element, which increases the overall corrosion potential of the matrix. Therefore, the corrosion potential of WNZ is higher than that of BM. For HAZ, the solid-solution theory is also applicable, except that it is not subjected to severe mechanical stirring, and the remaining precipitates are coarsened but not refined. These differences make the corrosion potential of HAZ lower than that of WNZ.

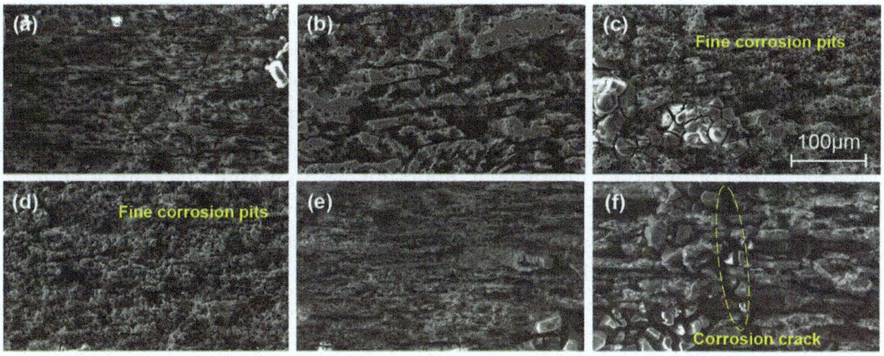

Fig. 3.44 Typical corroded surfaces at different microzones with 48 h corrosion time: **a** BM-2219, **b** HAZ-2219, **c** WNZ-2219, **d** WNZ-2195, **e** HAZ-2195, and **f** BM-2195

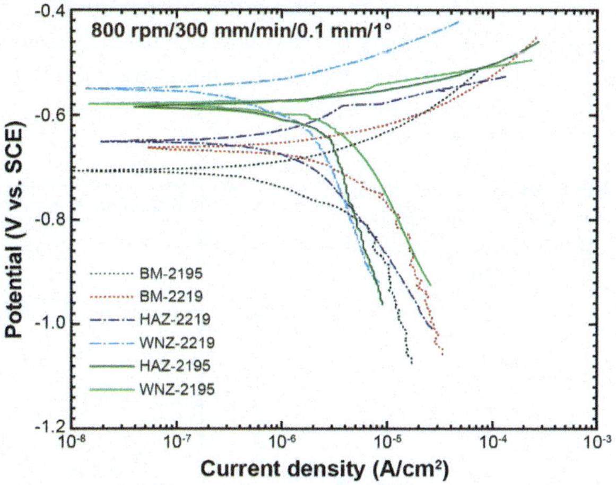

Fig. 3.45 Potentiodynamic polarization curves of each microzone in the heterogeneous joint

Table 3.4 Electrochemical parameters of potentiodynamic polarization curves calculated from each microzone of the heterogeneous joint

Area	Corrosion potential E_{corr} (V vs. SCE)	Current density I_{corr} (A/cm^2)
BM-2195	-0.713 ± 0.055	$1.67 \pm 0.22 \times 10^{-6}$
BM-2219	-0.659 ± 0.037	$3.66 \pm 0.49 \times 10^{-6}$
HAZ-2219	-0.629 ± 0.039	$2.52 \pm 0.30 \times 10^{-6}$
WNZ-2219	-0.550 ± 0.033	$1.62 \pm 0.31 \times 10^{-6}$
HAZ-2195	-0.576 ± 0.028	$2.72 \pm 0.19 \times 10^{-6}$
WNZ-2195	-0.591 ± 0.040	$2.06 \pm 0.38 \times 10^{-6}$

The entire joints at the welding speed of 100 and 300 mm/min were also chosen to conduct the Tafel curves to access the macroscopic galvanic effect, as shown in Fig. 3.46 and Table 3.5. The corrosion potential is enhanced at both welding speeds, indicating the improvement of the corrosion resistance. Higher welding speed contributes to higher corrosion potential, which shows that the corrosion resistance has a negative correlation with the heat input. It is worth noting that both at high and low welding speeds, the corrosion current density has increased compared with the BMs. This implies that the heterogeneous joint has a certain degree of microstructural heterogeneity due to the differences between microzones. This heterogeneity results in the formation of a macroscopic galvanic effect on the joint, accelerating the corrosion behaviors. However, the corrosion current density at 300 mm/min is only slightly higher than that of the BMs. As such, a summary can be drawn that under the premise of ensuring welding formation, increasing the welding speed and reducing the heat input of the joint can improve the corrosion resistance of the joint without significantly increasing the corrosion rate.

Figure 3.47 shows the Nyquist electrochemical impedance spectroscopy (EIS) results of each microzones under the corrosion potential of 3.5 wt.% NaCl aqueous solution. Each area exhibits similar electrochemical behaviors. Capacitive loop and

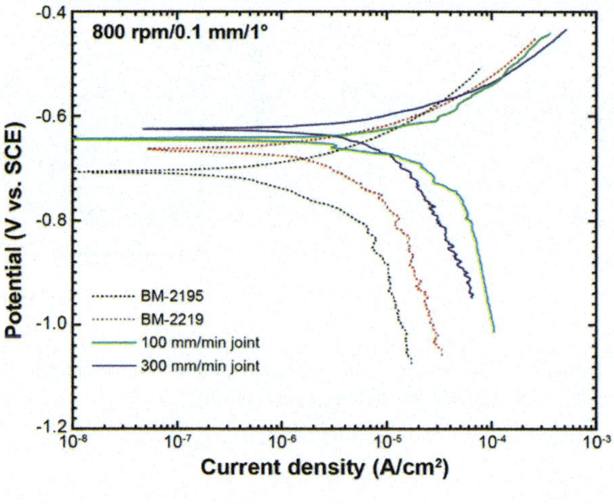

Fig. 3.46 Potentiodynamic polarization curves of the entire joints at different welding speeds

Table 3.5 Electrochemical parameters of potentiodynamic polarization curves calculated from the joint at different welding speeds

Welding speed/base metals	Corrosion potential E_{corr} (V vs. SCE)	Current density I_{corr} (A/cm^2)
AA2195-T8	-0.713 ± 0.055	$1.67 \pm 0.22 \times 10^{-6}$
AA2219-CS	-0.659 ± 0.037	$3.66 \pm 0.49 \times 10^{-6}$
100 mm/min	-0.643 ± 0.031	$4.41 \pm 0.55 \times 10^{-6}$
300 mm/min	-0.614 ± 0.042	$3.92 \pm 0.26 \times 10^{-6}$

inductive loop can be observed due to the passive film of aluminum oxide, the effect of porous corrosion products containing hydrate aluminum oxide/chlorides, and partial unsolvable precipitates. The inductive loop may be attributed to the change in the corrosion potential during EIS tests owing to the time need to obtain data at low frequency. According to the above circuit characteristics, an equivalent circuit is established to quantitatively fit the electrochemical impedance response of all microzones, as shown in Fig. 3.48. Among these parameters, R_s is the resistance of the corrosive electrolyte, that is, 3.5 wt.% NaCl aqueous solution. The parallel connection of R_1 and constant phase element (CPE) indicates the charge transfer impedance at the interface and corresponding electric double-layer capacitance. The CPE represents the capacitive element related to the non-ideal capacitive behavior, which characterizes the non-uniform distribution of the surficial reaction, that is, the microscopic galvanic couple between the precipitates and the matrix in each microzones. Its impedance is related to the frequency as follows: $Z_{CPE} = 1/T(i\omega)^n$, where T is the magnitude and n is the exponent. The parallel connection of R_2 and C is used to describe the resistance and capacitance of the surface oxide film. The inductance element L is used to explain the changes in the active area of the anode. R_3 represents the resistance associated with changes in the local corrosion environment near the anode area. Since these electrochemical/chemical reactions occur on the surface of the alloy simultaneously, they are all connected in parallel with CPE to reflect their coupled influence on the corrosion behavior. In addition, there is another critical parameter, polarization resistance R_p, whose reciprocal is considered proportional to the corrosion rate [166]. According to the equivalent circuit, we can derive its calculation equation as:

$$\frac{1}{R_p} = \frac{1}{R_1 + R_2} + \frac{1}{R_3} \tag{3.40}$$

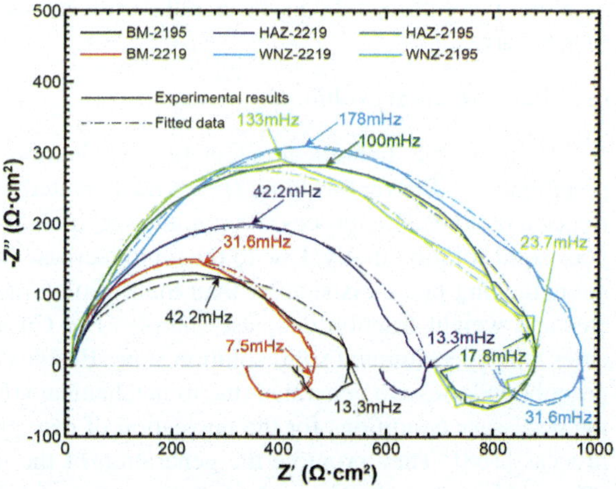

Fig. 3.47 Nyquist spectra of each microzone in the heterogeneous joint

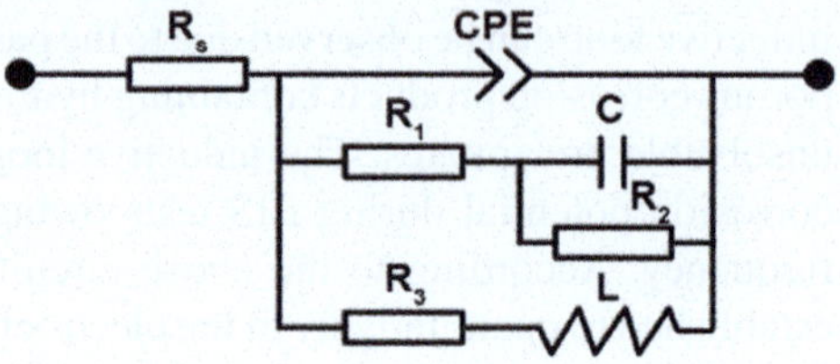

Fig. 3.48 Equivalent circuit for modeling the electrochemical impedance response of each microzone

Besides, it is worth noting that the CPE has the dimension of s^n, which does not represent the actual and pure capacitance of the electric double layer. Therefore, converting it into capacitance is critical to evaluate its actual impact on corrosion behavior. Brug et al. [167] introduced a model to describe the behavior of CPE as:

$$C_{Brug} = T_{CPE}^{\frac{1}{n}}\left(\frac{R_p R_s}{R_p + R_s}\right)^{\frac{1-n}{n}} \tag{3.41}$$

Table 3.6 lists the fitted results extracted from the EIS results in Fig. 3.47. The resistance R_s of all corrosive mediums has little difference, indicating that the resistance of the electrolyte changes little and the tests are reliable. Comparing with the welded joints, the polarization resistance R_p of each WNZ and HAZ has increased. HAZ-2219 amid them has the lowest increase in corrosion resistance, and the WNZ-2219 has the highest resistance. Comparing the CPE magnitude and exponent in difference microzones, it can be found that the heterogeneity affected by welding heat input and mechanical stirring has increased, while the value of R_2 related to the oxide film and corrosion products has reduced in the welding-affected zones, indicating that the density of corrosion products produced is less. It proves that the corrosion resistance of the welded joint is enhanced. The representative value L of the surface anode active area becomes higher in the welding-affected zone. It elucidates that the WNZ and HAZ have a larger reactive size during the corrosion process; that is, more exposed metallic surfaces reflect the enhancement of corrosion resistance in these areas.

6. Microstructural evolution

According to the theoretical calculation by the above modeling, the variations of precipitates at different welding speeds are depicted in Fig. 3.49. WNZs firstly experience a rapid heating process. Precipitations, including T_1 and θ''/θ' phases, undergoes solid solution firstly. Due to the rapid increase in temperature, the size of most strengthening phases is smaller than their critical radius [161]. The average diameter and weight fraction decrease sharply. The Cu and Li elements in the matrix enter the supersaturated solid solution state. Besides, high-density dislocations and grain boundaries are generated due to mechanical stirring [95]. These sites provide the necessary conditions for the nucleation of new precipitations in the subsequent process [168]. They promote the generation of the uniform and fine T_1 and θ''/θ' phases. Moreover, the Cu content in the matrix of WNZ increases slightly with the

Table 3.6 Equivalent circuit parameters by fitting the experimental EIS results

Area	R_s (Ω cm^2)	CPE		R_1 (Ω cm^2)	R_2 (Ω cm^2)	C (μF cm^{-2})	R_3 (Ω cm^2)	L (H cm^{-2})	R_p (Ω cm^2)
		T ($\mu\Omega^{-1}$ cm^{-2} s^n)	n						
BM-2195	8.2	39.7	0.92	413.5	217.1	145.0	856.9	198.0	388.5
BM-2219	8.9	52.0	0.95	328.7	232.9	132.2	1032.8	370.5	363.8
HAZ-2219	6.8	42.3	0.91	583.3	185.5	86.5	857.0	2350.4	405.3
WNZ-2219	5.9	11.8	0.89	801.3	82.5	13.7	1292.7	1135.4	524.9
HAZ-2195	6.6	21.3	0.90	687.9	66.8	107.2	1025.3	1798.7	434.7
WNZ-2195	7.1	12.9	0.90	735.2	93.9	76.3	1358.9	3558.9	514.9

increase of the welding speed. Since the solid-soluted Cu greatly affects the corrosion potential of the matrix, a higher content usually means better corrosion resistance due to the alleviation of electrochemical heterogeneity [169]. In addition, the average size of the precipitates in both WNZ and HAZ decreases with the increase of welding speed. The coarsening tendency of precipitates in HAZ is significantly at a higher welding speed. Larger precipitates contribute to the deterioration of corrosion resistance. However, due to the solid-solutioned Cu element in the matrix increase the corrosion potential of the matrix, HAZ shows an increase in corrosion resistance.

7. Precipitation-driven corrosion behavior on macro- and microscale

The calculation of the precipitation evolution by the modified Wagner–Kampmann model shows that the solid-soluted high corrosion potential elements in the matrices of WNZ and HAZ have increased during the welding process, and the content of Cu in the matrix of WNZ is much higher. As depicted in Fig. 3.50, this phenomenon makes all the welding-affected zones acquire the improvement of corrosion resistance on the macro scale. The BMs on both sides are corroded first and gradually covered by corrosion products due to their lowest corrosion potential. Then the major corroded area transferred to HAZ and finally extended to the noblest WNZ, which

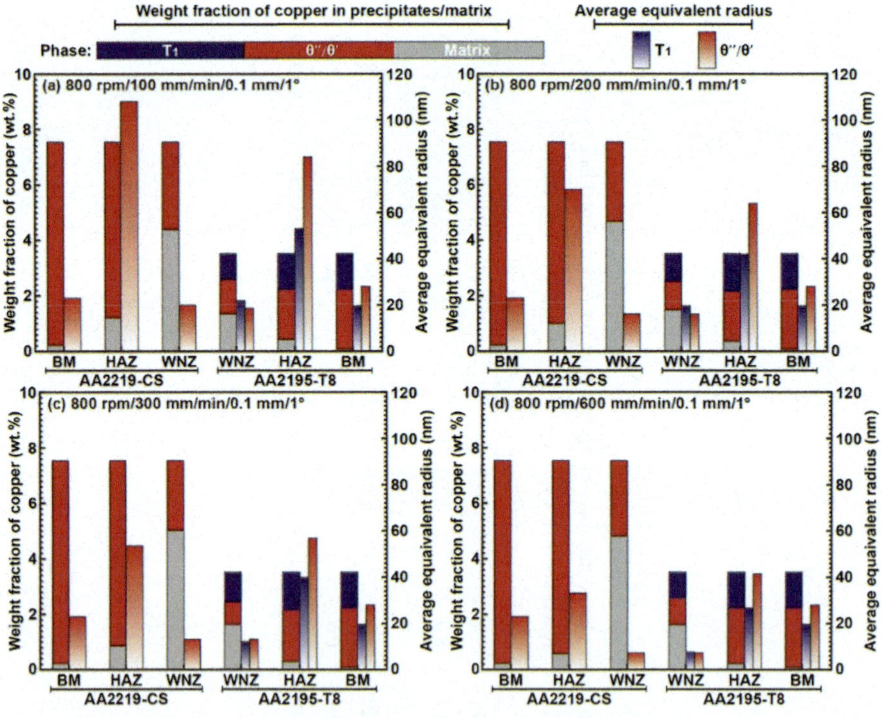

Fig. 3.49 Variation of precipitations during FSW at different welding speeds: **a** 100 mm/min, **b** 200 mm/min, **c** 300 mm/min, and **d** 600 mm/min

is consistent with relevant open literature [170]. It is the macroscopic part of the precipitation-driven corrosion behavior of the heterogeneous joint. The heat input induced by the FSW process contributes to the dissolution of the precipitates, and part of them re-precipitates in the subsequent cooling stage with a more refined shape [171]. The content of the element nobler than the aluminum matrix increases due to the reduction of volume fraction of the precipitates [172]. Besides, the severe plastic deformation results in the broken of remaining precipitates, forming a more "electrochemically homogeneous" microstructure [173], and eventually improving the corrosion resistance of the entire joint.

In addition, the sizes of the precipitates also play a critical role in the corrosion process [174]. Due to the severe plastic deformation and low heat input of FSW, the precipitations in the WNZ tend to re-precipitate more uniform and finer [175], which inhibits the initiation speed of local pitting corrosion and enhances the corrosion resistance of WNZ. For HAZ, an obvious coarsening of precipitates occurs [128]. Combined with the effect of the solid-soluted Cu element in the matrix, the overall corrosion resistance is slightly improved.

Therefore, we proposed a microscopic corrosion behavior of the heterogeneous FSWed joint between AA2195-T8 and AA2219-CS, as shown in Fig. 3.51. In the initial stage of corrosion, due to the corrosion potential difference between the precipitates and the matrix, the adjacent matrix is preferentially dissolved as an anode [176]. Relatively porous corrosion products are formed, which gradually cover and the surface neat the corrosion pits. As time goes on, the precipitate separates from

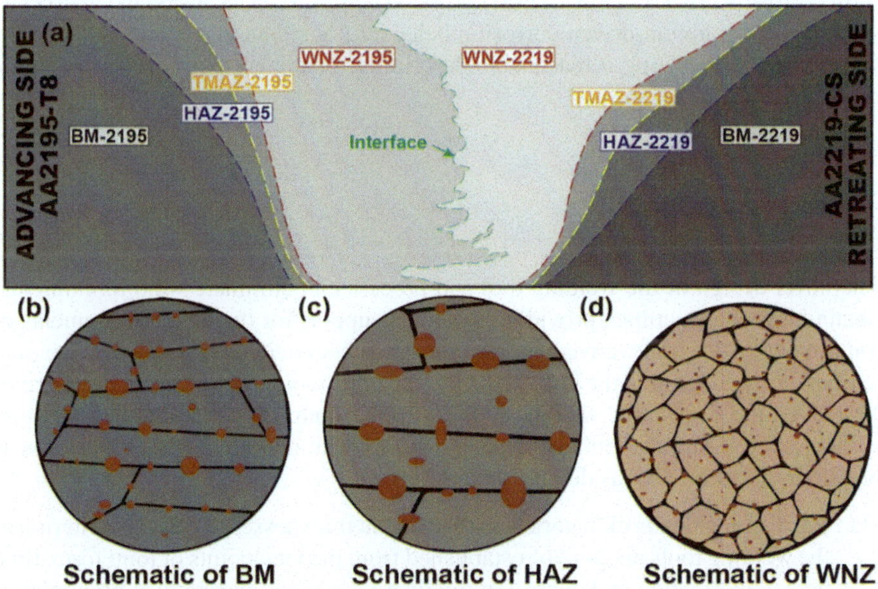

Fig. 3.50 Macroscopic corrosion characteristics of each microzone at the heterogeneous FSWed joint: **a** heterogeneous joint, **b** BM, **c** HAZ, and **d** WNZ

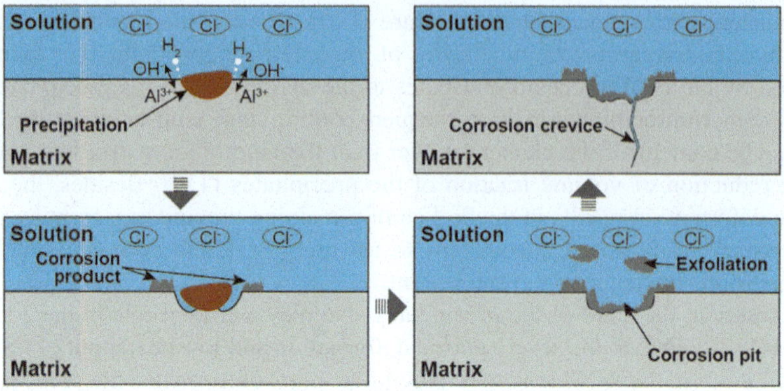

Fig. 3.51 Microscopic corrosion behavior between the precipitate and the matrix

the metal surface due to the lack of connection and eventually forms an unevenly distributed surface corrosion product film [177]. Since the adhesion of these corrosion products with the matrix is usually not dense enough, they tend to exfoliate from the surface and enter the corrosive medium but still cover a partial surface surrounding the original corrosion pit. Local protection is formed and inhibits further corrosion behavior. Moreover, due to the uneven distribution of the oxidation corrosion products covered on the surface, some of the spalling areas allow the passage of electrolyte, causing the corrosion to expand along the depth direction and finally form corrosion crevices [178]. Thus, corrosion begins to occur inside the matrix. As such, the precipitation-driven corrosion behavior co-occurs inside and between the microzones, completing a reasonable description of the corrosion behavior in the heterogeneous joints.

3.3.6 Prospects

The novel design of the welding tool and process can eliminate weld thinning and obtain high-quality joints, providing technical support for the high requirements of joint integrity. The innovative welding concept is beneficial to reducing the welding temperature and refining the grain size, thereby improving the mechanical properties. Moreover, the detachable welding tool is simpler than the stationary shoulder tool system and had more potentials in the industrial manufacturing fields. Some aspects remain to be addressed, as detailed hereafter:

(1) For different kinds of materials with different thicknesses, the design criterion of the welding tools needs to be established from the viewpoints of joint formation, defect elimination and control of microstructures and mechanical properties.
(2) The design of the MPD-FSW tool can potentially be combined with the lower shoulder in bobbin tool friction stir welding (BT-FSW) or $SSFSW_2$, realizing

non-weld-thinning of hollow or enclosed structures, subsequently expanding the industrial applications.

(3) Technical and fundamental issues need to be elucidated, including the design of the welding tool or structure, heat generation, material flow behaviour, microstructural evolution and mechanical properties.

(4) Hybrid joining processes, such as electricity-assisted FSW, heat-assisted FSW, ultrasonic-assisted FSW and underwater FSW, can be combined with advanced NWT-FSW to simultaneously improve the microstructure and mechanical properties by utilizing their individual advantages.

References

1. Nandan R, Debroy T, Bhadeshia HKDH. Recent advances in friction-stir welding—process, weldment structure and properties. Prog Mater Sci. 2008;53(6):980–1023.
2. Wang BB, Chen FF, Liu F, et al. Enhanced mechanical properties of friction stir welded 5083Al-H19 joints with additional water cooling. J Mater Sci Technol. 2017;33(9):1009–14.
3. Sree Sabari S, Malarvizhi S, Balasubramanian V. Characteristics of FSW and UWFSW joints of AA2519-T87 aluminium alloy: effect of tool rotation speed. J Manuf Process. 2016;22:278–89.
4. Lin SB, Zhao YH, Wu L, et al. Integral and layered mechanical properties of friction stir welded joints of 2014 aluminium alloy. Met Sci J. 2014;22(8):995–8.
5. Sattari S, Bisadi H, Sajed M. Mechanical properties and temperature distributions of thin friction stir welded sheets of AA5083. Int J Mech Appl. 2012;2(1):1–6.
6. Kumar M, Kailas SV, Narayanan RG. Influence of external weld flash on the in-plane plane-strain formability of friction stir welded sheets. J Strain Anal Eng Des. 2013;48(6):376–85.
7. Gangwar K, Ramulu M. Friction stir welding of titanium alloys: a review. Mater Des. 2018;141:230–55.
8. Ji S, Li Z, Wang Y, et al. Joint formation and mechanical properties of back heating assisted friction stir welded Ti–6Al–4V alloy. Mater Des. 2017;113:37–46.
9. Huang Y, Meng X, Xie Y, et al. Friction spot welding of carbon fiber-reinforced polyetherimide laminate. Compos Struct. 2018;189:627–34.
10. Azarsa E, Mostafapour A. Experimental investigation on flexural behavior of friction stir welded high density polyethylene sheets. J Manuf Process. 2014;16(1):149–55.
11. Eslami S, Ramos T, Tavares PJ, et al. Shoulder design developments for FSW lap joints of dissimilar polymers. J Manuf Process. 2015;20:15–23.
12. Davies PS, Wynne BP, Rainforth WM, et al. Development of microstructure and crystallographic texture during stationary shoulder friction stir welding of Ti-6Al-4V. Metall Mater Trans A. 2011;42(8):2278–89.
13. Delta-N FS system for friction-stir welding. 2014-02-03. https://conceptsystemsinc.com/tag/deltan-fs/
14. Wu H, Chen YC, Strong D, et al. Stationary shoulder FSW for joining high strength aluminum alloys. J Mater Process Technol. 2015;221:187–96.
15. Russell MJ, Threadgill PL, Thomas MJ, et al. Static shoulder friction stir welding of Ti-6Al-4V: process and evaluation. In: Proceedings of 11th world conference on titanium. Kyoto; 2007.
16. Russell MJ, Blignault C, Horrex NL, et al. Recent developments in the friction stir welding of titanium alloys. Weld World. 2008;52(9–10):12–5.

17. Brassington WDP, Colegrove PA. Alternative friction stir welding technology for titanium–6Al–4V propellant tanks within the space industry. Sci Technol Weld Join. 2017;22(4):300–18.
18. Rai R, De A, Bhadeshia HKDH, et al. Review: friction stir welding tools. Sci Technol Weld Join. 2011;16(4):325–42.
19. Liu HJ, Li JQ, Duan WJ. Friction stir welding characteristics of 2219-T6 aluminum alloy assisted by external non-rotational shoulder. Int J Adv Manuf Technol. 2013;64(9–12):1685–94. https://doi.org/10.1007/s00170-012-4132-1.
20. Ji SD, Meng XC, Liu JG, et al. Formation and mechanical properties of stationary shoulder friction stir welded 6005A-T6 aluminum alloy. Mater Des. 2014;62:113–7.
21. Li JQ, Liu HJ. Effects of tool rotation speed on microstructures and mechanical properties of AA2219-T6 welded by the external non-rotational shoulder assisted friction stir welding. Mater Des. 2013;43:299–306.
22. Zafar A, Awang M, Khan SR, et al. Investigating friction stir welding on thick Nylon 6 plates. Weld Res. 2016;95:210–8.
23. Barmouz M, Shahi P, Asadi P. Friction stir welding/processing of polymeric materials. Adv Frict Weld Process. 2014:601–70.
24. Mendes N, Loureiro A, Martins C, et al. Effect of friction stir welding parameters on morphology and strength of acrylonitrile butadiene styrene plate welds. Mater Des. 2014;58:457–64.
25. Mostafapour A, Taghizad Asad F. Investigations on joining of Nylon 6 plates via novel method of heat assisted friction stir welding to find the optimum process parameters. Sci Technol Weld Join. 2016;21(8):660–9.
26. Li D, Yang X, Cui L, et al. Fatigue property of stationary shoulder friction stir welded additive and non-additive T joints. Sci Technol Weld Join. 2015;20(8):650–4.
27. Wei S, Martin J. New techniques: robotic friction stir welding. In: Proceedings of 10th international symposium on friction stir welding. Beijing; 2014. p. 10.
28. Maltin CA, Nolton LJ, Scott JL, et al. The potential adaptation of stationary shoulder friction stir welding technology to steel. Mater Des. 2014;64:614–24.
29. Martin J, Stanhope C, Gascoyne S. Novel techniques for corner joints using friction stir welding. In: TMS annual meeting; 2011. p. 177–86.
30. Jesus JS, Costa JM, Loureiro A, et al. Assessment of friction stir welding aluminium T-joints. J Mater Process Technol. 2018;255:387–99.
31. Feistauer EE, Bergmann LA, dos Santos JF. Effect of reverse material flow on the microstructure and performance of friction stir welded T-joints of an Al–Mg alloy. Mater Sci Eng A. 2018;731:454–64.
32. Texier D, Atmani F, Bocher P, et al. Fatigue performances of FSW and GMAW aluminum alloys welded joints: competition between microstructural and structural-contact-fretting crack initiation. Int J Fatigue. 2018;116:220–33. https://doi.org/10.1016/j.ijfatigue.2018.06.020.
33. Li D, Yang X, Cui L, et al. Investigation of stationary shoulder friction stir welding of aluminum alloy 7075-T651. J Mater Process Technol. 2015;222:391–8.
34. Meng X, Jin Y, Ji S, et al. Improving friction stir weldability of Al/Mg alloys via ultrasonically diminishing pin adhesion. J Mater Sci Technol. 2018;34(10):1817–22. https://doi.org/10.1016/j.jmst.2018.02.022.
35. Liu Z, Meng X, Ji S, et al. Improving tensile properties of Al/Mg joint by smashing intermetallic compounds via ultrasonic-assisted stationary shoulder friction stir welding. J Manuf Process. 2018;31:552–9.
36. Huang Y, Meng X, Xie Y, et al. Joining of carbon fiber reinforced thermoplastic and metal via friction stir welding with co-controlling shape and performance. Compos Part A Appl Sci Manuf. 2018;112:328–36.
37. Ji SD, Meng XC, Li ZW, et al. Experimental study of stationary shoulder friction stir welded 7N01-T4 aluminum alloy. J Mater Eng Perform. 2016;25(3):1228–36.

38. Shah LH, Othman NH, Gerlich A. Review of research progress on aluminium–magnesium dissimilar friction stir welding. Sci Technol Weld Join. 2017;1718:1–15.
39. Strand S. Effects of friction stir welding on polymer microstructure. 2004.
40. Ji S, Huang R, Meng X, et al. Enhancing friction stir weldability of 6061-T6 Al and AZ31B Mg alloys assisted by external non-rotational shoulder. J Mater Eng Perform. 2017;26(5):2359–67.
41. Ji S, Li Z, Zhang L, et al. Effect of lap configuration on magnesium to aluminum friction stir lap welding assisted by external stationary shoulder. Mater Des. 2016;103:160–70.
42. Ji S, Meng X, Liu Z, et al. Dissimilar friction stir welding of 6061 aluminum alloy and AZ31 magnesium alloy assisted with ultrasonic. Mater Lett. 2017;201:173–6.
43. Meng X, Jin Y, Ji S, et al. Improving friction stir weldability of Al/Mg alloys via ultrasonically diminishing pin adhesion. J Mater Sci Technol. 2018;34(10).
44. Meng X, Xu Z, Huang Y, et al. Interface characteristic and tensile property of friction stir lap welding of dissimilar aircraft 2060-T8 and 2099-T83 Al–Li alloys. Int J Adv Manuf Technol. 2018;94(1–4):1253–61.
45. Ji S, Li Z. Microstructure and mechanical properties of friction stir lap welded Mg/Al joint assisted by stationary shoulder. Met Mater Int. 2017;23(6):1158–67.
46. Yue Y, Zhou Z, Ji S, et al. Effect of welding speed on joint feature and mechanical properties of friction stir lap welding assisted by external stationary shoulders. Int J Adv Manuf Technol. 2017:1691–8.
47. Li Z, Yue Y, Ji S, et al. Joint features and mechanical properties of friction stir lap welded alclad 2024 aluminum alloy assisted by external stationary shoulder. Mater Des. 2016;90:238–47.
48. Wen Q, Li WY, Wang WB, et al. Experimental and numerical investigations of bonding interface behavior in stationary shoulder friction stir lap welding. J Mater Sci Technol. 2019;35(1):192–200.
49. Ahmed MMZ, Wynne BP, Rainforth WM, et al. Through-thickness crystallographic texture of stationary shoulder friction stir welded aluminium. Scr Mater. 2011;64(1):45–8.
50. Li D, Yang X, Cui L, et al. Effect of welding parameters on microstructure and mechanical properties of AA6061-T6 butt welded joints by stationary shoulder friction stir welding. Mater Des. 2014;64:251–60.
51. Sun T, Roy MJ, Strong D, et al. Comparison of residual stress distributions in conventional and stationary shoulder high-strength aluminum alloy friction stir welds. J Mater Process Technol. 2017;242:92–100.
52. Sun T, Tremsin AS, Roy MJ, et al. Investigation of residual stress distribution and texture evolution in AA7050 stationary shoulder friction stir welded joints. Mater Sci Eng A. 2018;712:531–8.
53. Barbini A, Carstensen J, dos Santos JF. Influence of a non-rotating shoulder on heat generation, microstructure and mechanical properties of dissimilar AA2024/AA7050 FSW joints. J Mater Sci Technol. 2017;34(1):119–27.
54. Regensburg A, Schürer R, Weigl M, et al. Influence of pin length and electrochemical platings on the mechanical strength and macroscopic defect formation in stationary shoulder friction stir welding of aluminium to copper. Metals (Basel). 2018;8(85):2–9.
55. Mostafapour A. A study on the role of processing parameters in joining polyethylene sheets via heat assisted friction stir welding: investigating microstructure, tensile and flexural properties. Int J Phys Sci. 2012;7(4).
56. Kiss Z, Czigany T. Microscopic analysis of the morphology of seams in friction stir welded polypropylene. Express Polym Lett. 2012;6(1):54–62.
57. Bagheri A, Azdast T, Doniavi A. An experimental study on mechanical properties of friction stir welded ABS sheets. Mater Des. 2013;43:402–9.
58. Rahbarpour R, Azdast T, Rahbarpour H, et al. Feasibility study of friction stir welding of wood–plastic composites. Sci Technol Weld Join. 2014;19(8):673–81.
59. Mendes N, Loureiro A, Martins C, et al. Morphology and strength of acrylonitrile butadiene styrene welds performed by robotic friction stir welding. Mater Des. 2014;64:81–90.
60. Czigány T, Kiss Z. Friction stir welding of fiber reinforced polymer composites. In: Proceedings of the 18th international conference on composite materials. Jeju, South Korea: ICCM; 2011. p. 21–6.

61. Eslami S, Mourão L, Viriato N, et al. Multi-axis force measurements of polymer friction stir welding. J Mater Process Technol. 2017;2018(256):51–6.
62. Huang Y, Wang Y, Wan L, et al. Material-flow behavior during friction-stir welding of 6082-T6 aluminum alloy. Int J Adv Manuf Technol. 2016;87(1–4):1115–23.
63. Ji S, Li Z, Zhang L, et al. Eliminating the tearing defect in Ti-6Al-4V alloy joint by back heating assisted friction stir welding. Mater Lett. 2017;188:21–4.
64. Zhou L, Liu HJ, Liu QW. Effect of process parameters on stir zone microstructure in Ti-6Al-4V friction stir welds. J Mater Sci. 2010;45(1):39–45. https://doi.org/10.1007/s10853-009-3881-1.
65. Pilchak AL, Juhas MC, Williams JC. Microstructural changes due to friction stir processing of investment-cast Ti-6Al-4V. Metall Mater Trans A Phys Metall Mater Sci. 2007;38(2):401–8.
66. Knipling KE, Fonda RW. Microstructural evolution in Ti-5111 friction stir welds. Metall Mater Trans A. 2011;42(8):2312–22.
67. Avettand-Fènoël MN, Taillard R. Effect of a pre or postweld heat treatment on microstructure and mechanical properties of an AA2050 weld obtained by SSFSW. Mater Des. 2016;89:348–61.
68. Ma ZY, Feng AH, Chen DL, et al. Recent advances in friction stir welding/processing of aluminum alloys: microstructural evolution and mechanical properties. Crit Rev Solid State Mater Sci. 2017;8436:1–65.
69. Leng L, Zhang ZJ, Duan QQ, et al. Improving the fatigue strength of 7075 alloy through aging. Mater Sci Eng A. 2018;738:24–30.
70. Ferragut R, Somoza A, Tolley A. Microstructural evolution of 7012 alloy during the early stages of artificial ageing. Acta Mater. 1999;47(17):4355–64.
71. Berg LK, Gjoønnes J, Hansen V, et al. GP-zones in Al–Zn–Mg alloys and their role in artificial aging. Acta Mater. 2001;49(17):3443–51.
72. Hu Y, Liu H, Li S, et al. Improving mechanical properties of a joint through tilt probe penetrating friction stir welding. Mater Sci Eng A. 2018;731:107–18.
73. Sun Z, Yang X, Li D, et al. The local strength and toughness for stationary shoulder friction stir weld on AA6061-T6 alloy. Mater Charact. 2016;111:114–21.
74. Arici A, Sinmazçelýk T. Effects of double passes of the tool on friction stir welding of polyethylene. J Mater Sci. 2005;40(12):3313–6.
75. Arici A, Selale S. Effects of tool tilt angle on tensile strength and fracture locations of friction stir welding of polyethylene. Sci Technol Weld Join. 2007;12(6):536–9.
76. Huang Y, Meng X, Xie Y, et al. Friction stir welding/processing of polymers and polymer matrix composites. Compos Part A Appl Sci Manuf. 2018;105:235–57.
77. Gao J, Shen Y, Zhang J, et al. Submerged friction stir weld of polyethylene sheets. J Appl Polym Sci. 2014;41059:1–8.
78. He X, Gu F, Ball A. A review of numerical analysis of friction stir welding. Prog Mater Sci. 2014;65:1–66.
79. Chen G, Li H, Wang G, et al. Effects of pin thread on the in-process material flow behavior during friction stir welding: a computational fluid dynamics study. Int J Mach Tools Manuf. 2018;124:12–21.
80. Chen G, Ma Q, Zhang S, et al. Computational fluid dynamics simulation of friction stir welding: a comparative study on different frictional boundary conditions. J Mater Sci Technol. 2017;34(1):128–34.
81. Li JQ, Liu HJ. Design of tool system for the external nonrotational shoulder assisted friction stir welding and its experimental validations on 2219-T6 aluminum alloy. Int J Adv Manuf Technol. 2013;66(5–8):623–34.
82. Chu K, Wang X, Wang F, et al. Largely enhanced thermal conductivity of graphene/copper composites with highly aligned graphene network. Carbon N Y. 2018;127:102–12.
83. Li H, Dai S, Miao J, et al. Enhanced thermal conductivity of graphene/polyimide hybrid film via a novel "molecular welding" strategy. Carbon N Y. 2018;126:319–27.
84. Chung SH, Kim H, Jeong SW. Improved thermal conductivity of carbon-based thermal interface materials by high-magnetic-field alignment. Carbon N Y. 2018;140:24–9.

85. Aydin M. Effects of welding parameters and pre-heating on the friction stir welding of UHMW-polyethylene. Polym Plast Technol Eng. 2010;49(6):595–601.
86. Vijendra B, Sharma A. Induction heated tool assisted friction-stir welding (i-FSW): a novel hybrid process for joining of thermoplastics. J Manuf Process. 2015;20:234–44.
87. Liu Z, Ji S, Meng X. Joining of magnesium and aluminum alloys via ultrasonic assisted friction stir welding at low temperature. Int J Adv Manuf Technol. 2018;97(9–12):4127–36.
88. Ji S, Li Z, Zhou Z, et al. Microstructure and mechanical property differences between friction stir lap welded joints using rotating and stationary shoulders. Int J Adv Manuf Technol. 2017;90(9–12):3045–53.
89. Jayaraman N, Prevéy P, Mahoney M. Fatigue life improvement of an aluminum alloy FSW with low plasticity burnishing. In: TMS annual meeting; 2003. p. 259–69.
90. Huang Y, Wan L, Lv S, et al. In situ rolling friction stir welding for joining AA2219. Mater Des. 2013;50:810–6.
91. Huang Y, Wan L, Lv S, et al. Gradient micro-structured surface layer on aluminum alloy fabricated by in situ rolling friction stir welding. Mater Des. 2013;52:821–7.
92. Huang YX, Wan L, Lv SX, et al. New technique of *in situ* rolling friction stir welding. Sci Technol Weld Join. 2012;17(8):636–42.
93. Ji S, Meng X, Ma L, et al. Vertical compensation friction stir welding assisted by external stationary shoulder. Mater Des. 2015;68:72–9.
94. Sandnes L, Grong Ø, Torgersen J, et al. Exploring the hybrid metal extrusion and bonding process for butt welding of Al–Mg–Si alloys. Int J Adv Manuf Technol. 2018;98(5–8):1067.
95. Li N, Li W, Yang X, et al. Corrosion characteristics and wear performance of cold sprayed coatings of reinforced Al deposited onto friction stir welded AA2024-T3 joints. Surf Coat Technol. 2018;349:1069–76.
96. Li N, Li WY, Yang XW, et al. An investigation into the mechanism for enhanced mechanical properties in friction stir welded AA2024-T3 joints coated with cold spraying. Appl Surf Sci. 2018;439:623–31.
97. Huang Y, Han B, Lv S, et al. In-situ additive friction stir welding method for eliminating weld-thinning: 中国, CN103157903A; China; 2013. p. 7.
98. Li W, Assadi H, Gaertner F, et al. A review of advanced composite and nanostructured coatings by solid-state cold spraying process. Crit Rev Solid State Mater Sci. 2018;8436:1–48.
99. Ogura T, Saito Y, Nishida T, et al. Partitioning evaluation of mechanical properties and the interfacial microstructure in a friction stir welded aluminum alloy/stainless steel lap joint. Scr Mater. 2012;66(8):531–4.
100. Meng X, Gao S, Ma L, et al. Effects of rotational velocity on microstructures and mechanical properties of surface compensation friction stir welded 6005A-T6 aluminum alloy. Eng Rev. 2016;36(2):107–13.
101. Wen Q, Yue Y, Ji S, et al. Effect of welding speeds on mechanical properties of level compensation friction stir welded 6061-T6 aluminum alloy. High Temp Mater Process. 2016;35(4):375–9.
102. Liu H, Fujii H, Maeda M, et al. Heterogeneity of mechanical properties of friction stir welded joints of 1050-H24 aluminum alloy. J Mater Sci Technol. 2003;22:441–4.
103. Xu W, Liu J, Luan G, et al. Temperature evolution, microstructure and mechanical properties of friction stir welded thick 2219-O aluminum alloy joints. Mater Des. 2009;30(6):1886–93.
104. Korayem AH, Chen SJ, Zhang QH, et al. Failure of CFRP-to-steel double strap joint bonded using carbon nanotubes modified epoxy adhesive at moderately elevated temperatures. Compos Part B Eng. 2016;94:95–101.
105. İpekoğlu G, Akçam O, Çam G. Effect of plate thickness on weld speed in friction stir welding of AA6061-T6 Al-alloys plates. In: Proceedings of the 10th welding technology national congress and exhibition; 2017. p. 17–8.
106. Kuang B, Shen Y, Chen W, et al. The dissimilar friction stir lap welding of 1A99 Al to pure Cu using Zn as filler metal with "pinless" tool configuration. Mater Des. 2015;68:54–62.
107. Ji SD, Meng XC, Ma L, et al. Effect of groove distribution in shoulder on formation, macrostructures, and mechanical properties of pinless friction stir welding of 6061-O aluminum alloy. Int J Adv Manuf Technol. 2016;87(9–12):3051–8.

108. Yang XW, Fu T, Li WY. Friction stir spot welding: a review on joint macro- and microstructure, property, and process modelling. Adv Mater Sci Eng. 2014;2014:1–11.
109. Acerra F, Buffa G, Fratini L, et al. On the FSW of AA2024-T4 and AA7075-T6 T-joints: an industrial case study. Int J Adv Manuf Technol. 2010;48(9–12):1149–57.
110. Li B, Shen Y. A feasibility research on friction stir welding of a new-typed lap-butt joint of dissimilar Al alloys. Mater Des. 2012;34:725–31.
111. Zhang H, Wang M, Zhou W, et al. Microstructure–property characteristics of a novel non-weld-thinning friction stir welding process of aluminum alloys. Mater Des. 2015;86:379–87.
112. Wang M, Zhang H, Yu T, et al. Influence and affecting mechanism of tool pin geometrical configuration on weld formation during non-tilting friction stir welding. J Mech Eng. 2017;53(18):63.
113. Guan M, Wang Y, Huang Y, et al. Non-weld-thinning friction stir welding. Mater Lett. 2019;255:126506.
114. Zhang HJ, Wang M, Zhu Z, et al. Improving the structure-property of aluminum alloy friction stir weld by using a non-shoulder-plunge welding tool. Int J Adv Manuf Technol. 2016;87(1–4):1095–104.
115. Huang Y, Meng X, Xie Y, et al. Improving mechanical properties of composite/metal friction stir lap welding joints via a taper-screwed pin with triple facets. J Mater Process Technol. 2019;268:80–6.
116. Çam G, Mistikoglu S. Recent developments in friction stir welding of Al-alloys. J Mater Eng Perform. 2014;23(6):1936–53.
117. Xie Y, Meng X, Li Y, et al. Insight into ultra-refined grains of aluminum matrix composites via deformation-driven metallurgy. Compos Commun. 2021;26:100776.
118. Xie Y, Meng X, Mao D, et al. Homogeneously dispersed graphene nanoplatelets as long-term corrosion inhibitors for aluminum matrix composites. ACS Appl Mater Interfaces. 2021;13(27):32161–74.
119. Ma Y, Yang B, Lou M, et al. Effect of mechanical and solid-state joining characteristics on tensile-shear performance of friction self-piercing riveted aluminum alloy AA7075-T6 joints. J Mater Process Technol. 2020;278:116543.
120. Zeng XH, Xue P, Wang D, et al. Material flow and void defect formation in friction stir welding of aluminium alloys. Sci Technol Weld Join. 2018;23(8):677–86.
121. Doude H, Schneider J, Patton B, et al. Optimizing weld quality of a friction stir welded aluminum alloy. J Mater Process Technol. 2015;222:188–96.
122. Du B, Cui L, Yang X, et al. Weakening mechanism and tensile fracture behavior of AA 2219-T87 friction plug welds. Mater Sci Eng A. 2017;693:129–35.
123. Ma Y, Niu S, Liu H, et al. Microstructural evolution in friction self-piercing riveted aluminum alloy AA7075-T6 joints. J Mater Sci Technol. 2021;82:80–95.
124. Du Q, Tang K, Marioara CD, et al. Modeling over-ageing in Al–Mg–Si alloys by a multi-phase CALPHAD-coupled Kampmann–Wagner numerical model. Acta Mater. 2017;122:178–86.
125. dos Santos JF, Staron P, Fischer T, et al. Understanding precipitate evolution during friction stir welding of Al–Zn–Mg–Cu alloy through in-situ measurement coupled with simulation. Acta Mater. 2018;148:163–72.
126. Khan NZ, Siddiquee AN, Khan ZA, et al. Mechanical and microstructural behavior of friction stir welded similar and dissimilar sheets of AA2219 and AA7475 aluminium alloys. J Alloys Compd. 2017;695:2902–8.
127. Chen S, Zhang H, Jiang X, et al. Mechanical properties of electric assisted friction stir welded 2219 aluminum alloy. J Manuf Process. 2019;44:197–206.
128. Meng X, Huang Y, Cao J, et al. Recent progress on control strategies for inherent issues in friction stir welding. Prog Mater Sci. 2021;115:100706.
129. Entringer J, Reimann M, Norman A, et al. Influence of Cu/Li ratio on the microstructure evolution of bobbin-tool friction stir welded Al–Cu–Li alloys. J Mater Res Technol. 2019;8(2):2031–40.
130. Zhang C, Liu M, Meng Z, et al. Microstructure evolution and precipitation characteristics of spray-formed and subsequently extruded 2195 Al–Li alloy plate during solution and aging process. J Mater Process Technol. 2020;283:116718.

131. Rafieazad M, Mohammadi M, Gerlich A, et al. Enhancing the corrosion properties of additively manufactured AlSi10Mg using friction stir processing. Corros Sci. 2021;178:109073.
132. Niu PL, Li WY, Li N, et al. Exfoliation corrosion of friction stir welded dissimilar 2024-to-7075 aluminum alloys. Mater Charact. 2019;147:93–100.
133. Rambabu G, Balaji Naik D, Venkata Rao CH, et al. Optimization of friction stir welding parameters for improved corrosion resistance of AA2219 aluminum alloy joints. Def Technol. 2015;11(4):330–7.
134. Mishra RS, Ma ZY. Friction stir welding and processing. Mater Sci Eng R Rep. 2005;50(1–2):1–78.
135. Hu W, Ma Z, Ji S, et al. Improving the mechanical property of dissimilar Al/Mg hybrid friction stir welding joint by PIO-ANN. J Mater Sci Technol. 2020;53:41–52.
136. Liu Z, Cui H, Ji S, et al. Improving joint features and mechanical properties of pinless fiction stir welding of alcald 2A12-T4 aluminum alloy. J Mater Sci Technol. 2016;32(12):1372–7.
137. Khodabakhshi F, Marzbanrad B, Jahed H, et al. Interfacial bonding mechanisms between aluminum and titanium during cold gas spraying followed by friction-stir modification. Appl Surf Sci. 2018;462:739–52.
138. Huang Y, Wan L, Meng X, et al. Probe shape design for eliminating the defects of friction stir lap welded dissimilar materials. J Manuf Process. 2018;35:420–7.
139. Cho HH, Hong ST, Roh JH, et al. Three-dimensional numerical and experimental investigation on friction stir welding processes of ferritic stainless steel. Acta Mater. 2013;61(7):2649–61.
140. Zienkiewicz OC, Cormeau IC. Visco-plasticity-plasticity and creep in elastic solids—a unified numerical solution approach. Int J Numer Methods Eng. 1974;8(4):821–45.
141. Su H, Wu CS, Pittner A, et al. Thermal energy generation and distribution in friction stir welding of aluminum alloys. Energy. 2014;77:720–31.
142. Mendez PF, Tello KE, Lienert TJ. Scaling of coupled heat transfer and plastic deformation around the pin in friction stir welding. Acta Mater. 2010;58(18):6012–26.
143. Huang Y, Xie Y, Meng X, et al. Numerical design of high depth-to-width ratio friction stir welding. J Mater Process Technol. 2018;252:233–41.
144. Li S, Khan HA, Hihara LH, et al. Corrosion behavior of friction stir blind riveted Al/CFRP and Mg/CFRP joints exposed to a marine environment. Corros Sci. 2018;132:300–9.
145. Ly R, Hartwig KT, Castaneda H. Effects of strain localization on the corrosion behavior of ultra-fine grained aluminum alloy AA6061. Corros Sci. 2018;139:47–57.
146. Fukuda H, Szpunar JA, Kondoh K, et al. The influence of carbon nanotubes on the corrosion behaviour of AZ31B magnesium alloy. Corros Sci. 2010;52(12):3917–23.
147. Sinhmar S, Dwivedi DK. A study on corrosion behavior of friction stir welded and tungsten inert gas welded AA2014 aluminium alloy. Corros Sci. 2018;133:25–35.
148. Decreus B, Deschamps A, de Geuser F, et al. The influence of Cu/Li ratio on precipitation in Al–Cu–Li–x alloys. Acta Mater. 2013;61(6):2207–18.
149. Deschamps A, Brechet Y. Influence of predeformation and ageing of an Al–Zn–Mg alloy-II. Modeling of precipitation kinetics and yield stress. Acta Mater. 1998;47(1):293–305.
150. Han J, Thomas SL, Srolovitz DJ. Grain-boundary kinetics: a unified approach. Prog Mater Sci. 2018;98:386–476.
151. Ruoff AL, Balluffi RW. Strain-enhanced diffusion in metals. II. Dislocation and grain-boundary short-circuiting models. J Appl Phys. 1963;34(7):1848–53.
152. Simar A, Bréchet Y, de Meester B, et al. Integrated modeling of friction stir welding of 6xxx series Al alloys: process, microstructure and properties. Prog Mater Sci. 2012;57(1):95–183.
153. Atz Dick P, Knörnschild GH, Dick LFP. Anodising and corrosion resistance of AA 7050 friction stir welds. Corros Sci. 2017;114:28–36.
154. Hasani BM, Hedaiatmofidi H, Zarebidaki A. Effect of friction stir process on the microstructure and corrosion behavior of AZ91 Mg alloy. Mater Chem Phys. 2021;267:124672.
155. Jin B, Xiong DB, Tan Z, et al. Enhanced corrosion resistance in metal matrix composites assembled from graphene encapsulated copper nanoflakes. Carbon N Y. 2019;142:482–90.
156. Zhou YT, Zan YN, Wang QZ, et al. Grain boundary segregation of alloying Cu induced intergranular corrosion of B4C-6061Al composite. Mater Charact. 2021;173:110930.

157. Ba J, Ji X, Wang B, et al. Root-like C/SiC surface structure fabricated by the thermal and electrochemical corrosion for brazing to Nb. Compos Part B Eng. 2021;218:108942.
158. De Souza Carvalho Machado C, Donatus U, Milagre MX, et al. How microstructure affects localized corrosion resistance of stir zone of the AA2198-T8 alloy after friction stir welding. Mater Charact. 2021;174:111025.
159. Surekha K, Murty BS, Prasad Rao K. Effect of processing parameters on the corrosion behaviour of friction stir processed AA 2219 aluminum alloy. Solid State Sci. 2009;11(4):907–17.
160. Li J, Meng X, Li Y, et al. Friction stir extrusion for fabricating Mg-RE alloys with high strength and ductility. Mater Lett. 2021:129414.
161. Huang Y, Xie Y, Meng X, et al. Atypical grain coarsening of friction stir welded AA6082-T6: characterization and modeling. Mater Sci Eng A. 2019;740–741:211–7.
162. Buchheit RG. A compilation of corrosion potentials reported for intermetallic phases in aluminum alloys. J Electrochem Soc. 1995;142(11):3994–6.
163. Zander D, Schnatterer C, Altenbach C, et al. Microstructural impact on intergranular corrosion and the mechanical properties of industrial drawn 6056 aluminum wires. Mater Des. 2015;83:49–59.
164. Bellezze T, Giuliani G, Viceré A, et al. Study of stainless steels corrosion in a strong acid mixture. Part 2: anodic selective dissolution, weight loss and electrochemical impedance spectroscopy tests. Corros Sci. 2018;130:12–21.
165. Widener C, Lam TJ, Burford D. Corrosion in 2XXX-T8 aluminum alloys. In: TMS 2009 friction stir welding and processing. San Francisco; 2009.
166. Zhang F, Örnek C, Nilsson J-O, et al. Anodisation of aluminium alloy AA7075—influence of intermetallic particles on anodic oxide growth. Corros Sci. 2020;164:108319.
167. Brug GJ, van den Eeden ALG, Sluyters-Rehbach M, et al. The analysis of electrode impedances complicated by the presence of a constant phase element. J Electroanal Chem Interfacial Electrochem. 1984;176(1–2):275–95.
168. Liu H, Papadimitriou I, Lin FX, et al. Precipitation during high temperature aging of Al–Cu alloys: a multiscale analysis based on first principles calculations. Acta Mater. 2019;167:121–35.
169. Li JF, Li CX, Peng ZW, et al. Corrosion mechanism associated with T1 and T2 precipitates of Al–Cu–Li alloys in NaCl solution. J Alloys Compd. 2008;460(1–2):688–93.
170. Xu W, Ma J, Wang M, et al. Effect of cooling conditions on corrosion resistance of friction stir welded 2219-T62 aluminum alloy thick plate joint. Trans Nonferrous Met Soc China. 2020;30(6):1491–9.
171. Kamp N, Sullivan A, Tomasi R, et al. Modelling of heterogeneous precipitate distribution evolution during friction stir welding process. Acta Mater. 2006;54(8):2003–14.
172. Sinhmar S, Dwivedi DK. Effect of weld thermal cycle on metallurgical and corrosion behavior of friction stir weld joint of AA2014 aluminium alloy. J Manuf Process. 2019;37:305–20.
173. Zheng X, Zhang T, Yang H, et al. Friction stir processing induced elctrochemical performance improvement of commercial Al for Al-air battery. Electrochim Acta. 2020;354:136635.
174. Ralston KD, Birbilis N, Weyland M, et al. The effect of precipitate size on the yield strength-pitting corrosion correlation in Al–Cu–Mg alloys. Acta Mater. 2010;58(18):5941–8.
175. Steuwer A, Dumont M, Altenkirch J, et al. A combined approach to microstructure mapping of an Al–Li AA2199 friction stir weld. Acta Mater. 2011;59(8):3002–11.
176. Acosta G, Veleva L, López JL, et al. Contrasting initial events of localized corrosion on surfaces of 2219-T42 and 6061-T6 aluminum alloys exposed in Caribbean seawater. Trans Nonferrous Met Soc China. 2019;29(1):34–42.
177. Zhou L, Chen K, Chen S, et al. Correlation between stress corrosion cracking resistance and grain-boundary precipitates of a new generation high Zn-containing 7056 aluminum alloy by non-isothermal aging and re-aging heat treatment. J Alloys Compd. 2021;850:156717.
178. Li J, Wei H, Zhao K, et al. Effect of anodizing temperature and organic acid addition on the structure and corrosion resistance of anodic aluminum oxide films. Thin Solid Films. 2020;713:138359.

Chapter 4
Friction Stir-Based Remanufacturing

4.1 Background

Metallic materials have been extensively applied in several manufacturing fields, including aerospace, railway, and shipping transportation, which have drawn increasing attention to the development of welding techniques. Conventional fusion techniques, due to coarse grains, hot cracks and pore defects, have difficulty in obtaining high-quality welding joints. Friction stir welding (FSW) is a mature solid-state joining technique involving temperature, mechanics and metallurgy and their interactions. FSW has the advantages of low temperatures, high-quality joints, small residual stress and distortion, and no pollution, which can effectively avoid the problems associated with conventional fusion welding. When performed with inappropriate welding tools, process parameters or technological conditions, welding defects easily form and deteriorate the mechanical properties. Moreover, aiming at some structural parts after long service, corrosion, wear and other severe environments are prone to causing defects, which are detrimental to the service life of the parts, as described hereafter [1]:

(1) Failure of processes or procedures during part fabrication.
(2) In-service defects were missed during fabrication or grew during operation.
(3) Degradation of parts and components through unforeseen environmental factors or extended operation, such as corrosion and wear.
(4) Damage resulting from incorrect use or external factors.

All the service lifetimes of these structural parts involve "manufacturing-servicing-scrapping" and then result in economic loss and material waste, which are detrimental to the development of the recycling economy. Introducing a repairing process into the service lifetime of the parts can enable recyclability, such as "manufacturing-servicing-scrapping-repairing", which can significantly improve material utilization, resource conservation, energy savings and emission reduction. Green remanufacturing, with the characteristics of high efficiency, low cost and

Y. Huang et al., *Friction Stir Welding and Processing*, Materials Forming, Machining and Tribology, https://doi.org/10.1007/978-981-99-8688-0_4

little pollution, can restore and upgrade the properties of the parts via advanced welding techniques [2–4]. It is necessary to perform green remanufacturing techniques to enhance all service lifetimes. Solid-state green remanufacturing techniques, based on friction stir-based technologies, can achieve high-quality joints due to low temperature and severe plastic deformation.

4.2 Principle and Advantages

Friction stir repairing (FSR), based on the characteristics of FSW, involves temperature, mechanics, metallurgy and interactions, which has the feasibility and potential to repair the damaged structural parts. During FSR, an unconsumable or semi-consumable tool plunges into the workpieces to be repaired, and generates frictional heat and severe plastic deformation. Under the peak temperature and material flow, the defects can be repaired by filling with plasticized materials. Owing to the advantages, including low temperature, high-quality joints, and little pollution, friction stir-based remanufacturing techniques have been extensively developed and investigated, which are mainly divided into repetitive FSW and additive FSW. Figure 4.1 shows the key factors affecting joint integrity and quality, which consists of equipment parameters, welding tool parameters, and material properties. The welding tool and the equipment parameters are mainly dependent on the mechanical properties of the workpieces, such as yield strength, ductility, and hardness, which play significant influences on plastic deformation [5]. High heat input is required for a high melting point or constant-pressure specific heat materials, while materials with a low melting point or constant-pressure specific heat need relatively lower heat input [6]. The thermal properties of the workpieces primarily affect the welding peak temperature and thermal transfer. For materials with high thermal conductivity that easily result in heat loss, high heat input is necessary to guarantee sufficient heat input, obtaining a sound joints [7]. In contrast, the thermal conductivity of polymers is lower than that of Al alloys, which makes it difficult to preheat the materials at the front of the rotating tool. Higher heat input is also required to attain sound joints. Reasonable selections of these parameters are the key and prerequisite to achieve excellent repaired joints.

Compared with conventional fusion repairing techniques, FSR has many advantages.

(1) Low peak temperature can reduce the degree of grain coarsening, thereby improving mechanical properties.
(2) Low heat input and severe plastic deformation can eliminate the pore and crack defects associated with fusion welding techniques.
(3) Low residual stress and small distortion improve the structural integrity.
(4) A smooth surface avoids subsequent surface treatment, reducing the complexity of the manufacturing processes.

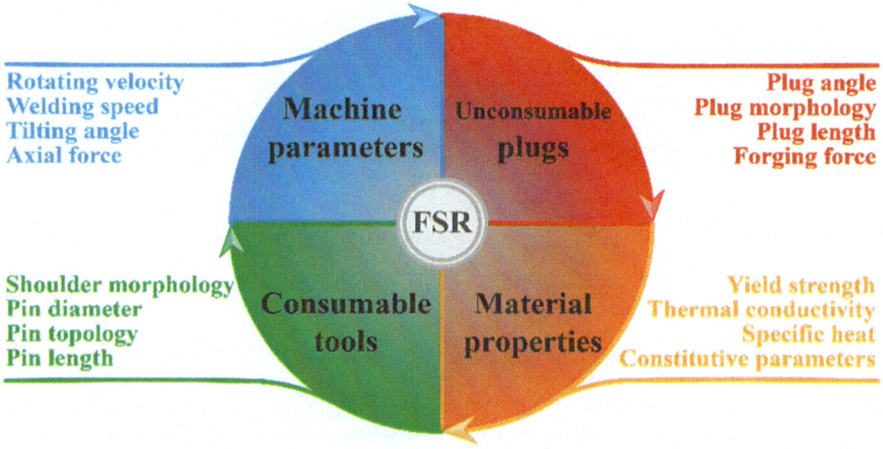

Fig. 4.1 Process variables of FSR [8]

(5) Strong material compatibility: FSR can also realize the repairing of high-strength Al alloys, such as 2xxx and 7xxx alloys.
(6) FSR has the potential to repair the parts of dissimilar metals according to engineering requirements.
(7) FSW equipment extensively used in the automobile and aerospace industries, which is characterized by low cost and high energy efficiency, has universal properties.

4.3 Types of FSW Defects

During the FSW process, a rotating tool with a shoulder and a pin rotates and plunges into the welding workpieces, which can produce frictional heat and material flow due to the friction between the rotating tool and the welding workpieces. After completing FSW, the rotating tool retracts, and then a keyhole forms at the end of the weld, which easily deteriorates mechanical properties [9]. Moreover, other welding defects, such as cavities [10], tunnels [11], grooves [12] and kissing bonds [13], caused by improper process parameters or technological conditions are also detrimental to mechanical performance [14]. Several characteristic defects are identified as the effect of either flow or geometric. The flow-related defects occur outside the acceptable processing window with parameters that are considered either too hot or too cold [15], as shown in Fig. 4.2. For example, a higher welding heat input or a larger plunge depth easily forms flash defects and then results in weld thinning and stress concentration. The insufficient material flow induced by lower heat input leads to inner joint defects, such as cavities, tunnels and grooves. Khan et al. [11] stated that the kissing bond and the tunnel defects were easily influenced by the plunge depth and offset during FSW of 5083-H116 and 6063-T6 Al alloys. Tao et al. [16] indicated that the fracture

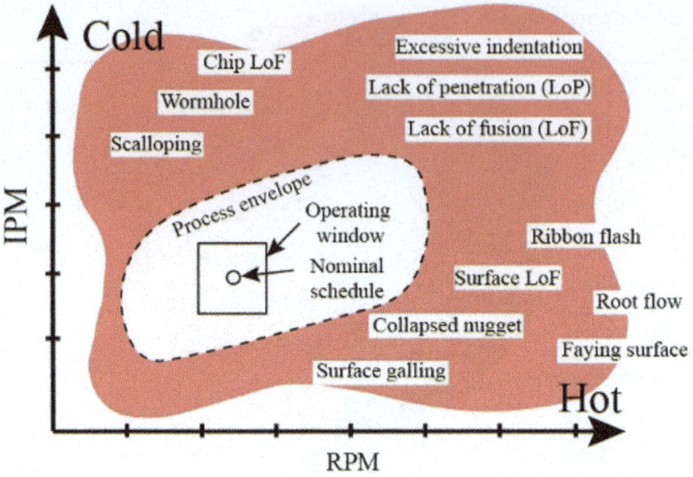

Fig. 4.2 Forming conditions of several defects [15]

behaviour of FSW joints was strongly influenced by the kissing bond featuring cracks from the lack of root penetration. Kainuma et al. [17] and Zhou et al. [18] reported that the existence of a kissing bond would seriously deteriorate the fatigue lifetimes of a joint in both low-cycle and high-cycle tests. Arici et al. [19] stated that the FSW joints of polymers could be broken by hand due to low strength resulting from kissing bond defects. Tunnelling and kissing bond defects are formed at pin offsets towards stronger materials at the AS and without offset due to insufficient mixing.

Without proper treatment or design, the flashes on the surface may cause significant problems in terms of stress concentrations, corrosion and surface finish [20]. Moreover, the kissing bond defect severely affects the mechanical properties of the welded joints compared with the tunnel defect. When the gap between the welding workpieces induced by the extrusion effect during the FSW process or accumulation of some manufacturing errors of large structures is mismatched, the kissing bond and the lack of root penetration are easily formed. Shultz et al. [21] found that increasing the gap width decreased the attainable maximum joint efficiency, while increasing the welding tilt angle was beneficial to increasing the joint efficiency to a certain degree, but the weld thinning was severe. It is difficult to obtain defect-free joints if the gap is too wide. Wanjara et al. [22] reported that when the thickness of the 6061 Al alloy was 3.18 mm, a gap greater than 0.5 mm resulted in the formation of cavity defects in FSW joints, which was determined to be the critical limit for practical application using serial industrial robots. Yang et al. [23] demonstrated that a gap between the workpieces reduced the effective area of the cross-section around the gap and led to an unsuitable welds. According to the characteristics and types of welding defects, they can be divided into global and local defects. The detailed features, definitions and causes of these defects are listed in Table 4.1.

Table 4.1 Typical defects appearing at the FSW joints

Defect	Typical view	Definition	Possible causes
Keyhole [14]		A hole matching the shape of the rotating pin	A hole left by the retraction of the rotating pin
Flash [20]	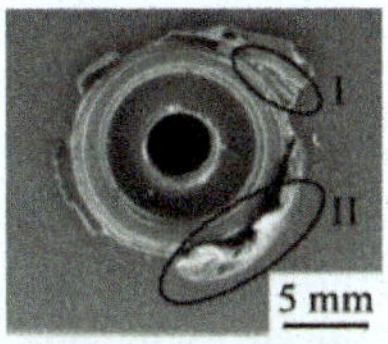	Excessive overflow of materials on top surface leaving a corrugated or ribbonlike effect along the RS or AS	Excessive forge load or plunge depth; excessively hot weld; Smaller shoulder diameter
Tunnel [24]	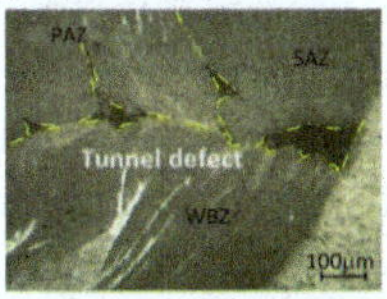	Internal cavity along the welding direction	Insufficient material flow; improper welding tool; overflow of materials; smaller plunge depth
Weld thinning [25]	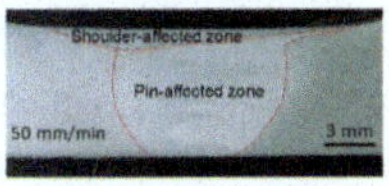	The thickness of NZ is lower than that of the BM	Larger plunge depth; overflow of materials; improper back support
Groove [26]	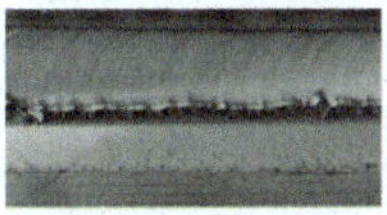	A continuous dented surface characterized by insufficient fill with plasticized materials	Insufficient material flow; improper back support; insufficient forge pressure; overflow of materials
Cavity [27]	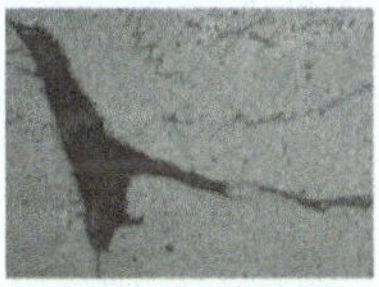	Incomplete joining induced by insufficient consolidated and forged materials at the RS	Lower heat input; excessive travel speed; insufficient material flow; improper welding tool
Lack of root penetration [28]	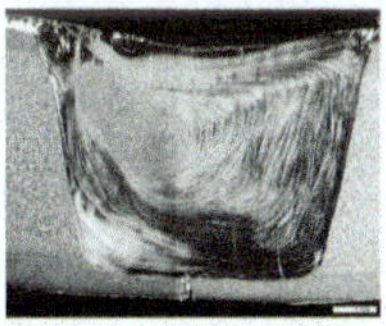	Incomplete joining at the back area of butt plate	Shorter pin length; smaller plunge depth; larger fitting gap or manufacturing errors

Table 4.2 lists the recent progress for solid-state repairing of welding defects or structural inner defects. The main research objects are focused on lightweight metals, such as Al alloys and Mg alloys, among which the 2xxx, 6xxx and 7xxx series are the most common Al alloys in aerospace and railways, and AZ31 is the most typical alloy. FSR has special advantages for these lightweight metals. Moreover, the FSR of high melting point materials is rare, and few studies are associated with high melting point alloys, which are only 316L stainless steel and CuNi 70/30. This is because the wear of the FSW tool is the critical issue for harder metals with high melting temperatures [29, 30] or metal matrix composites (MMCs) [31, 32], which is characterized by the deformation and reduction in the pin diameter and morphology. According to results of Rai et al. [33], plastic deformation, abrasion, diffusion and reaction between the environment and the tool materials were the major wear mechanisms that occurred in FSW tools. The welding tool requires higher mechanical properties at high peak temperatures, such as high compressive yield strength, good strength, dimensional stability, creep resistance and good thermal fatigue strength [34–36]. Moreover, these materials are usually expensive and difficult to be machined, which are detrimental to the economic benefits and repairing ability. It follows that FSR can be used for many kinds of light alloys and has strong material compatibilities for the main welding or repairing damaged defects in structural parts. From Table 4.2, it can be concluded that the strength coefficients of the FSR joints all reach 80% of the optimum welding joints or base material (BM), indicating the high-quality repairing. The above-mentioned welding defects or exogenous source-induced defects are prone to causing micro cracks which propagate and severely reduce the load bearing capacity and service life. Scholars in the scientific community and industry have performed extensive investigations in this area. According to the characteristics of the defects, the repairing process is mainly composed of defect detection, location, geometrization, and repairing. In this section, highly efficient and high-quality solid-state repairing is reviewed, which provides technical support and references for the repair of metallic structural parts.

4.4 Repetitive Friction Stir Remanufacturing

The global defects mainly contain surface peeling, grooves, tunnels, kissing bond, lack of root penetration, surface damage and long cracks induced by the exogenous source. The above defects always appear along a certain direction, curve and depth, which are primarily repaired by repetitive friction stir remanufacturing technologies.

A concept of combining friction stir spot welding and end milling was proposed to remove flashes simultaneously. The removal process of flashes in general has little effect on the mechanical properties of friction stir spot welds, except under very low indentation rates [20]. Li et al. [38] employed repetitive FSW to reduce or eliminate the tunnel defect of Al–Zn–Mg alloy FSW joints and then avoided the stress concentrations, in which the grain size in the repaired region was reduced compared with that in the original nugget zone (NZ). The maximum tensile strength

Table 4.2 Research progress in FSR of materials with main defects and their repairing techniques

Materials	T (mm)	Type of defects	Repairing methods	Optimum parameters	Maximum tensile strength/ load	Ref
Al–Cu Al alloy	3	Shrinkage/ porosity	Repetitive FSR	$w = 900$ rpm; $v = 150$ mm/min; PD $= 0.1$ mm; $T = 2.5°$	301.6 MPa	[37]
Al–Zn–Mg alloy	6	Tunnel	Repetitive FSR	$w = 800$ rpm; $v = 300$ mm/min; $F = 9$ kN	207.4 MPa	[38]
304L steel	12	Cavity (gap)	FSR	$w = 250$ rpm; $v = 100$ mm/min		[39]
AA6061-T6	3.2	Flashes	Friction stir spot welding	$w = 3000$ rpm; $v = 0.2$ mm/s; DT $= 0.5$ s	4678N	[20]
AA2219-T87	6	Kissing bond	Repetitive FSR	$w = 600$ rpm; $v = 200$ mm/min; PD $= 0.2$ mm; $T = 3°$	248 MPa	[40]
AA2219-T87	6	Cavity	Repetitive FSR	$w = 800$ rpm; $v = 400$ mm/min; PD $= 0.3$ mm; $T = 3°$	235 MPa	[40]
AA2219-T6	7	Groove	Offset FSR	$w = 600$ rpm; $v = 200$ mm/min; $F = 4.6$ kN; $T = 2.5°$	335 MPa	[41]
AA2219-T6	7.5	Groove	Gas tungsten arc welding (GTAW) + FSW	GTAW: $I = 200$ A; $v = 120$ mm/min FSW: $w = 600$ rpm; $v = 200$ mm/min; $F = 4.6$ kN; $T = 2.5°$	305 BM	[42]
AA6061-T6	4	Gap	Vertical compensation FSW (VCFSW)	$w = 2000$ rpm; $v = 50$ mm/min; $P = 0.1$ mm; $T = 2.5°$	214 MPa	[26]
AA7N01-T4	4	Gap	VCFSW	$w = 2000$ rpm; $v = 50$ mm/min; $P = 0.1$ mm; $T = 2.5°$	295.7	[43]
AA2024/ AA7075	3.5	Gap	VCFSW	$w = 2000$ rpm; $v = 20$ mm/min; $P = 0.2$ mm; $T = 2.5°$; FD $= 1.5$ mm	307.2 MPa	[44]

(continued)

Table 4.2 (continued)

Materials	T (mm)	Type of defects	Repairing methods	Optimum parameters	Maximum tensile strength/load	Ref
AA2219-T6	7.8	Keyhole	Filling friction stir welding (FFSW)	FFSW: w = 800 rpm; PD = 0.1 mm; T = 2.5° FSP: w = 800 rpm; v = 200 mm/min; PD = 0.1 mm; T = 2.5°	179.6 MPa	[45]
AA2219-T87	10	Through hole	Friction plug welding (FPW)	w = 7500 rpm; AF = 50 kN; F = 60 kN; Push amount = 5 mm; FT = 5 s	329 MPa	[46]
AA2195-T8	6.35	Through hole	FPW	w = 4250 rpm; A plug with a diameter of 19.1 mm	342 MPa	[47]
AA5456	7.5	Keyhole	FFSW	w = 400 rpm; v = 50 mm/min	91% of defect joint	[48]
316L steel	10	Keyhole	FFSW	FFSW: w = 1500 rpm; F = 35 kN; DT = 9 s	92% of BM	[49]
316L steel	10	Keyhole	Self-refilling FSW	SRFSW1-4: w = 1200 rpm; F = 30 kN; DT = 5 s; SRFSW5-8: w = 1500 rpm; F = 30 kN; DT = 5 s;	643 MPa	[50]
API X52 steel	16	Blind hole	Friction taper plug welding (FTPW)	w = 7500 rpm; F = 35 kN; FT = 5 s; F = 40 kN	Equal to BM	[51]
API 5L X65 steel	30	Blind hole	FTPW	w = 7000 rpm; F = 50 kN; FT = 5 s; F = 40 kN; B = 14 mm	Equal to BM	[52]
CuNi 70/30	3.13	Corrosion hole	Repetitive FSR	w = 1200 rpm; v = 12.7 mm/min; T = 3°	339.4 MPa	[53]
AZ31B	4	Keyhole	Active–passive filling friction stir repairing (A-PFFSR)	w = 1300 rpm; v = 1 mm/min; T = 2.5°; DT = 5 s	189.7 MPa	[9]
AA7N01-T4	4	Keyhole	A-PFFSR	w = 1600 rpm; v = 2 mm/min; T = 2.5°; DT = 5 s	311.1 MPa	[54]

(continued)

Table 4.2 (continued)

Materials	*T* (mm)	Type of defects	Repairing methods	Optimum parameters	Maximum tensile strength/load	Ref
AZ31B	1.5	Circular hole	Drilling-filling FSR (DFFSR)	$w = 1600$ rpm; $T = 2.5°$; DT = 20 s; PD = 0.4 mm	217 MPa	[55, 56]
AA6061-T6	4.8	Through hole	Refill friction stir spot welding (RFSSW)	$w = 2290$ rpm; PS = 1.93 mm/s; $F = 16$kN	213 MPa	[57]
AA2198–T851	3	Through hole	RFSSW	$w = 1000$ rpm; PS = 0.85 mm/s; PD = 2.9 mm	78%	[58]
AA7075-T761	6	Through hole	RFSSW	$w = 1500$ ~ 2100 rpm; PS = 0.7 ~ 1 mm/s; DT = 2, 4, and 6 s	425 MPa	[59]

Note Thickness: *T* (mm); Rotational velocity: *w* (rpm); Welding speed: *v* (mm/min); Plunge depth: *PD* (mm); Forging force: *F* (kN); Tilt angle: *T* (°); Dwelling time: *DT* (s); Current: *I* (A); Filler width: *FD* (mm); Axial force: *AF* (kN); Forging time: *FT* (s); Plunge speed: *PS* (mm/s); Burn-off: *B* (mm)

of the repaired joint reached 180.2% higher than that of the original joint with the tunnel defects. Li et al. [40] also used repetitive FSW to repair the lack of root penetration and cavity defects in AA2219-T87 FSW joints. To repair the surface groove and tunnel defects, Liu et al. [41] proposed offset FSW to eliminate groove defects in 7.5-mm-thick AA2219 FSW joints, and the offset was used so that the repairing pin was at the location of the defect rather than at the welding centreline. They stated that the offset FSR could effectively eliminate the cavity defect and an obvious aggregation of the hard-brittle phase Al_2Cu associated with the symmetrical FSR, as shown in Fig. 4.3. A high-quality joint repaired by offset FSR with a tensile strength of 335 MPa was achieved and fractured near the interface between the NZ and the thermo-mechanically affected zone (TMAZ), presenting ductile fracture characterized by large dimples. Subsequently, Liu et al. [42] employed tungsten inert gas welding (TIG) and FSW to remove large groove defects that cannot be eliminated by repetitive FSW, in which a high tensile strength equivalent to 70% of that of the BM was achieved and the fracture feature of the repaired joint was partially plastic and partially brittle. Arici et al. [19] and Arici and Sinmazçelýk [60] expounded that the lack of root penetration induced by the low thermal conductivity of medium-density polyethylene was eliminated by double-pass FSW, whereas weld thinning was increased. Saeedy et al. [61] also repaired the lack of root penetration of FSWed high-density polyethylene joints by double-pass FSW and then stated that its joint strength was twice that achieved by single-pass FSW. Cabibbo et al. [62] welded AA6082 sheets by double-pass FSW to eliminate kissing bond, in which the second

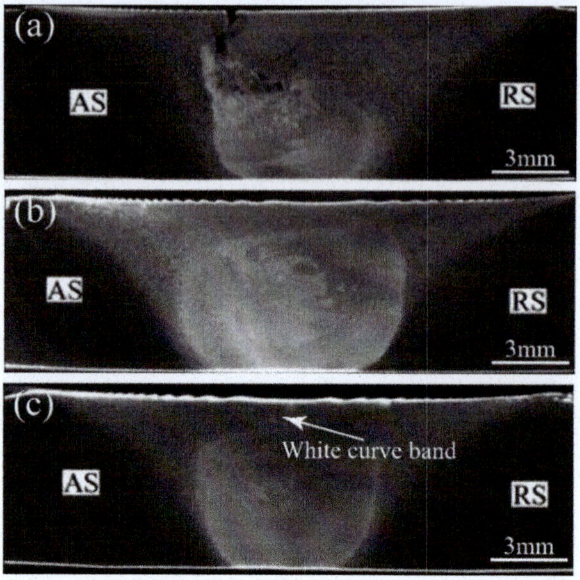

Fig. 4.3 Morphologies of the initial joints with groove defects and their repaired joints: **a** initial, **b** offset FSR and **c** symmetrical FSR joints [64]

pass was performed on the other side of the sheets. Compared to conventional FSW, the grains and precipitates in the TMAZ of double-pass FSW joints were coarser due to a twice overlapping heat input, thereby leading to poor tensile properties. Although double-pass FSW can solve the lack of root penetration, the surface formation is poor, and weld thinning is serious. Furthermore, the welding process still needs to be further optimized. Widener et al. [63] demonstrated that stationary shoulder friction stir welding ($SSFSW_2$) with a high rotational velocity can be performed in thin-gauge Al alloys with relatively low process forces, making smaller portable devices possible for repairing applications and achieving high-quality repaired joints without weld thinning.

In addition, when the welding or exogenous defects are relatively large, they easily result in the loss of extremely mass materials, such as large tunnel and groove defects and gap defects induced by either the extrusion effect during the FSW process or the accumulation of some manufacturing errors of large structures that are mismatched. Introducing additional filling materials into these locations is extremely essential. Ji et al. [26, 43, 65] proposed vertical compensation FSW (VCFSW) technique to repair these defects and then selected the gap defect as the main research object, as shown in Fig. 4.4. Additional compensation materials were added to these defects, which were homogeneous materials similar to the workpieces to be repaired. During VCFSW, the rotating tool simultaneously rotated, stirred and mixed the welding workpieces and the compensation materials, realizing high-strength solid-state repairing of defects with large sizes for 6061–T6 and 7N01 Al alloys, respectively. Moreover, they expounded that employing materials with low melting points and high strength easily achieved high arrangement repairing of materials with high melting points, such as the workpieces to be welded AA6061–T6 and the compensation materials

AA2024–T4 [26]. Two vortex-like structures appeared in the SAZ and PAZ, which was similar to the "onion ring" pattern reported in the literature [66]. The forming mechanism of "onion ring" patterns was formed due to loads from both the external and the internal screws at each rotation [67–69]. The formation of the "onion ring" pattern markedly improved the mechanical interlocking and metallurgical bonding, enhancing the mechanical properties. Abu-Okail et al. [44] stated that the quality of the joints welded by VCFSW depended on the ability of compensation materials to fill in and mix with the BM through VCFSW. Sandnes et al. [70] used the hybrid metal extrusion and bonding (HYB) process to eliminate the insufficient material feeding induced by the lack of filler materials when strict base plate and profile tolerances were required.

To eliminate the potential defects and severe joint softening induced by conventional fusion welding and the back-supported body that is needed in additive filling FSW, a hybrid between TIG welding and pinless friction stir processing (FSP) is proposed by our workgroup, as shown in Fig. 4.5. TIG welding is responsible for the weld reinforcement in the welding line, eliminating possible defects at the ending stage. Here, a "multi-strand welding wire" technique is used, and the shape and performance control of the filling joints can be realized by several factors, including the composition design, diameter selection, and wire assembly modes. Subsequently, a pinless tool with a special shoulder morphology and diameter is used to modify the microstructure produced by TIG welding, which transforms the coarse casting microstructures into fine and equiaxed forging microstructures, eliminating fusion defects and strengthening the repaired joints. Additionally, by

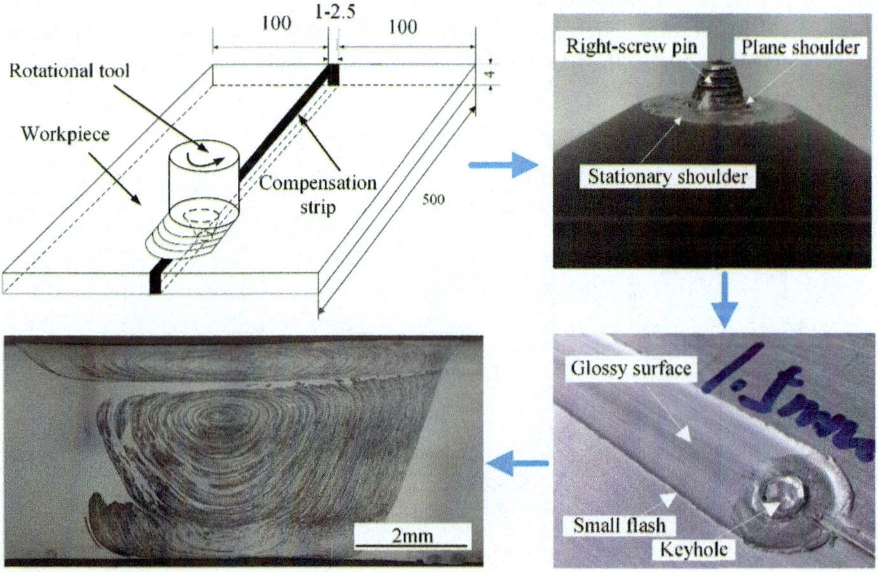

Fig. 4.4 Schematic of VCFSW [26]

regulating the shoulder topology (shoulder diameter, morphology and angle) and processing parameters (plunge depth/force, rotational velocity, processing speed and tilt angle), the microstructures and mechanical properties can be optimized, which realize the excellent repairing of volume defects.

For materials that have higher melting points than Al alloys, few studies exist due to the high cost and wear of the rotating tool. Mukherjee and Ghosh [53] combined laser-assisted direct metal deposition with FSW to repair corrosion holes in a CuNi 70/30 alloy. This additive FSW process allowed for site-specific filling and repairing of the harmful features in laser-assisted direct metal deposition microstructures, including porosity, large grain size, residual stress and solute segregation. Gunter et al. [39] employed FSW to repair crack defects in 304L stainless steel using a PCBN/W–Re tool, as shown in Fig. 4.6. Through this process, it was possible to heal a crack that began narrow and then progressively grew up to a width of 2 mm.

During FSW, the material intersecting the path of the pin deformed in a similar manner. As the pin approached, the materials were initially compressed, rotated around the pin, and finally deposited behind the pin at the same transverse position as its beginning (Fig. 4.7). The typical microstructural evolution of stainless steel during the deformation process could be presented through a microstructural analysis along the centre of the weld path. The final microstructure in the NZ was a result of dynamic recrystallization (DRX) and some grain growth. FSW of 304L stainless steel resulted in similar levels of dislocation density, as shown in Fig. 4.8.

For the solid-state repairing methods, the materials around the repairing interfaces experienced severe thermo-mechanical behaviours, which further resulted in the formation of chemical and mechanical bonds at the repairing interfaces. The microstructural evolution and interfacial behaviour significantly influenced the mechanical properties. The macrostructure of the FSW welds was closely related to the workpiece materials, tool design, and offset, as well as welding variables and

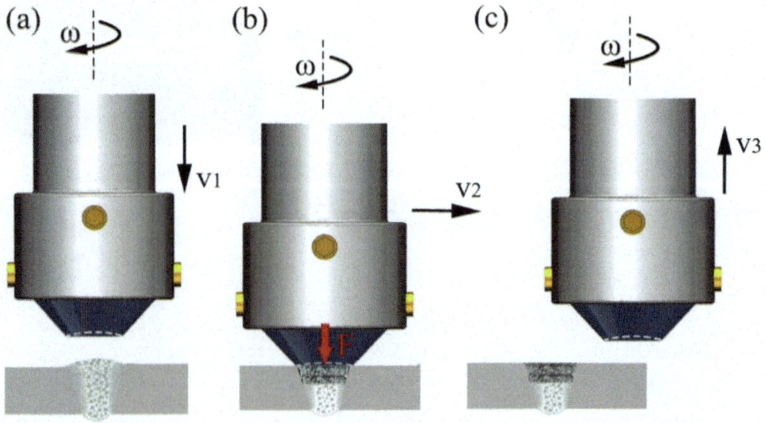

Fig. 4.5 Keyhole closure technique of TIG filler wire welding with subsequent FSP: **a** TIG filler wire welding, **b** FSP and **c** completion of keyhole repairing

Fig. 4.6 FSR tool (70% PCBN/30% W–Re) [39]

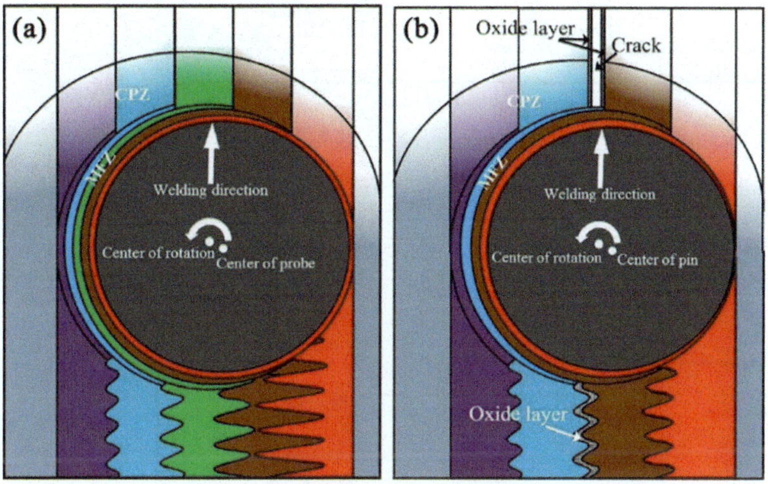

Fig. 4.7 Pin-driven material flow around the tool: **a** schematic of FSP performed on crack-free stainless steel and **b** schematic of FSP performed along a crack [39, 71]

setup. Good metallurgical bonding may reduce the presence of defects at the repairing interface, while complex mechanical bonding promotes better stress distribution during loading and a more complex crack propagation path [73]

Investigating the interfacial microstructures and healing of the repairing interfaces can provide important references and technical support for the regulation and control of mechanical properties. Current knowledge on the interface characteristics and microstructural evolution of the repaired joint is presented in this section. During the FSR process, the materials in the repairing zone always experienced high peak

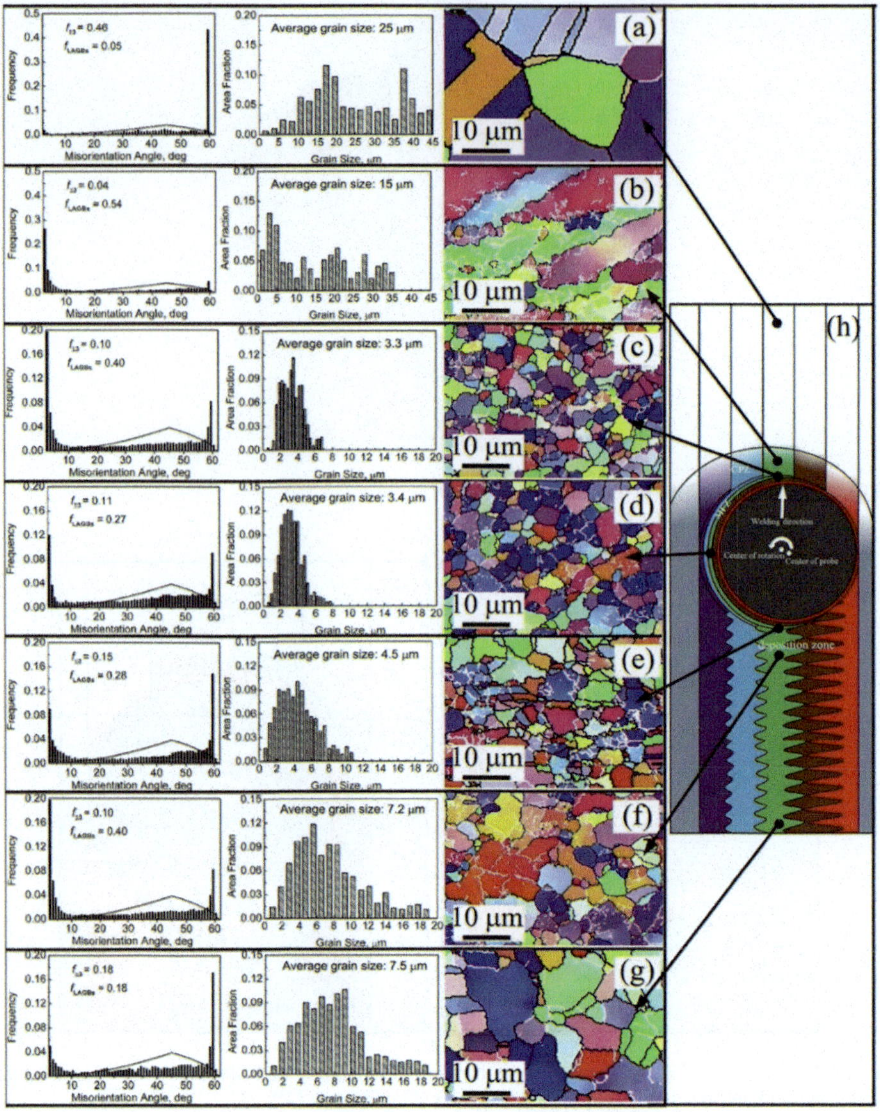

Fig. 4.8 Grain structure evolution in 304L stainless steel during FSW [72]

temperatures and high strain rates, and then DRX occurred, which further resulted in grain refinement or phase transformation. Moreover, severe plastic deformation can improve the mechanical mixing between the filling materials and the workpieces. Ji et al. [37] stated that the casting structure in Al-Cu casting alloy was transformed into fine and equiaxed grains due to DRX, thereby improving mechanical properties.

The main aim behind solid-state repairing is to improve the mechanical properties, corrosion and wear resistance of joints. The mechanical properties mainly include microhardness, tensile strength, yield strength and elongation in addition to impact and fatigue strength. Various strengthening mechanisms contribute to the enhancement of mechanical properties. Parameter optimization and optimal welding setup are essential to produce high-quality repaired joints. Estimation of the microhardness of the repaired joints requires the consideration of contrary parameters, which further influences the tensile properties. On the one hand, severe plastic deformation refines the grain size and then increases the microhardness according to the Hall–Petch relation. On the other hand, high temperatures easily result in annealing of materials [74]. Dislocation density and residual compressive stress are decreased in the annealed materials, leading to a reduction in microhardness. The grain refinement effect dominates the annealing effect [75]. Consequently, the grains are refined during FSR, which increases the microhardness values. Moreover, second-phase particle strengthening is another main factor besides grain refinement contribution. It can be concluded that strengthening is a combined effect of grain refinement and strengthening phases [76]. For Al alloys and interstitial-free (IF) steel, the Hall–Petch relationship fit the data quite well. For commercially pure Ti, the relationship breaks down when the grain sizes are smaller than 6.4 μm, wherein the measured microhardness is higher than that predicted by the Hall–Petch relationship. The lack of correlation to the Hall–Petch effect for fine Ti grains was attributed to the high dislocation density within the grains.

4.5 Additive Friction Stir Remanufacturing

For local volume defects, such as keyhole, cavity and pitting defects, digging and geometrization are the essential procedures due to the irregular distribution inside the parts, in which the loss of materials can be compensated by filling additional materials. In recent years, a series of friction stir-based closure techniques are reported that attempt to transfer the advantages of FSR to closure methods of volume defects. Additive FSR techniques mainly consist of friction plug welding (FPW), friction taper plug welding (FTPW), filling FSW (FFSW), and progressive FSW.

4.5.1 Friction Plug Welding

FPW is proposed to repair keyhole defects, which has been widely reported by relevant researchers. FPW is primarily used by TWI [77] and NASA Marshall Space Flight Center [78, 79]. In this process, a taper plug is coaxially forced into a keyhole that has a similar taper. The keyhole is filled with the taper plug by the friction between the taper plug and the surface of the hole. For the FPW joints, post machining is essential to remove the redundant materials of the plug and the materials that flow

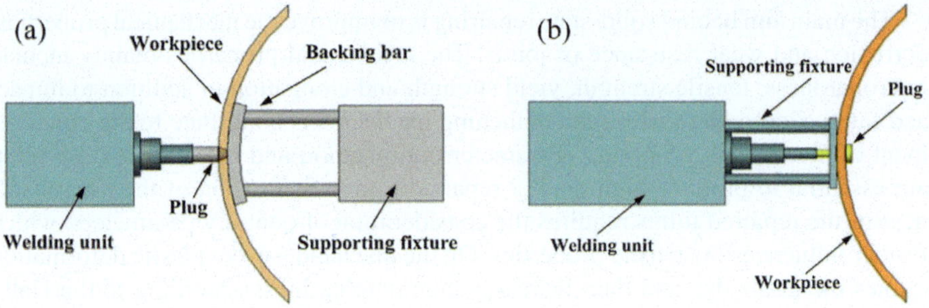

Fig. 4.9 Schematics of FPW: **a** friction push plug welding and **b** friction pull plug welding [64]

out of the weld. According to the load pressure, FPW is divided into friction push plug welding and friction pull plug welding, as shown in Fig. 4.9 [64].

Friction push plug welding was first introduced by Lockheed Martin to repair the defects of AA2195 FSW joints in 1995, replacing manual TIG welding. Subsequently, Lockheed Martin, TWI and NASA optimized the welding parameters for a more robust process [78–80]. Friction push plug welding has become an effective repairing method in rocket tanks. During welding, excessive force on the taper plug pushes the plug into the hole, forming a compact joint. Friction push plug welding has many advantages characterized by a plug unit located in front of the workpieces and withdrawing compressive stress and a wide process window, high welding stability and qualified welding rate. This method is unsuitable for enclosed structures because the rigid backing plate has to withstand a great welding force. The published references have reported that a larger plug diameter in relation to the hole diameter is much more effective in repairing defects. The main process parameters affecting the friction push plug welding process and its welded joints are rotational velocity, axial force, forging force, push amount, forging time and plug morphology. A tapered plug can make the flow direction change instantly from the upper surface to the lower surface of the sheet. Metz and Barkey [47] and Metz et al. [81] studied the microstructures, hardness and fatigue properties of 2195-T8 FPW welds. A typical weld can be classified into six regions: plug metal, plug thermo-mechanically affected zone, plug recrystallized zone (PRZ), TMAZ, heat affected zone (HAZ) and BM (Fig. 4.10). The hardness variations in these zones may control the fatigue behaviour of the joints. The TMAZ that is close to the bonding interface is the weakest location of the joint due to the dissolution of precipitates and the redistribution of constituent particles. The repaired joints sustained reductions in both the tensile properties and the fatigue life when compared with the FSW joints. The reduction in fatigue life is most likely due to the complication of the weld geometry, the interacting HAZ and the strength mismatch between the BM, the FSW joint and the plug materials [47].

Lange and Schwenker [83] studied the bending, tensile and thermal characterization of AA2024–T3 plug welds. Defect-free joints were successfully achieved with optimized parameters. At the beginning of the tensile stress application ($t \leq 20$ s),

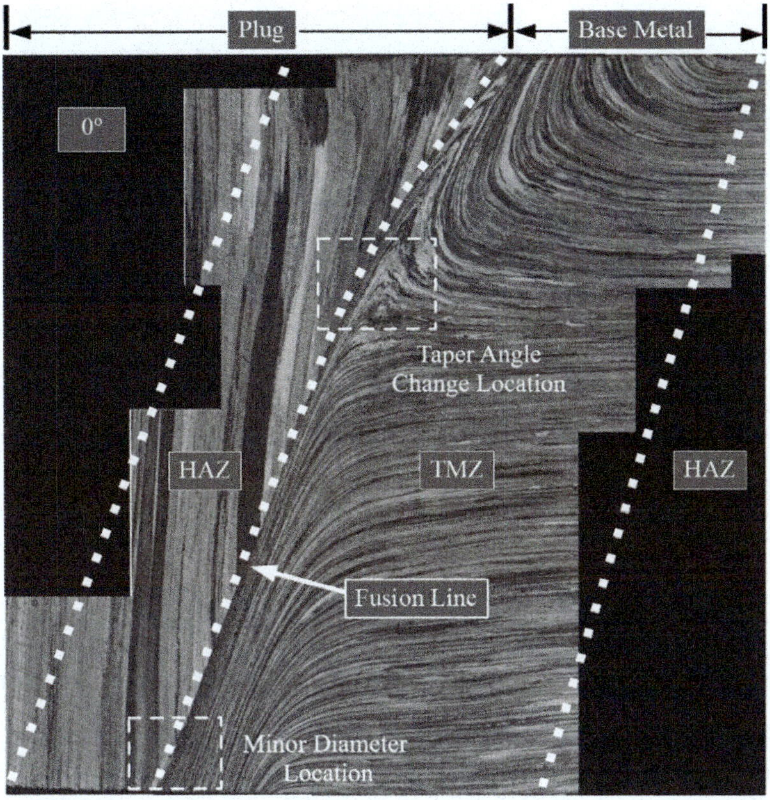

Fig. 4.10 A typical weld [82]

the joint was under the elastic deformation stage, exhibiting uniform elastic strain across the plug joint (Fig. 4.11).

When the global strain reached 0.18% (t = 35 s), the strain concentration could be examined initially near the TMAZ. A slight plastic deformation occurred in the HAZ but was not obvious in the BM. With conducting the tensile process, the stress showed a fast and linear increase accompanied by a slow extension of the plastic deformation zone. When the global strain reached 0.86%, yield occurred in the joint with necking. When the global strain reached 5.08%, the joint fractured with a local strain of 6.8% in the TMAZ. By optimizing the design of joint geometry, the maximum tensile strength of the repaired joint of AA2219–T87 reached 329 MPa, which is ~ 72.3% of that of the BM [46].

Friction pull plug welding was proposed and patented by Paula Hartley and Riki Takeshita from Lockheed Martin [64]. It was originally received as a key technique for the external tank of the space shuttle and the keyhole in BT-FSW [85], as indicated in Fig. 4.12 [86]. The welding equipment of friction pull plug welding is located on one side of the workpieces, which requires a simple fixture without rigid backing plates,

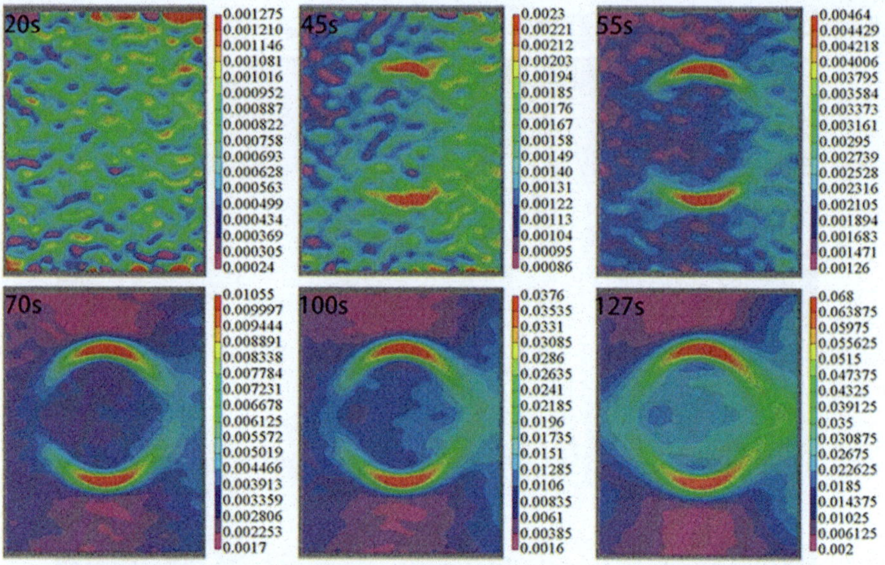

Fig. 4.11 Strain maps of a plug joint at different times during a tensile test [84]

indicating that friction pull plug welding is more suitable for enclosed complicated structures than friction push plug welding. The narrower process window in the friction push plug welding results in lower joint quality and less welding stability than that in the friction push plug welding. The friction pull plug welding process was applied to 4-mm-thick AA2219. By appropriate selection of the welding parameters and joint configuration, the tensile strength reached 330 MPa or above, which was equal to or even greater than that of the FSW joints.

Fig. 4.12 Friction pull plug weld operations [86]

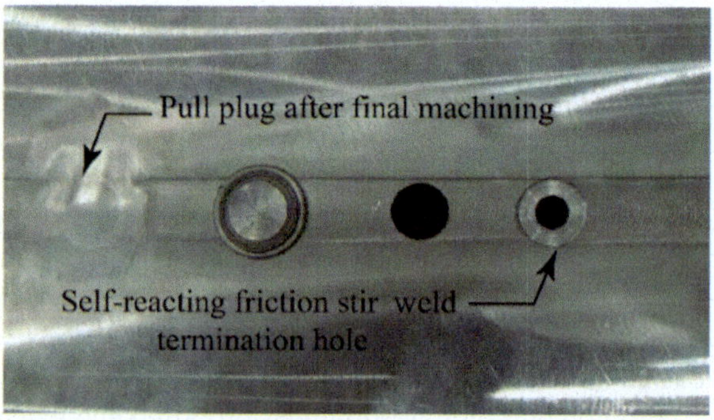

4.5.2 Friction Taper Plug Welding

As a new and promising solid-state repairing technique, FTPW was proposed and has been widely investigated, which has the feasibility and potential to repair underwater metal structures, pipes and thick section materials or seam thick plates or components [87, 88]. Pauly [89] and Nicholas [90] defined that FTPW was a hole filling process that involved drilling a blind hole and then forcing a consumable plug into the hole coaxially, similar to FPW. A schematic of the FTPW process is shown in Fig. 4.13. Synergistic effects between frictional heat and forging force led to the plug materials to be softened and flow continuously, and then the blind hole could be filled. After completing the plug consumption process, another axial forging force was applied immediately to forge the materials in the weld [91]. FTPW joints were mostly fabricated in underwater wet conditions, and the repairing interfaces exhibit as-forged microstructures without hydrogen-induced cold cracks [92–94]. Characteristics of the basic welding parameters have been described in previous investigations and are reviewed in Meyer and Ambroziak's works [95, 96].

Thorough insight into material flow and heat generation in different welding conditions is of practical importance for the comprehension and elimination of welding defects. Yin et al. [52] showed the typical plug material flow behaviour with the variation in weld torque during the FTPW process, which contained a heat-dominated phase (HDP), heat-shear transition phase (HSTP) and shear dominated phase (SDP), as shown in Fig. 4.14. The HDP was defined as the time between the start of the weld and the point where substantial plastic shear occurred, which resulted in a substantial drop in peak torque. In the SDP, the torque stabilized at a considerably lower plateau level for the remainder of the weld. Under the severe heat dissipation caused by both cold workpieces and the water environment, a stable temperature field and heat input could hardly be established. The plug material flow was not sufficient

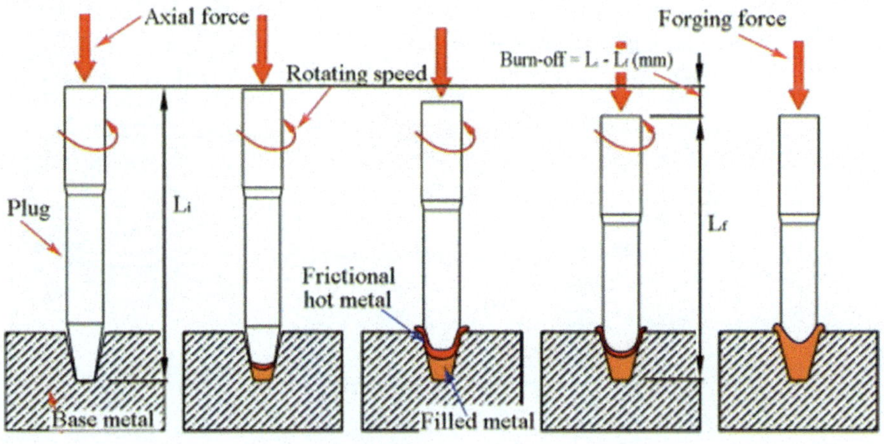

Fig. 4.13 Schematic of the FTPW process [97]

to fill the hole, further forming effective bonding with the hole sidewall. Nevertheless, at the end of HDP, a certain amount of plasticized material accumulated at the plug frictional plane and pressed to the gap between the plug and BM at the start of HSTP. This "pressing behaviour" was reflected by the drop in weld torque and demonstrated by the Ni tracer since a significant amount of Ni was removed from the centre over a relatively larger area, especially under higher axial force. As the SDP began, a regular plug consumption process that was dominated by regular shear off of the plug and deposition was established. The weld torque was maintained at a quasi-stable level that was smaller than that in the other phases. Considering the severe heat dissipation and poor material flow at the HDP and HSTP, increasing material flow and the heat input during both phases was believed to be practical to eliminate root defects. Moreover, the material flow in the SDP was uniform and extensive. This should be the reason why incomplete filling defects never occur at the sidewall when the axial force is sufficient (higher than 20 kN).

The main process parameters affecting the FTPW process and its welded joints are rotational velocity, axial force, forging force, push amount, forging time, burn-off and plug morphology. Meyer and co-workers [95] found that the bonding qualities and properties of FTPW welds were significantly influenced by the weld geometry, rotational velocity, and axial force. However, the influences of the forging force and forging time were less substantial. Hattingh et al. [94] found that for AISI 4140 steel, a maximum tensile strength of 94% of that of the BM could be obtained with the following parameter combination: 6000 rpm rotational velocity, 11.7 kN axial force and 9.0 mm burn-off. The measured tensile strength and ductility were nearly equivalent to those of the BM owing to the overmatching of the welds. In Charpy impact tests, the properties would match the requirements of the standard for marine applications [95]. Chludzinski et al. [98] found that the fracture toughness of

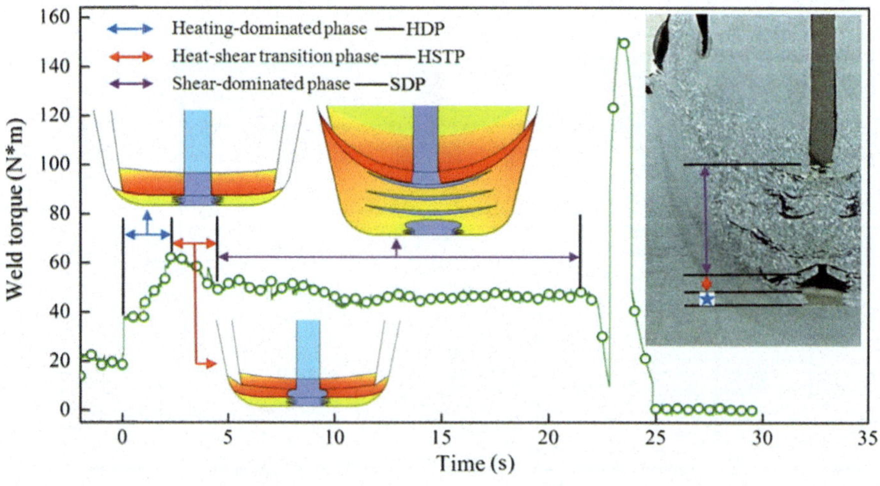

Fig. 4.14 Relationship between material flow behaviour and torque variation [52]

the weld was much lower than that of the BM. The varied axial forces could rarely influence the fracture toughness. Cui et al. [99] determined that a lack of bonding and incomplete filling defects were found, which easily formed under the combination of low rotational velocity and low welding force. High-quality joints were obtained at rotational velocities up to 7000 rpm and welding forces ranging from 25 to 40 kN. It can be concluded that superior FTPW joints can be achieved by regulating and controlling the welding parameters. A high rotational velocity and a larger axial force than conventional FSW are necessary.

4.5.3 Progressive Friction Stir Welding

Progressive FSW is different from the other filling FSW methods employing a consumable bit, whose principle is that the materials surrounding the keyhole are used to fill the volume defects and reduce the depth of the keyhole. To improve the filling of the volume defect completely, an additional filling piece is necessary.

Pinless FSW is a solid-state joining technique based on the material transfer induced by severe plastic deformation that can realize the high-quality joining of thin sheets lower than 2 mm [100–102]. Based on the concept of friction-induced plastic flow, the keyhole of FSW joints in thin-walled sheets could be repaired. The main factors affecting the joint strength are shoulder morphology, rotational velocity, plunge depth and dwelling time. Venukumar et al. [103] reported that a convex shoulder tool was better than a concave shoulder tool due to the complete elimination of insufficient refilling. Sajet et al. [104] reported that rotational velocity has the strongest effect on welding strength. Dwelling time has a minimum effect on the strength and maximum displacement of conventional joints and nearly no effect on refilled joints. The order of importance for parameters that affect conventional joints is rotational velocity > tool plunge depth > dwelling time. Huang et al. [55] proposed a drilling-filling FSR (DFFSR) technique to repair the volume defects in lightweight AZ31B Mg alloy with an additional filling plug and a pinless tool. An optimized joint with a maximum tensile strength of 217 MPa was produced, which fractured at the TMAZ rather than at the interface between the filler and the BM. Niu et al. [56] employed a non-rotating shoulder to assist the PFFSR and then obtained a uniform microhardness distribution along the whole repaired region. However, the effective depth of the pinless FSW is limited to 2 mm or even smaller.

Self-refilling friction stir welding ($SRFSW_2$) was proposed by Zhou et al. [50] to seal the keyhole left by FSW in stainless steel, as shown in Fig. 4.15. This multistage process used a series of non-consumable tools with gradual changes in pin geometry and size. Similar to the results of Ji et al. [9], based on the constancy of the material volume, a narrow and deep keyhole was filled and changed to a wide and shallow exit hole remaining at the surface due to the lack of filler materials. SRFSW is not applicable to through holes and has not been verified to work in Al alloys. This approach can also avoid preparing a specific tapered through hole in a plate, such as that required in the FPW/FTPW processes, thereby providing better adaptability for the actual working conditions.

Based on the advantages of $SRFSW_2$ [50] and pinless FSW [55], Ji et al. [9, 54] proposed the active–passive filling FSR (A-PFFSR) method to fill keyhole-based volume defects based on a series of non-consumable pinless tools and additional filler materials. A schematic diagram of the A-PFFSR process is shown in Fig. 4.16. The A-PFFSR process was divided into two parts: active filling (AF) and passive filling (PF). The keyhole could be repaired in several steps using different pinless rotating tools with different diameters. The keyhole in the AF process was filled with the materials surrounding the keyhole. The additional filler materials used in the PF process were the same as those used in the BM process. The diameter of the additional filler materials was the same as that of the rotating tool used in the last step, and the height was higher than the depth of the unfilled keyhole. During the AF and PF processes, the filler materials were heated and plasticized by frictional heat generated by the pinless tool. Under the stirring effect of the rotating tool, sufficient material flow behaviour can be produced, which can effectively fill the keyhole. In addition, the forging force provided by the rotating tool was beneficial to solid-state bonding between both filling layers. Significantly, to repair keyholes successfully and obtain high-quality joints, the optimum selections of the welding parameters (rotational velocity, plunge depth, tilt angle, plunge depth) and other variables (shoulder diameter and morphology and the filling piece size) were very important. Ji et al. [105] stated that a six-spiral-flute shoulder was designed to improve the material flow and eliminate the interfacial defects, in which six-spiral flutes were distributed across a greater proportion of the shoulder. During the PF process, to obtain surface formation and reduce or even eliminate the possibility of kissing bonds, the pinless rotating tool

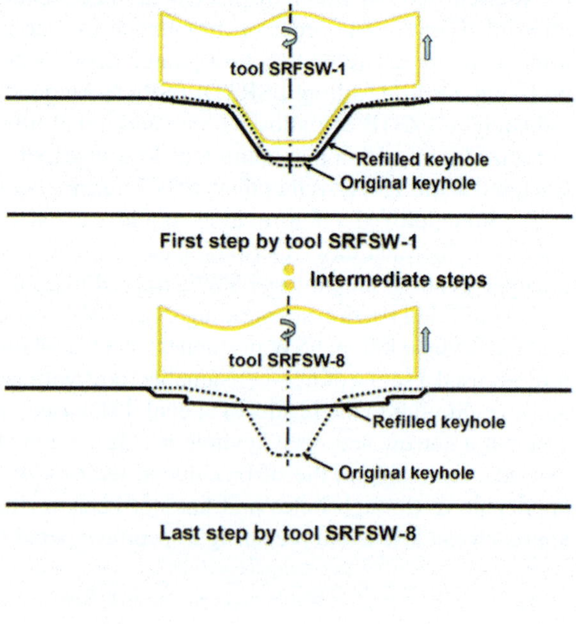

Fig. 4.15 New technique of $SRFSW_2$ [50]

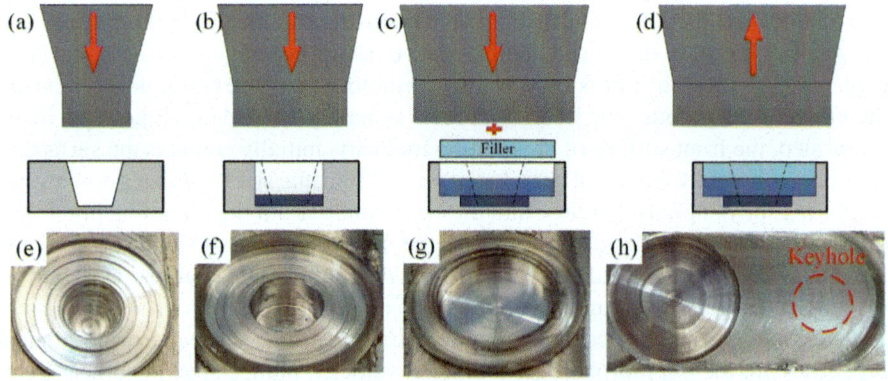

Fig. 4.16 Schematics and joint formations of the A-PFFSR process: schematics of the **a** first AF, **b** second AF, **c** PF and **d** completion of A-PFFSR; joint formations of the **e** keyhole, **f** after first AF, **g** after second AF and **h** after PF [9]

moved forward along the welding direction after dwelling for 5 s. They reported that the tensile strength and elongation of the repaired joint reached 189.7 MPa and 7.6% at a constant rotational velocity of 1300 rpm and a plunge speed of 1 mm/min, which correspond to 96.3 and 98% of those of the defect-free FSW joint, respectively. The diminished plastic deformability after repairing the FSW keyhole in the 4-mm-thick AZ31B Mg alloy sheet occurred due to the limitation of the number of slip systems of the hexagonal close packed crystal lattice. The mechanical properties of the A-PFFSR joints increased as the rotational velocity increased from 1000 to 1600 rpm. When the rotational velocity was 1600 rpm and the constant plunge speed was 2 mm/min, the tensile strength and elongation of the repaired joint reached maximum values of 311.1 MPa and 7.6%, which correspond to 82.1 and 95.8% of those of the defect-free FSW joint, respectively. It can be concluded that the solid-state repairing techniques, the severe plastic deformation and low peak temperature occur in the repairing interface, which can improve the interfacial bonding, and then achieve the repaired joints with quasi-equivalent strengths.

4.5.4 Refill Friction Stir Spot Welding

RFSSW, a solid-state spot welding process invented and patented by Helmholtz-Zentrum Geesthacht [106], was developed as an alternative to riveting or conventional fusion welding techniques to fabricate similar or dissimilar material joints, such as Al/Al alloys [107–110], Mg/Mg alloys [111], Al/Mg alloys [112], polymers/polymers [113–115] and polymers/metals [116–118]. The RFSSW tool consists of a clamping ring, a sleeve, and a pin, as shown in Fig. 4.17. The clamping ring does not rotate during welding. The main function of the clamping ring is to keep the sheets tightly joined and prevent the plasticized materials from escaping. The clamping ring always has a large diameter. The main components stirring the materials are the pin and the sleeve. Hence, both the pin and the sleeve have threads on their outer surfaces (Fig. 4.17). A clearance fit is used when assembling the three components to avoid their direct connections. The RFSSW process is mainly divided into four steps. In the initial step, the front surface of the rotating tool parts initially contacts the surface of the plate, generating frictional heat. In the second step, the sleeve plunges downward, displacing the materials that flow into the cavity underneath the retracting pin. In the third step, the pin moves downward and the sleeve retracts, pushing the softened materials back into the weld. In the final step, the tool retracts from the workpieces, leaving the plug and surrounding workpieces joined.

The main parameters controlling microstructures and subsequent mechanical properties are the pin morphology, rotational velocity, plunge speed and heat treatment for precipitation-hardened Al alloys. Defect-free joints were fabricated using a modified tool with three grooves or notches in the tool sleeve, which improved metallurgical bonding, material intermixing and mechanical interlocking at the interface compared with the standard sleeve with a flat surface, as shown in Fig. 4.18 [107]. With increasing tool wear, more materials are pressed into the gaps between the tool parts. This leads to larger required torques and forces and higher temperatures [119]. Based on numerical simulations, Ji et al. [120] studied the effects of welding tool morphology on heat generation and material flow, as shown in Fig. 4.19. They found that a large sleeve diameter increased the heat input and joining area, while an extremely high heat input was detrimental to the joint quality. The groove on the sleeve improved the material flow behaviour. The threaded groove on the sleeve walls or the unclosed groove on the sleeve bottom is more suitable for increasing the flow velocity along the vertical direction.

Reimann et al. [57] employed RFSSW and a plug as filler materials to seal through holes in 6-mm-thick AA6061-T6 workpieces. The RFSSW joints reached 68 and 55% of the ultimate tensile strength and yield strength of the BM, respectively. Later, Reimann et al. [58] applied RFSSW for termination hole closure in BT-FSW joints in 3-mm-thick AA2198–T8 workpieces. The mechanical performance of the closed hole is similar to that of the RFSSW keyhole closure welds in the base metal. Coefficient of 67% for hardness, 57% for yield strength and 78% for ultimate tensile

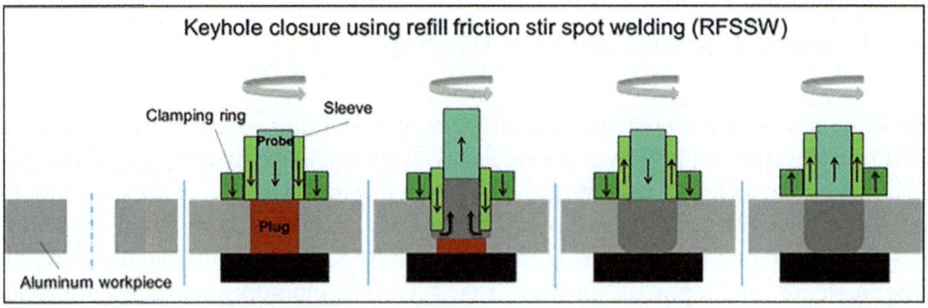

Fig. 4.17 RFSSW keyhole closure process [57]

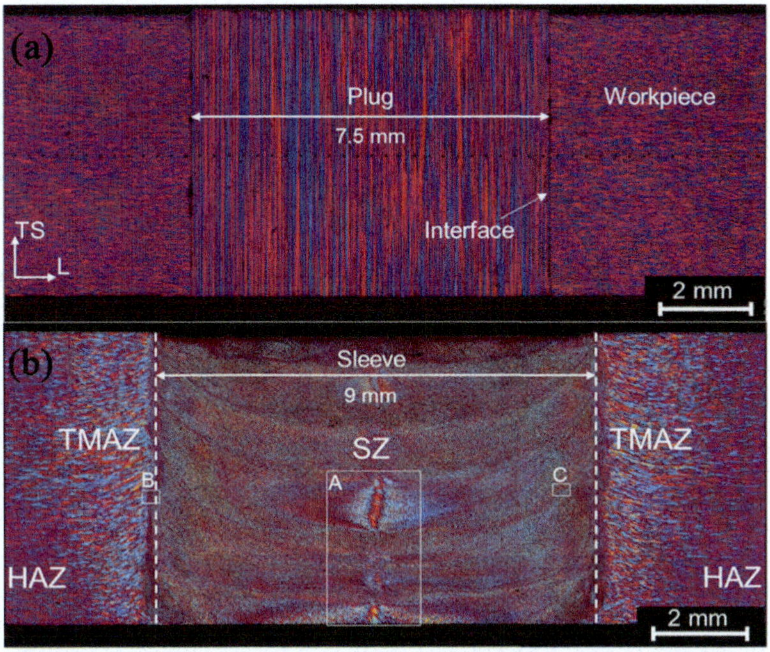

Fig. 4.18 Low-magnification overviews of unwelded and welded samples: **a** unwelded cross-section with extruded plug and surrounding workpieces and **b** typical appearance of the welded sample [59]

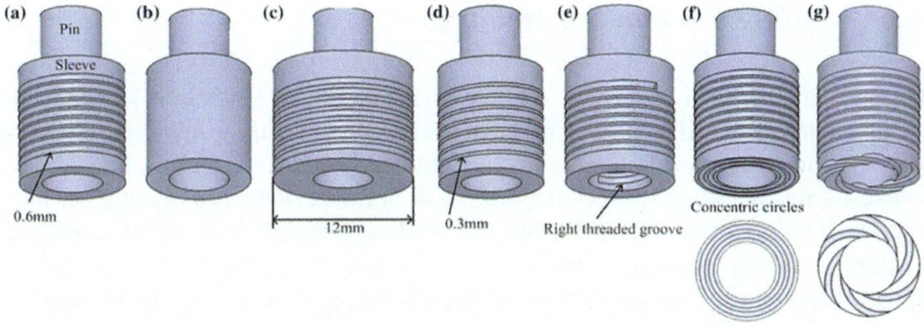

Fig. 4.19 Tool models in the simulation process: **a** regular tool, **b** tool with a smooth surface, **c** tool with a large-diameter sleeve, **d** tool with a smaller groove width, **e** tool with a groove on the inner sleeve surface, **f** tool with concentric circles on the sleeve bottom and **g** tool with a scrolled groove on the sleeve bottom [120]

strength were achieved. They also filled the through hole of an AA7075–T761 workpiece by RFSSW [59]. Reimann et al. [57–59] stated that post-weld natural ageing was shown to be highly significant for the mechanical properties. During ageing, the strength increased from 58 to 70% of the BM hardness in the area of lowest hardness in the HAZ and from 46 to 55% of the yield strength of the BM. The post-weld natural ageing stabilized after 4 weeks [59]. Keyhole closure using RFSSW was demonstrated as a universal through-hole closure method that offers advantages such as defect-free welds and superior surface appearance on both sides of the weld. Fox [121] developed a pre-cracking procedure to provide simulated fatigue cracks on which repairing could be performed by depositing AA2024 powders using cold spraying to fill the pre-cracks. Good repairing was also characterized and showed the improvement in fatigue life over cracks by $SSFSW_2$.

Additionally, this method does not require any surface preparation before or after the welding process and seals the keyhole in a single-step process. This method is not suitable for closed structures due to the need to install a rigid backing bar that has to withstand a large welding force. Some cross-sections of the RFSSW joints show cracks in the lower parts of the NZ, which initiated from the oxide remnants of the interface between the plug and surrounding workpieces. Moreover, the common defects are lack of mixing, incomplete refilling [122], voids [123] and tearing [124] at the TMAZ/NZ interface. Liu et al. [125] introduced ultrasonic oscillation into the RFSSW process to join AA5A06 and stated that lateral ultrasonic oscillation was more effective in improving the tensile strength of the joint.

4.5.5 Filling Friction Stir Welding

Huang et al. [14, 45, 126] proposed FFSW to repair similar keyhole volume defects, as shown in Fig. 4.20. The FFSW process was performed in three main steps: friction stage, stirring stage, and joining stage. First, the FFSW depends on a friction process, where a semi-consumable bit contacts the wall of the keyhole, rubs against the inner surface of the keyhole and self-cleans the bonding surfaces (removal of oxidation). Second, a stirring step, where the semi-consumable bit is plunged into the keyhole and the rotating shoulder made of steel contacts the upper surface of the workpieces, generates frictional heating at the interface between the bit and the wall of the keyhole. The materials surrounding the keyhole and the bit are plasticized and stirred, while severe plastic deformation and flow of the plasticized materials occur. The joining step follows, while the bit and the materials surrounding the keyhole are heated by frictional heat. With the combined functions of softening and torque, the joining bit is sheared and fractured at the corner of the shoulder, becoming the filler material to repair the keyhole.

Figures 4.21 and 4.22 show detailed images of the FFSW tool and joints. The welding tool is composed of a semi-consumable bit and a steel shoulder (Fig. 4.21).

The keyhole defect can be successfully filled (Fig. 4.22). In the following FSP, a non-consumable tool consisting of a pinless shoulder provides mechanical mixing

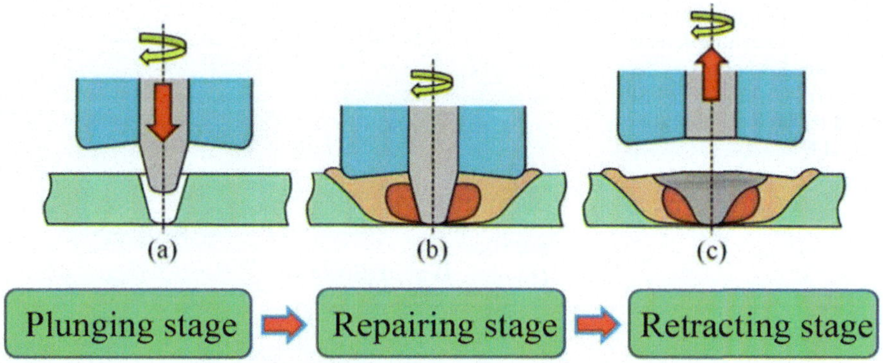

Fig. 4.20 Schematics of FFSW: **a** plunging stage, **b** repairing stage and **c** retreating stage

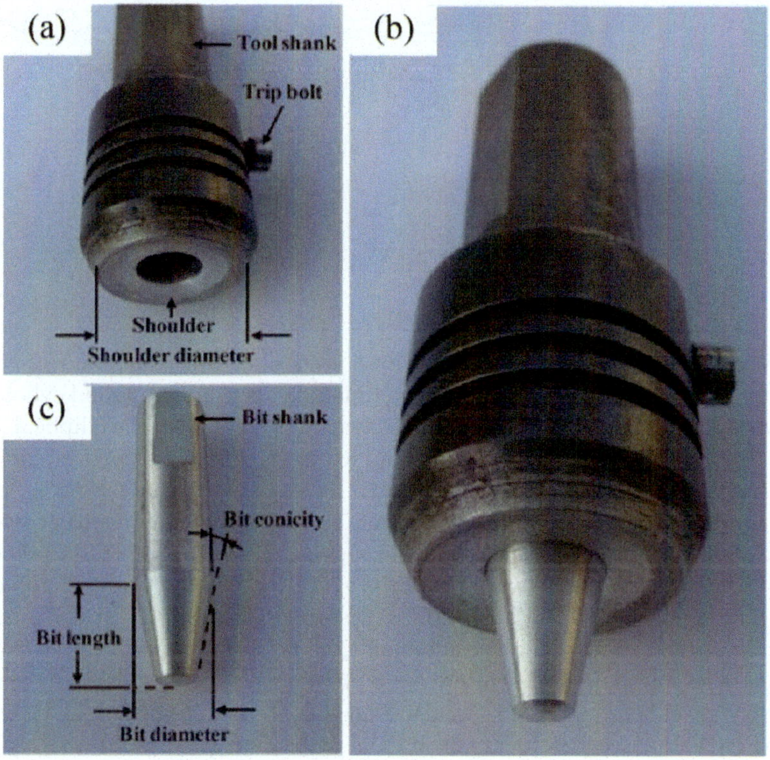

Fig. 4.21 Features of the FFSW tool: **a** steel shoulder, **b** AA7075 bit and **c** assembled semi-consumable FFSW tool [14]

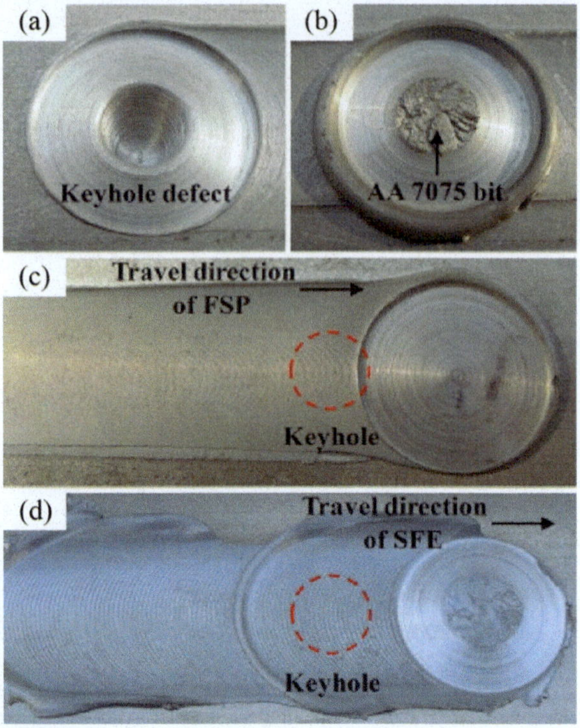

Fig. 4.22 Detailed images of FFSW: **a** keyhole defects, **b** repaired keyhole, **c** repaired keyhole after FSP and **d** repaired keyhole after SFE [14]

and plastic deformation. Moreover, the flash around the filled keyhole was abraded and stirred. The semi-consumable bit and materials surrounding the keyhole have been metallurgically bonded very well. A classical formation of the elliptical "onion ring" structure parallel to the interface formed in the transition between the filling zone (FZ) and the TMAZ [14]. Mechanical interlocking is also the main bonding mechanism in dissimilar FSW joints that rely on forming a complicated geometry at the bonding interface to enhance mechanical strength. During the FFSW process, as the bit keeps inserting downward into the keyhole, the bit head begins to contact the zigzag structures on the keyhole surface. Material deformation and fierce attrition occurred at the contact region. Additionally, some of the superficial bit materials were stripped off and fell down into the bottom of the keyhole, resulting from the relatively lower hardness (96 HV) and the keyhole friction. As a result of the features of the FFSW tool, the materials at the bottom of the FZ experienced relatively gentle stirring and insufficient heating, especially compared with the NZ, and defects such as cavities and flaws are easily formed at the interlayer or interface between two adjacent layers. These problems can be solved by regulating and controlling the welding parameters in the FFSW.

A macroscopic overview of the cross section of the FFSW joint is presented in Fig. 4.23a. Macroscopic observation confirms that a defect-free joint is successfully produced by FFSW. The joining bit remains in the joint. Figure 4.23b shows the

Vicker's hardness profile along the line below the top surface 2 mm cross the FFSW joint. The hardness is higher than that of the base metal due to the combined functions of FFSW and FSP. Figure 4.24a–e presents the details of the bonding interface between the joining bit and the keyhole at five different locations (see labels in Fig. 4.24). Since the interface region is located beside the joining bit, it would experience both frictional heating and several plastic deformations during the joining step of FFSW. Based on microstructural characterization, five distinct zones, FZ, quasi-nugget zone (QNZ), TMAZ, HAZ, and NZ, have been identified. During the FFSW process, the heating is accomplished by both friction between the tool and the keyhole and plastic deformation of the joining bit and the keyhole wall. The FZ has formed through the bonding and self-breaking of consumable joining bit due to the function of heating and shearing. The center zone of FZ does not experience the plastic deformation. Near the FZ, the QNZ is defined which undergoes extreme thermal–mechanical deformation and material flow of the keyhole wall and the consumable joining bit. The TMAZ experiences both elevated temperature and plastic deformation. Heat is generated due to friction and plastic deformation at the tool-work-piece interface and due to plastic deformation in the TMAZ. The TMAZ is observed over the entire width of the weld region. Beyond the TMAZ, there is an HAZ which experiences a thermal cycle but no plastic deformation. The nugget of FSP intersected with zones of FFSW is formed. The upper surface experiences extreme deformation and frictional heating by contact with shoulder during FFSW and following FSP. The dynamic recrystallization occurs in the NZ due to the severe thermo-mechanical processing, resulting in homogenous fine equiaxed recrystallized grain structure. And the FSP can modify successfully some properties such as yield strength, fatigue and superplasticity. In the initial stage of tool plunge, the heating is primarily resulted from the friction between the joining bit and the keyhole wall. Meanwhile, some additional heating is generated through plastic deformation of abutted material. The tool is plunged till the shoulder intimately contacts the workpiece. The friction between the shoulder and the workpiece results in the majority of heating. The localized heating softens the material around the joining bit and leads to movement of the softened material combining with rotation of the partly plastic-deformed bit. With the combined functions of softening and torque, the joining bit is sheared at the corner of shoulder. Therefore, the keyhole is filled by the sheared joining bit. The plasticized material is forged by the intimate contact of the tool shoulder and then consolidated. It leaves a solid phase bond between the keyhole and the joining bit. As a result of this FFSW process a joint is produced on 'solid state'.

The welded joint is fundamentally defect free and displays excellent mechanical properties when compared to conventional fusion welds. Tensile specimens were machined from the filled zone in the direction normal to weld. The tensile testing was carried out on specimens that were 15 mm wide and 150 mm long. And tensile tests were performed at room temperature at a cross-head speed of 1 mm/min. Tensile specimens in the transverse orientation cover six different microstructural zones, i.e., parent material, HAZ, TMAZ, QNZ, NZ, and FZ. The observed ductility is an average strain over the gage length including various zones. Figure 4.25a shows the

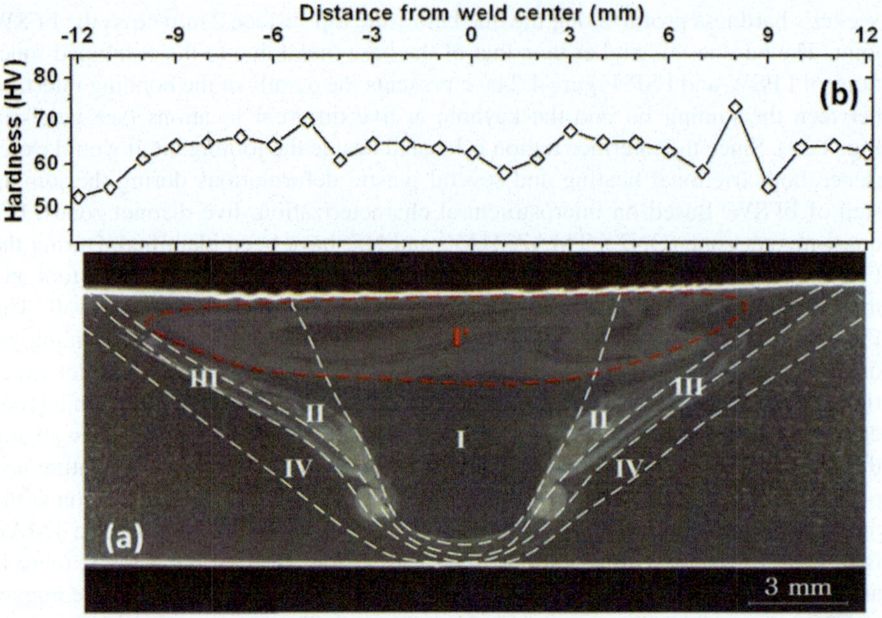

Fig. 4.23 Optical macrograph and hardness profile of FFSW joint for 7.8 mm aluminum plate. **a** Five distinct zones: I—FZ, II—QNZ, III—TMAZ, IV—HAZ, I—NZ of FFSW and following FSP, and **b** hardness profile in cross section [45]

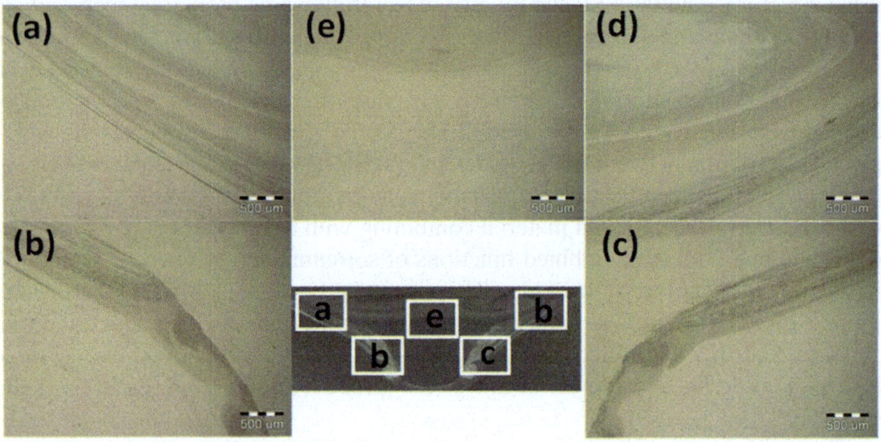

Fig. 4.24 Bonding interface detail between the joining bit and the keyhole (refer to five different locations) [45]

top and fracture surface views of the FFSW weld, and fracture occurs near FZ. The original interface between the wall of the keyhole and consumable aluminum alloy joining bit is a weak region due to inadequate material plastic deformation and flow. Specimen extension was measured by means of a 50 mm extensometer attached to the reduced section. The elongation was determined by scribing marks with known separation within the reduced section before testing and measuring their separation after testing. Figure 4.25b shows the tensile test results of the base weld without keyhole and repairing joints by FSP without filler material, by TIG fusion welding and by FFSW. It can be seen that the joints by FSP and TIG have considerably poor mechanical properties. Particularly, the relative elongation rate is only 11.5 and 12.2% of base weld, respectively. However, when the FFSW process is utilized, the relative tensile strength is 84.3% of the base weld, and the relative elongation is 98.9% of the base weld. The FFSW process increased the effective cross sectional area of the nugget, resulting in higher tensile strength. The fractured surface is characterized by dimples and tearing edges, indicating that plastic deformation occurs during tensile test. Obviously, the FFSW process is able to repair the keyhole and overcome the 'weakest link effect' caused by keyhole.

For FFSW, two parameters are very important: tool rotation rate and tool plunge speed along the axial line of keyhole. The rotation of tool results in stirring and mixing of material around the rotating joining bit. Furthermore, the plunge depth of joining bit into the keyhole is important for producing sound welds. The insertion depth is associated with the joining bit height and the keyhole depth. When the insertion depth is too small, the pin does not contact the bottom of keyhole adequately, and the shoulder does not contact firmly original workpiece surface. Thus, rotating shoulder can't move the stirred material efficiently, producing welds with inner hole or surface groove. When the insertion depth is too large, the shoulder plunges into the workpiece creating excessive flash.

As FSW, the FFSW advantages result from the fact that the process takes place in the solid phase below the melting point of the material to be joined. The benefits therefore include the ability to join materials that are difficult to be welded by conventional fusion welding, such as high strength aluminum alloy. However, several key problems and issues about the new technology of FFSW remain to study. Firstly, the fundamental knowledge of the FFSW process and the evolution of the structure and properties need to be combined to build process control models with a goal to achieve, defect free, structurally sound and reliable welds. The framework has been provided to understand material flow accounting for all of the observed phenomena, and in particular the cyclic variation in the microstructure and process response variables. Secondly, during FFSW process, the heat produced by friction and stirring may be not sufficient to soften and plasticize the material around the rotating bit. It is difficult to produce continuous defect-free weld with abundant material plastic deformation and flow. Thus, preheating of the keyhole is important for the FFSW process, and preheating or additional external heating source can facilitate the material flow and increase the process window.

The main process parameters in the FFSW process are rotational velocity, plunge speed and bit conicity degree for semi-consumable bits, though other variables such

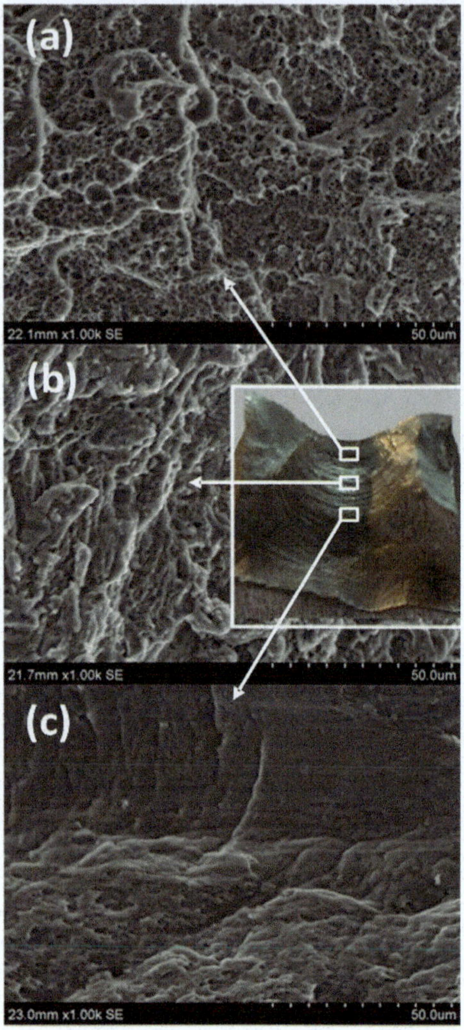

Fig. 4.25 Tensile fractured view and performance. **a** the view of failed FFSW joint, and **b** the relative strength and elongation rate of the FFSW, FSP without filler material, and TIG welds [45]

as welding speed, shoulder diameter, tilt angle, deep indentation depth for the rotating shoulder, and plate conditions may also influence the repair quality. The length of the bit should be 2 ~ 5 mm higher than the depth of the keyhole, which can guarantee sufficient filling of the keyhole. The bit conicity degree is always lower than that of the keyhole defects. Behmand et al. [48] discussed that an unsuitable geometry caused incorrect contact at the pin/exit hole faying surface, while sufficient friction and stirring were not attained. A low plunge speed is beneficial to sufficient plastic deformation at the repairing interface and improves metallurgical bonding. Moreover, the harder the semi-consumable bit is, the more severe the plastic deformation in the repairing interface and the higher the joint quality. Huang et al. [126] reported that when employing an AA2219 bit to repair the keyhole defect in AA2219 FSW joints, the average ultimate tensile strength and elongation were 172 MPa and 11.22%, which corresponded to 90 and 82% of those of the base weld without defects, respectively. Huang et al. [14, 45] also repaired the keyhole defect of AA2219–T6 FSW joints by choosing an AA7075–T6 bit whose hardness was higher than the workpieces to be repaired. The maximum ultimate tensile strength and elongation of the FFSW joint were 179.6 MPa and 13.7%, which corresponded to 96.6 and 99% of those of the original defect-free FSW joint, respectively [14]. Behmand et al. [48] applied FFSW to repair a 6.5-mm-deep exit hole in AA5456 FSW lap joints. The failure load on the coupons with the refilled keyhole reached 91% of that of a corresponding defect-free FSW joint. It can be concluded that quasi-equivalent mechanical performances without weld defects have been achieved by the FFSW joints with excellent interface and mechanical properties. Based on the technical advantage of FFSW, Yan et al. [127] successfully joined acrylonitrile butadiene styrene (ABS) sheets and stated that high joint strength was achieved when using a large plug diameter, a moderate plug length and a blind hole depth. In contrast, spot welds with small plug diameters, excessive plug lengths or large blind hole depths exhibited pores and cavities in the vicinity of the joining interface.

Zhou et al. [49] reported that void defects were formed at the bottom of the refilled original conical keyholes for all the tool design and process parameters used for 316L stainless steel. A refilled joint with a defect-free interface could be obtained on the modified spherical keyhole using the corresponding filling tool under processing parameters of 35 kN, 1500 rpm and 9 s. The microstructures surrounding the interface and in the refilled zone were significantly refined for all the refilled joints. Tensile test results showed that the tensile specimen fractured at the BM. The relative tensile strength and elongation were 105 and 92% of those of the as-received plate, respectively. Zhang et al. [128] modified the FFSW method using a pinless tool and a T-shaped filler bit to reduce the setup time for replacement of the tool between the filling and reprocessing operations. For a cylindrical bit, the failure loads were sensitive to the in-situ friction time. When the in-situ friction time exceeded 5 s, the filled samples exhibited favourable failure loads. They sealed keyholes left by FSW in 4.7-mm-thick AA1060 sheets. To eliminate voids in the lower portion of the joint, a Zn brazing foil was preset in the keyhole. The ultimate tensile strength of the keyhole closure welds reached 67.3% of the BM strength [128]. Until now, FFSW has been shown to repair keyholes left by a conical FSW tool, for which the geometry of the filler bit must be adapted and has not yet been applied to through holes. Moreover, FFSW is still in its infancy, and further investigations are necessary for its evaluation and application in manufacturing fields.

For dissimilar materials without metallurgical reactions, our research group proposed friction-based filling stacking joining (FFSJ), which is similar to the principle of FFSW, to fill the keyholes of polymer and metal FSLW joints, as shown in Fig. 4.26 [129]. The FFSJ was completed by filling a prefabricated hole with an additional filling stud. An excellent FFSJ joint was successfully achieved under the combined effects of frictional heat and thermo-mechanical behaviour at the interface between the filling stud and the polymer sheet. Intimate contact formed at the interface between the filling stud, the polymer and the metal, and the metal hole was completely filled with the stud. The maximum tensile shear strength of 13 MPa for the FFSJ was comparable to that produced by state-of-the-art welding. The fracture location of the FFSJ joint is located at the joining interface between the filling stud and the re-solidified polymer sheet.

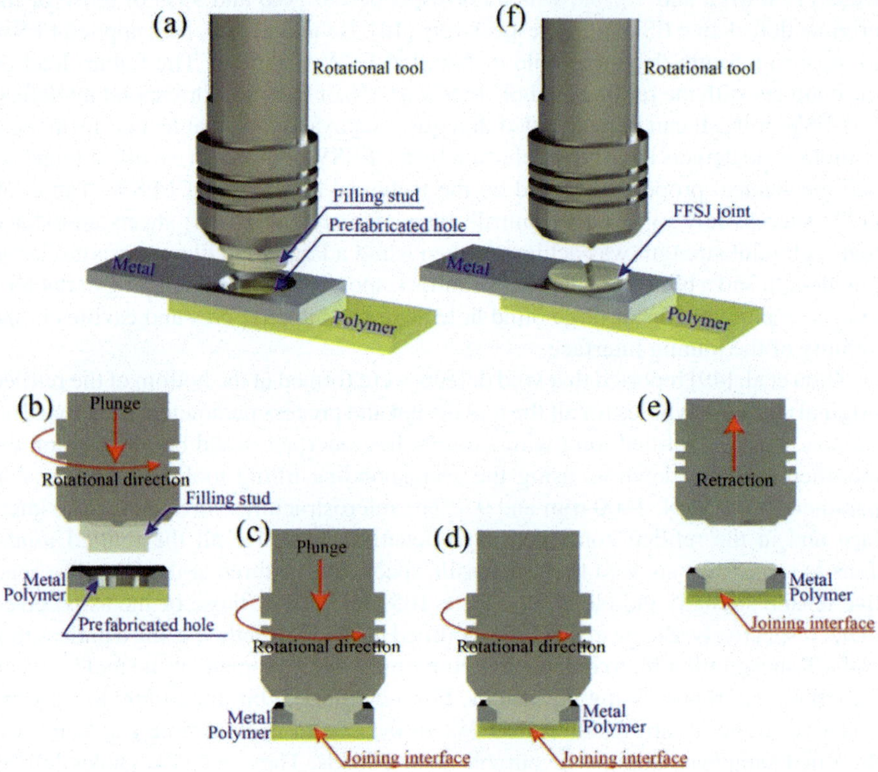

Fig. 4.26 Schematic of the FFSJ process: **a** positioning of the filling stud in the prefabricated hole; **b** rotation and plunge of the rotating tool and the filling stud; **c** the filling stud approaches the polymer sheet while the filling stud and polymer sheet are melted near the joining interface; **d** the rotating tool stops plunging and continues rotating for 1 s, then the tool stops rotating and the molten polymer begins to solidify; **e** the rotating tool retracts and the molten polymer continues to solidify; and **f** the hybrid FFSJ joint is achieved [129]

4.6 Prospects

Friction stir-based remanufacturing, a high-efficiency and green joining technique based on low frictional heat and severe plastic deformation, has great potential for high-quality repairing of metallic material. However, some issues need to be addressed.

(1) Aiming at the typical defect repairing of the outer environment or complex structural parts, the portable and multiple-degree-of-freedom FSW robot combined with FSR will certainly be a research hotspot. Recently, robotic FSW systems have been utilized in various industrial applications.
(2) Although FSR has the outstanding advantages to repair light materials of Al alloys and Mg alloys, it should be developed into the production of high melting materials, such as steels, Ti or Ti-based alloys and MMCs.
(3) FSR integrates advanced manufacturing processes such as additive manufacturing to obtain variants such as friction stir additive manufacturing and additive friction stir, resulting in landmark accomplishments towards the goal of innovative and modern prototyping/tooling/manufacturing engineering applications. It is challenging to extend its applications to include tubes, in particular small-diameter tubes, and other complex geometries.

References

1. TWI. Weld repair, reclamation and rectification[EB/OL].
2. Feng C, Liang J, Gong C, et al. Repair volume extraction method for damaged parts in remanufacturing repaira. Int J Adv Manuf Technol. 2018;98(5–8):1523–36.
3. Chai Q, Xiao Z, Lai K, et al. Can carbon cap and trade mechanism be beneficial for remanufacturing? Int J Prod Econ. 2018;203:311–21.
4. Wu X. Research on design management based on green remanufacturing engineering. Syst Eng Proced. 2011;2012(4):448–54.
5. Balasubramanian V. Relationship between base metal properties and friction stir welding process parameters. Mater Sci Eng A. 2008;480(1–2):397–403.
6. Heidarzadeh A, Jabbari M, Esmaily M. Prediction of grain size and mechanical properties in friction stir welded pure copper joints using a thermal model. Int J Adv Manuf Technol. 2015;77(9–12):1819–29.
7. Sharma V, Prakash U, Kumar BVM. Surface composites by friction stir processing: a review. J Mater Process Technol. 2015;224:117–34.
8. Huang Y, Meng X, Xie Y, et al. Friction stir welding/processing of polymers and polymer matrix composites. Compos A Appl Sci Manuf. 2018;105:235–57.
9. Ji S, Meng X, Zeng Y, et al. New technique for eliminating keyhole by active-passive filling friction stir repairing. Mater Des. 2016;97:175–82.
10. Teimurnezhad J, Pashazadeh H, Masumi A. Effect of shoulder plunge depth on the weld morphology, macrograph and microstructure of copper FSW joints. J Manuf Process. 2016;22:254–9.
11. Khan NZ, Siddiquee AN, Khan ZA, et al. Investigations on tunneling and kissing bond defects in FSW joints for dissimilar aluminum alloys. J Alloys Compd. 2015;648:360–7.

12. Zheng Q, Feng X, Shen Y, et al. Effect of plunge depth on microstructure and mechanical properties of FSW lap joint between aluminum alloy and nickel-base alloy. J Alloys Compd. 2017;695:952–61.
13. Zhou N, Song D, Qi W, et al. Influence of the kissing bond on the mechanical properties and fracture behaviour of AA5083-H112 friction stir welds. Mater Sci Eng A. 2018;719:12–20.
14. Han B, Huang Y, Lv S, et al. AA7075 bit for repairing AA2219 keyhole by filling friction stir welding. Mater Des. 2013;51:25–33.
15. Arbegast WJ. A flow-partitioned deformation zone model for defect formation during friction stir welding. Scr Mater. 2008;58(5):372–6.
16. Tao Y, Zhang Z, Ni DR, et al. Influence of welding parameter on mechanical properties and fracture behavior of friction stir welded Al–Mg–Sc joints. Mater Sci Eng A. 2014;612:236–45.
17. Kainuma S, Katsuki H, Iwai I, et al. Evaluation of fatigue strength of friction stir butt-welded aluminum alloy joints inclined to applied cyclic stress. Int J Fatigue. 2008;30(5):870–6.
18. Zhou C, Yang X, Luan G. Effect of kissing bond on fatigue behavior of friction stir welds on Al 5083 alloy. J Mater Sci. 2006;41(10):2771–7.
19. Arici A, Selale S. Effects of tool tilt angle on tensile strength and fracture locations of friction stir welding of polyethylene. Sci Technol Weld Join. 2007;12(6):536–9.
20. Zeng KW, Su ZM, Luo SM, et al. Removing approach for flashes of friction stir spot welds. J Mater Process Technol. 2013;213(10):1725–33.
21. Shultz EF, Cole EG, Smith CB, et al. Effect of compliance and travel angle on friction stir welding with gaps. J Manuf Sci Eng. 2010;132(4):041010.
22. Wanjara P, Monsarrat B, Larose S. Gap tolerance allowance and robotic operational window for friction stir butt welding of AA6061. J Mater Process Technol. 2013;213(4):631–40.
23. Yang Y, Kalya P, Landers RG, et al. Automatic gap detection in friction stir butt welding operations. Int J Mach Tools Manuf. 2008;48(10):1161–9.
24. Liu XC, Wu CS. Elimination of tunnel defect in ultrasonic vibration enhanced friction stir welding. Mater Des. 2016;90:350–8.
25. Zhang H, Wang M, Zhou W, et al. Microstructure-property characteristics of a novel non-weld-thinning friction stir welding process of aluminum alloys. Mater Des. 2015;86:379–87.
26. Ji S, Meng X, Ma L, et al. Vertical compensation friction stir welding assisted by external stationary shoulder. Mater Des. 2015;68:72–9.
27. Ji SD, Meng XC, Li ZW, et al. Experimental study of stationary shoulder friction stir welded 7N01-T4 aluminum alloy. J Mater Eng Perform. 2016;25(3):1228–36.
28. Strand S. Effects of friction stir welding on polymer microstructure. Provo: Brigham Young University; 2004.
29. Sahlot P, Jha K, Dey GK, et al. Quantitative wear analysis of H13 steel tool during friction stir welding of Cu–0.8%Cr–0.1%Zr alloy. Wear. 2017;378–379:82–9.
30. Sahlot P, Arora A. Numerical model for prediction of tool wear and worn-out pin profile during friction stir welding. Wear. 2018;408–409:96–107.
31. Salih OS, Ou H, Sun W, et al. A review of friction stir welding of aluminium matrix composites. Mater Des. 2015;86:61–71.
32. Bist A, Saini JS, Sharma B. A review of tool wear prediction during friction stir welding of aluminium matrix composite. Trans Nonferrous Met Soc China. 2016;26(8):2003–18.
33. Rai R, De A, Bhadeshia HKDH, et al. Review: friction stir welding tools. Sci Technol Weld Join. 2011;16(4):325–42.
34. Zhang YN, Cao X, Larose S, et al. Review of tools for friction stir welding and processing. Can Metall Q. 2012;51(3):250–61.
35. Avettand-Fènoël MN, Simar A. A review about friction stir welding of metal matrix composites. Mater Charact. 2016;120:1–17.
36. Mironov S, Sato YS, Kokawa H. Friction-stir welding and processing of Ti–6Al–4V titanium alloy: a review. J Mater Sci Technol. 2017;34(1):58–72.
37. Ji S, Huang R, Zhang L, et al. Microstructure and mechanical properties of friction stir repaired Al–Cu casting alloy. Trans Indian Inst Met. 2018;71(8):2057–65.

38. Li W, Diao G, Liang Z, et al. Repair welding of tunnel defect in friction stir weld of Al–Zn–Mg alloys. In: International conference on mechanical engineering and control automation; 2017, pp. 202–207.
39. Gunter C, Miles MP, Liu FC, et al. Solid state crack repair by friction stir processing in 304L stainless steel. J Mater Sci Technol. 2017;34(1):140–7.
40. Li Z, Han K, Liu C, et al. Repair welding of defects in 2219 Al alloy friction stir welded joints. J Aeronaut Mater. 2016;36(1):26–32.
41. Liu H, Zhang H. Repair welding process of friction stir welding groove defect. Trans Nonferrous Met Soc China. 2009;19(3):563–7.
42. Liu HJ, Zhang HJ. Study of hybrid welding repair process of friction stir welding grove defect. Sci Technol Weld Join. 2012;17(3):169–73.
43. Ji SD, Meng XC, Li ZW, et al. Investigation of vertical compensation friction stir-welded 7N01-T4 aluminum alloy. Int J Adv Manuf Technol. 2016;84(9–12):2391–9.
44. Abu-Okail M, Ata MH, Abu-Oqail A, et al. Production of tailor-welded blanks by vertical compensation friction stir welding technique. Mater Sci Technol. 2018;34(16):2030–41.
45. Huang YX, Han B, Tian Y, et al. New technique of filling friction stir welding. Sci Technol Weld Join. 2011;16(6):497–501.
46. Du B, Sun Z, Yang X, et al. Characteristics of friction plug welding to 10 mm thick AA2219-T87 sheet: weld formation, microstructure and mechanical property. Mater Sci Eng A. 2016;654:21–9.
47. Metz DF, Barkey ME. Fatigue behavior of friction plug welds in 2195 Al–Li alloy. Int J Fatigue. 2012;43:178–87.
48. Behmand SA, Mirsalehi SE, Omidvar H, et al. Filling exit holes of friction stir welding lap joints using consumable pin tools. Sci Technol Weld Join. 2015;20(4):330–6.
49. Zhou L, Nakata K, Tsumura T, et al. Microstructure and mechanical properties of 316L stainless steel filling friction stir-welded joints. J Mater Eng Perform. 2014;23(10):3718–26.
50. Zhou L, Liu D, Nakata K, et al. New technique of self-refilling friction stir welding to repair keyhole. Sci Technol Weld Join. 2012;17(8):649–55.
51. Xiong J, Yang X, Lin W, et al. Evaluation of inhomogeneity in tensile strength and fracture toughness of underwater wet friction taper plug welded joints for low-alloy pipeline steels. J Manuf Process. 2018;32:280–7.
52. Yin Y, Yang X, Cui L, et al. Material flow influence on the weld formation and mechanical performance in underwater friction taper plug welds for pipeline steel. Mater Des. 2015;88:990–8.
53. Mukherjee S, Ghosh AK. Friction stir processing of direct metal deposited copper-nickel 70/30. Mater Sci Eng A. 2011;528(9):3289–94.
54. Ji SD, Meng XC, Huang RF, et al. Microstructures and mechanical properties of 7N01-T4 aluminum alloy joints by active-passive filling friction stir repairing. Mater Sci Eng A. 2016;664:94–102.
55. Huang R, Ji S, Meng X, et al. Drilling-filling friction stir repairing of AZ31B magnesium alloy. J Mater Process Technol. 2018;255:765–72.
56. Niu S, Wu B, Ma L, et al. Passive filling friction stir repairing AZ31-B magnesium alloy by external stationary shoulder. Int J Adv Manuf Technol. 2018;97(5–8):2461–8.
57. Reimann M, Gartner T, Suhuddin U, et al. Keyhole closure using friction spot welding in aluminum alloy 6061–T6. J Mater Process Technol. 2016;237:12–8.
58. Reimann M, Goebel J, Gartner TM, et al. Refilling termination hole in AA 2198–T851 by refill friction stir spot welding. J Mater Process Technol. 2017;245:157–66.
59. Reimann M, Goebel J, dos Santos JF. Microstructure and mechanical properties of keyhole repair welds in AA 7075–T651 using refill friction stir spot welding. Mater Des. 2017;132:283–94.
60. Arici A, Sinmazçelýk T. Effects of double passes of the tool on friction stir welding of polyethylene. J Mater Sci. 2005;40(12):3313–6.
61. Saeedy S, Besharati Givi MK. Experimental investigation of double side friction stir welding (FSW) on high density polyethylene blanks. In: ASME 2010 10th Biennial Conference on Engineering Systems Design and Analysis; 2010, pp. 845–848

62. Cabibbo M, Forcellese A, el Mehtedi M, et al. Double side friction stir welding of AA6082 sheets: microstructure and nanoindentation characterization. Mater Sci Eng A. 2014;590:209–17.
63. Widener CA. High-rotational speed friction stir welding with a fixed shoulder. In: Proceedings of the 6th international symposium on friction stir welding. Montreal, 2006, pp. 1–45
64. Wang G, Zhao Y, Hao Y. Friction stir welding of high-strength aerospace aluminum alloy and application in rocket tank manufacturing. J Mater Sci Technol. 2017;34(1):73–91.
65. Ji S, Meng X, Xing J, et al. Vertical compensation friction stir welding of 6061–T6 aluminum alloy. High Temp Mater Process. 2016;35(8):843–51.
66. Park SK, Hong ST, Park JH, et al. Effect of material locations on properties of friction stir welding joints of dissimilar aluminium alloys. Sci Technol Weld Join. 2010;15(4):331–6.
67. Yoon TJ, Kang CY. Observations on metallurgical phenomena and formation of onion ring nugget during friction stir lap welding of dissimilar aluminum by a new 3D technique. Mater Lett. 2015;142:253–7.
68. Yoon TJ, Yun JG, Kang CY. Formation mechanism of typical onion ring structures and void defects in friction stir lap welded dissimilar aluminum alloys. Mater Des. 2016;90:568–78.
69. Krishnan KN. On the formation of onion rings in friction stir welds. Mater Sci Eng A. 2002;327(2):246–51.
70. Sandnes L, Grong Ø, Torgersen J, et al. Exploring the hybrid metal extrusion and bonding process for butt welding of Al–Mg–Si alloys. Int J Adv Manuf Technol. 2018;98(5–8):1059–65.
71. Liu FC, Nelson TW. In-situ material flow pattern around probe during friction stir welding of austenitic stainless steel. Mater Des. 2016;110:354–64.
72. Liu FC, Nelson TW. In-situ grain structure and texture evolution during friction stir welding of austenite stainless steel. Mater Des. 2017;115:467–78.
73. Venkateswaran P, Reynolds AP. Factors affecting the properties of friction stir welds between aluminum and magnesium alloys. Mater Sci Eng A. 2012;545:26–37.
74. McNelley TR, Swaminathan S, Su JQ. Recrystallization mechanisms during friction stir welding/processing of aluminum alloys. Scr Mater. 2008;58(5):349–54.
75. Dolatkhah A, Golbabaei P, Besharati-Givi MK, et al. Investigating effects of process parameters on microstructural and mechanical properties of Al5052/SiC metal matrix composite fabricated via friction stir processing. Mater Des. 2012;37:458–64.
76. Izadi H, Sandstrom R, Gerlich AP. Grain growth behavior and Hall-Petch strengthening in friction stir processed Al 5059. Metall Mater Trans A Phys Metall Mater Sci. 2014;45(12):5635–44.
77. Dunkerton S, Nicholas D, Sketchley P. Repairing defective metal workpiece-by friction welding with a metal plug. Britain. 1991;127:9125978.
78. Coletta ER, Cantrell MA, Mccool A. Friction plug weld repair geometric innovations [EB/OL]. https://ntrs.nasa.gov/search.jsp?R=20000112928; 2017
79. Li Z, Cantrell M, Brown R, et al. Process development and microstrutural characterization plug welded 2195 and 2219 alloys [EB/OL]. https://ntrs.nasa.gov/search.jsp?R=20000089963; 2000
80. Hartley PJ. Frition plug weld repair geometric innovations [EB/OL]. https://ntrs.nasa.gov/search.jsp?R=20000093962; 2017
81. Metz DF, Weishaupt ER, Barkey ME, et al. A microstructure and microhardness characterization of a friction plug weld in friction stir welded 2195 Al–Li. J Eng Mater Technol. 2012;134(2):021005.
82. Metz DF, Barkey ME, Bradt RC, et al. Experimental and numerical characterization of the fatigue and fracture properties of friction plug welds in 2195-T8 aluminum lithium alloy; 2013
83. Lange M, Schwenker S. Friction plug welding of 2024-T3 aluminum. In: Proceedings of aircraft structural integrity conference. San Antoni; 2014
84. Du B, Cui L, Yang X, et al. Weakening mechanism and tensile fracture behavior of AA 2219–T87 friction plug welds. Mater Sci Eng A. 2017;693:129–35.

85. Gregersen HB, Dyer JH. Lockheed martin chairman and CEO vance coffman on achieving mission success. Acad Manag Perspect. 2002;16(3):31–41.
86. Brooke SA, Bradford V. Friction pull plug welding in aluminum alloys [EB/OL]. https://ntrs.nasa.gov/search.jsp?R=20120014529; 2014
87. Gibson D, Meyer A, Vennemann O, et al. Engineering applications of friction stitch welding. Proc Int Confer Offshore Mech Arct Eng OMAE. 2001;14:139–43.
88. Costa Pereira DA, Cunha PH, Szabo R, Bergmann L, et al. Preliminary study on the application of friction welding in structural repairs; 1998
89. PAULY D. Process development on an experimental friction hydro pillar processing system. GKSS Int Rep; 1999
90. Nicholas E. Friction hydro pillar processing. In: Proceedings of the 11th annual North American welding research conference; 1995, pp. 7–9
91. Hattingh D, Van Zyl C. Temperature distribution for a friction taper stud weld in thick walled 10CrMo910 steel. South Afr Inst Mech Eng. 2012;28:37–45.
92. Meyer A, Pauly D, Dos Santos J, et al. Considerations on robotic friction stitch welding for the repair of marine structures. Proc Int Confer Offshore Mech Arct Eng. 2001;3:145–51.
93. Wu F, Yeh T, Henrique P, et al. Evaluation of discontinuities in A36 steel repairs with friction hydro pillar processing using different axial forces. ISIJ Int. 2013;53:2269–71.
94. Hattingh DG, Bulbring DLH, Els-Botes A, et al. Process parameter influence on performance of friction taper stud welds in AISI 4140 steel. Mater Des. 2011;32(6):3421–30.
95. Meyer A. Friction hydro pillar processing: bonding mechanism and properties; 2003
96. Ambroziak A, Gul B. Investigations of underwater FHPP for welding steel overlap joints. Arch Civ Mech Eng. 2007;7(2):67–76.
97. Cui L, Yang X, Wang D, et al. Friction taper plug welding for S355 steel in underwater wet conditions: welding performance, microstructures and mechanical properties. Mater Sci Eng A. 2014;611:15–28.
98. Chludzinski M, Paes MP, Bastian FL, et al. Fracture toughness of friction hydro-pillar processing welding in C-Mn steel. Mater Des. 2012;33(1):340–4.
99. Cui L, Yang X, Wang D, et al. Experimental study of friction taper plug welding for low alloy structure steel: welding process, defects, microstructures and mechanical properties. Mater Des. 2014;62:271–81.
100. Li W, Li J, Zhang Z, et al. Improving mechanical properties of pinless friction stir spot welded joints by eliminating hook defect. Mater Des. 2014;62:247–54.
101. Liu Z, Cui H, Ji S, et al. Improving joint features and mechanical properties of pinless fiction stir welding of alcald 2A12–T4 aluminum alloy. J Mater Sci Technol. 2016;32(12):1372–7.
102. Liu Z, Ji S, Meng X, et al. Improving joint formation and tensile properties of friction stir welded ultra-thin Al/Mg alloy sheets using a pinless tool assisted by a stationary shoulder. Int J Adv Manuf Technol. 2017;93(5–8):2071–9.
103. Venukumar S, Yalagi SG, Muthukumaran S, et al. Static shear strength and fatigue life of refill friction stir spot welded AA 6061–T6 sheets. Sci Technol Weld Join. 2014;19(3):214–23.
104. Sajed M. Parametric study of two-stage refilled friction stir spot welding. J Manuf Process. 2016;24:307–17.
105. Ji SD, Meng XC, Ma L, et al. Effect of groove distribution in shoulder on formation, macrostructures, and mechanical properties of pinless friction stir welding of 6061-O aluminum alloy. Int J Adv Manuf Technol. 2016;87(9–12):3051–8.
106. Schilling C, Dos Santos J. Method and device for linking at least two adjoining work pieces by friction welding; 2001
107. Shen Z, Ding Y, Gopkalo O, et al. Effects of tool design on the microstructure and mechanical properties of refill friction stir spot welding of dissimilar Al alloys. J Mater Process Technol. 2018;252:751–9.
108. Xu Z, Li Z, Ji S, et al. Refill friction stir spot welding of 5083-O aluminum alloy. J Mater Sci Technol. 2018;34(5):878–85.
109. Kubit A, Wydrzynski D, Trzepiecinski T. Refill friction stir spot welding of 7075–T6 aluminium alloy single-lap joints with polymer sealant interlayer. Compos Struct. 2018;201:389–97.

110. Cao JY, Wang M, Kong L, et al. Microstructure, texture and mechanical properties during refill friction stir spot welding of 6061–T6 alloy. Mater Charact. 2017;128:54–62.
111. Shen Z, Ding Y, Chen J, et al. Comparison of fatigue behavior in Mg/Mg similar and Mg/steel dissimilar refill friction stir spot welds. Int J Fatigue. 2016;92:78–86.
112. Suhuddin U, Fischer V, Kroeff F, et al. Microstructure and mechanical properties of friction spot welds of dissimilar AA5754 Al and AZ31 Mg alloys. Mater Sci Eng A. 2014;590:384–9.
113. Huang Y, Meng X, Xie Y, et al. Friction spot welding of carbon fiber-reinforced polyetherimide laminate. Compos Struct. 2018;189:627–34.
114. Junior WS, Emmler T, Abetz C, et al. Friction spot welding of PMMA with PMMA/silica and PMMA/silica-g-PMMA nanocomposites functionalized via ATRP. Polymer. 2014;55(20):5146–59.
115. Gonçalves J, dos Santos JF, Canto LB, et al. Friction spot welding of carbon fiber-reinforced polyamide 66 laminate. Mater Lett. 2015;159:506–9.
116. Goushegir SM, dos Santos JF, Amancio-Filho ST. Failure and fracture micro-mechanisms in metal-composite single lap joints produced by welding-based joining techniques. Compos A Appl Sci Manuf. 2016;81:121–8.
117. André NM, Goushegir SM, dos Santos JF, et al. Friction spot joining of aluminum alloy 2024–T3 and carbon-fiber-reinforced poly(phenylene sulfide) laminate with additional PPS film interlayer: microstructure, mechanical strength and failure mechanisms. Compos B Eng. 2016;94:197–208.
118. Esteves JV, Goushegir SM, dos Santos JF, et al. Friction spot joining of aluminum AA6181-T4 and carbon fiber-reinforced poly (phenylene sulfide): effects of process parameters on the microstructure and mechanical strength. Mater Des. 2015;66:437–45.
119. Montag T, Wulfsberg JP, Hameister H, et al. Influence of tool wear on quality criteria for refill friction stir spot welding (RFSSW) process. Proced CIRP. 2014;24:108–13.
120. Ji S, Wang Y, Li Z, et al. Effect of tool geometry on material flow behavior of refill friction stir spot welding. Trans Indian Inst Met. 2017;70(6):1417–30.
121. Fox SL. Refill friction stir spot weld repair of a fatigue crack; 2010
122. Parra B, Saccon VT, Alcântara NG, et al. An Investigation on friction spot welding in AA6181–T4 alloy. Tecnol Metal Mater. 2011;8(3):184–90.
123. Shi Y, Yue Y, Zhang L, et al. Refill friction stir spot welding of 2198–T8 aluminum alloy. Trans Indian Inst Met. 2018;71(1):139–45.
124. Li Z, Ji S, Ma Y, et al. Fracture mechanism of refill friction stir spot-welded 2024–T4 aluminum alloy. Int J Adv Manuf Technol. 2016;86(5–8):1925–32.
125. Liu X, Qiao F, Guo L, et al. Metallographic structure, mechanical properties, and process parameter optimization of 5A06 joints formed by ultrasonic-assisted refill friction stir spot welding. Int J Miner Metall Mater. 2017;24(2):164–70.
126. Huang YX, Han B, Lv SX, et al. Interface behaviours and mechanical properties of filling friction stir weld joining AA 2219. Sci Technol Weld Join. 2012;17(3):225–30.
127. Yan Y, Shen Y, Guo C, et al. Friction plug welding acrylonitrile butadiene styrene sheets: the investigation of welding process, joint morphology and mechanical property. Int J Mater Form. 2018;27:68–79.
128. Zhang GF, Jiao WM, Zhang JX. Filling friction stir weld keyhole using pin free tool and T shaped filler bit. Sci Technol Weld Join. 2014;19(2):98–104.
129. Huang Y, Meng X, Xie Y, et al. New technique of friction-based filling stacking joining for metal and polymer. Compos B Eng. 2019;163:217–23.

Chapter 5
High Depth-to-Width Ratio Friction Stir Welding

5.1 Numerical Design of High Depth-to-Width Ratio Friction Stir Welding

5.1.1 Introduction

For conventional friction stir welding (FSW), welding tool consists of shoulder and pin, in which the ratio of pin to shoulder is about 0.33. Zhang et al. [1] showed that the shoulder was responsible for main part of heat generation. Larger shoulder diameter resulted in that higher heat input concentrated on the surface of joint, which broadened the width of the heat affected zone (HAZ) and reduced mechanical properties. The rotational pin is the main driven force of material flow around the tool. In order to obtain high-quality joint, decreasing shoulder diameter and optimizing pin geometrical structure are of vital importance. However, decreasing heat input by simply reducing the shoulder diameter easily results in the fracture of the pin and formation of defect. Obviously, it is not an acceptable choice to attempt massive possibilities one by one due to its potential risk of pin fracture. Thus, numerical methods were proposed to provide a better alternative to welding tool design, which saved time cost and improve the efficiency of the design process.

Up to present, computer aided engineering (CAE) has potential to elucidate joining mechanism without limitation of objective conditions. The numerical methods CAE are divided into two classical categories: computational solid mechanics (CSM) and computational fluid dynamics (CFD). Due to its difficulty to guarantee convergence and calculate time of CSM methods, researchers utilize CFD methods to make the numerical process. Feulvarch et al. [2] proposed a robust moving mesh technique based on a Eulerian formalism. They divided the mesh into two parts: a first one fixed around the stirring zone and a second one including the base material near the tool and moving with a rotational solid motion. Mesh distortions were avoided and computing time was reduced effectively. Zhu et al. [3] assumed that the materials around the

Y. Huang et al., *Friction Stir Welding and Processing*, Materials Forming, Machining and Tribology, https://doi.org/10.1007/978-981-99-8688-0_5

tool flow non-uniformly distributed on the interface under the driving force of friction. The non-uniform friction force model provided a potential tool for predicting defects. Moreover, Arora et al. [4] proposed a numerical criterion for the topology of shoulder diameter based on the principle of maximum utilization of supplied torque for traction, which provided a novel approach to developing tool design. Shi et al. [5] developed a model to evaluate the effect of the welding parameters and shoulder size in reverse dual-rotation FSW.

To summarize, conventional computational methods have been widely accepted during FSW. However, when designing a new tool from nonexistence to pass into existence, conventional method is difficult to realize. High-throughput screening method, based on the marriage between massively parallel computational methods and existing database containing the calculated properties, is capable to explore hypothetical candidates. High depth-to-width ratio FSW was proposed to reduce the width of HAZ and then increase mechanical performances. A high-throughput geometry design method was proposed to design high depth-to-width ratio FSW tools. From the aspects of tool fracture, defect prediction, joint formation and width of HAZ, high depth-to-width ratio FSW was mainly investigated in detail.

5.1.2 Experimental Procedures

Al–Mg–Si alloy rolled sheets were used in this high depth-to-width ratio friction stir butt welding, of which the microhardness was about 95 HV. A plunge depth of 0.15 mm and a tilt angle of 2.5° were constant. Particular tools with features of high depth-to-width ratio, made of HS6-5-2C high speed steel, were used as illustrated in Fig. 5.1. The sizes of the tools are respectively 8.0 mm and 5.8 mm in the shoulder diameter and the pin length, which have a depth-to-width ratio of 0.6. The ratio of pin length to shoulder diameter used in conventional FSW situations of aluminum alloy is relatively low than 0.33, as counted in Fig. 5.2. The workpiece temperature field was recorded by four thermocouples which were fixed in blind holes with 1 mm diameter. Couples were named from T1 to T4, and were placed in their coordinates, as shown in Fig. 5.1. Specimens were cut from the welded workpiece transverse to the welding direction to carry out microstructural characterization. To analyze the microstructural evolution, the section of joint was observed by optical microscope (OM).

5.1.3 Numerical Modeling

A computational domain with dimensions of 5 mm × 55 mm × 50 mm was used in the mathematical evaluation model. Four kinds of tool geometry design, consisting of thread and milling facets, equal to the geometry parameters and plunging depth, were used to simulate the welding process. As illustrated in Fig. 5.3, the domain

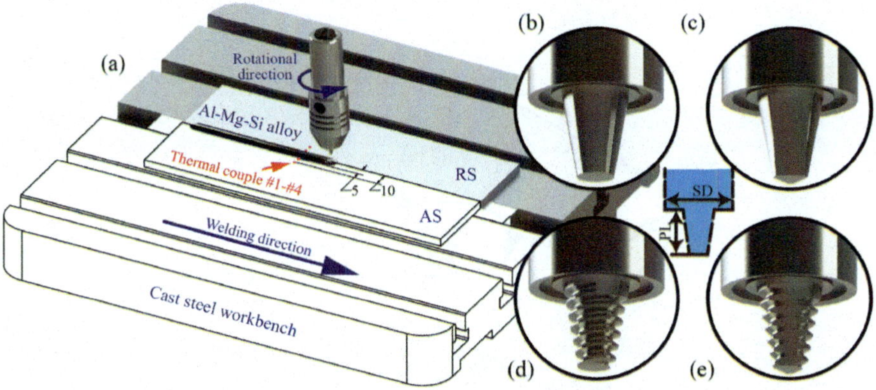

Fig. 5.1 Illustrations: **a** welding process, **b** conventional tapered pin, **c** tapered pin with triple facets, **d** tapered thread pin, and **e** tapered thread pin with triple facets

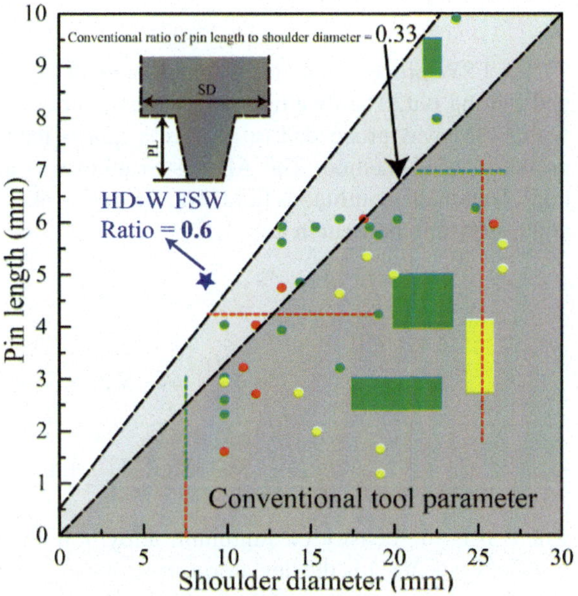

Fig. 5.2 Pin length and shoulder diameter from relevant references for FSW of aluminum alloy (the green, red and yellow shape means that these were counted from journal papers between 2012 and 2017, journal papers before 2012 and conference papers, respectively. The points meant single data while the squares and lines meant range data)

was divided into two relative zones by sliding mesh model (SMM) with the moving mesh and the fixed mesh. The origin of Cartesian coordinate located at center of the pin root, which was at rest relative to the fixed mesh. The moving mesh rotated synchronously with the welding tool. Though the diameter of the moving mesh has no effect on the numerical results, the interface between two zones could cause slight discontinuity on strain rate and temperature. The diameter was set as 13 mm in order to exceed the confines of the banded structures in the model.

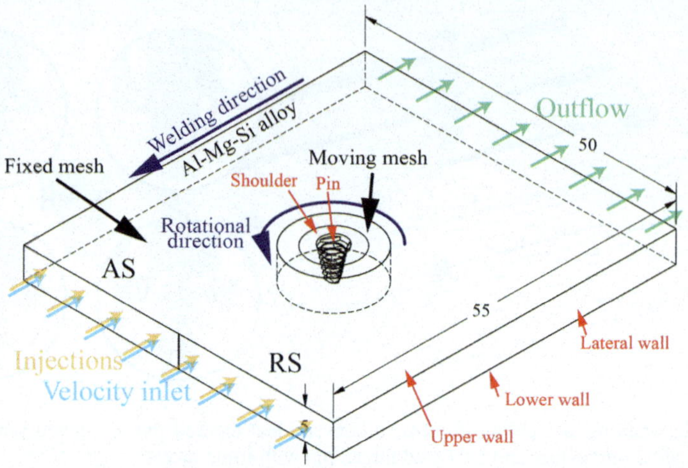

Fig. 5.3 Schematic view of high depth-to-width ratio FSW model

The FSW process includes three subsequent procedures with plunging, dwelling and pulling out. The dwelling period was mainly considered. The transient model with SMM was proceeded to achieve a quasi-steady state. The solver was set as pressure-based method. The Al–Mg–Si alloy was assumed as an incompressible, non-Newtonian, continuous fluid. The Navier–Stokes equations of mass, momentum and energy can be written as:

$$\nabla \cdot v = 0 \tag{5.1}$$

$$\rho \frac{dv}{dt} = -\nabla P + \nabla \cdot \sigma \tag{5.2}$$

$$\rho c_p \frac{dT}{dt} = -\nabla(K \nabla T) + (\sigma : \nabla v) - \phi \tag{5.3}$$

where ρ is the density of aluminum alloy, t is the flow time, v is flow velocity, $\sigma = \mu\left(\nabla v + \nabla v^t\right)$ is the deviatoric stress tensor, μ is the non-Newtonian viscosity, P is the relative pressure, T is the absolute temperature, ϕ is the sink term which represents the heat loss of convection and radiation, c_p is the specific heat capacity and k is the thermal conductivity. Temperature-dependent thermal conductivity and specific heat capacity of the Al–Mg–Si alloy rolled sheets are given in Fig. 5.4.

Attested by Zienkiewicz and Cormeau [6], the non-Newtonian viscosity of metal materials μ is dependent of temperature and effective strain rate, as written below:

$$\mu = \frac{\sigma_e}{3\,\dot{\varepsilon_e}} \tag{5.4}$$

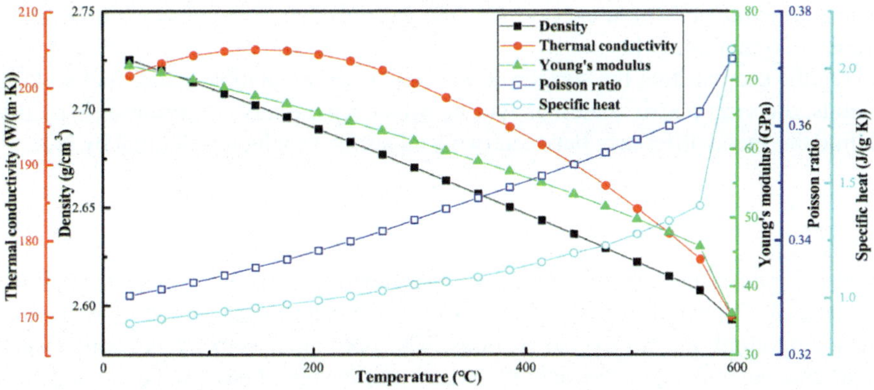

Fig. 5.4 Temperature-dependent thermal conductivity and specific heat capacity of Al–Mg–Si alloy

where σ_e is the effective stress which is regarded as the driving force to sustain the non-Newtonian plastic deformation, $\dot{\varepsilon_e}$ is the effective strain rate:

$$\sigma_e = \sqrt{\frac{3}{2}\sigma_{ij}\sigma_{ij}} \tag{5.5}$$

$$\dot{\varepsilon_e} = \sqrt{\frac{2}{3}\dot{\varepsilon_{ij}}\,\dot{\varepsilon_{ij}}} \tag{5.6}$$

where σ_{ij} and $\dot{\varepsilon_{ij}}$ are the components of the deviatoric stress tensor and the strain rate tensor, respectively. The strain rate tensor is a symmetric second-order tensor, which represents deformation parts of Helmholtz velocity decomposing theorem:

$$\dot{\varepsilon_{ij}} = \frac{1}{2}\left(v_{i,j} + v_{j,i}\right) \quad i, j = 1, 2, 3 \tag{5.7}$$

The Sellars–Tegart constitutive model used by Sheppard and Jackson [7] and ameliorated by Su et al. [8] determines the stress to sustain the plastic deformation:

$$\sigma_e = \left(1 - \sqrt{\frac{T - 273.15}{T_{melt} - 273.15}}\right)\frac{1}{\alpha}\sinh^{-1}\left[\left(\frac{Z}{A}\right)^{\frac{1}{n}}\right] + \sigma_{melt} \tag{5.8}$$

where α, A and n are material constants, T_{melt} is the melt temperature of Al–Mg–Si alloy, σ_{melt} is the yield stress above the melting point. Z is the Zener–Hollomon parameter, which is defined as:

$$Z = \dot{\varepsilon_e}\, e^{\frac{Q}{RT}} \tag{5.9}$$

where Q is the activation energy, R is the gas constant. The constants used above were listed in Table 5.1.

On the premise that the surface contact exists between tool surface and welding metals, the viscoplastic materials will be driven by the tools. Slip rate η is used to define the speed difference between two wall surfaces, which is defined by [9]:

$$\delta = 1 - e^{-\frac{\omega R_{ijk}}{\delta_0 \omega_0 R_s}} \tag{5.10}$$

where δ_0 is a constant taken as 0.4, ω_0 is a constant to nondimensionalize the rotational velocity, R_s is the radius of the shoulder. Figure 5.5 limns the velocity distributions on longitudinal and transverse sections. The horizontal velocity is proportion to rotational velocity ω, so the velocity on any infinitesimal body can be defined as:

$$\begin{cases} v_{horizon} = \eta\omega R_{ijk} \\ v_{shear} = v_{horizon}\cos\theta_{ijk} \\ v_{normal} = v_{horizon}\sin\theta_{ijk} \\ v_{vertical} = v_{horizon}\sin\xi_{ijk} \end{cases} \tag{5.11}$$

The other boundaries except inlet and outlet boundaries were considered as stationary wall with zero slip shear resistance. The velocities of inlet and outlet boundaries were in accordance with the welding speed, while the injections were released from the section of the inlet boundary.

Table 5.1 Material constants and other constants

Constant	α	ln A	n	T_{melt}	σ_{melt}	Q	R
Value	1.7×10^{-8} Pa^{-1}	27.418	6.66	908 K	20 MPa	181.53 kJ/mol	8.314 J/(mol K)

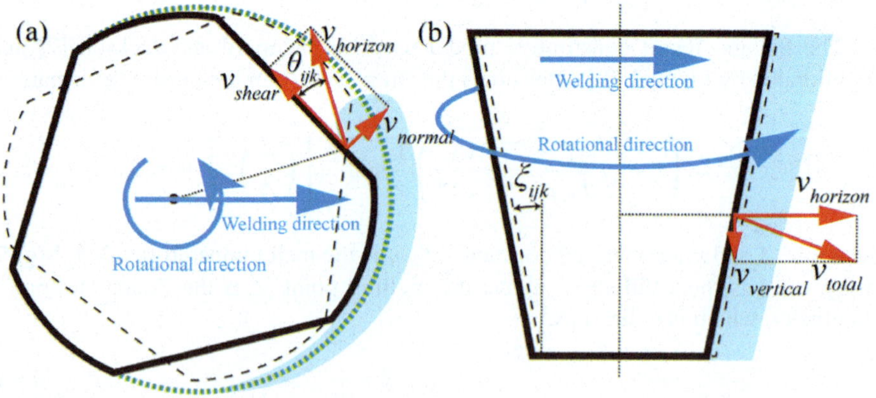

Fig. 5.5 Velocity distributions near-tool-surface: **a** horizontal section and **b** transverse section

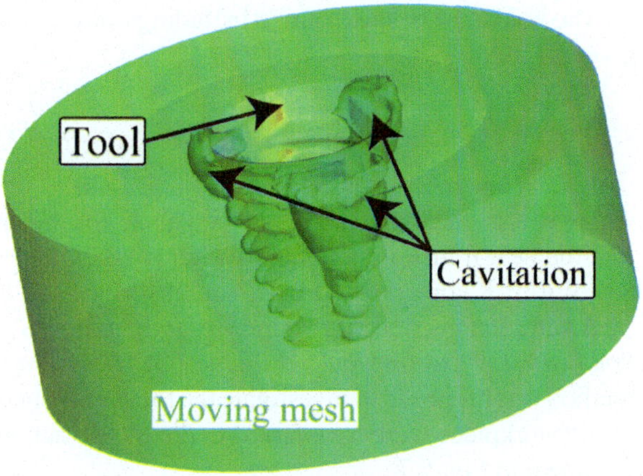

Fig. 5.6 Illustration of negative pressure cavitation

Friction and plastic deformation heats were synchronous considered. The friction heat source term F_1 was surficial heat source and represented the friction between tool and workpiece. F_1 equaled to zero when the surfaces were separated. Related numerical results like Tutunchilar et al. [10] attested that there was cavitation phenomenon occurring due to negative pressure, as shown in Fig. 5.6. The friction heat was governed as follows:

$$\begin{cases} F_1 = \beta_1 R_{ijk}\omega \cdot \left[\frac{\delta\sigma_e}{\sqrt{3}} + (1-\delta)\mu P_{ijk}\sin\xi_{ijk}\right] \cdot \cos\theta_{ijk} & P_{ijk} > 0 \\ F_1 = 0 & P_{ijk} \leq 0 \end{cases} \tag{5.12}$$

The plastic deformation heat source term F_2 is volumetric heat source, which is given as:

$$F_2 = \beta_2 \cdot \sigma_e \cdot \dot{\varepsilon}_e \tag{5.13}$$

where β_1 and β_2 are the empirical coefficient which represents thermal conversion rate, μ is the friction coefficient which is taken as 0.5.

The tool is made of high speed steel, and heat will be conducted to the tool and dissipate in the end. It is of vital importance to calculate the percentage of heat transported into the tool, which can be defined by:

$$\lambda = \frac{J_w}{J_w + J_T} \tag{5.14}$$

$$\mathrm{J} = \sqrt{k\rho C_p} \tag{5.15}$$

where J_w and J_T denote the effusivity of the workpiece and the tool, respectively. k, ρ, and C_p are the thermal conductivity, the density and the heat capacity, respectively.

The boundary heat exchange including convection and radiation of the surface is given as follows:

$$-k\frac{\partial T}{\partial z} = h(T - T_{\infty}) + \sigma\varepsilon\left(T^4 - T_{\infty}^4\right) \tag{5.16}$$

where σ is the Stefan–Boltzmann constant, ε is the emissivity which is taken as 0.4, h is the heat transfer coefficient which is taken as 15 W/(m^2 K), 15 W/(m^2 K) and 200 W/(m^2 K) on upper surface, lateral surface and lower surface, respectively.

As aforementioned, high depth-to-width ratio FSW tools face risk of pin fracture. It is necessary to find out when the pin fracture will occur. Fluid–solid interface method was proposed to investigate the tool stress state. The pressure and temperature of the workpiece were transferred to the tool to achieve coupled fluid–solid surface, as shown in Fig. 5.7.

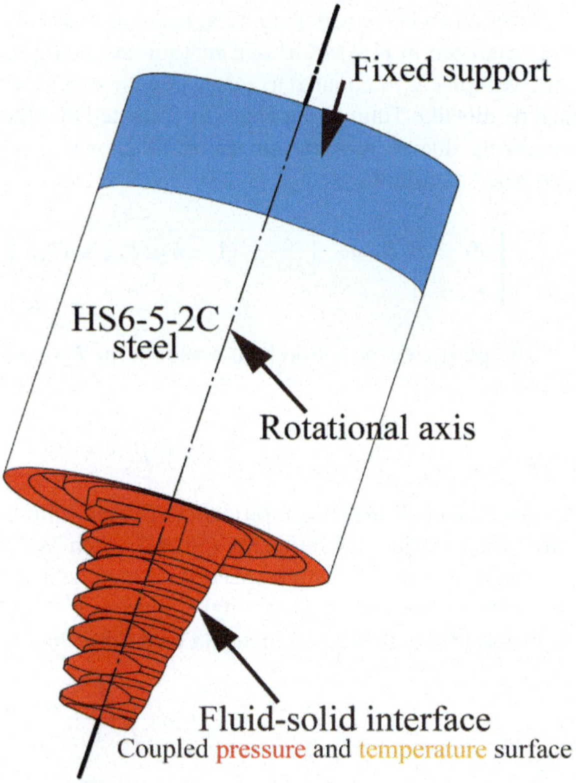

Fig. 5.7 Schematic illustration of fluid–solid interface method

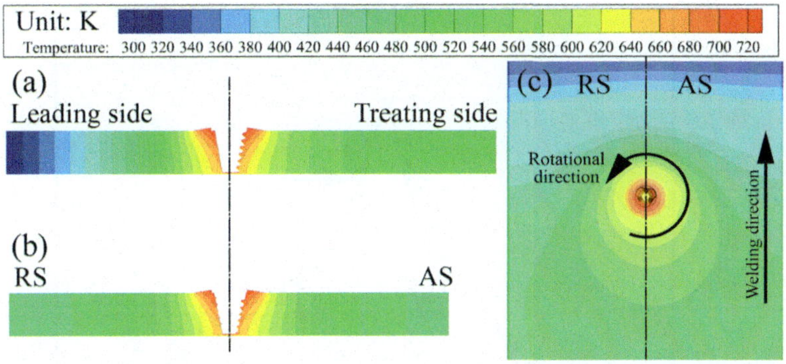

Fig. 5.8 The computed temperature contour: **a** longitudinal section, **b** transverse section, and **c** upper surface

5.1.4 Temperature Distribution and Validation

To validate the reliability of numerical model, a high depth-to-width ratio FSW joint was welded by a tapered thread pin with triple facets at a welding speed of 50 mm/min and a rotational velocity of 1200 r/m. The simulated temperature distributions along the longitudinal, transverse section around the tool and on the upper surface of the fluid domain are shown in Fig. 5.8, respectively. The computed peak temperature is about 719 K. The temperature on advancing side (AS) is slightly higher than that on retreating side (RS), which agrees with Mishra and Ma [11]. Figure 5.9 shows the history curves of temperature. The thermal curve corresponds well with that of experimental data measured by thermocouples. The average relative deviation of the prediction temperature is 1.7%.

5.1.5 Fracture Criteria

Figure 5.10 shows evaluation criteria of fracture potential possibility by fluid–solid interface method. To examine the tool fracture due to the lack of the strength, curves of tensile strength versus temperature and the computed maximal effective stress are plotted. The green shades indicate that the tool can work without risk of fracture, while the yellow shades indicate that the tool is likely to crack. The red zone on tools means the distribution position of maximal effective stress, which is the potential crack initiation position. For example, the tapered thread pin with triple facets at a welding speed of 30 mm/min and a rotational velocity of 800 r/m locates on the position of the blue pentagram, which means it will work regularly. The tapered thread pin with triple facets at a welding speed of 100 mm/min and a rotational velocity of 800 r/m locates on the position of the red hexagon, which is the sign of fracture.

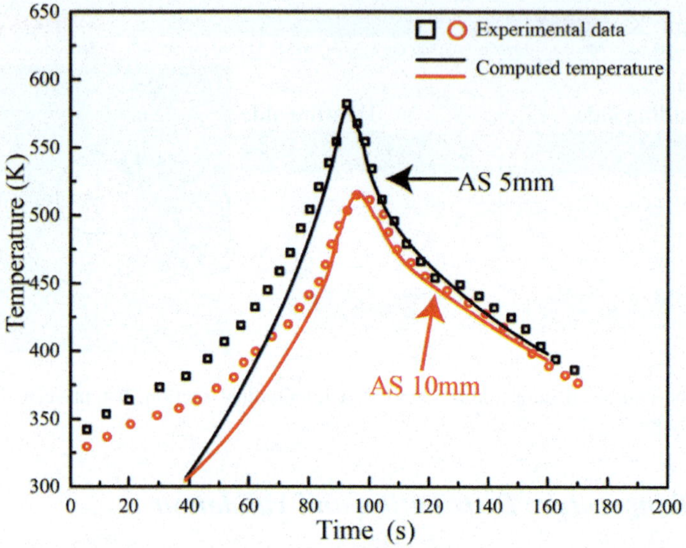

Fig. 5.9 Thermal curves of experiment and numerical simulation

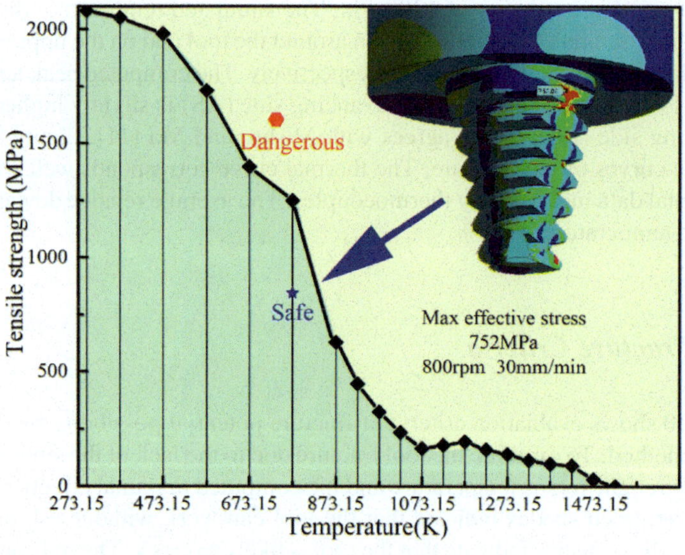

Fig. 5.10 Evaluation criteria of fracture potential possibility

Table 5.2 Fracture state evaluated criteria

Geometrical structure		A	B	C
Tapered pin	Parameter (r/m, mm/min)	800, 50	800, 30	800, 10
	Maximal effective stress (MPa)	2050	1940	1446
	State	Fracture	Fracture	Fracture
Tapered pin with triple facets	Parameter (r/m, mm/min)	800, 50	800, 30	800, 10
	Maximal effective stress (MPa)	1225	830	588
	State	Fracture	Fracture	**Safe**
Tapered thread pin	Parameter (r/m, mm/min)	800, 50	800, 30	800, 10
	Maximal effective stress (MPa)	1535	1386	900
	State	Fracture	Fracture	**Safe**
Tapered thread pin with triple facets	Parameter (r/m, mm/min)	800, 50	800, 30	800, 10
	Maximal effective stress (MPa)	1179	750	656
	State	Fracture	**Safe**	**Safe**

Table 5.2 shows the potential possibility of fracture. At a certain rotational velocity, lower welding speed reduces the maximal effective stress and the potential risk of fracture. The tapered thread pin with triple facets is the only geometrical structure to remain intact when welding speed reaches to 30 mm/min. The riveting discovery is that the maximal effective stress of 587.6 MPa by tapered pin with triple facets is significantly lower than those of thread pins at a welding speed of 10 mm/min, while the stress of thread pins is relatively lower than conventional tapered pin. It is conjectured that the thread structure can promote the material flow but increase the fracture risk.

5.1.6 Defect Prediction

The particle tracing technique is applied to predict the formation of high depth-to-width ratio FSW influenced by material flow and related factors comprehensively. The injections are set as massless particles, which are superimposed on the inlet surface with a random radius of 0.3 mm. The boundary conditions for injections are identical to the matrix metal. As shown in Fig. 5.11, the particles appear grey and the grey mass means dense particle distribution. The region where the red background exposes, means the occurrence of potential defects. Red background indicates the infeasibilities of this geometry with certain parameter. Defects mainly distribute in the welding nugget zone (WNZ) of AS and the root of joint. The defects at AS can be attributed to the insufficient flow, and the defects in the root of joint results from extreme low heat input. It is necessary to improve the effect of stir and raise root heat input appropriately.

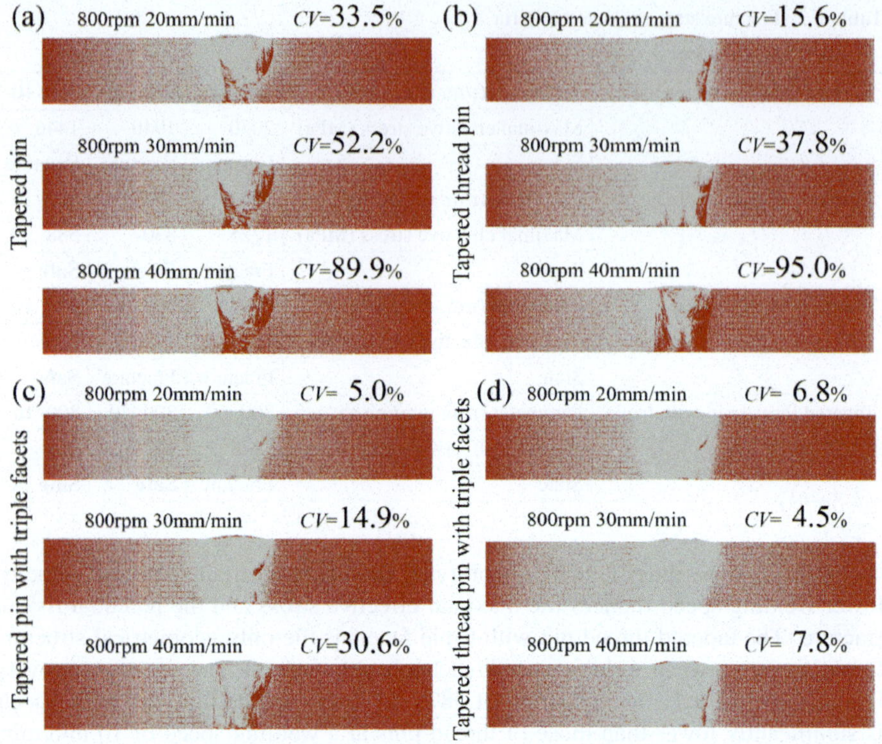

Fig. 5.11 Particle distribution at outlet plane and its CV value: **a** tapered pin, **b** tapered thread pin, **c** tapered pin with triple facets, and **d** tapered thread pin with triple facets

To describe the joint formation, an index to evaluate the distribution evenness of the particles was defined:

$$CV = \frac{SDd}{X} \times 100\% \tag{5.17}$$

where SDd is the color standard deviation within 5 mm from the weldline, X is the color average value within 5 mm from the weldline, CV is the evaluation coefficient. The formation comes better when CV decreases. According the function, tapered thread pin with triple facets comes the best results of which CV value is 5.512%, as shown in Fig. 5.11. Both tapered pin with triple facets and tapered thread pin with triple facets which have the geometrical structure of milling facets are more likely to perform high-quality joint for the lower CV value.

5.1.7 High-Throughput Screening

Both tool fracture analyses and joint formation are of vital importance to realize sound joints. These two influence factors into an evaluation system were taken which is processed by a massively parallel way. Fracture evaluation and defect prediction high-throughput maps comprising all 1600 systems considered are shown in Fig. 5.12. All those systems for which CV value is less than 5% without fracture are included in the list of potential candidates. The parameters are arranged linear scale in increasing order. Each circle, pentagram, hexagram and square represent a parameter combination. A grey square indicates that there are risks of fracture, which should be avoided. A colored circle or a colored pentagram indicates that it can be welded without tool fracture. The color contrast from yellow to red indicates increase of CV value and deterioration of the joint formation. Those yellow pentagrams are the systems of which CV value is less than 5%. There are 0, 0, 1 and 6 candidates in the geometry assembly of tapered pin, tapered thread pin, tapered pin with triple facets, tapered thread pin with triple facets, respectively. A lower CV value means a sounder joint. The system of tapered thread pin with triple facets at a welding speed of 30 mm/min and a rotational velocity of 800 r/m is regarded as the optimized geometry and parameters.

Tapered thread pin with triple facets has a geometry of thread structure and milling facets. In spite of potential weakening of the tools, the thread structure effectively promotes the material flow. Under the assistance of milling facets to promote the flow and soften the materials by increasing deformation heat, the structure is beneficial to realize the joint formation successfully without crack possibilities. These parameters give a feasible environment to ensure the realization of the dense and sound joint.

5.1.8 Joint Formation

Based on the optimized system evaluated by high-throughput screening, the joint produced by tapered thread pin with triple facets at a welding speed of 30 mm/min and a rotational velocity of 800 r/m was obtained. As shown in Fig. 5.13, the surface presents smooth which means lower heat input. Three distinct zones of HAZ, thermo-mechanically affected zone (TMAZ) and WNZ are identified, as shown in Fig. 5.15. The macrostructure displays lanky shape due to the high depth-to-width geometry of the welding tool. The interface between TMAZ and WNZ on the AS is obviously clear than conventional FSW joint obtained by Cabibbo et al. [12]. However, the interface on the RS becomes obscure. It is inferred that only a few of grains in the TMAZ on the RS flows into the region behind the pin on the AS, while metal is driven by the shear force of the rotational tool to fill the gap behind on the RS resulting in a clear interface between equiaxed WNZ grains and elongated TMAZ grains. Interestingly, the multiple onion ring patterns are observed in the WNZ stacked vertically through the thickness and close to the AS. Onion rings have alternate bright and dark rings.

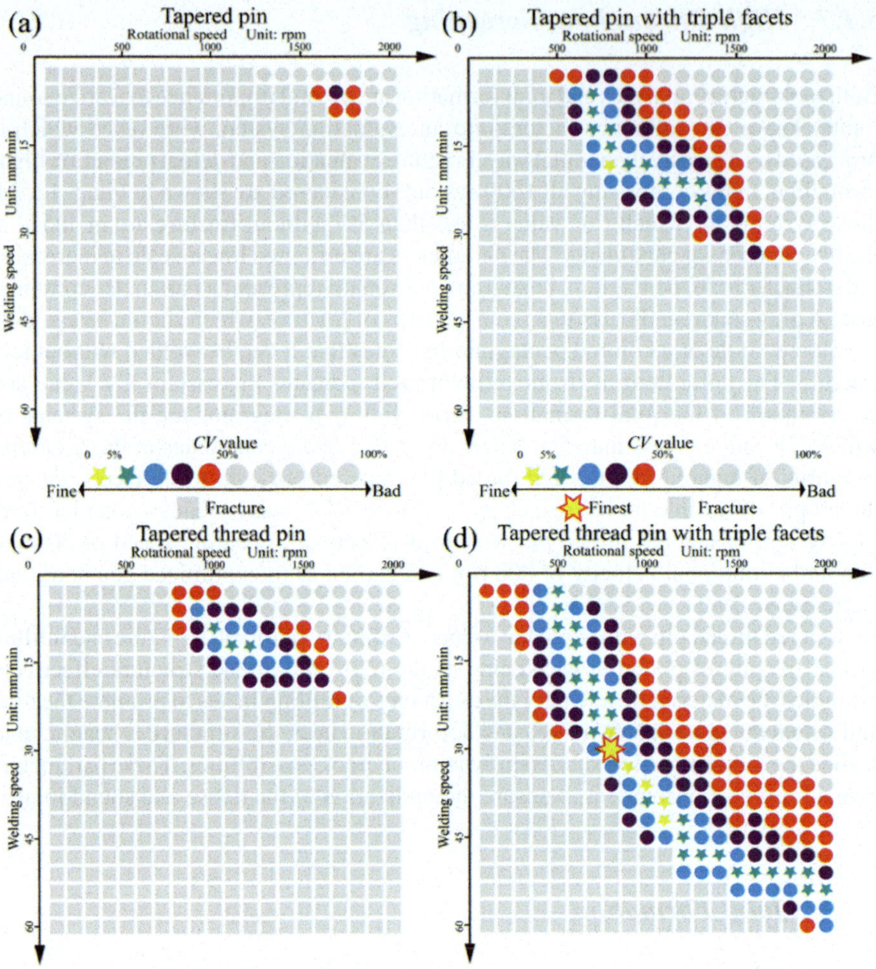

图中 rpm 改为 r/m

Fig. 5.12 The parameters are arranged linear scale in increasing order. Each circle, pentagram, hexagram and square represents a parameter combination. A grey square indicates that there are risks of fracture, and this combination will be ignored. A colored circle, a hexagram or a colored pentagram indicates that it can be welded without fracture. The color contrast from yellow to red indicates increasing *CV* value and deterioration of the joint formation. The hexagram indicates the optimized parameter combination: **a** tapered pin, **b** tapered pin with triple facets, **c** tapered thread pin, and **d** tapered thread pin with triple facets

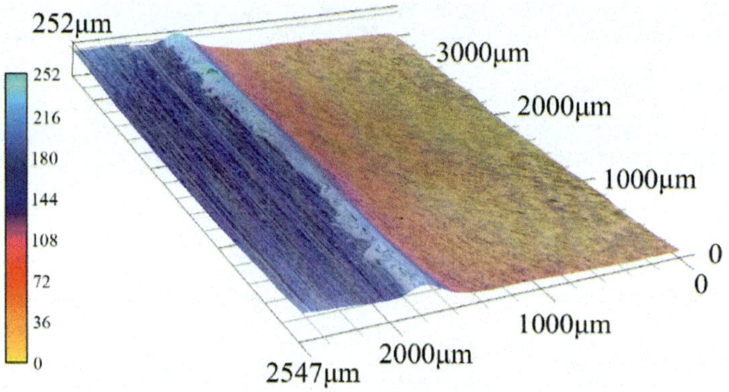

Fig. 5.13 Optical surface macrograph of high depth-to-width ratio FSW joint for 5 mm Al–Mg–Si alloy rolled sheet

Since the formation of onion rings is believed to be the results of maximal effective plastic strain rate by Tongne et al. [13], the multiple onion rings found in the WNZ are a direct evidence of thread-driven material transport phenomena during high depth-to-width ratio FSW. The flow velocity near-tool-surface can be defined as:

$$v_{horizon} = \eta \omega R_{ijk} \tag{5.18}$$

There is a great difference of flow velocity between the top and the bottom of the thread, which results in periodic drastic changes of strain rate tensor. The multiple onion rings pattern stacked vertically through the thickness emerge. For the Al–Mg–Si alloys, the zigzag line easily appears at the WNZ, which is detrimental to tensile property as described by Dai et al. [14]. The optimized tapered thread pin with triple facets can produce the multiple onion rings, which results in the elimination of zigzag line and mechanical interlocking, improving mechanical properties.

The temperature distribution of transverse section is shown in Fig. 5.15. The attenuation of the peak temperature occurs rapidly from welding center to sides, which generates less heat effect to the joints. In Fig. 5.14, the width of the HAZ is lower than conventional FSW notably, which is attributed to the lower heat input. Microhardness distribution of the high depth-to-width ratio FSW joint is shown in Fig. 5.16. The hardness gap of the WNZ between upper line and others implies that the shoulder affected zone is softened by more heat accumulation than the pin affected zone. The hardness of the TMAZ at the middle is approximately equivalent to that at the upper due to the similar accumulated strain and thermal cycling, as shown in Fig. 5.15. The width of HAZ ranges from 270 to 660 μm, while the average value is about 440 μm. HAZ is usually to be the weakness of FSW joint [15]. This technique can reduce the width of HAZ significantly, which effectively diminishes the disadvantages of wider HAZ. Compared with underwater FSW by Zhang et al.

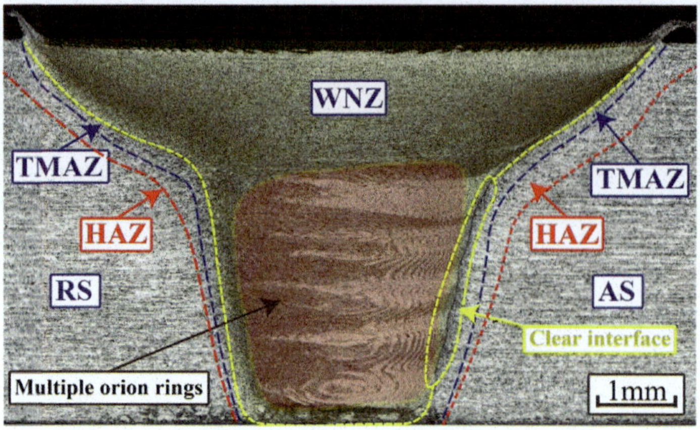

Fig. 5.14 Microstructure in transverse section

[16], the mechanism of narrowed HAZ induced by high depth-to-width ratio FSW is reducing heat input significantly while that of underwater FSW is increasing heat dissipation rate. High depth-to-width ratio FSW reduces the width of HAZ further, which means the potential promotion of mechanical performance.

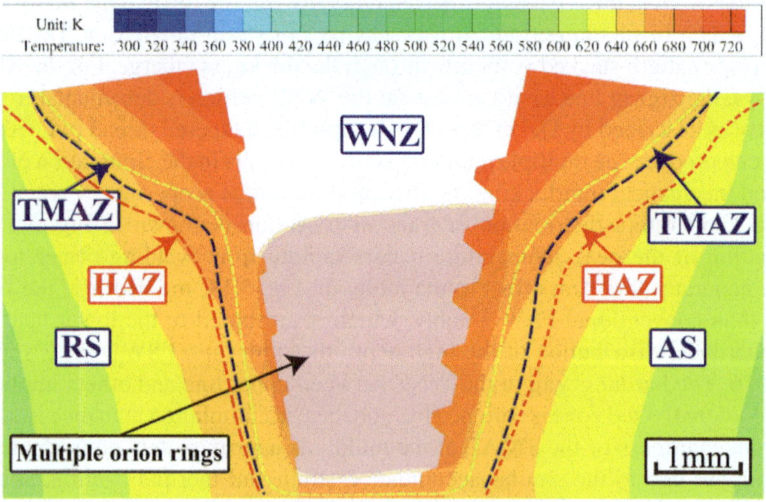

Fig. 5.15 Temperature distribution on transverse section of high depth-to-width ratio FSW joint

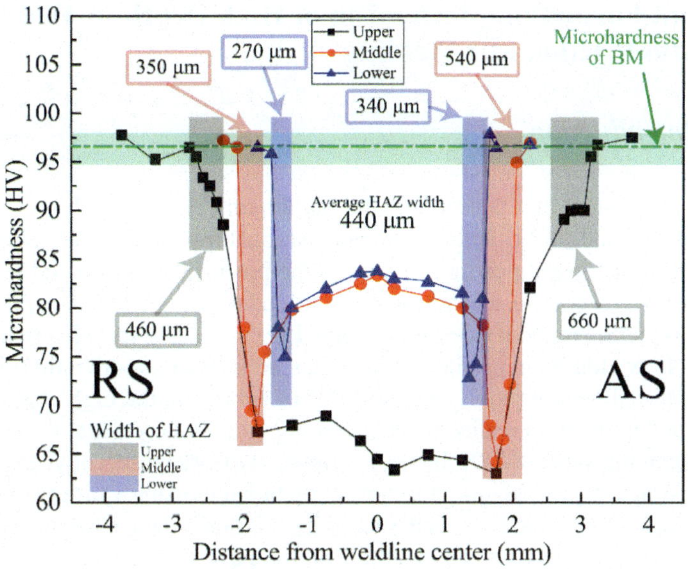

Fig. 5.16 Microhardness profiles of the high depth-to-width ratio FSW joint

5.1.9 Summary

(1) The numerical evaluation model was proved to be accurate and practical for high-throughput screening.
(2) The thread structure could promote the material flow but increase the fracture risk. The CV value revealed that pins with milling facets were more beneficial to achieve sound joint.
(3) Optimized geometry of tapered thread pin with triple facets at a welding speed of 30 mm/min and a rotational velocity of 800 r/m was obtained. Sound joint without obvious thickness reduction and flashed was attained based on the optimal geometrical structure.
(4) High depth-to-width ratio FSW has potential to reduce HAZ and improve joint quality. The average width of HAZ was 440 μm, which was attributed to the lower heat input generated by high depth-to-width structure.

5.2 Joint Formation Mechanism of High Depth-to-Width Ratio Friction Stir Welding

5.2.1 Introduction

Liu et al. [17] stated that the ratio of the pin length (PL) to the shoulder diameter (SD) should be modified to obtain sufficient material flow and heat input. Unrestricted increase in the PL with a certain SD led to the pin fracture or caused the welding defects. For over a decade, an empirical cognition has formed that the optimum ratio of the PL to the SD is about 0.33, as shown in Fig. 5.17a. For specific applications of FSW of hollow aluminum profiles with a large thickness-to-width ratio (Fig. 5.17b), the conventional welding tools are incapable of achieving the high-quality joining. An important strategy to realize this challenging goal is the design of the pin based on numerical analysis [18–20]. Indeed, significant efforts in recent years have been aimed at extracting process–structure–property linkages using a variety of numerical approaches [21–24]. Colegrove and Shercliff [25] established a 3-dimensional CFD model in FSW. The framework gained a preliminary understanding of the plasticized flow around a complex FSW tool. Hoyos et al. [26] presented a phenomenological based semi-physical model to produce an interpretable flow model near the tool surface. Liu et al. [27] evaluated the volume of the deformed materials around the pin, the material flow velocity, and the strain/stress rate based on numerical methods. Non-continuous low and high speed regions around a cylinder pin were revealed. Fagan et al. [28] proposed a friction stir forming model to analyze the complex physical mechanisms, which visualized the surface flow on an analytical circle tool using a particle streamline tracing approach. Tongne et al. [13] studied the material flow induced by the tool pin based on a 2-dimensional Coupled Eulerian–Lagrangian (CEL) formulation. Nevertheless, the comprehensive analysis about the 3-dimensional flow near the tool with the complex topological structure is seldom. A credible plasticized material flow model corresponding to the exploration of the process–structure–property linkages is necessary [29–32].

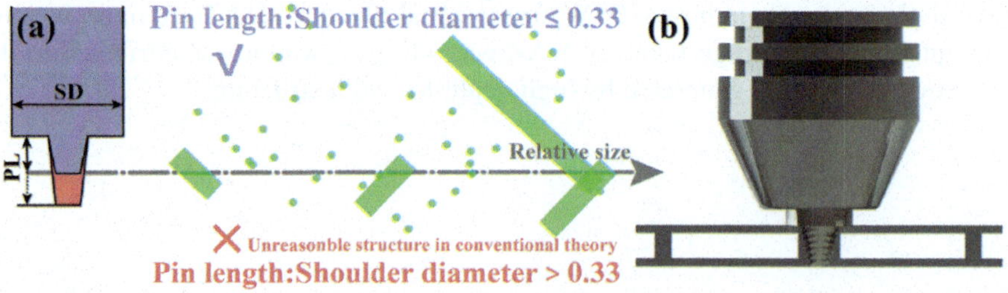

Fig. 5.17 Design of friction stir welding tools: **a** depth-to-width ratio data of welding tools (the green points and the green blocks were the depth-to-width ratio data in the published open literatures) and **b** hollow aluminum profiles with a large thickness-to-width ratio

In the present paper, high depth-to-width ratio FSW, a method suitable for the specific application (Fig. 5.17b), was considered. A fluid–solid interaction (FS) algorithm was proposed to establish the coupling model consisting of CFD and CSM. The plasticized material flow near a designed tool with a tapered thread triple-facet pin was studied. Joint formation mechanisms including flow feature, defect analysis, precipitation evolution and joint characterization were discussed in detail.

5.2.2 Experimental Procedures

The base material (BM) was 6082-T6 Al–Mg–Si precipitation hardenable alloy, whose sheet dimensions were 300 mm × 75 mm × 5 mm. The chemical composition and mechanical properties are listed in Table 5.3. Schematic of the high depth-to-width ratio FSW is shown in Fig. 5.18. The welding tool is made of H13 high speed steel and has the topologies of the thread and the three milling facets. It has the sizes of 8, 6 and 5.8 mm in the SD, the pin root diameter (PRD) and the PL, of which the depth-to-width ratio is 0.6. In contrast, the ratio of conventional welding tool is about 0.33, as shown in Fig. 5.17a. A rotational velocity of 800 r/m, a shoulder plunge depth of 0.15 mm and a tool tilt angle of 2.5°, were identical in the high depth-to-width ratio FSW.

Weld surface appearances were observed by a surface reconstruction method. Metallographic and mechanical specimens were cut perpendicular to the welding line using an electrical discharge cutting equipment. Metallographic specimens were etched by anodic coating with 43 ml H_2SO_4, 38 ml H_3PO_4 and 19 ml H_2O. Tensile tests were carried out at room temperature at a crosshead speed of 2.0 mm/min using a computer-controlled testing machine. Tensile properties were evaluated with three tensile specimens. Electron backscattered diffraction (EBSD) pattern and fractographs were analyzed by a scanning electron microscope (SEM). X-ray diffraction (XRD) using a Cu-Kα radiation was made from 30 to 80° at a scan rate of 2°/min, to investigate the precipitates on the fracture surface.

Table 5.3 Chemical composition and mechanical properties of Al–Mg–Si alloy sheets

Chemical composition (wt.%)									Mechanical properties	
Mg	Si	Cu	Fe	Mn	Zn	Cr	Ti	Al	Tensile strength (MPa)	Elongation (%)
1.10	0.81	0.23	0.75	0.15	0.31	0.07	0.19	Bal.	310	15.6

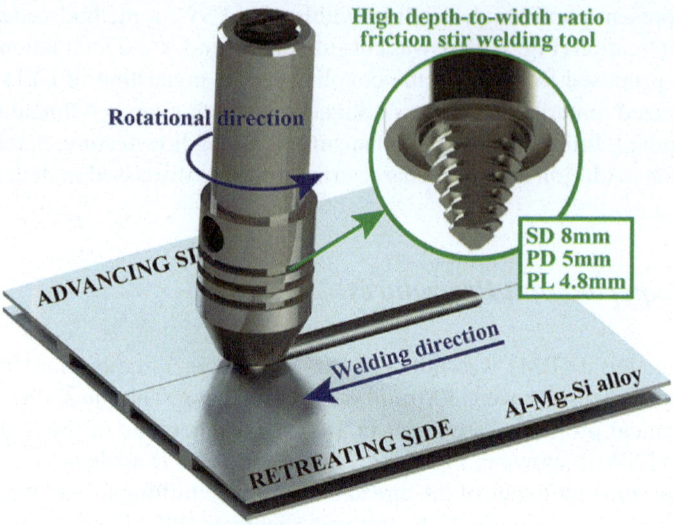

Fig. 5.18 Schematic of high depth-to-width ratio FSW

5.2.3 *Numerical Modeling*

Figure 5.19 shows the components of the high depth-to-width ratio FSW model. The model consists of three key parts: a CFD method of the fluid domain to simulate the heat generation and the plasticized material flow, a CSM method of the solid structure to describe the deformation and the heat dissipation of the welding tools, and a coupling FSI algorithm to integrate the front two parts and transfer the flux data. Full details of three parts and other essential points are described here.

A computational domain with the dimensions of 5 mm × 50 mm × 50 mm was used in the numerical model. The dynamic mesh technique was adopted to achieve the FSI coupling. Only the dwelling procedure during the FSW was calculated. The calculation time lasted for more than 15 s to achieve a quasi-steady state. The solver was set as a pressure-based mode. The Al alloy to be welded was assumed as an incompressible, continuous fluid with Navier–Stokes descriptions as:

$$\nabla \cdot v = 0 \tag{5.19}$$

$$\rho \frac{dv}{dt} = -\nabla P + \nabla \cdot \sigma \tag{5.20}$$

$$\rho c_p \frac{dT}{dt} = -\nabla(K\nabla T) + (\sigma : \nabla v) - \phi \tag{5.21}$$

where ρ is the density of the Al–Mg–Si alloy, t is the flow time, v is flow velocity, $\sigma = \mu\left(\nabla v + \nabla v^t\right)$ is the deviatoric stress tensor, μ is the non-Newtonian viscosity,

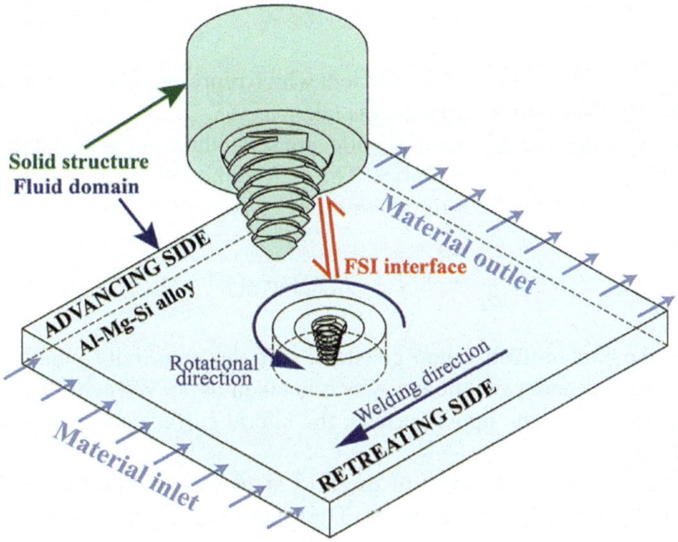

Fig. 5.19 Schematic view of high depth-to-width ratio FSW numerical model including solid structure and fluid domain

P is the relative pressure, T is the absolute temperature, ϕ is the sink term which represents the heat loss of convection and radiation, c_p is the specific heat capacity and k is the thermal conductivity.

The thermal conductivity and the specific heat capacity of the alloy were given in Sect. 5.1.3. The yield stress σ_e of the Al alloy equals to the flow stress approximately [33]. It is highly temperature and strain rate dependent according to the constitutive equation introduced by the Sellars–Tegart model:

$$\sigma_e = \left(1 - \sqrt{\frac{T - 273.15}{T_{melt} - 273.15}}\right)\frac{1}{\alpha}\sinh^{-1}\left[\left(\frac{Z}{A}\right)^{\frac{1}{n}}\right] + \sigma_{melt} \tag{5.22}$$

where α, A, and n are the material constants of 1.7×10^{-8} Pa^{-1}, 8.081×10^{11} and 6.66, respectively, T_{melt} is the melt temperature of the Al–Mg–Si alloy with 908 K, σ_{melt} is the yield stress with 15.4 MPa above the melt temperature. Z is the Zener–Hollomon parameter, which is defined as:

$$Z = \dot{\varepsilon_e}\, e^{\frac{Q}{RT}} \tag{5.23}$$

where Q is the activation energy with 181.53 kJ/mol, R is the gas constant with 8.314 J/(mol K), and $\dot{\varepsilon_e}$ is the effective strain rate.

Two parts of the heat input were taken into consideration. The surficial heat source term F represents the friction heat at the fluid–solid coupling interface:

$$\phi = \beta_2 \cdot \sigma_e \cdot \dot{\varepsilon}_e \tag{5.24}$$

where β_1 and β_2 are the empirical coefficient which represents the thermal conversion rate, μ is the friction coefficient which is taken as 0.5.

The material inlet and the material outlet are set at the front side and the back side of the model as the velocity boundaries. The thermal boundary condition for the heat exchange between the surfaces of the workpieces is given by:

$$-k\frac{\partial T}{\partial z} = h(T - T_\infty) + \sigma\varepsilon\left(T^4 - T_\infty^4\right) \tag{5.25}$$

where σ is the Stefan–Boltzmann constant, ε is the emissivity which is taken as 0.4, h is the heat transfer coefficient which is taken as 15 W/(m^2 K), 15 W/(m^2 K) and 200 W/(m^2 K) on the upper surface, the lateral surface and the lower surface, respectively.

Johnson–Cook material model of the H13 steel was chosen as the constitutive equation for the elastic–plastic analysis. The equation can be written as:

$$\sigma = \left(A + B\varepsilon^n\right)\left[1 + C\left(\ln\frac{\dot{\varepsilon}}{\dot{\varepsilon}_0}\right)\right]\left[1 - \left(\frac{T - T_0}{T_{meltSteel} - T_0}\right)^m\right] \tag{5.26}$$

where A, B, C, m, and n are material constants, T_0 is the reference temperature and $T_{meltSteel}$ is the melting temperature of the H13 steel. The values of the Johnson–Cook model and the other physical properties are shown in Table 5.4.

The coupling algorithm defines the interaction between the tool surface and the plasticized materials, which can be written as:

$$\begin{bmatrix} K_{11} & K_{12} & 0 & K_{14} \\ K_{21} & K_{22} & 0 & 0 \\ 0 & 0 & K_{33} & K_{34} \\ K_{41} & 0 & K_{43} & 0 \end{bmatrix} \begin{Bmatrix} \tilde{U} \\ \tilde{p} \\ \tilde{d} \\ \tilde{\lambda} \end{Bmatrix}^{n+1} = \begin{Bmatrix} f_1 \\ f_2 \\ f_3 \\ f_4 \end{Bmatrix} \tag{5.27}$$

where f_1, f_2, f_3, and f_4 terms are the forcing terms, and the other nonzero entries of the equation can be written as:

Table 5.4 Constitutive properties of H13 high speed steel

Density (kg/m^3)	Young's modulus (GPa)	Poisson's ratio	Specific heat (J/(kg K))	Thermal expansion coefficient (10^{-6} K^{-1})	Thermal conductivity (W/(m K))
7800	211	0.28	560	37	9.1
A (MPa)	B (MPa)	C	m	n	T_0 (K)
908.5	321.4	0.28	1.18	0.278	300

$$K_{11} = M_u + \frac{1}{\rho} K_\tau + C_u - \Delta t K_u \tag{5.28}$$

$$K_{12} = G^T + \frac{\Delta t}{2} P \tag{5.29}$$

$$K_{21} = G \tag{5.30}$$

$$K_{22} = \Delta t \left[G M_u^{-1} \left(G^T + \frac{\Delta t}{2} P \right) - H \right] \tag{5.31}$$

$$K_{33} = c_3 M_d + c_2 C_d + c_1 K_d \tag{5.32}$$

$$K_{14} = M_{\lambda u}^{\Gamma,T} \tag{5.33}$$

$$K_{34} = M_{\lambda d}^{\Gamma,T} \tag{5.34}$$

$$K_{41} = M_{\lambda u}^{\Gamma} \tag{5.35}$$

$$K_{43} = M_{\lambda d}^{\Gamma} \tag{5.36}$$

In addition, the thermodynamic properties of the Al–Mg–Si alloy used in this model were calculated using the JMatPro, a CALPHAD based thermodynamic software.

5.2.4 Joint Formation

Surface appearances of the high depth-to-width ratio FSW joints at different welding speeds are shown in Fig. 5.20. The shoulder marks with concave troughs and convex crests are clearly visible. The distance of the two neighboring troughs or two neighboring crests is proportional to the advance per rotation at certain welding parameters. The thickness reductions decrease with the increase of welding speed, which are below 100 μm at the welding speeds higher than 300 mm/min. Specifically, the joints in Fig. 5.20 exhibit the regularly fluctuant surfaces with tiny flashes, indicating the adequate plasticized materials flow under the driving effect of high depth-to-width ratio tool.

The typical macrostructure is shown in Fig. 5.21. A sound joint has been realized under a low heat input generated by a tiny SD. Apart from BM, WNZ and TMAZ can be identified in Fig. 5.21a. No HAZ is observed, as the EBSD pattern depicted in Fig. 5.21g. The shape of the WNZ differs from that of the conventional FSW [34] and appears to be lanky but non-axisymmetric between the advancing side and the

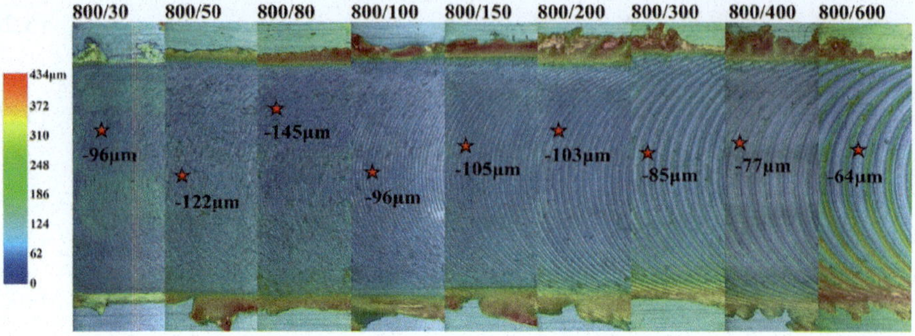

Fig. 5.20 Characteristics of weld surface appearances by high depth-to-width ratio FSW (the pentagons were the trough thickness reduction points at different welding speeds)

retreating side. The interface between the TMAZ and the WNZ at the advancing side is clear, but the interface at the retreating side is obscure as shown in Fig. 5.21b, f. The material domain at the leading side is driven by the shear force to refill the cavity behind the tool [9]. The converged materials exhibit a periodic fluctuant structure due to the deep thread feature. There is a great difference in the flow velocities between the crest and the root of the thread, which results in the drastic changes of the strain rate along the interface. A lamellar structure inside the WNZ is observed as shown in Fig. 5.21c. According to Tongne et al. [35], the lamellar band is the result of the periodical deposited layers. The space between the two bands is equal to one third of the advance per tool rotation, which accords with the three milling facets. The milling facets increase the extent of the turbulent flow, promoting the joint formation. Figure 5.21d, e show the fluctuant interface between the WNZ and the TMAZ at the advancing side. The materials of the WNZ was converged to the interface according to these red arrows, while the materials of the TMAZ was converged to the intervals of the interface. Based on the tool topology in Fig. 5.18, the downward inclined plane of the thread transforms the materials to move downward, while the upward inclined plane pushes the materials to move upward. Subsequently, the downward and upward materials are accumulated at the converged domains, of which the number is equal to the rounds of the thread.

5.2.5 *Numerical Evaluation*

Numerical modeling is applied to simulate the flow field near the tool surface for the precise description of the flow pattern. Octree contours of the equivalent strain rate, the velocity and their comparison are displayed in Fig. 5.22. The distributions of the strain rate and the velocity show the similar feature as the metallographic images (Fig. 5.21). The differences of the velocity at the thread crest and root are obvious with

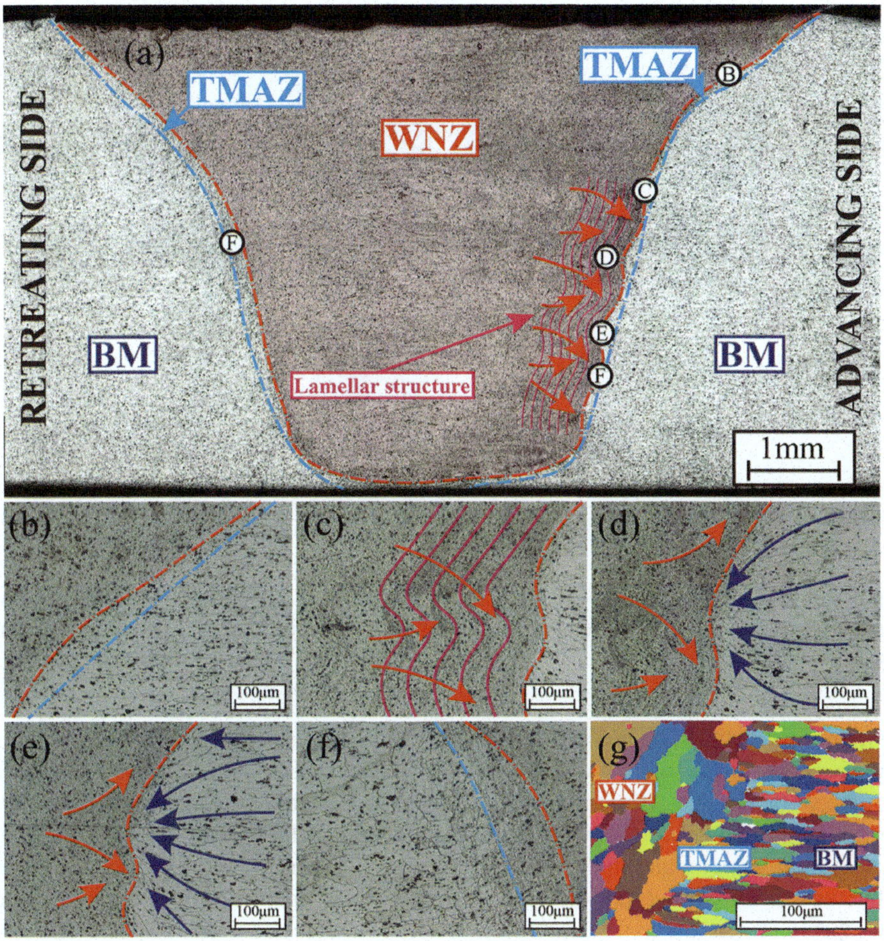

Fig. 5.21 Macrostructure and microstructures of the typical joint: **a** joint macrostructure; microstructures of **b** "B", **c** "C", **d** "D", **e** "E" and **f** "F" in **a**; **g** EBSD pattern at advancing side

a fluctuant amplitude of about 2.5. There are opposite pushing directions between the downward and the upward inclined planes of the thread.

A quantitative curve analysis is carried out by the simulation data from the sampling points, as shown in Fig. 5.23a. The 200 sampling points distribute uniformly along the red edge of the pin at the advancing side, of which the sampling direction is from the bottom to the top. Undulation curves of the equivalent strain rate at different welding speeds can be attained in Fig. 5.23b. The strain rate of the thread crest reaches 300 s^{-1}, while that of the thread root is only 70 s^{-1}. The drastic change of the strain rate promotes the plasticized turbulence near the pin surface. Additionally, the double-apex structure on single thread crest is an another feature on the

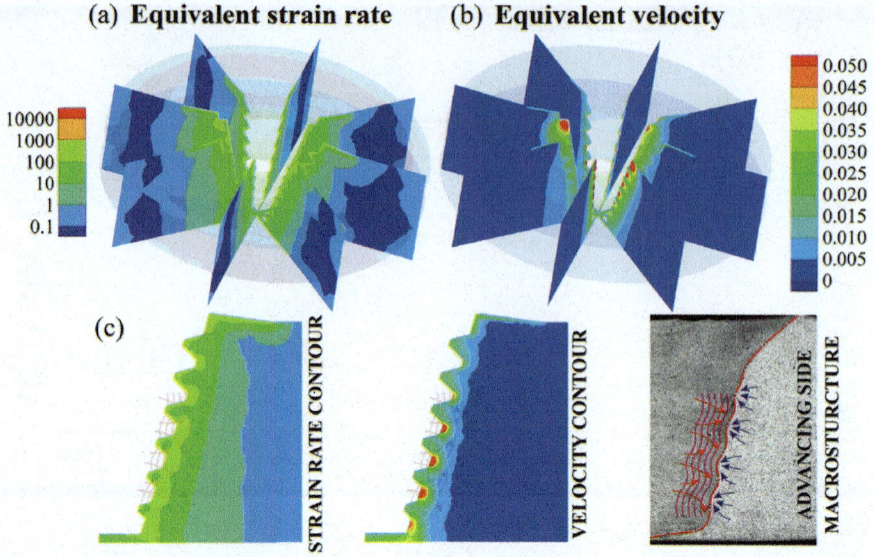

Fig. 5.22 Distributions of equivalent strain rate, velocity and their comparison: **a** equivalent strain rate contour, **b** equivalent velocity contour, and **c** comparison between simulation and experiment at advancing side

strain rate curves. The strain rate gradients at the upward and downward inclined planes of the thread are extremely high, which lead to the relatively lower strain rate at the vertical plane of the thread. The average strain rate of the high depth-to-width ratio FSW is much higher than that of conventional FSW [36]. The plots of the equivalent velocity and the peak temperature are shown in Fig. 5.23c, d. The same fluctuant feature induced by the thread structure are obtained. The velocity and the temperature near the tool surface decrease with the raise of welding speed. The low heat generation per length and the less nominal flow velocity at the high welding speed have almost no influence on the severe material flow. An effective promotion of the plasticized materials is obtained.

Figure 5.24a exhibits the plasticized material flow calculated by the numerical model. In general, the material flow moves downward near the pin surface, while the other flow from the bottom of the joint moves upward near the sub pin surface and refills the cavity induced by the downward stream. In the local area near the thread, the plasticized flow is greatly enhanced by the converging effect of the thread and the dynamic flow by the milling facets. According to Fig. 5.23d, the heat input becomes lower obviously compared to the conventional method [1]. The great plastic strain energy induced by the drastic plasticized turbulence compensates the extremely low heat input, which contributes to the sound joint formation during the relatively low-thermal process.

The velocity distributions on three Cartesian axes as defined in Fig. 5.24a are shown in Fig. 5.24b. The material flow mainly exhibits the characteristics of the

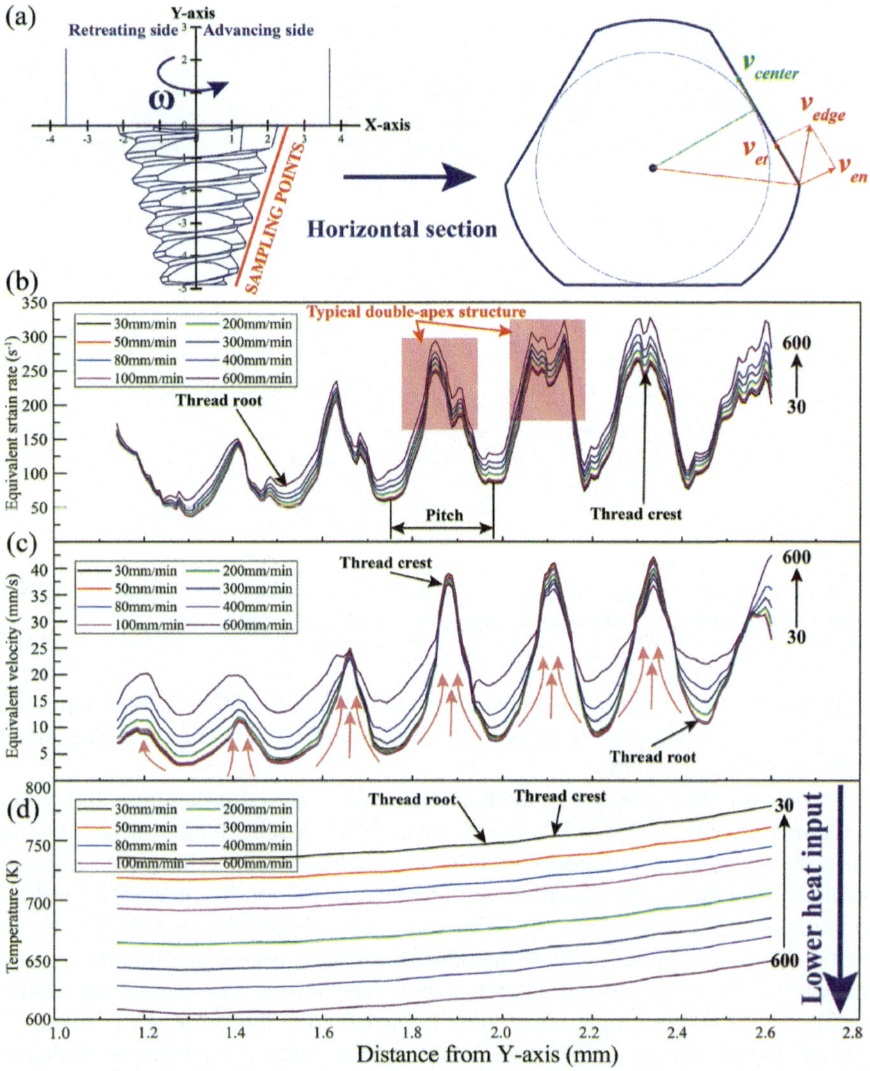

Fig. 5.23 Flow characteristics: **a** distributions of the 200 sampling points at the red line, **b** curves of equivalent strain rate, **c** curves of equivalent velocity, and **d** curves of peak temperature at different welding speeds

downward movement along the thickness direction, and moves backward along the direction of the weld synchronously. However, there is also forward stream existing along the welding direction due to the thread pin, increasing the complexity of the material flow during the high depth-to-width FSW. In the lateral direction of the weld, the plasticized materials are mainly characterized by the flow from the advancing side to the retreating side at the leading side of the tool, while the materials flow from the

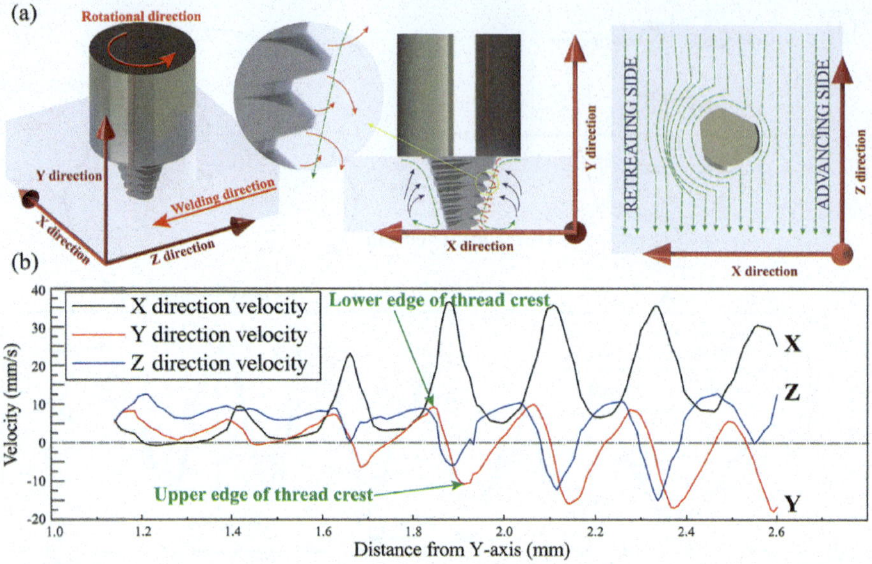

Fig. 5.24 Flow characteristics on three axes: **a** illustration of plasticized material flow via high depth-to-width ratio FSW and **b** velocity distributions on three Cartesian axes

retreating side to the advancing side at the treating side to fulfill the joining process. In conclusion, the flows on the three axes are jointly affected by the fluctuating effects of the deep-thread feature.

Particle tracing technique is applied to trace the material streamlines and predict the formation of the welds to validate the reliability. The injections are set as massless particles, which are superimposed on the inlet surface with a radius of 0.5 mm. The boundary conditions for injections are identical to the matrix metal. The welding formation predictions and their experimental contrasts are depicted in Fig. 5.25. The grey clusters of particles indicate the dense particle distribution, while the exposed red region means the occurrence of the defects. The defects appear when the welding speed reaches 400 mm/min, which mainly distribute in the WNZ at the advancing side and the root of the joints. From the comparison data, it can be inferred that the numerical modeling can simulate and predict the defects effectively, which proves the accuracy of the flow analyses powerfully [37].

5.2.6 *Fractrography*

Figure 5.26 depicts the tensile fracture morphologies of the joints at different welding speeds. The voids with an average diameter of 19 μm distribute at the root of the WNZ in 300 mm/min. Many voids appear and penetrate each other at the welding speeds of 400 and 600 mm/min. However, no welding defect can be observed from

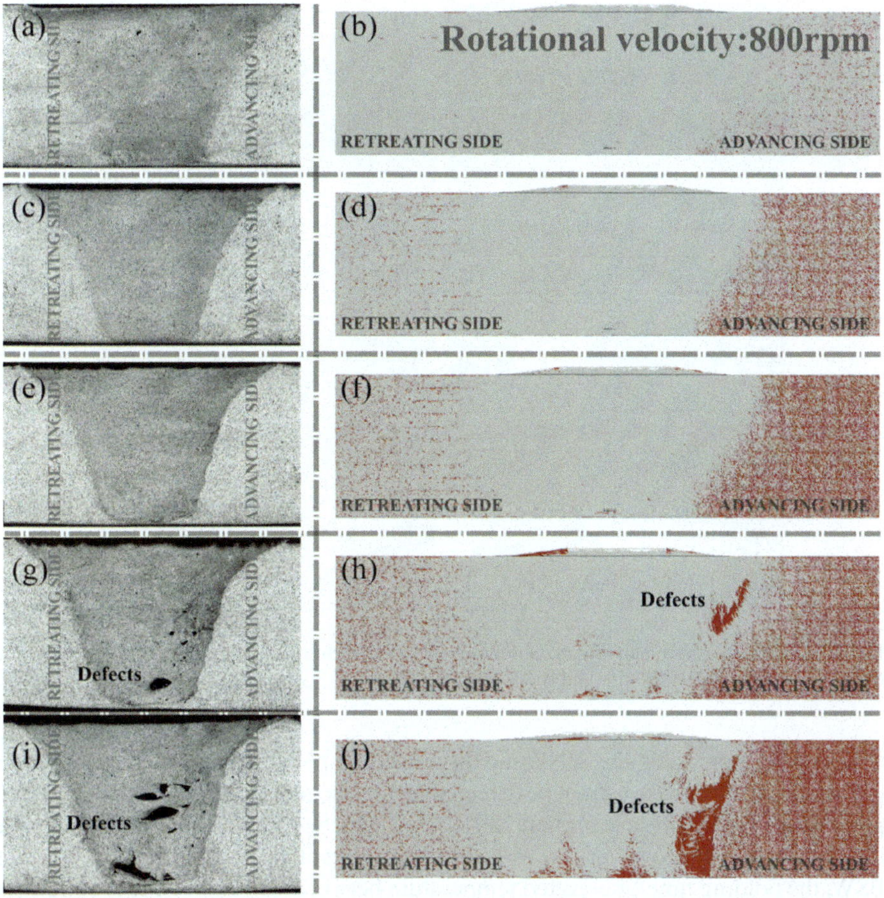

Fig. 5.25 Comparison of defect distributions between experiments and simulations: **a**, **b** 30 mm/min; **c**, **d** 100 mm/min; **e**, **f** 300 mm/min; **g**, **h** 400 mm/min; **i**, **j** 600 mm/min

the metallographic image in Fig. 5.25e, which implies that the nucleation of the voids occurs during the tensile process rather than the welding process. The heat generation per length increases with the decrease of welding speed, resulting in the increase of the peak temperature near the tool surface [31]. According to Fig. 5.23b, the equivalent strain rate becomes bigger when getting closer to the shoulder along the thickness direction. Therefore, the diameters of the voids decrease gradually and diminish eventually with the increases of the temperature and the strain rate. Additionally, the distance between each row of the voids is approximately equal to one third of the advance per rotation, as depicted in Fig. 5.26b, c.

Precipitates can be observed in the dimples of the fracture morphologies in Fig. 5.26a. From the XRD results of Fig. 5.27, the phase is concluded to be Mg_2Si and $Al_{12}(Fe,Mn)_3Si$. The needle-like β″ phase ($Al_{x+y}Mg_{5-x}Si_{6-y}$) is unstable against the

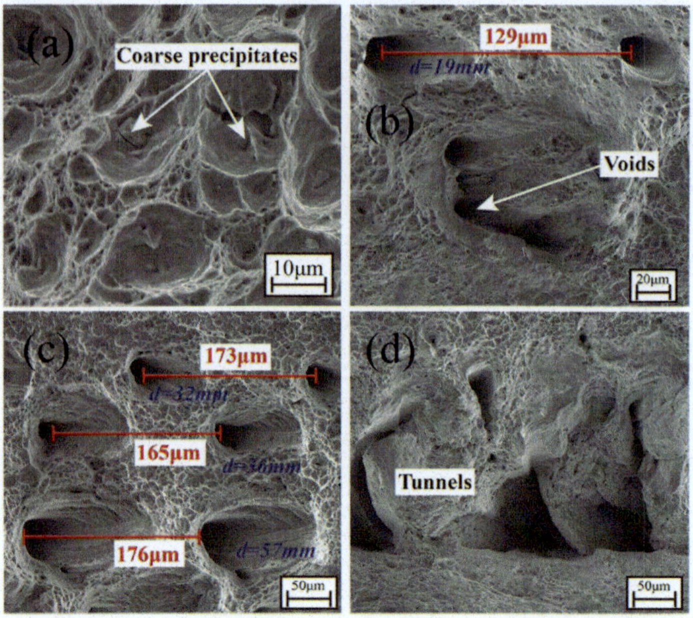

Fig. 5.26 Partly fracture surface morphologies of high depth-to-width ratio FSW: **a** 30 mm/min, **b** 300 mm/min, **c** 400 mm/min, and **d** 600 mm/min

temperature increase in the WNZ and dissolves completely during the FSW process [38]. Thus, the β″ phase cannot be detected in the fracture surface of the WNZ. The β phase (Mg_2Si) dissolves at around 720–770 K, which is significantly affected by the thermal cycle [39]. However, due to the low heat input of the high depth-to-width ratio FSW, the holding time of elevated temperature between 720 and 770 K is extremely short and even non-existing. A large amount of the β phases remain undissolved. Subsequently, the β phase is coarsened in accordance with the Wagner–Kampmann (KWN) model [40], which results in coarsened β phase existing in the WNZ and becomes the nuclei of the initial cracks. The rate law of KWN model is defined as:

$$v = \frac{\partial r}{\partial t} = \frac{C_{ss} - C_{int}}{C_p - C_{int}} \frac{D}{r} \tag{5.37}$$

$$D = D_0 \exp\left(-\frac{Q_d}{R_g T}\right) \tag{5.38}$$

where v is the growth rate of the β phase, C_{ss} is the concentration in the precipitation controlling element Mg within the matrix, C_{int} is the concentration of the Mg element at the matrix/precipitate interface given by the Gibbs–Thomson equation, C_p is the concentration in the precipitate of the Mg element, D is the diffusion coefficient, D_0 is a constant, Q_d is the activation energy, R_g is the gas constant and T is the temperature.

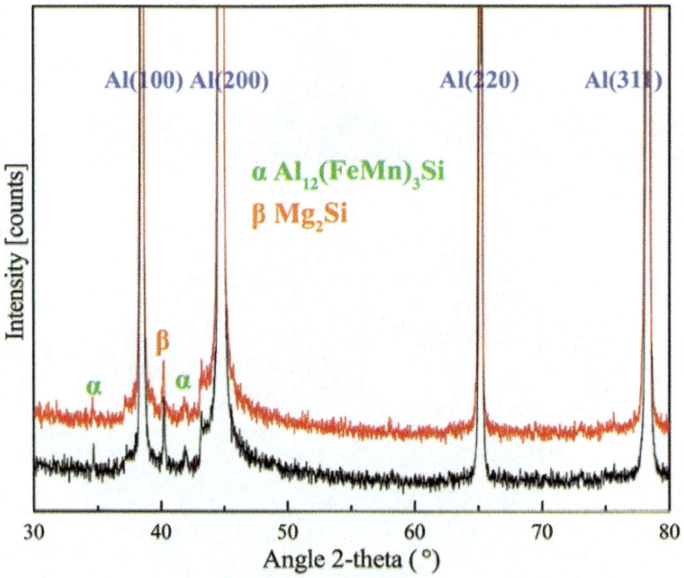

Fig. 5.27 XRD pattern showing precipitations on fracture surfaces

According to the formulae, the peak temperature and the holding time of elevated temperature drop with the increasing welding speeds, resulting in the decrease of the diffusion coefficient. The growth rate of the β phase reduces then, leading to the relatively small radii of the β phase. In Fig. 5.26, these hard precipitates become the nuclei of the cracks during the tensile process. The material necking occurs and the cracks form with the increasing plastic deformation [41]. The micro voids grow uniformly and eventually form the dimples, which result in the crack of the joint. Thus, the mechanical properties of the joints are improved effectively by hindering the coarsening of the precipitates, under the condition of extremely low heat input.

5.2.7 Mechanical Properties

Increasing welding speed can improve the mechanical performances of the joint through reducing heat input, while it is also responsible for the welding defects. It is of vital importance to find the best welding parameters for high depth-to-width ratio FSW. A comprehensive evaluation based on the response surface method is applied to optimize the process window. As shown in Fig. 5.28, when the peak temperature is 648 K and the strain rate is 151 s^{-1}, the void defects are eliminated. Compared to Fig. 5.23b, d, the welding parameters of 800 r/m and 300 mm/min are the optimal parameters, which correspond to the best mechanical performances.

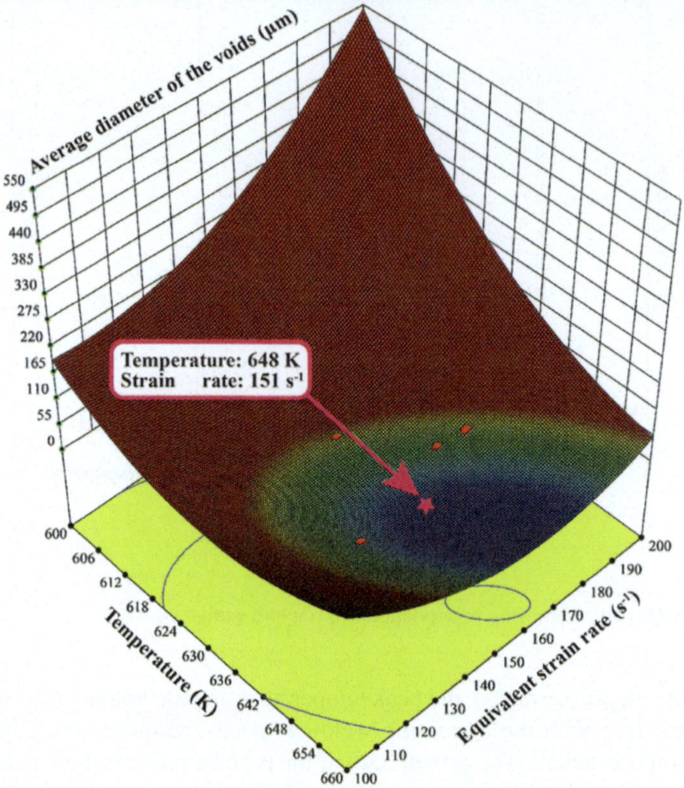

Fig. 5.28 Response surface of the average void diameter based on the temperature and the equivalent strain rate

Figure 5.29 shows the tensile properties of the joints at different welding speeds. The tensile strength gradually increases firstly and then decreases as the welding speed increases. The elongation decreases with the increase of welding speeds. The tensile properties change drastically due to the strengthening effect of the precipitates [42]. The coarsening degree of the precipitates decreases with the increase of welding speed. Meanwhile, the degree of the dynamic grain refinement is promoted due to the low heat generation. However, the tensile properties decrease drastically due to the welding defects with the welding speeds exceeding 300 mm/min. The maximum tensile strength of 265 MPa is achieved at the welding speed of 300 mm/min and rotational velocity of 800 r/m, which is obviously higher than that of the conventional FSW.

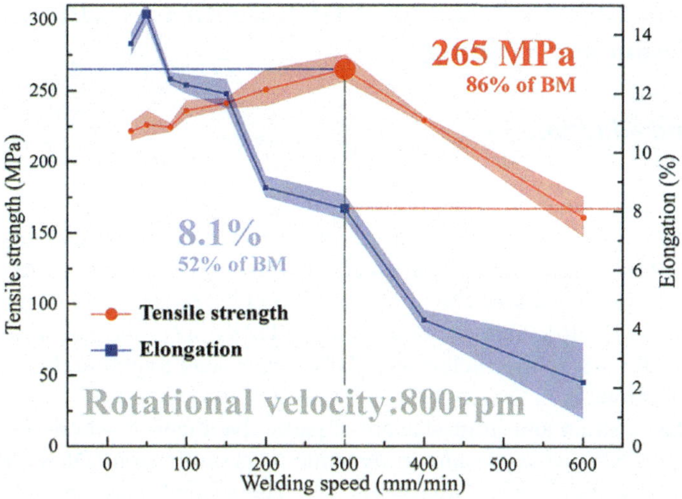

Fig. 5.29 Tensile properties of the high depth-to-width ratio FSW joints at different welding speeds

5.2.8 Summary

(1) The material flow model consisting of the CFD method, the CSM method and the FSI algorithm was proposed to investigate the joint formation mechanism. The numerical model was proved to be accurate and practical for high depth-to-width ratio FSW.

(2) The heat input decreased effectively and the plasticized material flow was enhanced compared to the conventional FSW. The tiny shoulder diameter reduced the heat generation. The thread structure and the milling facets increased the strain rate greatly under the extremely low heat generation, avoiding the welding defects.

(3) The process–structure–property linkage was achieved via the comprehensive evaluation. The peak temperature and the strain rate of 648 K and 151 s^{-1} eliminated the void defects and achieved the lowest coarsening degree of precipitate.

(4) The optimum tensile strength and elongation of the joint at the welding speed of 300 mm/min and rotational velocity of 800 r/m were 265 MPa and 8.1%, equivalent to 86 and 52% of the BM.

5.3 Grain Growth Behavior of High Depth-to-Width Ratio Friction Stir Welding

5.3.1 Introduction

Severe plastic deformation (SPD) is a state-of-the-art process to achieve strong grain refinement by imposing severe shear deformation and extrusion pressure on materials [11]. Ultrafine-grained (UFG) materials with an average grain size lower than 500 nm can be produced based on SPD, like FSW [17, 43–45], equal-channel angular pressing (ECAP) [46], high-pressure torsion (HPT) [47], and single-roll angular rolling (SRAR) [48]. Nevertheless, can SPD always lead to grain refinement in any process conditions?

High plastic strain and strain rate are of particular attention among various characteristics of FSW, which results in dynamic recrystallization (DRX) and heterogeneous precipitation synchronously [8, 49]. The second-phase particles dissolve during the heating stage, and then the dissolving induced supersaturated solid solution (SSS) precipitates heterogeneously at grain boundaries [50] and dislocations [51] caused by the high plastic strain. These regions usually have a higher free energy such that the formation of second-phase particle there minimizes the increase in surface energy [52]. At the same time, the DRX takes place during straining as long as the temperature is above 0.5 melting point [53]. The new grains appear at the nucleation strain, grow up under the driving force of strain energy, and then replace the initial microstructure completely [54]. During the recrystallization process, the heterogeneously distributed precipitates retard or even suppress recrystallization, which limited the recrystallized grain size [55]. The microstructure of the joint is rebuilt under the combined effects of the DRX and the heterogeneous precipitation. Gu and Tóth [56] reported that grain refinement took place due to high strain during the SPD. Sitdikov et al. [57] suggested that the kinetics of the grain refinement were accelerated but the degree of the grain refinement decreased with increasing deformation temperature. It appears that increasing plastic strain and decreasing processing temperature are capable of refining the microstructure further [58].

High depth-to-width ratio (up to 0.6) FSW was proposed by us for the specific application like hollow aluminum profiles with a large thickness-to-width ratio. The tool design [59] and the joint formation mechanism were investigated. The results showed that the tiny shoulder generated much lower heat input, while the pin with complex surface morphology brought about enhanced material flow, which resulted in greater plastic strain and strain rate [60]. According to the conventional theory, UFG microstructure should be found in the WNZ. However, we obtained an atypical grain coarsening phenomenon in the high depth-to-width ratio FSW without post-processing heat treatment, which was inconsistent with our previous cognition. The objective is to explore (a) the evolution of the heterogeneous precipitation via a size class approach combined with CFD method, and (b) the process of the DRX associated with different second-phase particle sizes and volume fractions. From the

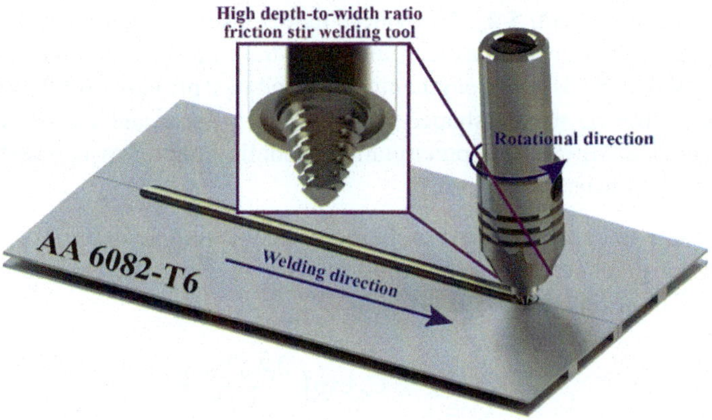

Fig. 5.30 Illustration of the high depth-to-width ratio FSW

aspects of numerical analyses and experiments, the atypical grain phenomenon was investigated in detail.

5.3.2 Experimental Procedures

The as-received material was 5-mm-thick 6082-T6 Al–Mg–Si alloy. The sheets were friction stir welded at a rotational velocity of 800 r/m and welding speeds of 30 and 300 mm/min. A H13 steel welding tool with the features of thread and three milling facets was used, as shown in Fig. 5.30. The sizes of the tool are 8 mm, 6 mm and 5.8 mm in the shoulder diameter, the pin root diameter and the pin length, respectively.

Microstructural characterization was carried out by OM (Keyence VHX-1000E), EBSD (FEI Quanta 200FEG), transmission electron microscope (TEM, FEI Talos F200X), and scanning transmission electron microscope (STEM, FEI Talos F200X). Metallographic specimens were etched by anodic coating with 43 ml H_2SO_4, 38 ml H_3PO_4 and 19 ml H_2O. The coating voltage and current density were 25 V and 0.7 A/cm^2. Specimens for EBSD and TEM were prepared by argon ion polishing (GATAN 685) and ion milling (GATAN 695), respectively. EBSD was operated at 20 kV. The TEM and the STEM were operated at 200 kV. All the distributions of different phases were counted by computational pixel filtering.

5.3.3 Numerical Modeling

The numerical model consists of two components: a macro CFD model to calculate the thermal and flow field, and a micro structure evolution model to describe the heterogeneous precipitation and the recrystallization. The full details of the CFD model were given in Sect. 5.2.2. The micro model takes the outputs from the macro CFD model (thermal cycle, plastic strain, and strain rate) as its inputs. The model tracks two major kinds of strengthening precipitates present in Al–Mg–Si alloy. These phases are the needle-shaped β''-$Al_{x+y}Mg_{5-x}Si_{6-y}$ and the heterogeneous, platelet-shaped β-Mg_2Si. The precipitation modeling approach is based on KWN method [40]. It consists of four parts:

(1) A rate law to calculate the dissolution or coarsening velocity v of the precipitates, as given by the Whelan formalism in Eqs. (5.37) and (5.38). The Gibbs–Thomson equation is given by:

$$C_{int} = C_e \exp\left(\frac{2\gamma_{int} V_m}{r_i R_g T}\right) \tag{5.39}$$

where C_e is the equilibrium solute content at the interface as given by Du et al. [42], γ_{int} is the modified interface energy, and V_m is the molar volume of the precipitate.
According to Eqs. (5.37)–(5.39), the critical rate radius controlling the dissolution or the coarsening of the precipitate can be given when $C_{ss} = C_{int}$, as:

$$r^* = 2\gamma_{int} V_m \left[R_g T \ln\left(\frac{C_{ss}}{C_{e0}}\right) + Q_e\right]^{-1} \tag{5.40}$$

When the actual precipitate radius is larger than the critical rate value, the precipitate continues to coarsening, and vice versa.

(2) A nucleation law to predict the number of stable precipitate nuclei that form at each time step, as:

$$J_{nucl} = \frac{J_0}{\sqrt{T}} \exp\left\{-\frac{A_0}{(R_g T)^3 [\ln(C_{ss}/C_e)]^2} - \frac{Q_d}{R_g T}\right\} \tag{5.41}$$

$$J_0 = \frac{2\rho V_a D_0 C_{at} \sqrt{\gamma_{int}}}{a^4 b \sqrt{k}} \tag{5.42}$$

$$A_0 = \frac{16\pi \varphi_i \gamma_{int}^3 N_A V_m^2}{3C_p^2} \tag{5.43}$$

where ρ is the dislocation density which provides the position for nucleation, V_a is the volume of Al atom, k is the Boltzmann constant, C_{at} is the atom fraction of Mg in matrix, a is the lattice constant of Al matrix, b is the burgers vector, φ_i is the wetting constant at dislocation, and N_A is the Avogadro's number.
The critical nucleation radius of nucleation can be defined as:

$$r_c = -\frac{2\gamma_{int}}{\Delta G_v + \Delta G_e} \tag{5.44}$$

where ΔG_v is the volume free energy change and ΔG_e is the strain energy change associated to the precipitate nucleation.

(3) A continuity equation to record the amount of solute being involved in the precipitates, as:

$$C_{ss} = \frac{C_0 - C_p \sum f_v^i}{1 - \sum f_v^i} \tag{5.45}$$

$$f_v^i = \sum \frac{4}{3}\pi r_i^3 N_i \Delta r \tag{5.46}$$

where C_0 is the solute content of element in alloy, f_v is the volume fraction of the precipitates, N_i is the density and Δr is the discrete step.

Thus, the final radius of the precipitates can be expressed as:

$$r = \frac{1}{f_v} \sum f_v \frac{\sum r_i N_i}{\sum N_i} \tag{5.47}$$

(4) A geometric correction equation [61]. All the model is based on the assumption that the second-phase has the spherical shape, so the needle-like β'' need a geometric correction given as:

$$\frac{4}{3}\pi r^3 = \pi \left(\frac{l}{16}\right)^2 l \tag{5.48}$$

where l is the length of the needle-like β''.

Table 5.5 lists the typical values of the parameters used in the modified micro structure evolution model.

5.3.4 *Microstructural Factors*

Figure 5.31 shows the joint macrostructures at the two welding speeds of 300 and 30 mm/min. The joint at 30 mm/min was divided into WNZ, TMAZ, HAZ and BM, as shown in Fig. 5.32e. However, the TMAZ and the HAZ could not be seen at 300 mm/min (Fig. 5.32c), suggesting that the heat input was extremely low. This is the first report on the FSW joint without the HAZ and TMAZ so far.

The microstructures of the WNZ at different welding speeds are shown in Fig. 5.32b, d. Notable grain coarsening can be obtained at 300 mm/min. The average grain size and the aspect ratio were determined to be respectively 266.3 μm and 2.85. The aspect ratio was greater than the value 1.47 at 30 mm/min. As shown in Fig. 5.33, the average grain size at 300 mm/min was approximately twice larger than that in BM (Fig. 5.32a), while that at the lower welding speed was just about one twentieth of the BM. However, the fine equiaxed grain should have been obtained via

Table 5.5 Summary of parameter values used in the model

Nomenclature	Symbol	Value (unit)
Concentration of β''	$C_{p,\beta''}$	0.419
Concentration of β	$C_{p,\beta}$	0.634
Constant in Eq. (5.38)	D_0	3.71×10^{-4} m^2/s
Gas constant	R_g	8.314 J/(K mol)
Diffusion activation energy	Q_d	130,632 J/mol
Modified interface energy of β''	$\gamma_{int,\beta''}$	0.298 J/m^2
Modified interface energy of β	$\gamma_{int,\beta}$	0.257 J/m^2
Molar volume	V_m	1×10^{-5} m^3/mol
Volume of Al atom	V_a	3.16×10^{-30} m^3
Boltzmann constant	k	1.38×10^{-23} J/K
Atom fraction of Mg in matrix	C_{at}	0.01
Lattice constant of Al matrix	a	5.04×10^{-10} m
Burgers vector	b	2.85×10^{-10} m
Wetting constant at dislocation	φ_i	2.2×10^{-3}
Avogadro's number	N_A	6.02214×10^{23} mol^{-1}
Strain energy change	ΔG_e	1.2×10^9 J/m^3

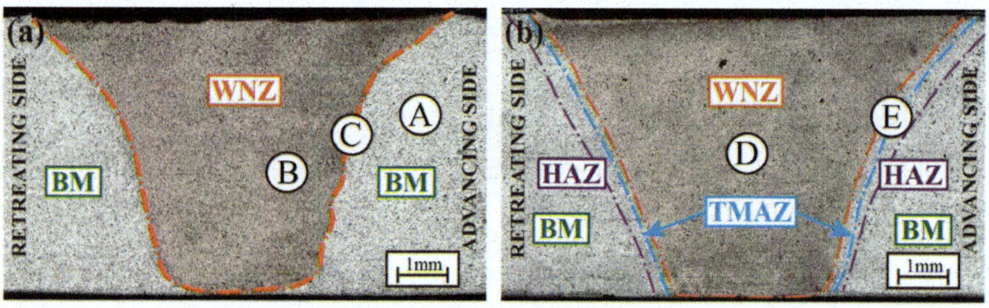

Fig. 5.31 Macrostructures at different welding speeds: **a** 300 mm/min and **b** 30 mm/min

FSW in empirical cognition [62]. The coarse microstructure occurred without post-processing heat treatment [63, 64] or superplastic forming [65], which was unlike abnormal grain growth (AGG) caused by inherently instable fine-grained structures by the theory of Humphreys [66, 67]. To our best knowledge, this atypical coarsening phenomenon has not been reported in other researches until now.

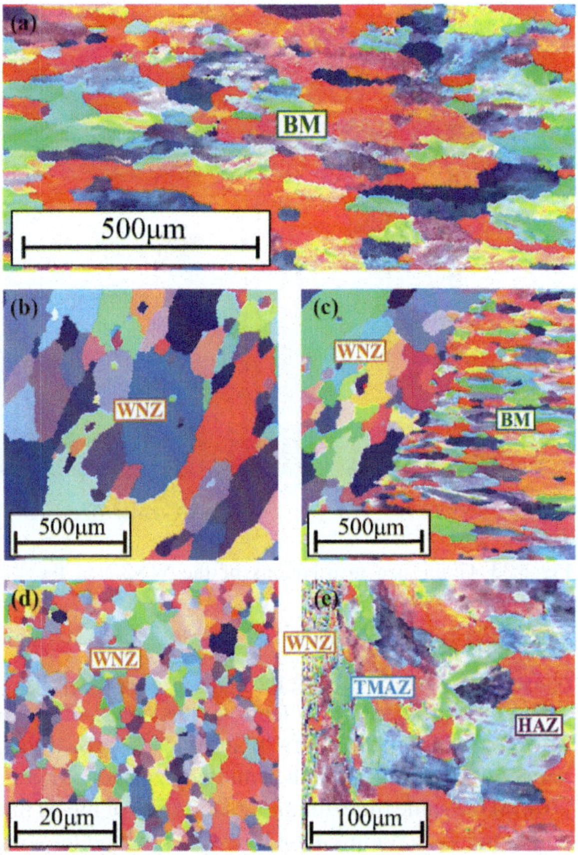

Fig. 5.32 Inverse pole figure (IPF) maps: **a** BM; **b** WNZ, and **c** interface at 300 mm/min; **d** WNZ and **e** interface at 30 mm/min marked by "A", "B", "C", "D" and "E", respectively

5.3.5 Precipitation Transformation

To check the main factor influencing the grain evolution, the modified KWN model was applied to investigate the mechanism of the heterogeneous precipitation based on our previous numerical simulation results [59]. Figure 5.34 shows the distribution of β and β'' phases in the BM. Large amounts of sampling statistics were counted to calculate the initial volume fractions and average sizes. The initial volume fractions were 0.71 and 0.82% of the β and the β'' phases. The initial average radius of the β phase was 32 nm, while the initial average length of the β'' phase was 14 nm. Figure 5.35a, b reflect the variations of the two precipitates at 300 mm/min. The temperature related to the heat input decreased greatly with enhanced material flow due to the tiny shoulder and the pin with complex surface morphology of high depth-to-width ratio FSW. The precipitates dissolved during the heating stage at first. The β'' phase dissolved completely while some β phase still remained, which was consistent with other researches [49, 68]. However, the major radii of this precipitates were smaller than the critical rate radius, which resulted in the shrink of the residual β

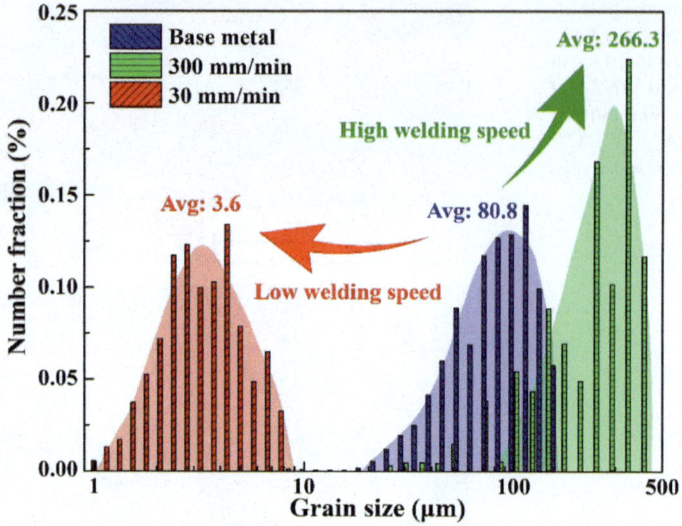

Fig. 5.33 Grain size distributions of the WNZ and the BM at different welding speeds (in logarithmic coordinate)

phase. SSS formed during the heating process [69]. During the subsequent cooling stage, lots of potential nucleation sites emerged due to the high dislocation density. Lower critical nucleation radii and a lower growth rate were achieved due to the low temperature, resulting in the smaller size and less volume fraction of the precipitates [70]. At the same time, the supersaturation of the Al matrix remained relatively higher due to the fast cooling, which resulted in a better solution strengthening effect [71]. The final volume fraction and the average size of the nanoscale β phase were 0.09% and 6 nm as shown in Fig. 5.36. As a comparison, the precipitation variation with a higher heat input at 30 mm/min was shown in Fig. 5.35c, d. The heating stage also contributed to the dissolving of the precipitates. However, both the β'' phase and the β phase dissolved completely due to the higher temperature. Then, the precipitates began to nucleate heterogeneously from the SSS during the subsequent cooling stage. The critical nucleation radius was much larger and the grain growth rate increased notably due to the higher heat input. More second-phase with the larger radii formed subsequently. The final volume fraction and the average size of the β phase were respectively 1.26% and 33 nm.

5.3.6 Dynamic Recrystallization and Pinning Effect

The DRX took place during the extreme straining, which was greatly influenced by the heterogeneous precipitation [72]. The new grains appeared at the nucleation

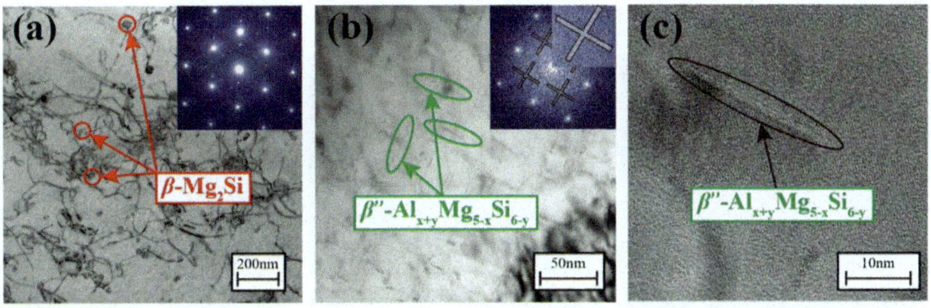

Fig. 5.34 TEM bright field images of **a** β phase and **b** β'' phase in the BM; **c** high resolution TEM image of the β'' phase in the BM

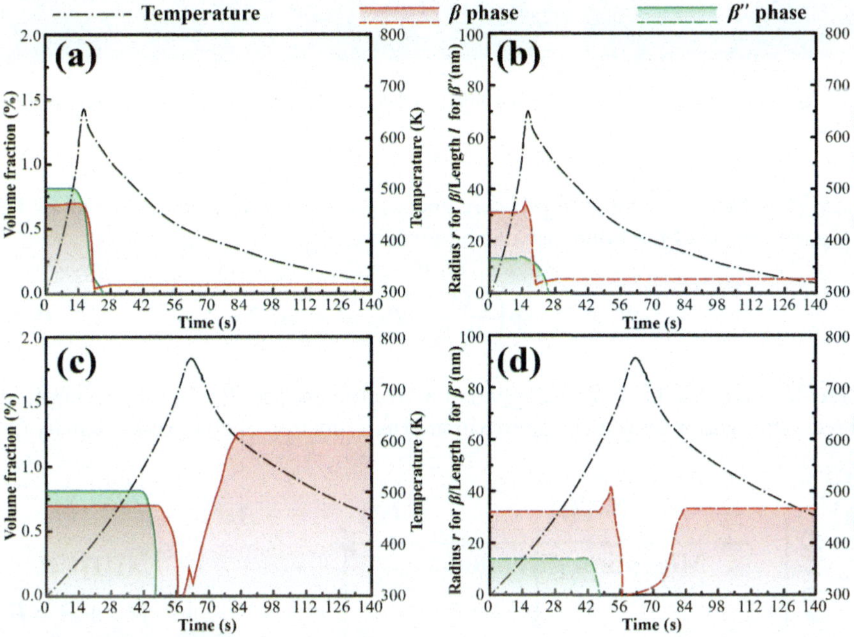

Fig. 5.35 Variations of precipitates during FSW: **a** volume fractions and **b** geometrical sizes at 300 mm/min; **c** volume fractions and **d** geometrical sizes at 30 mm/min (the temperature cycles were taken from the outputs of the macro CFD model)

strain and grew to replace the initial microstructure. The precipitate pinned the rearrangement of dislocations and grain boundary migration (Fig. 5.37) according to the Smith–Zener pinning principle in Eq. (5.49) [73]:

$$P_z = \frac{3 f_v \gamma_{AB}}{4r} \tag{5.49}$$

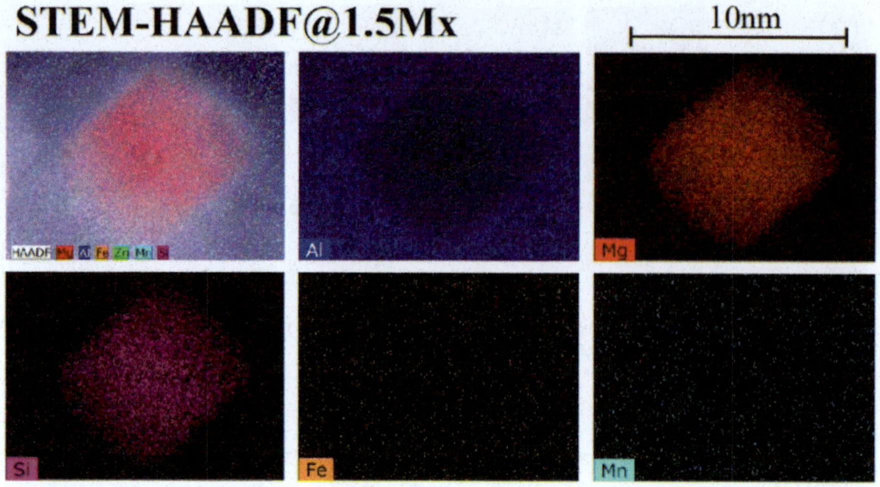

Fig. 5.36 Super energy dispersive spectrometer results of the nano β particle in the WNZ at 300 mm/min

where P_z is the Zener pinning pressure and γ_{AB} is the grain boundary tension. The migration rate V of the boundary can be given as:

$$V = M \exp\left(\frac{Q}{R_g T}\right)(P_{dis} - P_z \pm P_c) \quad (5.50)$$

where M is a constant, Q is the apparent activation energy, P_{dis} is the driving force for recrystallization and can be approximated as a function of dislocation density (ρ),

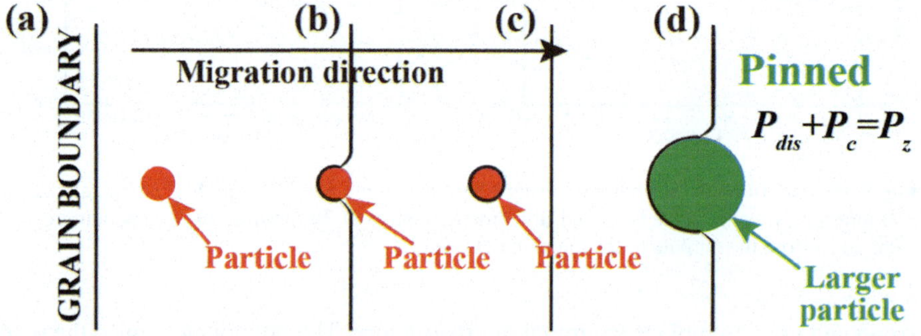

Fig. 5.37 Schematic representation of the precipitate/grain boundary interaction and the pinning effect. The grain boundary is migrating from the left to the right. Black solid line is the precipitate/grain boundary interface: **a** before migration, **b** interaction between a small precipitate and the grain boundary, **c** after migration, and **d** pinning effect between a relatively larger precipitate and the grain boundary

shear modulus (G) and Burgers vector (b) as $0.5\rho Gb^2$, and P_c is the geometrical driving force from the curvature of the grain boundaries.

As shown in Fig. 5.37, when a precipitation intersects with a migrating grain boundary, it forms a new interface with the growing grain, replacing the original interface with the shrinking grain [74]. A balance needs to be reached between the line tension of the grain-precipitation interface and the grain boundary along the migration direction [75]. Because the grain sizes are much larger than the precipitations, curvature effect is the major driving force of the grain boundary motion. Stable condition of the migrating grain boundary has the bow-out shape shown in Fig. 5.37b. For the smaller precipitations shown in Fig. 5.37c, the increasing grain curvature results in higher driving force and the migrating grain boundaries cannot be pinned entirely. For the larger precipitations shown in Fig. 5.37d, the driving force becomes lower and the pinning force is high enough to stop the grain boundary migration.

The ratio of the volume fraction to the average grain size at 300 mm/min was significantly smaller than that at 30 mm/min. Thus, the pinning effect P_z decreased greatly due to the smaller precipitation size (Fig. 5.38a), while the driving force P_{dis} of the boundary migration was not changed basically according to our previous numerical work [59]. The driving force of the migration rate V became higher than that at the lower welding speed [76]. The following happened due to the low heat input: recrystallization grains had fewer nucleation sites and almost no dissolving of AlFeMnSi phase occurred [77], resulting in strong pinning effects on partial grains. At the same time, the temperature became lower than the DRX temperature before the tool completely left due to the larger cooling rate [59]. Partial plastic deformation occurred after DRX, which resulted in the slightly elongated grains. Finally, the microstructure of mainly unequiaxed coarse grains doped with a small amount of small grains (Fig. 5.32b) formed. As a contrast, the DRX occurred completely due to the ample heat input and the pinning effect induced by larger precipitates (Fig. 5.38b) at 30 mm/min, leading to the microstructure of homogeneously small equiaxed grains shown in Fig. 5.32d. However, due to the better solution strengthening effect [71] and the better precipitation strengthening effect [78], the mechanical properties at 300 mm/min were still better than that at 30 mm/min.

The coarsening tendency was calculated to analyze the relationship between grain coarsening and welding parameters at different welding speeds in Fig. 5.39. When the heat input became higher (lower welding speed), the DRX occurred completely and the precipitates had significant pinning effects on boundary migration, implying a microstructure of small equiaxed grains. When the heat input was much lower (higher welding speed), the DRX could not occur because the temperature was below 0.5 melting point, inhibiting the coarsening tendency greatly. The grains were refined by the shear effect notably. As shown in Fig. 5.39, the atypical grain coarsening could only be obtained at about 224–358 mm/min.

It's important to point out that although the heat input seems to dominate the coarsening tendency, the high strain rate and plastic strain are also prerequisites for the atypical grain coarsening. For example, the grain refinement occurred due to the relatively smaller strain rate during the low temperature process like ECAP [79].

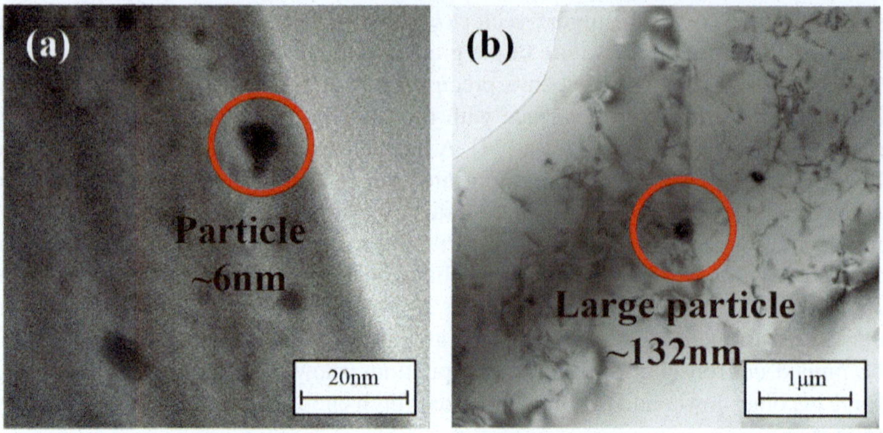

Fig. 5.38 Pinning effects observed by bright field images: **a** nano particle and **b** large particle

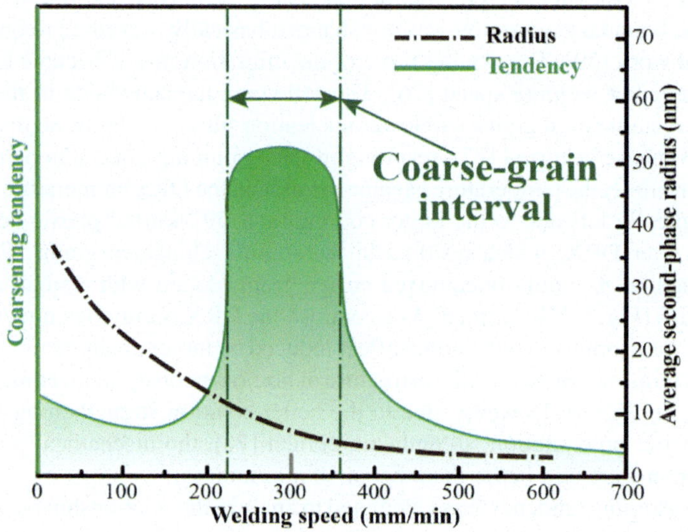

Fig. 5.39 Calculated results for the average second-phase radius and the coarsening tendency at different welding speeds

5.3.7 *Summary*

In conclusion, an atypical grain coarsening phenomenon was obtained in the WNZ without post-processing heat treatment for the first time. The low thermal cycle brought about the low-volume-fraction, low-average-radius nanoscale β phase. The DRX with fewer nucleation sites and lower Smith–Zener pinning effects were confirmed. A microstructure of mainly unequiaxed coarse grains doped with a small amount of small grains was obtained under the extremely low heat input and the high plastic strain rate. The atypical coarsening phenomenon only occurred at certain conditions: extremely high plastic strain and strain rate by friction stir welding; an elevated temperature slightly above 0.5 melting point to inhibit the DRX refinement under higher temperature and shear induced refinement under lower temperature.

References

1. Zhang XX, Xiao BL, Ma ZY. A transient thermal model for friction stir weld. Part II: effects of weld conditions. Metall Mater Trans A. 2011;42(10):3229–39.
2. Feulvarch E, Roux JC, Bergheau JM. A simple and robust moving mesh technique for the finite element simulation of friction stir welding. J Comput Appl Math. 2013;246:269–77.
3. Zhu Y, Chen G, Chen Q, et al. Simulation of material plastic flow driven by non-uniform friction force during friction stir welding and related defect prediction. Mater Des. 2016;108:400–10.
4. Arora A, De A, Debroy T. Toward optimum friction stir welding tool shoulder diameter. Scr Mater. 2011;64(1):9–12.
5. Shi L, Wu CS, Liu HJ. The effect of the welding parameters and tool size on the thermal process and tool torque in reverse dual-rotation friction stir welding. Int J Mach Tools Manuf. 2015;91:1–11.
6. Zienkiewicz OC, Cormeau IC. Visco-plasticity-plasticity and creep in elastic solids—a unified numerical solution approach. Int J Numer Methods Eng. 1974;8(4):821–45.
7. Sheppard T, Jackson A. Constitutive equations for use in prediction of flow stress during extrusion of aluminium alloys. Mater Sci Technol. 1997;13(3):203–9.
8. Su H, Wu CS, Bachmann M, et al. Numerical modeling for the effect of pin profiles on thermal and material flow characteristics in friction stir welding. Mater Des. 2015;77:114–25.
9. Nandan R, Roy GG, Lienert TJ, et al. Three-dimensional heat and material flow during friction stir welding of mild steel. Acta Mater. 2007;55(3):883–95.
10. Tutunchilar S, Haghpanahi M, Besharati Givi MK, et al. Simulation of material flow in friction stir processing of a cast Al–Si alloy. Mater Des. 2012;40:415–26.
11. Mishra RS, Ma ZY. Friction stir welding and processing. Mater Sci Eng R Rep. 2005;50(1–2):1–78.
12. Cabibbo M, McQueen HJ, Evangelista E, et al. Microstructure and mechanical property studies of AA6056 friction stir welded plate. Mater Sci Eng A. 2007;460–461:86–94.
13. Tongne A, Desrayaud C, Jahazi M, et al. On material flow in friction stir welded Al alloys. J Mater Process Technol. 2017;239:284–96.
14. Dai QL, Wang XY, Hou ZG, et al. Microcavities accompanying a zigzag line in friction stir welded A6082-T6 alloy joint. Sci Technol Weld Join. 2015;20(1):68–74.
15. Meng X, Huang Y, Cao J, et al. Recent progress on control strategies for inherent issues in friction stir welding. Prog Mater Sci. 2021;115:100706.
16. Zhang HJ, Liu HJ, Yu L. Microstructure and mechanical properties as a function of rotation speed in underwater friction stir welded aluminum alloy joints. Mater Des. 2011;32(8–9):4402–7.
17. Liu HJ, Fujii H, Maeda M, et al. Tensile properties and fracture locations of friction-stir-welded joints of 2017-T351 aluminum alloy. J Mater Process Technol. 2003;142(3):692–6.
18. Martinez N, Kumar N, Mishra RS, et al. Microstructural comparison of friction-stir-welded aluminum alloy 7449 starting from different tempers. J Mater Sci. 2018;53(12):9273–86.
19. Xu Z, Li Z, Ji S, et al. Refill friction stir spot welding of 5083-O aluminum alloy. J Mater Sci Technol. 2018;34(5):878–85.
20. Aziz SB, Dewan MW, Huggett DJ, et al. Impact of friction stir welding (FSW) process parameters on thermal modeling and heat generation of aluminum alloy joints. Acta Metall Sin (Engl Lett). 2016;29(9):869–83.

21. Chen G, Ma Q, Zhang S, et al. Computational fluid dynamics simulation of friction stir welding: a comparative study on different frictional boundary conditions. J Mater Sci Technol. 2018;34(1):128–34.
22. Ji SD, Jin YY, Yue YM, et al. Effect of temperature on material transfer behavior at different stages of friction stir welded 7075-T6 aluminum alloy. J Mater Sci Technol. 2013;29(10):955–60.
23. Liu CY, Qu B, Xue P, et al. Fabrication of large-bulk ultrafine grained 6061 aluminum alloy by rolling and low-heat-input friction stir welding. J Mater Sci Technol. 2018;34(1):112–8.
24. Wang BB, Chen FF, Liu F, et al. Enhanced mechanical properties of friction stir welded 5083Al-H19 joints with additional water cooling. J Mater Sci Technol. 2017;33(9):1009–14.
25. Colegrove PA, Shercliff HR. 3-Dimensional CFD modelling of flow round a threaded friction stir welding tool profile. J Mater Process Technol. 2005;169(2):320–7.
26. Hoyos E, López D, Alvarez H. A phenomenologically based material flow model for friction stir welding. Mater Des. 2016;111:321–30.
27. Liu X, Wu C, Padhy GK. Characterization of plastic deformation and material flow in ultrasonic vibration enhanced friction stir welding. Scr Mater. 2015;102:95–8.
28. Fagan T, Lemiale V, Nairn J, et al. Detailed thermal and material flow analyses of friction stir forming using a three-dimensional particle based model. J Mater Process Technol. 2016;231:422–30.
29. Zhang Z, Wu Q, Grujicic M, et al. Monte Carlo simulation of grain growth and welding zones in friction stir welding of AA6082-T6. J Mater Sci. 2016;51(4):1882–95.
30. Barbini A, Carstensen J, dos Santos JF. Influence of a non-rotating shoulder on heat generation, microstructure and mechanical properties of dissimilar AA2024/AA7050 FSW joints. J Mater Sci Technol. 2017;34(1):119–27.
31. Niu P, Li W, Zhang Z, et al. Global and local constitutive behaviors of friction stir welded AA2024 joints. J Mater Sci Technol. 2017;33(9):987–90.
32. Schneider J, Cobb J, Carpenter JS, et al. Maintaining nano-lamellar microstructure in friction stir welding (FSW) of accumulative roll bonded (ARB) Cu–Nb nano-lamellar composites (NLC). J Mater Sci Technol. 2018;34(1):92–101.
33. Su H, Wu CS, Pittner A, et al. Thermal energy generation and distribution in friction stir welding of aluminum alloys. Energy. 2014;77:720–31.
34. Hannard F, Castin S, Maire E, et al. Ductilization of aluminium alloy 6056 by friction stir processing. Acta Mater. 2017;130:121–36.
35. Tongne A, Jahazi M, Feulvarch E, et al. Banded structures in friction stir welded Al alloys. J Mater Process Technol. 2015;221:269–78.
36. Shi L, Wu CS, Gao S, et al. Modified constitutive equation for use in modeling the ultrasonic vibration enhanced friction stir welding process. Scr Mater. 2016;119:21–6.
37. Zhao K, Liu Z, Xiao B, et al. Friction stir welding of carbon nanotubes reinforced Al–Cu–Mg alloy composite plates. J Mater Sci Technol. 2017;33(9):1004–8.
38. Sauvage X, Dédé A, Muñoz AC, et al. Precipitate stability and recrystallisation in the weld nuggets of friction stir welded Al–Mg–Si and Al–Mg–Sc alloys. Mater Sci Eng A. 2008;491(1–2):364–71.
39. Gupta AK, Lloyd DJ, Court SA. Precipitation hardening processes in an Al-0.4%Mg-1.3%Si-0.25%Fe aluminum alloy. Mater Sci Eng A. 2001;301(2):140–6.
40. Simar A, Bréchet Y, de Meester B, et al. Integrated modeling of friction stir welding of 6xxx series Al alloys: process, microstructure and properties. Prog Mater Sci. 2012;57(1):95–183.
41. Chu Q, Li WY, Yang XW, et al. Microstructure and mechanical optimization of probeless friction stir spot welded joint of an Al–Li alloy. J Mater Sci Technol. 2018;34(10):1739–46.
42. Du Q, Tang K, Marioara CD, et al. Modeling over-ageing in Al–Mg–Si alloys by a multi-phase CALPHAD-coupled Kampmann–Wagner numerical model. Acta Mater. 2017;122:178–86.
43. Huang Y, Meng X, Zhang Y, et al. Micro friction stir welding of ultra-thin Al-6061 sheets. J Mater Process Technol. 2017;250:313–9.
44. Padhy GK, Wu CS, Gao S. Precursor ultrasonic effect on grain structure development of AA6061-T6 friction stir weld. Mater Des. 2017;116:207–18.

45. Huang Y, Meng X, Xie Y, et al. Friction stir welding/processing of polymers and polymer matrix composites. Compos Part A Appl Sci Manuf. 2018;105:235–57.
46. Beyerlein IJ, Tóth LS. Texture evolution in equal-channel angular extrusion. Prog Mater Sci. 2009;54(4):427–510.
47. Čížek J, Janeček M, Krajňák T, et al. Structural characterization of ultrafine-grained interstitial-free steel prepared by severe plastic deformation. Acta Mater. 2016;105:258–72.
48. Lee HH, Yoon JI, Kim HS. Single-roll angular-rolling: a new continuous severe plastic deformation process for metal sheets. Scr Mater. 2018;146:204–7.
49. Ma ZY, Feng AH, Chen DL, et al. Recent advances in friction stir welding/processing of aluminum alloys: microstructural evolution and mechanical properties. Crit Rev Solid State Mater Sci. 2018;43(4):269–333.
50. Hin C, Bréchet Y, Maugis P, et al. Kinetics of heterogeneous grain boundary precipitation of NbC in α-iron: a Monte Carlo study. Acta Mater. 2008;56(19):5653–67.
51. Rainforth WM, Black MP, Higginson RL, et al. Precipitation of NbC in a model austenitic steel. Acta Mater. 2002;50(4):735–47.
52. Huang K, Marthinsen K, Zhao Q, et al. The double-edge effect of second-phase particles on the recrystallization behaviour and associated mechanical properties of metallic materials. Prog Mater Sci. 2018;92:284–359.
53. McQueen HJ, Kassner ME. Comments on "a model of continuous dynamic recrystallization" proposed for aluminum. Scr Mater. 2004;51(5):461–5.
54. Sakai T, Belyakov A, Kaibyshev R, et al. Dynamic and post-dynamic recrystallization under hot, cold and severe plastic deformation conditions. Prog Mater Sci. 2014;60(1):130–207.
55. Han J, Thomas SL, Srolovitz DJ. Grain-boundary kinetics: a unified approach. Prog Mater Sci. 2018;98:386–476.
56. Gu CF, Tóth LS. Texture development and grain refinement in non-equal-channel angular-pressed Al. Scr Mater. 2012;67(1):33–6.
57. Sitdikov O, Sakai T, Miura H, et al. Temperature effect on fine-grained structure formation in high-strength Al alloy 7475 during hot severe deformation. Mater Sci Eng A. 2009;516(1–2):180–8.
58. Vinogradov A, Estrin Y. Analytical and numerical approaches to modelling severe plastic deformation. Prog Mater Sci. 2018;95:172–242.
59. Huang Y, Xie Y, Meng X, et al. Numerical design of high depth-to-width ratio friction stir welding. J Mater Process Technol. 2018;252:233–41.
60. Rai R, De A, Bhadeshia HKDH, et al. Review: friction stir welding tools. Sci Technol Weld Join. 2011;16(4):325–42.
61. Kelly PM. The effect of particle shape on dispersion hardening. Scr Metall. 1972;6(8):647–56.
62. Wang B, Lei B, Zhu J, et al. EBSD study on microstructure and texture of friction stir welded AA5052-O and AA6061-T6 dissimilar joint. Mater Des. 2015;87:593–9.
63. Charit I, Mishra RS. Abnormal grain growth in friction stir processed alloys. Scr Mater. 2008;58(5):367–71.
64. Vysotskiy I, Malopheyev S, Mironov S, et al. Pre-strain rolling as an effective tool for suppression of abnormal grain growth in friction-stir welded 6061 aluminum alloy. Mater Sci Eng A. 2018;733:39–42.
65. Charit I, Mishra RS, Mahoney MW. Multi-sheet structures in 7475 aluminum by friction stir welding in concert with post-weld superplastic forming. Scr Mater. 2002;47(9):631–6.
66. Humphreys FJ. A unified theory of recovery, recrystallization and grain growth, based on the stability and growth of cellular microstructures—I. The basic model. Acta Mater. 1997;45(10):4231–40.
67. Humphreys FJ. A unified theory of recovery, recrystallization and grain growth, based on the stability and growth of cellular microstructures—II. The effect of second-phase particles. Acta Mater. 1997;45(12):5031–9.
68. Ma ZY, Sharma SR, Mishra RS. Microstructural modification of as-cast Al–Si–Mg alloy by friction stir processing. Metall Mater Trans A. 2006;37(11):3323–36.

69. Bayat N, Carlberg T, Cieslar M. In-situ study of phase transformations during homogenization of 6005 and 6082 Al alloys. J Alloys Compd. 2017;725:504–9.
70. Madanat M, Liu M, Banhart J. Reversion of natural ageing in Al–Mg–Si alloys. Acta Mater. 2018;159:163–72.
71. Huskins EL, Cao B, Ramesh KT. Strengthening mechanisms in an Al–Mg alloy. Mater Sci Eng A. 2010;527(6):1292–8.
72. Kim MS, Jung JY, Song YM, et al. Simulation of microtexture developments in the stir zone of friction stir-welded AZ31 Mg alloys. Int J Plast. 2017;94:24–43.
73. Agnoli A, Bozzolo N, Logé R, et al. Development of a level set methodology to simulate grain growth in the presence of real secondary phase particles and stored energy—application to a nickel-base superalloy. Comput Mater Sci. 2014;89:233–41.
74. Wang N, Ji Y, Wang Y, et al. Two modes of grain boundary pinning by coherent precipitates. Acta Mater. 2017;135:226–32.
75. Abdeljawad F, Foiles SM, Moore AP, et al. The role of the interface stiffness tensor on grain boundary dynamics. Acta Mater. 2018;158:440–53.
76. Vuppuluri A. Theory and simulation of microstructure evolution due to simultaneous grain boundary migration and grain rotation with misorientation dependent energy and mobility. Mater Sci Eng A. 2018;713:118–24.
77. Kumar N, Goel S, Jayaganthan R, et al. Effect of grain boundary misorientation, deformation temperature and AlFeMnSi-phase on fatigue life of 6082 Al alloy. Mater Charact. 2017;124:229–40.
78. Murugesan AP, Rajinikanth V, Mahato B, et al. Concurrent precipitation and associated texture evolution in AA 6082 alloy during high pressure torsion (HPT) processing. Mater Sci Eng A. 2017;700:487–94.
79. Ribbe J, Schmitz G, Rösner H, et al. Effect of back pressure during equal-channel angular pressing on deformation-induced porosity in copper. Scr Mater. 2013;68(12):925–8.

Chapter 6
Entire-Process Simulation of Friction Stir Welding

6.1 Experiments and Simulation

6.1.1 Introduction

Friction stir welding (FSW), a solid-state joining technology, has become an ideal welding method to join materials with low weldability [1, 2]. The heat input, including a surficial frictional heat source and a volumetric deformation heat source, is generated by the contact between welding tool and workpieces, which is inversely related to the transient flow stress of the material, ensuring the transient temperature just above the critical value to proceed dynamic recovery and recrystallization [3, 4]. Several variants of FSW further enhance material flow and heat input via the structural design of welding tools. These techniques, such as threaded pin [5], milling facets [6], external scribe [7], stationary shoulder [1], and external heat assistance [8], increase the flexibility of temperature-sensitive applications. However, the more complex deformation mechanisms have hindered the understand and industrial applications.

New strategies to analyze the entire welding process systematically are highly appreciated. Finite element methods are acceptable underlying frameworks for the calculation strategy since they can quantitatively describe the distribution of temperature, strain, and pressure during the welding process [9]. The current finite element method can be divided into two categories: computational solid mechanics (CSM) and computational fluid dynamics (CFD) [10]. Due to the difficulty in controlling the calculation time and convergence of CSM, researchers tend to adopt the CFD method to conduct the field calculation of the FSW process [11]. Su et al. [12] conducted an accurate prediction of thermal generation and distribution for parametric optimization. A modified constitutive equation was proposed to correct the numerical results. Arora et al. [13] proposed a criterion for the design of shoulder diameter based on the principle of maximum utilization of supplied torque for traction, which established a relationship between welding tools and joint performance. Wu et al. [14] utilized the finite element method to simulate the temperature field throughout the joint for

Y. Huang et al., *Friction Stir Welding and Processing*, Materials Forming, Machining and Tribology, https://doi.org/10.1007/978-981-99-8688-0_6

the microstructural explanation. A dynamic recrystallization analysis was conducted based on the numerical model. For the combined strategy, taking the CFD method as the underlying framework, relevant field data, including temperature and strain rate, can be provided based on various welding tools and welding parameters.

Microstructural evolution containing precipitation and recrystallization behavior can further accuracy the calculation of mechanical properties. The microstructural model inherits the outputs from the CFD model as its input. For Al-Cu-Li alloys, the precipitate model tracks three kinds of primary strengthening precipitates. dos Santos et al. [15] calculated the fraction and radius revolution of precipitates in FSWed Al–Zn–Mg–Cu alloys via the Wanger-Kampmann (KWN) model, proving that the KWN model is applicable for the FSW process. The dynamic recrystallization (DRX) model is also needed. Irani et al. [16] stated the application of the Johnson–Mehl–Avrami-Kolmogorov model combined with the finite element method in DRX and recovery. The accuracy between measured and predicted grain sizes was convincing. Combining the two micro models, the local and global mechanical properties can be subsequently calculated based on precipitation strengthening, grain boundary strengthening, dislocation strengthening, and solution strengthening [17].

We proposed an entire-process simulation strategy based on numerical simulation to predict the tensile strength of friction stir welded joints. Choosing the third-generation AA2195-T8 alloys owning the more complex phase composition and precipitations as the research object, which is the preferred material of rocket fuel tanks in aerospace. This combined model provides an accuracy strategy to mapping the macro and microstructural response and will provide sufficient datasets for the following NN (neural network) methods in Sect. 6.2.

6.1.2 Experimental Procedures

The chosen object was AA2195-T8 alloy with dimensions of 300 mm × 100 mm × 6 mm. FSW was performed at a rotational velocity range of 600–1500 r/min and a welding speed range of 100–500 mm/min. A plunge depth of 0.1 mm and a tilt angle of 1° were constant. The welding tool made of H13 steel was composed of two parts: a shoulder and a rotational pin. The rotating and stationary shoulder diameters in the experimental section were 14 mm and 18 mm, respectively. The sizes of the pin were respectively 8 mm and 6.9 mm in diameter and length, and the pin was characterized by threads and triple milling facets. The thermal cycle was recorded by several thermocouples fixed in the prefabricated blind holes with a diameter of 1 mm.

An optical microscope (OM, Keyence VHX-7000) was used to observe metallography morphologies in joint transverse sections. A scanning electron microscope (SEM, Zeiss Supra 55) at a working voltage of 20 kV was utilized to perform an electron backscattered diffraction (EBSD) characterization with an HKL detector. The specimens for OM and EBSD were polished by respectively mechanical polishing and a JEOL IB-09020CP cross-section polisher. The OM specimens were etched by

Keller's reagent (2.5 mL HNO_3, 1.5 mL HCl, 1 mL HF, and 95 mL H_2O). An Instron 5569 tester was applied to evaluate the tensile properties of the joints at a constant crosshead speed of 2 mm/min. The tensile samples had the dimensions of 100 mm × 25 mm × 6 mm. The microhardness required for the local strength conversion was obtained by the micro Vickers hardness test with a load of 200 g and a dwelling time of 10 s.

6.1.3 Finite Element Modeling

Figure 6.1 illustrates the schematic outline of the combined calculation strategy. As the underlying framework, the CFD model served as a converter from primitive causative variables to field variables, including strain rate and temperature. The materials to be welded and simulated, AA2195-T8 alloys, were treated as incompressible, non-Newtonian fluid. According to Boussinesq's approximation, the change of alloy density was ignored during the simulation process. The tilt angle and plunge depth were taken the same as the experimental settings. The calculation was chosen as the steady mode to obtain the quasi-steady state of the welding process. The domain for calculation with coordinates of 60 mm × 90 mm × 6 mm was used.

The materials in the domain were described as an incompressible fluid with a constant density [18], the deformation and flow behaviors followed the Navier-Stoke equations and Boussinesq's approximation [19], which were solved by finite volume method via FLUENT:

$$\nabla \cdot v = 0 \tag{6.1}$$

$$\rho \frac{dv}{dt} = -\nabla P + \nabla \cdot \sigma \tag{6.2}$$

$$\rho c_p \frac{dT}{dt} = -\nabla(K \nabla T) + (\sigma : \nabla v) - \phi \tag{6.3}$$

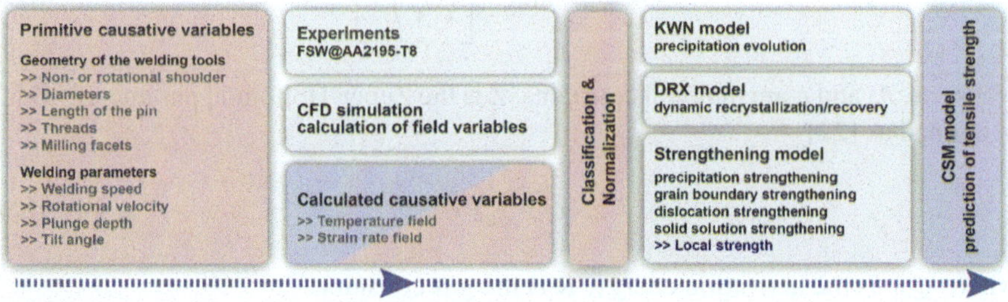

Fig. 6.1 Schematic outline of the combined strategy

where ρ is the density of AA2195-T8 alloy, t is the step time, v is transient velocity, $\sigma = \mu(\nabla v + \nabla v^t)$ is the deviatoric stress tensor, μ is the non-Newtonian viscosity, P is the structural pressure, T is the absolute temperature, ϕ is the source term which represents the heat generation of volumetric deformation heat and surficial frictional heat, and the heat loss of convection and radiation, c_p is the specific heat capacity, and k is the thermal conductivity.

Since the Al-Cu-Li alloy is not fluid, the relationship between flow stress and viscosity is essential to access the real-time non-Newtonian viscosity. The equation between viscosity, equivalent strain rate, and flow stress is defined by Perzyna's viscoplasticity model [20]:

$$\mu = \frac{\sigma_e}{3\dot{\varepsilon}_e} \tag{6.4}$$

where σ_e is the flow stress or called yield stress which can be defined by a constitutive equation of aluminum alloys, $\dot{\varepsilon}_e$ is the equivalent strain rate:

$$\sigma_e = \sqrt{\frac{3}{2}\sigma_{ij}\sigma_{ij}} \quad i, j = 1, 2, 3 \tag{6.5}$$

$$\dot{\varepsilon}_e = \sqrt{\frac{2}{3}\dot{\varepsilon}_{ij}\dot{\varepsilon}_{ij}} \quad i, j = 1, 2, 3 \tag{6.6}$$

where, σ_{ij} and $\dot{\varepsilon}_{ij}$ are the components of the second-order deviatoric stress tensor and the second-order strain rate tensor, respectively. The strain rate tensor can be defined by the Helmholtz velocity decomposing theorem:

$$\dot{\varepsilon}_{ij} = \frac{1}{2}\left(v_{i,j} + v_{j,i}\right) \quad i, j = 1, 2, 3 \tag{6.7}$$

The flow stress can be calculated by Sellars-Tegarts constitutive equations based on strain rate and temperature [21]:

$$\sigma_e = \frac{1}{\alpha}\sinh^{-1}\left[\left(\frac{Z}{A}\right)^{\frac{1}{n}}\right] \tag{6.8}$$

where, α, A, and n are material constants. Z is the Zener-Hollomon parameter, which is defined as [22]:

$$Z = \dot{\varepsilon}_e e^{\frac{Q_{zh}}{RT}} \tag{6.9}$$

where Q_{zh} is the activation energy, R is the gas constant.

However, there are limitations and drawbacks of Eq. (6.8) when the temperature rises near melting point [12]. Thus, a modified constitutive equation is proposed:

$$\sigma_e = \left[1 - \left(\frac{T}{T_s}\right)^{\gamma}\right]\frac{1}{\alpha}\sinh^{-1}\left[\left(\frac{Z}{A}\right)^{\frac{1}{n}}\right] \tag{6.10}$$

where T_s is the solidus temperature of AA2195-T8 alloy, and γ is the influencing factor.

For the heat boundary conditions, there are two parts of the heat source: frictional heat and deformation heat. Frictional heat is a surficial heat source applied to the interface between welding tool and workpiece, which is described as:

$$F = \beta_1 R\omega \cdot \left[\frac{\delta\sigma_e}{\sqrt{3}} + (1-\delta)\mu P \sin\xi\right] \cdot \cos\theta \tag{6.11}$$

The deformation heat is a volumetric heat source, which can be defined as:

$$\Phi = \beta_2\mu\left[2\left(\frac{\partial u}{\partial x}\right)^2 + 2\left(\frac{\partial v}{\partial y}\right)^2 + 2\left(\frac{\partial w}{\partial z}\right)^2 + \left(\frac{\partial u}{\partial y} + \frac{\partial v}{\partial x}\right)^2 + \left(\frac{\partial u}{\partial z} + \frac{\partial w}{\partial x}\right)^2 + \left(\frac{\partial w}{\partial y} + \frac{\partial v}{\partial w}\right)^2\right] \tag{6.12}$$

where β_1 and β_2 are the empirical coefficients, μ is the frictional coefficient, R is the radius of elements to the axis, ω is the rotational velocity, δ is the slip rate, P is the axial pressure, ξ and θ represent the orientation of the element. The slip rate can be defined as [23]:

$$\delta = 1 - e^{-\frac{\omega R}{\delta_0\omega_0 R_p}} \tag{6.13}$$

where δ_0 is an interfacial constant, ω_0 is a constant for nondimensionalization, and R_p is the radius of the pin. Besides, in consideration of heat dissipation by the welding tool, the interfacial thermal efficiency is defined by:

$$\lambda = \frac{J_d}{J_d + J_t} \tag{6.14}$$

$$J = \sqrt{k\rho C_p} \tag{6.15}$$

where J_d and J_t represent the domain and the tool, respectively. k, ρ, and C_p are the thermal conductivity, the density, and the heat capacity, respectively.

The boundary heat convection and radiation can be given as follows [24]:

$$-k\frac{\partial T}{\partial z} = h(T - T_\infty) + \sigma\varepsilon\left(T^4 - T_\infty^4\right) \tag{6.16}$$

where h is the heat transfer coefficient, σ is the Stefan-Boltzmann constant, and ε is the emissivity of the domain.

For the motion boundary conditions, the thermo-plasticized AA2195-T8 alloy, which is regarded as a non-Newtonian fluid, is driven by the velocity equation [10]:

$$\begin{cases} v_{horizon} = \eta\omega R \\ v_{shear} = v_{horizon}\cos\xi\cos\theta \\ v_{normal} = v_{horizon}\cos\xi\sin\theta \\ v_{vertical} = v_{horizon}\sin\xi \end{cases} \tag{6.17}$$

Particles tracing technique was applied to assess the defect formation during the FSW process. The injections were set to be massless, whose generating position had a random amplitude of 0.3 mm. The distribution of particles was obtained at the section of the outlet. Areas without particles were marked as crimson (#DC143C), while the area with particles was marked as grey (#808080). All grey implied that the joint was sound and defect-free. An index to evaluate the distribution of these particles was expressed as:

$$\chi = \frac{\mu_R + \mu_G + \mu_B - 240}{54} \tag{6.18}$$

where μ_R, μ_G, μ_G are the average value of the colored section in the decimal system. The higher the index, the better the joint formation.

6.1.4 Precipitation Evolution Modeling

The microstructural evolution models are the secondary framework of the combined strategy. The input data for these models were taken from the CFD simulation results, which were called calculated causative variables in Fig. 6.1. The first part of the microstructural evolution model is a KWN precipitation evolution model. This evolution algorithm is based on the discretization of time and particle size. The overall calculation process mainly includes three parts, as shown in Fig. 6.2a: ① Calculating the growth rate of precipitation particles including dissolving and coarsening; ② Calculating the nucleation rate of particles with different sizes at each time step; ③ Recording the continuum equation of each element between matrix and precipitations. Since δ′ is not the major precipitation for the alloys whose element ratio of Cu to Li is larger than 2.5, there are two kinds of major strengthening particles in AA2195 alloy [25]. Cu element and Li element, which constitute the T_1 and θ″/θ′ phases, are the binary diffusion controlling elements. Both two elements are involved in the calculation, and the element with a lower diffusion rate at the specific time shall prevail. Besides, the shapes of the two major phases are platelets rather than globular. Tremendous increases in the complexity and calculation time of the modified KWN model would occur if the influence of phase morphology were considered.

Therefore, for the convenience of calculation, all the particles precipitating during the welding process were treated as globular. Assuming that the ratio of radius to the thickness of the precipitations is a specific value k_s, then the equivalent radius can be extracted. There is a relationship between thickness and equivalent radius of the platelet-like precipitations.

$$r = \sqrt[3]{\frac{3k_s^2\delta^3}{4}} \tag{6.19}$$

where r is the equivalent radius, δ is the thickness.

The first part is to calculate the growth rate of precipitation particles. It depends on the solute concentration at the interface between the particles with a certain radius and the matrix. The growth rate can be described as [26]:

$$v_i = \frac{\partial r_i}{\partial t} = \frac{C_{ss} - C_{int}}{C_p - C_{int}}\frac{D_f}{r_i} \tag{6.20}$$

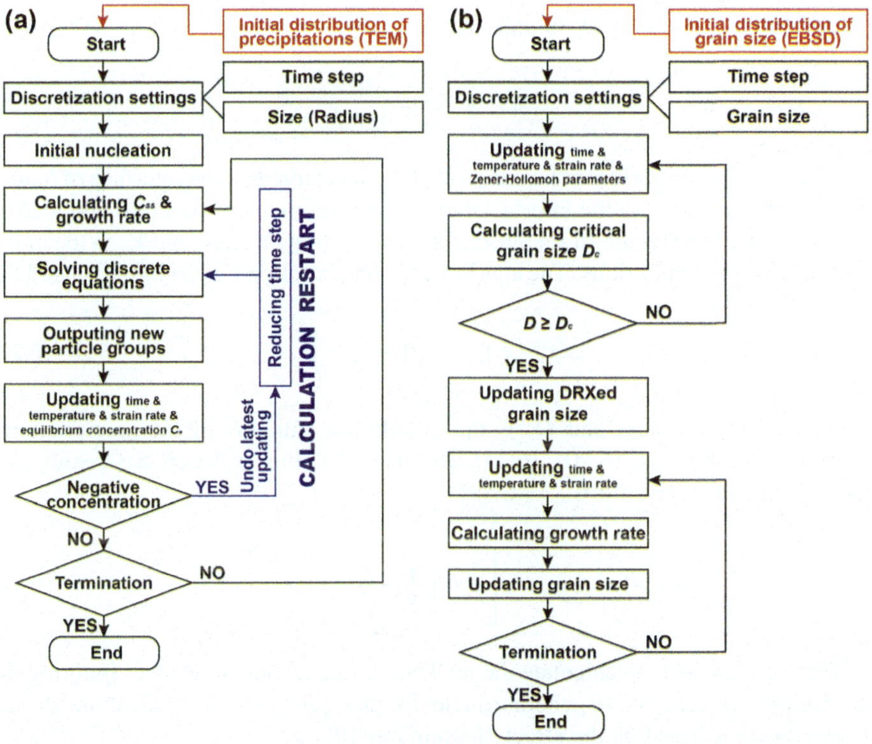

Fig. 6.2 Flow charts of **a** modified KWN model and **b** GDRX model

where C_{ss} is the solute concentration of diffusion controlling element far away from the adjacent area of precipitations, C_{int} is the concentration at the interface between precipitations and matrix, which can be calculated from the Gibbs–Thomson equation [27], C_p is the concentration of the diffusion controlling element in the precipitations. When $C_{ss} < C_{int}$, the radius of particles is gradually reduced, showing the state of continuous dissolution; when $C_{ss} > C_{int}$, the radius of particles increases gradually, showing the state of continuous coarsening conversely, D_f is the diffusion coefficient, and r_i is the radius of the particle groups. Due to the severe plastic deformation essence of FSW, the dislocation density increases in an order of magnitude during the welding process, contributing to the presence of pipe diffusion [28]. This phenomenon greatly improves the diffusion rate. In order to reflect this characteristic in the modified KWN model, an additional coefficient is introduced into the diffusion equation, which can be written as [29]:

$$D_f = D_0\sqrt{1+\dot{\varepsilon}}\exp\left(-\frac{Q_d}{RT}\right) \tag{6.21}$$

where D_0 is the diffusion factor; Q_d is the diffusion activation energy. The interfacial concentration is given by Du et al. [30]:

$$C_{\text{int}} = C_e \exp\frac{2\gamma_{\text{int}} V_m}{r_i RT} \tag{6.22}$$

where C_e is the equilibrium concentration of solute element at the interface of matrix and precipitation, γ_{int} is the interfacial energy per unit area of particles and matrix, which is influenced by the strain rate during the welding process, and V_m is the molar volume of the precipitations. The equilibrium concentration C_e can be calculated by:

$$C_e = C_{e0}\exp\left(-\frac{Q_e}{RT}\right) \tag{6.23}$$

where C_{e0} is a constant and Q_e is the dissolution enthalpy of the precipitations. According to the Eqs. (6.20), (6.23), the critical radius between coarsening and dissolution can be given if $C_{ss} = C_{int}$:

$$r^* = 2\gamma_{int} V_m\left[RT\prod_{i=1}^{n} ln\frac{C_{iss}}{C_{ie0}} + Q_e\right]^{-1} \tag{6.24}$$

The second part is to calculate the nucleation rate at each time step. Ignoring the incubation time required to generate nuclei for precipitations, the nucleation rate can be expressed as based on the effect of strain rate [9]:

$$J_{nucl} = \frac{J_0\sqrt{1+\dot{\varepsilon}}}{\sqrt{T}}\exp\left\{-\frac{A_0}{(RT)^3[\ln(C_{ss}/C_e)]^2} - \frac{Q_d}{RT}\right\} \tag{6.25}$$

$$J_0 = \frac{2\rho_{dis} V_a D_0 C_{at} \sqrt{\gamma_{int}}}{a^4 b \sqrt{k}} \tag{6.26}$$

$$A_0 = \frac{16\pi \varphi_i \gamma_{int}^3 N_A V_m^2}{3C_p^2} \tag{6.27}$$

where ρ_{dis} is the dislocation density which provides the position for nucleation, V_a is the volume of the matrix atom, k is the Boltzmann constant, C_{at} is the atom fraction of diffusion controlling element in the matrix, a is the lattice constant of the matrix, b is the burger's vector, φ_i is the wetting constant at dislocation, and N_A is Avogadro's number.

The last part is the continuum equation. The core of particle grouping strategy is to discretize particles according to radii, simultaneously simulating nucleation, dissolution, and coarsening. Various groups of particles with different numbers flow between the given control equation. Two kinds of precipitates were divided from small to large into groups with a fixed radius. Each particle group can be defined as a control body and be an instance of the group class. Therefore, the nucleation, dissolution, and coarsening of particles can be written as methods of the group class, which symbolizes the source term, inflow, and outflow of the group instance. The material flow equation based on the material flow can be written as:

$$\frac{\partial \mathrm{N}}{\partial t} = -\frac{\partial F_n}{\partial r} + \phi_n \tag{6.28}$$

where N is the number density of the particles in each particle group, F_n is the material flow, and ϕ_n is the source term, which refers to the nucleation rate at each time step. The material flow is described based on the equation of growth rate:

$$F_n = \Delta N v \tag{6.29}$$

Therefore, the first-order upwind material flow equation is further expressed as:

$$\frac{\partial \mathrm{N}_i}{\partial t} = \begin{cases} \frac{\partial F_{n,i-1}}{\partial r_{i-1}} - \frac{\partial F_{n,i+1}}{\partial r_{i+1}} + \phi_{n,i} & F_{n,i-1} > 0,\ F_{n,i+1} < 0 \\ -\frac{\partial F_{n,i+1}}{\partial r_{i+1}} + \phi_{n,i} & F_{n,i-1} \le 0,\ F_{n,i+1} < 0 \\ \frac{\partial F_{n,i-1}}{\partial r_{i-1}} + \phi_{n,i} & F_{n,i-1} > 0,\ F_{n,i+1} \ge 0 \\ \phi_{n,i} & F_{n,i-1} \le 0,\ F_{n,i+1} \ge 0 \end{cases} \tag{6.30}$$

The explicit continuum equation among all the particle groups and matrix can be given as:

$$C_{ss} = \frac{C_0 - C_p \sum f_v^i}{1 - \sum f_v^i} \tag{6.31}$$

$$f_v^i = \frac{4}{3}\pi r_i^3 N_i \tag{6.32}$$

where C_0 is the total concentration of the diffusion controlling element, f_v^i is the volume fraction of each particle group based on the globular assumption, and $\sum f_v^i$ is the volume fraction of the precipitation which the particles points to. An accuracy issue exists in the calculation: the matrix concentration is obtained by updating the previous time step. The value of the time step needs to be focused on to avoid negative concentration; that is, the change between two adjacent time steps should not be too large. An adaptive step time was applied by auto-restart strategy, as shown in Fig. 6.2a.

6.1.5 Dynamic Recrystallization Modeling

The existing researches usually indicate that DRX is mainly dominated by geometric DRX (GDRX) [31]. Therefore, the DRX behaviors in the joint, including welding nugget zone (WNZ), thermo-mechanically affected zone (TMAZ), and heat-affected zone (HAZ), were calculated based on GDRX mechanisms. The critical grain size of GDRX can be calculated by:

$$D_C = (a + b\ln Z)^{-1} \tag{6.33}$$

where a and b are the constants, and Z is the Zener-Hollomon parameters mentioned in Eq. (6.9). The criteria can be used to determine whether GDRX occurs: no recrystallization occurs when the current grain size is greater than D_C. Obviously, this is what happened to HAZ; recrystallization occurs when the current grain size is smaller than or equal to D_C. The initial DRXed grain size is equal to D_C. For the subsequent calculation of grain size evolution, the following assumptions were made: ① DRX occurs instantaneously when the critical recrystallization size is reached; ② the DRXed grains only undergo coarsening with subsequent deformation; ③ in the subsequent thermal process, the grains grow up statically with the formation of high angle grain boundaries. Besides, there are precipitations in Al-Cu-Li alloys, the pinning effect of these particles should be considered [32]. The growth equation of grains is expressed as:

$$\frac{dD}{dt} = M_0\left(\frac{2\gamma_b}{D} - \sum \frac{3\gamma_b \sum f_v^i}{2r}\right)\exp\left(-\frac{Q_G}{RT} - \dot{\varepsilon}\right) \tag{6.34}$$

where M_0 is the migration coefficient of grain boundaries, γ_b is the grain boundary energy, D is the diameter of grains, $\sum f_v^i$ is the volume fraction of each precipitation, $\sum \frac{3\gamma_b \sum f_v^i}{2r}$ is the pinning effect of the major precipitations, and Q_G is the migration activation energy of grain boundaries. One should note that this model adopted the discretization method of grain diameter, which is similar to the modified KWN model. The detailed calculating process is depicted in Fig. 6.2b.

6.1.6 Strengthening Modeling

As a precipitation strengthening alloy, Al-Cu-Li alloy has four major strengthening mechanisms: grain boundary strengthening, solid solution strengthening, precipitation strengthening, and dislocation strengthening. The total yield strength (The yield strength was used as the output value) can be calculated by the linear addition of inherent strength and these four strengthening mechanisms [33]:

$$\sigma_{total} = \chi^{P}\left[\sigma_{\text{in}} + \sigma_{\text{gb}} + M\left(\tau_{\text{ss}} + \tau_{\text{prec}} + \tau_{\text{dis}}\right)\right] \tag{6.35}$$

where χ^P is the sensitivity coefficient of defects in joints, σ_{in} is the intrinsic strength of aluminum alloys, σ_{gb} is the strength of grain boundary strengthening, M is the Taylor factor [34], τ_{ss} is the strength of solid solution strengthening, τ_{prec} is the strength of precipitation strengthening, and τ_{dis} is the strength of dislocation strengthening. The grain boundary strengthening follows Hall–Petch formula [35]:

$$\sigma_{\text{gb}} = k_{\text{gb}} D^{-\frac{1}{2}} \tag{6.36}$$

where k_{gb} is a constant. For the solid solution strengthening, when solute atoms dissolve into the matrix, distortion of the matrix lattice occurs, increasing the migration resistance of dislocations, contributing to the increasing strength. The degree of strengthening is affected by size difference, valence electrons, and moduli between the matrix and the solute atoms. The increase in critical shear stress of decomposition caused by the solid solution elements can be expressed as:

$$\tau_{\text{ss}} = \sum A_i c_{\text{ss},i}^{\frac{2}{3}} \tag{6.37}$$

where A_i is a constant. For the precipitation strengthening, it can be calculated by the Orowan strengthening theory. Among the many equations based on Orowan theory, the most famous one is the Orowan-Ashby equation [36]:

$$\tau_{\text{prec}} = \frac{0.13Gb}{\lambda} ln\frac{r}{b} \tag{6.38}$$

where G is the shear modulus of aluminum, and λ is the mean distance of precipitations which can be calculated from volume fraction and radius of the precipitations. For the dislocation strengthening, it can be calculated as [37]:

$$\tau_{\text{dis}} = \alpha Gb\sqrt{\rho} \tag{6.39}$$

where α is a constant.

6.1.7 Tensile Behavior Modeling

To determine the global mechanical properties of the joint, a CSM simulation was conducted to obtain the structural response of the virtual joint via ABAQUS. The local strength was calculated above, and the elongation was converted according to the ratio of the local strength to the strength of the base material (BM). The output value, global tensile strength, will be used for the final output of the subsequent NN in Sect. 6.2. All the parameters used in the modeling process are listed in Table 6.1.

6.1.8 Causative Variables and Experimental Validations

Figure 6.3 shows the surficial and macro morphologies of the FSWed obtained with typical welding parameters. The value of welding thinning is directly related to the tilt angle and plunge depth. Excessive welding thinning will contribute to a severe drop in load-bearing capacity at welding nugget zone (WNZ) [38]. The issue of weld thinning will be embodied in the subsequent computational solid mechanics (CSM) tensile model. The macrostructures reflect the formation characteristics of the joint with different combinations of welding parameters, including rotational or stationary shoulder, welding speed, and rotational velocity. The stationary shoulder, greater welding speed, and lower rotational velocity contribute to lower heat input, which is usually a necessary and sufficient condition for better joint performances under the premise of no welding defects. However, these parameters are highly coupled. It is difficult to isolate and analyze every single parameter without being dependent on pure experimental methods. Therefore, the CFD method was applied to access the static and dynamic distributions of temperature and flow behaviors in the welding process.

The calculated field data via CFD methods are shown in Fig. 6.4. The temperature around the welding tool is significantly high. The highest temperature is located at the rear side of the shoulder, which is caused by the plunge due to the tilt angle. This angle concentrates more axial pressure in the rear side of the welding tool, thereby increasing frictional heat. The temperature at the advancing side is higher than that at the retreating side. The overall distribution of the temperature field shows an ellipsoidal shape [24]. This unique temperature distribution affects the macro and microstructural evolution of the joint. Materials at high temperatures tend to migrate the grain boundaries fast and own high diffusion rates, resulting in significant DRX and precipitation evolution. Besides, the thermo-plasticized materials brought by high temperature play critical roles in joint formation. Sufficient heat input guarantees the fluidity of the materials, thus obtaining defect-free joints. Figure 6.4c shows the comparison between the calculated and experimental data. The relative error of peak temperature of less than 2% shows that the proposed CFD model is reasonable and accurate, proving that the CFD model and its field output can be reliably applied to subsequent microstructural models and NN in Sect. 6.2.

Table 6.1 Physical and empirical parameters used in numerical models

Parameter	Symbol	Value
Material constant #1 of AA2195-T8	α	1.1×10^{-8} Pa^{-1}
Material constant #2 of AA2195-T8	A	3.77×10^{8}
Material constant #3 of AA2195-T8	n	3.55
Influencing factor of AA2195-T8	γ	0.4
Activation energy of Sellars-Tegart constitutive equation of AA2195-T8	Q_{zh}	1.52×10^{5} J/mol
Gas constant	R	8.314 J/(mol K)
Solidus temperature of AA2195-T8	Ts	807 K
Empirical coefficient #1	β_1	0.8
Empirical coefficient #2	β_2	0.7
Friction coefficient	μ	0.4
Axial pressure	P	65 MPa
Interfacial constant	δ_0	0.4
Thermal conductivity of AA2195-T8	$k_D(T)$	92.5 + 0.17*T W/(m K)
Thermal conductivity of H13	k_T	29 W/(m K)
Thermal conductivity of M42	k_T	20 W/(m K)
Thermal conductivity of WC–Co	k_T	85 W/(m K)
Thermal conductivity of W–Re	k_T	21 W/(m K)
Thermal conductivity of PCBN	k_T	79 W/(m K)
Density of AA2195-T8	ρ_D	2690 kg/m^3
Density of steel	ρ_T	7800 kg/m^3
Density of WC–Co	ρ_T	15,630 kg/m^3
Density of W–Re	ρ_T	19,650 kg/m^3
Density of PCBN	ρ_T	2250 kg/m^3
Heat capacity of AA2195-T8	$c_{pD}(T)$	703 + 0.373*T + 0.00155*T^2 J/(kg K)
Heat capacity of H13	c_{pT}	470 J/(kg K)
Heat capacity of M42	c_{pT}	450 J/(kg K)
Heat capacity of WC–Co	c_{pT}	280 J/(kg K)
Heat capacity of W–Re	c_{pT}	120 J/(kg K)
Heat capacity of PCBN	c_{pT}	970 J/(kg K)
Heat transfer coefficient of upper and side wall	h_w	15 W/(m^2 K)
Heat transfer coefficient of lower wall	h_{lw}	130 W/(m^2 K)
Stefan-Boltzmann constant	σ	6.67×10^{-8} W/(m^2 K^4)
Emissivity of AA2195-T8	ε	0.35
Diffusion factor of Cu element	$D_{0,Cu}$	6.5×10^{-5} m^2/s
Diffusion factor of Li element	$D_{0,Li}$	4.7×10^{-5} m^2/s

(continued)

Table 6.1 (continued)

Parameter	Symbol	Value
Diffusion activation energy of Cu element	$Q_{d,Cu}$	1.35×10^5 J/mol
Diffusion activation energy of Li element	$Q_{d,Li}$	1.03×10^5 J/mol
Interfacial energy of T_1	$\gamma_{int,T1}$	0.23 J/m^2
Interfacial energy of θ variants	$\gamma_{int,\theta}$	0.2 J/m^2
Molar volume of T_1	$V_{m,T1}$	3.15×10^{-5} m^3/mol
Molar volume of θ variants	$V_{m,\theta}$	2.85×10^{-5} m^3/mol
Concentration constant of T_1	$C_{e0,T1}$	10.35
Concentration constant of θ variants	$C_{e0,\theta}$	9.68
Dissolution enthalpy of T_1	$Q_{e,T1}$	4.87×10^4 J/mol
Dissolution enthalpy of θ variants	$Q_{e,\theta}$	4.62×10^4 J/mol
Nucleation constant #1 of T_1	$J_{0,T1}$	1.035×10^{35} /(m^3 s)
Nucleation constant #1 of θ variants	$J_{0,\theta}$	9.67×10^{34} /(m^3·s)
Nucleation constant #2 of T_1	$A_{0,T1}$	1.92×10^4 J/mol
Nucleation constant #2 of θ variants	$A_{0,\theta}$	1.63×10^4 J/mol
Geometric constant	K_s	11.0
DRX constant #1	a	-3.43×10^{-9} /m
DRX constant #2	b	0.16×10^{-9} /m
Migration coefficient	M_0	5×10^{-6} m^2/s
Grain boundary energy	γ_b	0.3 J/m^2
Migration activation energy	Q_G	7.7×10^4 J/mol
Sensitivity coefficient of defects	χ^p	3.0
Intrinsic strength of aluminium alloys	σ_{in}	16 MPa
Taylor factor	M	3.06
Hall–Petch constant	k_{gb}	0.065 MPa s$^{1/2}$
Solid solution strengthening constant of Cu	A_{Cu}	12.431
Solid solution strengthening constant of Li	A_{Li}	23.772
Shear modulus	G	26 GPa
Burger's vector	b	2.84×10^{-10} m
Dislocation strengthening constant	α	0.3

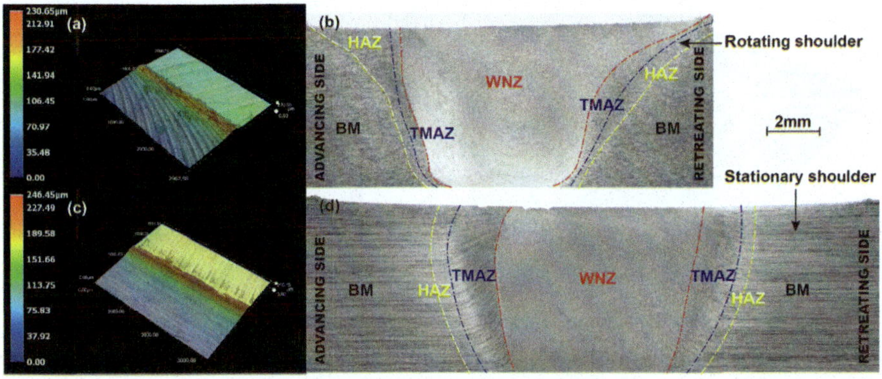

Fig. 6.3 Surficial morphologies and macrostructures of **a** and **b** rotating shoulder with 800 r/min (rotational velocity)-300 mm/min (welding speed)-0.1 mm (plunge depth)-1° (tilt angle); **c** and **d** stationary shoulder with 1500r/min (rotational velocity)-150 mm/min (welding speed)-0.1 mm (plunge depth)-1° (tilt angle)

Although the thermal cycle is one of the major prerequisites for sound joints, the joining of BM is not only obtained by thermal diffusion. Severe plastic deformation is another characteristic. The global and local material flow driven by the welding tool contributes to the main formation behaviors, including metallurgical reactions and mechanical interlocking. The material flow around the welding tool can be expressed via strain rate, a quantitative scalar of the velocity gradient tensor. A higher strain rate means results in several aspects. ① More uniform and more frequent blending of the thermo-plasticized materials occurs, improving the load-bearing capacity of the joints. ② A higher velocity gradient also brings spatial discontinuity in material flow, which potentially promotes the initiation of welding defects. ③ More heat input is introduced by the greater strain rate in the form of deformation heat. However, we tend to reduce heat input in most FSW processes. Higher heat input coarsens grains and precipitations, which deteriorates the mechanical properties of the joints. The stain rate of the thermo-plasticized AA2195 alloy is affected by the transient distributions of velocities around the welding tool. Figure 6.4b shows the contour of the strain rate around the welding tool. Strain rate decreases rapidly with increasing distance from the welding tool, consistent with the observations in other peer-reviewed literature [39]. In addition, the local strain rate is significantly affected the pin morphologies. The milling facets and threads increase the strain rate, thereby improving the joint formation and increasing the tendencies of welding defects.

The above discussion is based on the observation of the contour. However, quantitative data are necessary to directly make the outputs applicable for the subsequent NN in Sect. 6.2 to read such a highly nonlinearly-coupled field dataset. Figure 6.5a, b depict the evolution of temperature and strain rate along the streamlines of different models. These curves represent the thermal cycle and plastic deformation experienced by each cell throughout the welding process. By these changes in strain rate and temperature, we can speculate on the microstructural behaviors, including DRX

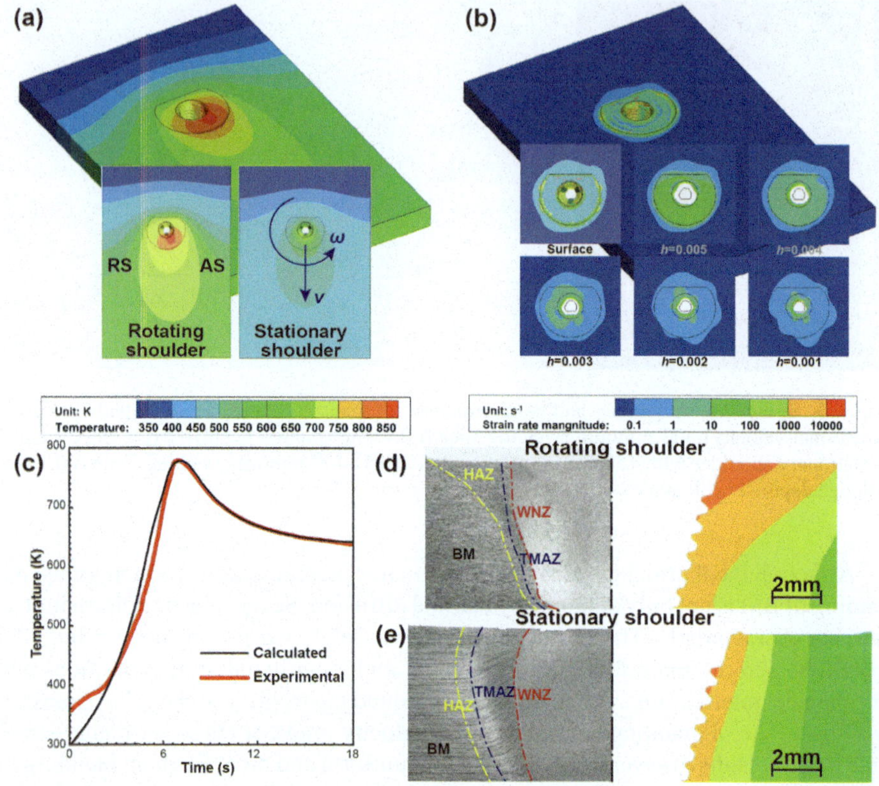

Fig. 6.4 Field outputs from the CFD models: **a** distribution of quasi-steady temperature, **b** distribution of quasi-steady strain rate, and **c** comparison of thermocouple-measured curves and numerical results. Comparisons between the calculated cross-sections of FSWed AA2195-T8 alloy with typical welding parameters: **d** rotating shoulder with rotating shoulder and **e** stationary shoulder

and precipitation evolutions. In addition, the curves extracted from the pin with complex structures bring a greater velocity gradient, as shown in Fig. 6.5c, d. The stationary shoulder is also capable of reducing the heat input. These prove that our extracted data is reliable.

It is worth noting that ameliorating the joint formation by increasing strain rate is a "double-edged sword", which has been mentioned above several times [40]. The initiation possibility of welding defects increases with the appearance of complex structures located on the welding tools. This possibility is difficult to evaluate with a simple CFD model. Lower heat input is equivalent to the better load-bearing capacity of welding joints if welding defects are ignored. This is ridiculous and unacceptable for a systematic calculation model. Therefore, a defect prediction model based on the discrete phase model was utilized to characterize the forming characteristics of the joint formation. Figure 6.6 shows the distribution of massless particles generated by a discrete phase model with different welding parameters. Compared with Fig. 6.3,

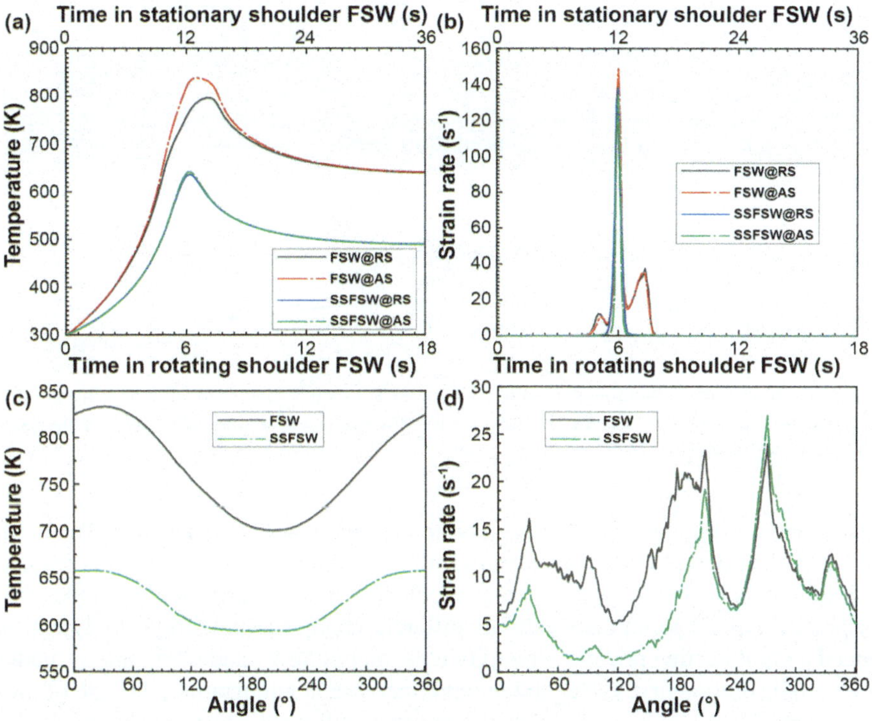

Fig. 6.5 Simulated curves of **a** temperature and **b** strain rate along the streamline. Simulated curves of **c** temperature and **d** strain rate around welding tool (SSFSW refers to stationary shoulder FSW; AS refers to advancing side; RS refers to retreating side)

the crimson area in WNZ corresponds to the welding defects. The formation index correlates well with the experimental distribution of welding defects. The performance of the joint with defects will drop sharply, which happens when the heat input is too low, and the strain rate is not sufficient. This index was applied to predict the global tensile strength of the joint, avoiding the appearance of ridiculous results. By combining the CFD model and discrete phase model, we have successfully implemented a model that can effectively predict the macrostructural evolution during the welding process.

6.1.9 Microstructural Evolutions

Since the weight fractions of Cu and Li are about 4% and 1%, no δ' phase but large amounts of T_1 and θ''/θ' phases exist [25]. However, these experimental results obtained by high-resolution transmission electron microscopy can only obtain the distributions of precipitations before and after welding. The microstructural evolution

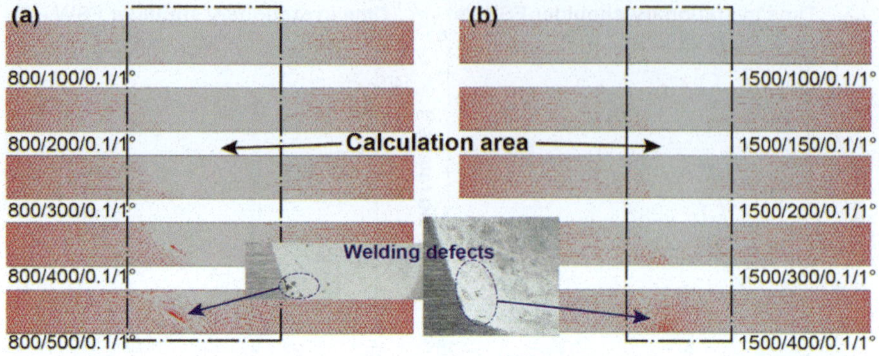

Fig. 6.6 Predictions of welding parameters via massless discrete phase model: **a** rotating shoulder and **b** stationary shoulder (The labels under each figure refer to rotational velocity/welding speed/plunge depth/tilt angle)

during the welding process cannot be monitored by experimental means. The modified KWN model provides a preferred solution. A shown in Fig. 6.7, the evolution of precipitations in the WNZ of rotating shoulder and stationary shoulder under their respective typical parameters exhibits different characteristics. Both firstly experienced a rapid heating process. Precipitations, including T_1 and θ''/θ' phases, undergoes solid solution firstly. Due to the rapid increase in temperature, the size of most strengthening phases is smaller than their critical radius, according to Eq. (6.24). The average diameter and volume fraction decrease sharply. The Cu and Li elements in the matrix enter the supersaturated solid solution state. Besides, high-density dislocations and DRXed grain boundaries are generated due to the severe plastic deformation of FSW. These sites provide the necessary conditions for the nucleation of new precipitations in the subsequent cooling process [29]. These promote the generation of the uniform and fine T_1 and θ''/θ' phases. Decreus et al. [25] pointed out that the evolution kinetics of these two strengthening phases were significantly affected by plastic deformation. Their distributions depended on the local concentration of the Li element. The T_1 phase preferentially nucleates in the area with high-density dislocations induced by the FSW process and gradually approaches the saturation state. Then, due to the local depletion of the Li element, the θ''/θ' phase appears in the area where the T_1 phase has not yet been formed. This explains why the formation of the T_1 phase is relatively earlier than the θ''/θ' phase, which proves the rationality of the calculates curve. In addition, in the case of the rotating shoulder, the precipitations dissolve entirely due to the higher heat input in the heating process. The higher temperature also brings a greater critical nucleation radius, making the initial size of precipitation larger and the nucleation rate lower. However, the volume fraction of the joint is still high. The coarser precipitations reduce their strengthening effect [33]. Furthermore, due to massive precipitations, the proportion of solid solution strengthening has also decreased, although solid solution strengthening is not the key strengthening mechanism for Al-Cu-Li alloys. By contrast, in the case of stationary

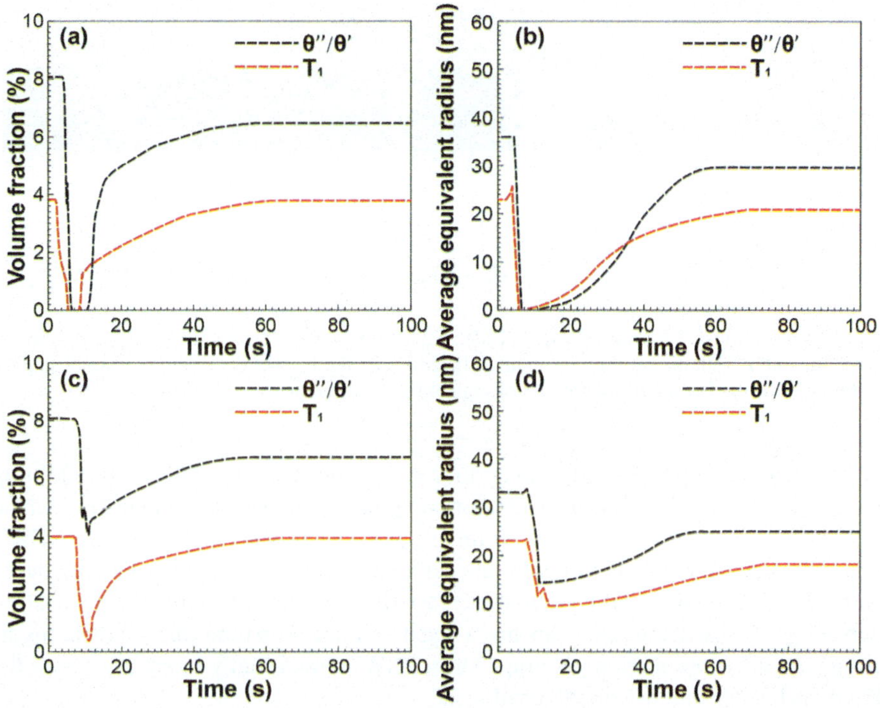

Fig. 6.7 Variations of precipitations during FSW: average radius and volume fraction from modified KWN model of **a**, **b** rotating shoulder, **c**, **d** stationary shoulder

shoulder, the great reduction in heat input causes the peak temperature to drop by about 20%. The precipitations only dissolve partially during the heating process. In the subsequent cooling process, the lower temperature contributes to a smaller critical nucleation radius, which effectively improves the precipitation strengthening effect.

The transformation of precipitations is the embodiment of the precipitation strengthening effect. However, the load-bearing capacity of the heat-treated strengthening aluminum alloy does not depend solely on this. Hall–Petch effect and dislocation strengthening are also not negligible [17]. Figure 6.8 shows the calculated and experimental results of DRX at the WNZ and HAZ based on the geometric DRX model. Due to severe plastic deformation, the materials in the WNZ quickly undergo DRX with sufficient strain rates and transform from a deformed microstructure to fine equiaxed crystal microstructures. We have assumed that the materials did not undergo the second DRX. These materials in the nugget gradually recover under the co-effects of subsequent heat and plastic deformation, resulting in a gradual increase in grain size. One can see from Eq. (6.34) that the enhancement of strain rate can effectively suppress the coarsening of the grains, while the increases in temperature will accelerate the growth of grains. Therefore, our gist is to achieve a combination

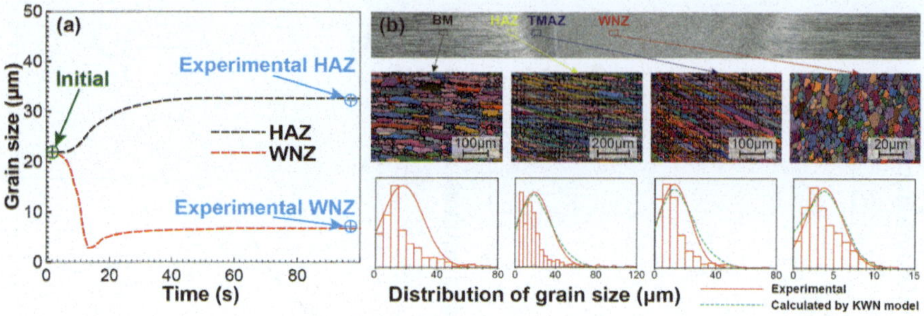

Fig. 6.8 Variations of grain sizes during FSW: **a** average diameter calculated by GDRX model of HAZ and WNZ with the stationary shoulder. Inverse pole figures and distribution of grain sizes in welding joints via EBSD, **b** distribution of grain sizes in different zones of the welds

of low peak temperature and high strain rate without welding defects. In addition, heat affected zone (HAZ) is also a major component, which usually manifests as the weakness of the joints. There is no strain rate and accumulated plastic strain, so DRX does not occur in HAZ. Throughout the calculation process, only the static recovery was observed, leading to the gradual increase of the grain sizes while the dislocation density gradually decreased. This has a negative impact on both fine-grain strengthening and dislocation strengthening. Thus, FSW joints usually break at or near the HAZ during the unidirectional tensile test.

The strengthening effects on the local area were obtained using the above analyses. The local strength of any locations at the joints can be calculated. These values consider various strengthening mechanisms and weakening effects of welding defects. However, the actual tensile strength of the whole joint is the reflection of all the cells in the welding joints. Appling these local values to a two-dimensional CSM tensile model is a good choice. The local elongation is generated by referring to the local strengthen in a linear ratio. Figure 6.9 shows the dynamic stress distribution and fracture position. This result is obtained from the typical parameters of the stationary shoulder. The stress–strain curves were obtained by calculating the stress evolution on the CSM model, as shown in Fig. 6.9b, c. It matches the experimental curves well, which shows that the above micro-evolution model is reasonable. As such, we have successfully integrated the macro and microstructural response of the FSWed joints, providing datasets for the following NN in Sect. 6.2.

6.1.10 Summary

An entire-process simulation strategy was proposed to map the macro- and microstructural response of friction stir welded aluminum alloys. AA2195-T8 alloys, owning more complex phase composition and precipitations, were chosen as the research object. This model adopts a series of numerical models, including the

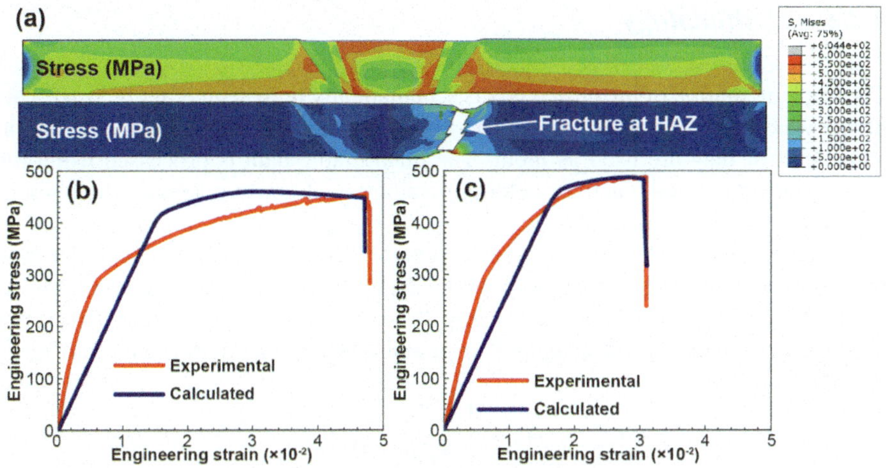

Fig. 6.9 Results of CSM tensile simulation: **a** typical results, comparison of calculated curves and tensile tests with **b** rotating shoulder and **c** stationary shoulder

computational fluid dynamics model, modified Wagner-Kampmann model, dynamic recrystallization and recovery model, and computation structural mechanics model. The comparison between experimental results and the combined model proved the rationality and accuracy of this numerical model. This combined model will provide necessary datasets for the following NN methods in the next Section.

6.2 Implementation of Neural Networks

6.2.1 Introduction

FSW, sparked with its high flow nonlinearity induced by severe plastic deformation, remains great challenges in mapping their mechanical and microstructural responses with various parameters [9]. Data-driven learning is beneficial to establish the weight tables among different datasets of parametric variables [41–43]. In machine learning methods, the primitive datasets can be obtained in peer-reviewed literature and experimental results, as well as numerical simulation in Sect. 6.1, avoiding the unacceptable cost for massive experimental results [44]. In this way, we can have the opportunity to realize a high-throughput numerical model and complete the quantitative control and optimization.

6.2.2 Methodology

The implementation of the nested NN is schematically represented in Fig. 6.10. This model was composed of two parts. The first part is a generator of field data output. Two types of data are used as input. The quantifiable data is processed using the normalization method, such as welding parameters, diameters, length, the pitch of the thread, and the number of milling facets. The unquantifiable data is processed by a one-hot method, such as thread type. This part of the NN was majorly composed of full connect and transposed convolution layers. The applied activation function was leaky ReLU [Eq. (6.40)] and ReLU [Eq. (6.41)]. The loss function and optimizer used were respectively mean squared error and ADAM [Eqs. (6.42) and (6.43)]:

$$y_i = \begin{cases} x_i, x_i \geq 0 \\ \frac{x_i}{5}, x_i < 0 \end{cases} \tag{6.40}$$

$$y_i = \begin{cases} x_i, x_i \geq 0 \\ 0, x_i < 0 \end{cases} \tag{6.41}$$

$$g_t = \nabla_\theta J(\theta_{t-1}) \tag{6.42}$$

$$\theta_{t+1} = \theta_t - \alpha g_t / \sqrt{\sum_{i=1}^{t} g_t^2} \tag{6.43}$$

where g_t represents the gradient of the t time step (a vector, including the partial derivative corresponding to each parameter), and g_t^2 represents the gradient squared at the t time step.

The output data were three field outputs, which had the same shape as the CFD results. The evaluation criterion R-squared was given as:

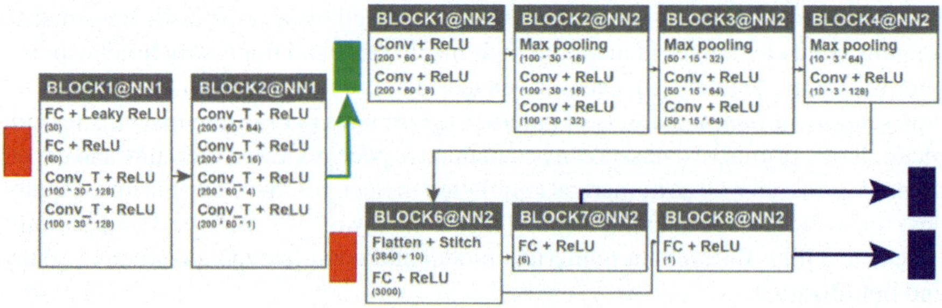

Fig. 6.10 Nested NN model

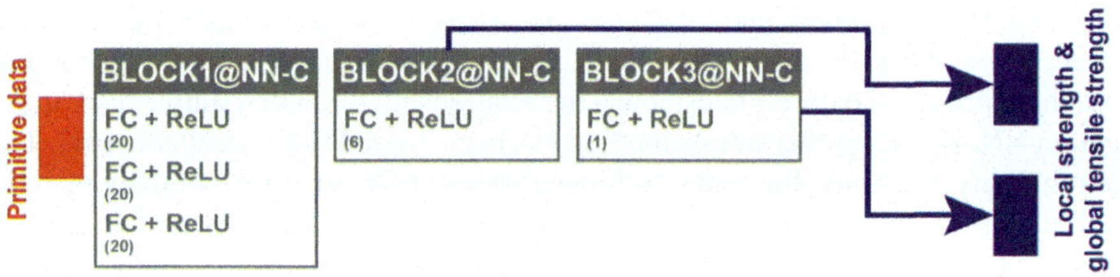

Fig. 6.11 Simple NN model for comparison

$$R^2 = 1 - \frac{\sum_i (\hat{y}_i - y_i)^2}{\sum_i (\overline{y} - y_i)^2} \tag{6.44}$$

The second part of the NN is a predictor of local and global strength. All the inputs of the first part were also used as input in this part. Moreover, the output from the first part was also used as another input data. This NN part was majorly composed of convolution, max pooling, and full connection layers. The applied activation function was ReLU. The loss function and optimizer used were respectively mean squared error and RMSprop [Eqs. (6.42) and (6.45)]. The evaluation criterion had the same form as Eq. (6.44). The output was a global tensile strength and six local strengths. When the training of the model is completed, the two NN are spliced to form the final model. Besides, a simple NN with only full connection layers and ReLU activation was also trained for contrast, as shown in Fig. 6.11.

$$\theta_t = \theta_{t-1} - \alpha * g_t / \left(10^{-8} + \sqrt{v_t}\right) \tag{6.45}$$

18 datasets from our experiments and 122 extra datasets from experiments in other peer-reviewed literature were used as test sets. 6400 CFD datasets and microstructural models validated by thermocouple temperature tests were used as the train sets and development sets with the ratio of 8:2. We adjusted the input corresponding to different sheet thicknesses in peer-reviewed papers, including length and diameter, by proportional scaling to standardize the data. The microstructural models were conducted based on the CFD results. The actual coordinates of each point were taken from the surface near the outlet surface of the domain. The temperature and strain rate cycle were obtained by retrospectively tracing the streamline. Tensorflow conducted all the training processes and validating processes.

6.2.3 Implementation Evaluation

To obtain a satisfactory NN model, great efforts are needed to provide sufficient training samples in experiments and numerical simulations. The primitive input variables include 13 types, as shown in Table 6.2. The output variables are composed of global tensile strength and local strength, where the local strength is obtained by linearly converting the microhardness in experiments. A simple method is to directly use these primitive datasets to establish the relationship through a simple backpropagation NN. However, this simple method has been proved to be unreliable [41]. This unreliability is mainly due to the following issues: FSW involves complex material nonlinearity, geometric nonlinearity, and boundary nonlinearity [4, 45, 46]. These facts and the essence of severe plastic deformation result in that the depth of simple NN cannot explain so much information. For such complex system engineering, we apply the CFD simulation and microstructural model as an effective upgrade in the complexity of the model, which was provided Sect. 6.1. This strategy converts the information contained in the numerical model into the NN model. Therefore, a nested NN was proposed. The primitive dataset is firstly converted into the required field output through a generator network and trained by the results from CFD simulation. Thus, we can embed the information in the quasi-steady temperature field and strain rate field in the welding process into the NN model, improving the rationality of massive parameters in the generator network. Then the corresponding field dataset is combined with the primitive data. The prediction of global and local strength is obtained by a predictor NN. KWN model, dynamic recrystallization model, and strengthening model provide the technical information required.

Based on the above systematic numerical model, we obtained the detailed evolution and relevant datasets of the FSWed joints. We trained these data through the nested NN method illustrated in Sect. 6.1. Figure 6.12 depicts the curves of the R-squared during the training of the nested model and a simple model for contrast as the increased iterations. The final R-squared of our generator model is about 0.967, while the R-squared of the predictor model is about 0.983. One should note that we did not pursue a lower error to avoid the occurrence of overfitting. Finally, we spliced the generator and the predictor to form the final model. The input of this model includes only the primitive datasets, while the output is local and global mechanical properties. Only the global tensile strength was used for the evaluation by test datasets. The final R-squared based on the experimental result for tests is 0.951, which can effectively predict the load-bearing capacity of the joints. In contrast, the simple NN illustrated in Sect. 6.1 cannot obtain satisfactory results. The R-squared curve fluctuated violently during training. This is because the primitive data cannot directly bear such high information density, including several field datasets. By splitting the NN into two parts and supplemented by numerical simulation results, we indirectly introduce these field data that affect the mechanical performance into the model. The nested NN model, including a generator and a predictor, has been independently trained to improve the reliability and accuracy of the model significantly.

Figure 6.13 shows the comparison of the joint performance based on the typical parameters generated by the model. Each input parameter except non- or rotating shoulder is drawn in a single figure, while this excepted parameter is depicted in each figure. The rotation velocity and welding speed are undoubtedly the most critical factors affecting the mechanical properties of the joints. Turning points are observed in both the figures, which means a compromise between low heat input and no welding defects. There are conditional influences on the load-bearing capacity for

Table 6.2 Standardization and normalization of raw inputs

Input	Method	Value
Diameter of the shoulder	Standardization	Input = Raw value/(1 mm)
Diameter of the pin	Standardization	Input = Raw value/(1 mm)
Length of the pin	Standardization	Input = Raw value/(1 mm)
Pitch of the thread	Standardization	Input = Raw value.1 mm)
Numbers of the facets	Standardization	Input = Raw value
Depth of the facets	Standardization	Input = Raw value/(0.1 mm)
Welding speed	Standardization	Input = Raw value/(100 mm/min)
Rotational velocity	Standardization	Input = Raw value/(100 r/min)[a]
Plunge depth	Standardization	Input = Raw value/(0.1 mm)
Tilt angle	Standardization	Input = Raw value/(1°)
Non- or rotational shoulder	One-hot method	Stationary shoulder (0) Rotational shoulder (1)
Type of the thread	One-hot method	Triangular (1,0,0) Trapezoidal (0,1,0) Circular (0,0,1)
Material of the tool	One-hot method	H13 (1,0,0,0,0) M42 (0,1,0,0,0) WC–Co (0,0,1,0,0) W–Re (0,0,0,1,0) PCBN (0,0,0,0,1)

Note [a] In the case of the rotating shoulder, this value is the rotational velocity of both the shoulder and pin, while in the case of the stationary shoulder, this value is the rotational velocity of the pin only

plunge depth and tilt angle. Only when the induced welding thinning is too larger than the load-bearing area of the WNZ decreases significantly, so the load-bearing area of WNZ has been lower than that of HAZ, the global tensile strength starts to decline. In addition, the tilt angle also has effects on the suppression of welding defects. Zero or much too small tilt angle tends to induce welding defects, which reduces the performance of the joints. Moreover, for various welding tools, their impacts on joint performance are mainly reflected in regulating heat input and changing material flow behavior. These behaviors are difficult to analyze through existing theories. Our prediction achieves the dissecting of valuable information. There is a turning point in the prediction of each parameter to provide design references for obtaining optimal local combinations of welding parameters. The nested NN has been built with a built-in generator and predictor, rather than a simple full connect NN using only primitive parameters. This allows our model to obtain more intrinsic information

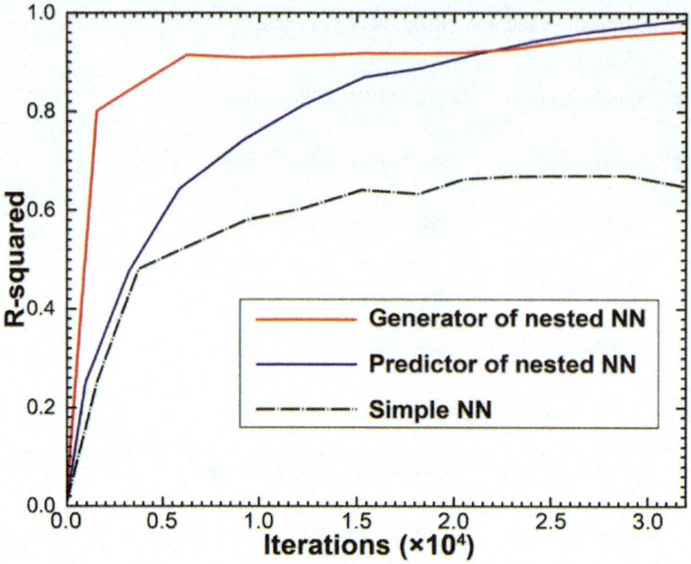

Fig. 6.12 Evolution of evaluation criterion R-squared during training of the nested NN

on the impact of various parameters on the load-bearing capacity of the joints. This combined strategy makes the model have better generalization performances and provides a potential for the transfer application.

6.2.4 Summary

(1) A nested NN, including a generator and a predictor, can accurately predict the tensile strength of the joint. The final R-squared based on the experimental result for tests is 0.951. The nested NN is better than the simple NN with low information density at forecasting the tensile performance of the welding joints. The introduction of the intermediate field data effectively improves the generalization performances.

(2) The primitive parameters affect the mechanical performances of the joints via changing the thermal cycle and the velocity gradient around the welding tools. Reducing heat input can decrease the critical nucleation size of the precipitations, increased the nucleation rate, and obtain more efficient precipitation strengthening. It also reduces the coarsening tendency of grains with or without recrystallization, which avoids the loss of grain boundary strengthening. The strain rate representing the velocity gradient improves the joint performances by increasing the diffusion rate, increasing the nucleation rate, and suppressing the grain coarsening.

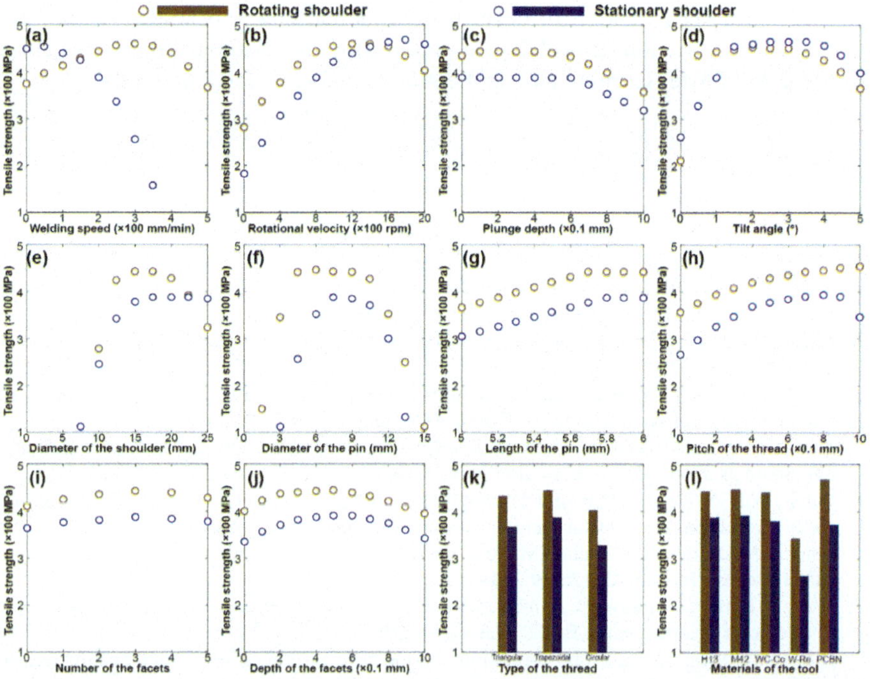

Fig. 6.13 Curves of the effect of primitive parameters on joint performances in uncoupled mode: **a** welding speed, **b** rotational velocity, **c** plunge depth, **d** tilt angle, **e** diameter of the shoulder, **f** diameter of the pin, **g** length of the pin, **h** pitch of the thread, **i** numbers of facets, **j** depth of the facets, **k** type of the thread, and **l** material of the welding tool. Default combination of parameters: 200 mm/min, 800 r/min, 0.1 mm, 1°, 16 mm, 8 mm, 6.8 mm, 0.75 mm, 3, 0.4 mm, trapezoidal, and H13, respectively

(3) Reducing heat input as much as possible and remaining an adequate load-bearing area of the welding nugget zone under the premise of no welding defects are sufficient and necessary conditions to obtain high performance of the joints. The strategy can be accessed based on the selection of welding parameter combinations and tailored to the geometric design of the welding tool to improve the material flow. Threads and milling facets on welding pin can increase the strain rate, but it also increases the tendencies of welding defects. Increasing the diameter of the welding tool enhances material flow but also increases the heat input. Therefore, a compromise needs to be realized to obtain the ideal joint performances.

(4) This strategy reveals the macro–micro dynamic evolution during the whole friction stir welding process. Combining classic numerical simulation and machine learning provides an efficient and accurate prediction of the joint load-bearing characteristics. Since the reliability of the NN can continue to increase with more datasets from both experiments and simulation, this combined strategy

has charming potential for development. In addition, this strategy is not only suitable for friction stir welded aluminum alloys but also can effectively achieve general transfer learning and application for more welded materials.

References

1. Meng X, Huang Y, Cao J, et al. Recent progress on control strategies for inherent issues in friction stir welding. Prog Mater Sci. 2021;115:100706.
2. Mishra RS, Ma ZY. Friction stir welding and processing. Mater Sci Eng R Reports. 2005;50(1–2):1–78.
3. Xie Y, Meng X, Li Y, et al. Insight into ultra-refined grains of aluminum matrix composites via deformation-driven metallurgy. Compos Commun. 2021;26:100776.
4. Xie Y, Meng X, Mao D, et al. Deformation-driven modification of Al-Li-Mg-Zn-Cu high-alloy aluminum as anodes for primary aluminum-air batteries. Scr Mater. 2022;212:114551.
5. Sun Z, Wu CS. A numerical model of pin thread effect on material flow and heat generation in shear layer during friction stir welding. J Manuf Process. 2018;36:10–21.
6. Tongne A, Desrayaud C, Jahazi M, et al. On material flow in friction stir welded Al alloys. J Mater Process Technol. 2017;239:284–96.
7. Wang T, Ramírez-Tamayo D, Jiang X, et al. Effect of interfacial characteristics on magnesium to steel joint obtained using FAST. Mater Des. 2020;192:108697.
8. Ji S, Li Z, Wang Y, et al. Joint formation and mechanical properties of back heating assisted friction stir welded Ti-6Al-4V alloy. Mater Des. 2017;113:37–46.
9. Simar A, Bréchet Y, de Meester B, et al. Integrated modeling of friction stir welding of 6xxx series Al alloys: Process, microstructure and properties. Prog Mater Sci. 2012;57(1):95–183.
10. Huang Y, Xie Y, Meng X, et al. Numerical design of high depth-to-width ratio friction stir welding. J Mater Process Technol. 2018;252:233–41.
11. Ji SD, Wen Q, Li ZW. A novel friction stir diffusion bonding process using convex-vortex pin tools. J Mater Sci Technol. 2020;48:23–30.
12. Su H, Wu CS, Pittner A, et al. Thermal energy generation and distribution in friction stir welding of aluminum alloys. Energy. 2014;77:720–31.
13. Arora A, De A, Debroy T. Toward optimum friction stir welding tool shoulder diameter. Scr Mater. 2011;64(1):9–12.
14. Wu LH, Hu XB, Zhang XX, et al. Fabrication of high-quality Ti joint with ultrafine grains using submerged friction stirring technology and its microstructural evolution mechanism. Acta Mater. 2019;166:371–85.
15. dos Santos JF, Staron P, Fischer T, et al. Understanding precipitate evolution during friction stir welding of Al-Zn-Mg-Cu alloy through in-situ measurement coupled with simulation. Acta Mater. 2018;148:163–72.
16. Irani M, Lim S, Joun M. Experimental and numerical study on the temperature sensitivity of the dynamic recrystallization activation energy and strain rate exponent in the JMAK model. J Mater Res Technol. 2019;8(2):1616–27.
17. Xie Y, Huang Y, Wang F, et al. Deformation-driven metallurgy of SiC nanoparticle reinforced aluminum matrix nanocomposites. J Alloys Compd. 2020;823:153741.
18. Huang Y, Wan L, Meng X, et al. Probe shape design for eliminating the defects of friction stir lap welded dissimilar materials. J Manuf Process. 2018;35:420–7.
19. Pei X, Dong P. A selectively-coupled shear localization model for friction stir welding process window estimation. Int J Mach Tools Manuf. 2017;123:89–104.
20. Cho HH, Hong ST, Roh JH, et al. Three-dimensional numerical and experimental investigation on friction stir welding processes of ferritic stainless steel. Acta Mater. 2013;61(7):2649–61.

21. Zienkiewicz OC, Cormeau IC. Visco-plasticity-plasticity and creep in elastic solids-A unified numerical solution approach. Int J Numer Methods Eng. 1974;8(4):821–45.
22. Zhao W, Wu C. Constitutive equation including acoustic stress work and plastic strain for modeling ultrasonic vibration assisted friction stir welding process. Int J Mach Tools Manuf. 2019;145:103434.
23. Mendez PF, Tello KE, Lienert TJ. Scaling of coupled heat transfer and plastic deformation around the pin in friction stir welding. Acta Mater. 2010;58(18):6012–26.
24. He X, Gu F, Ball A. A review of numerical analysis of friction stir welding. Prog Mater Sci. 2014;65:1–66.
25. Decreus B, Deschamps A, de Geuser F, et al. The influence of Cu/Li ratio on precipitation in Al-Cu-Li-x alloys. Acta Mater. 2013;61(6):2207–18.
26. Deschamps A, Brechet Y. Influence of predeformation and ageing of an Al-Zn-Mg Alloy-II. Modeling of precipitation kinetics and yield stress. Acta Mater. 1998;47(1):293–305.
27. Han J, Thomas SL, Srolovitz DJ. Grain-boundary kinetics: a unified approach. Prog Mater Sci. 2018;98:386–476.
28. Ruoff AL, Balluffi RW. Strain-enhanced diffusion in metals. II. Dislocation and grain-boundary short-circuiting models. J Appl Phys. 1963;34(7):1848–53.
29. Xie Y, Meng X, Wang F, et al. Insight on corrosion behavior of friction stir welded AA2219/AA2195 joints in astronautical engineering. Corros Sci. 2021;192:109800.
30. Du Q, Tang K, Marioara CD, et al. Modeling over-ageing in Al-Mg-Si alloys by a multi-phase CALPHAD-coupled Kampmann-Wagner numerical model. Acta Mater. 2017;122:178–86.
31. Ma ZY, Feng AH, Chen DL, et al. Recent advances in friction stir welding/processing of aluminum alloys: microstructural evolution and mechanical properties. Crit Rev Solid State Mater Sci. 2018;43(4):269–333.
32. Speight MV. Growth kinetics of grain-boundary precipitates. Acta Metall. 1968;16(1):133–5.
33. Xie Y, Meng X, Chang Y, et al. Ameliorating strength-ductility efficiency of graphene nanoplatelet-reinforced aluminum composites via deformation-driven metallurgy. Compos Sci Technol. 2022;219:109225.
34. Hu P, Liu Y, Zhu Y, et al. Crystal plasticity extended models based on thermal mechanism and damage functions: application to multiscale modeling of aluminum alloy tensile behavior. Int J Plast. 2016;86:1–25.
35. Hansen N. Hall-Petch relation and boundary strengthening. Scr Mater. 2004;51(8):801–6.
36. Chen B, Shen J, Ye X, et al. Length effect of carbon nanotubes on the strengthening mechanisms in metal matrix composites. Acta Mater. 2017;140:317–25.
37. Huskins EL, Cao B, Ramesh KT. Strengthening mechanisms in an Al-Mg alloy. Mater Sci Eng A. 2010;527(6):1292–8.
38. Guan M, Wang Y, Huang Y, et al. Non-weld-thinning friction stir welding. Mater Lett. 2019;255:126506.
39. Feulvarch E, Roux JC, Bergheau JM. A simple and robust moving mesh technique for the finite element simulation of friction stir welding. J Comput Appl Math. 2013;246:269–77.
40. Huang Y, Xie Y, Meng X, et al. Joint formation mechanism of high depth-to-width ratio friction stir welding. J Mater Sci Technol. 2019;35(7):1261–9.
41. Du Y, Mukherjee T, Mitra P, et al. Machine learning based hierarchy of causative variables for tool failure in friction stir welding. Acta Mater. 2020;192:67–77.
42. Hu W, Ma Z, Ji S, et al. Improving the mechanical property of dissimilar Al/Mg hybrid friction stir welding joint by PIO-ANN. J Mater Sci Technol. 2020;53:41–52.
43. Wang B, Hu SJ, Sun L, et al. Intelligent welding system technologies: state-of-the-art review and perspectives. J Manuf Syst. 2020;56:373–91.
44. Du Y, Mukherjee T, Debroy T. Conditions for void formation in friction stir welding from machine learning. Npj Comput Mater. 2019;5(1):68.
45. Mesmoudi S, Timesli A, Braikat B, et al. A 2D mechanical-thermal coupled model to simulate material mixing observed in friction stir welding process. Eng Comput. 2017;33(4):885–95.
46. Mesmoudi S, Braikat B, Lahmam H, et al. Three-dimensional numerical simulation of material mixing observed in FSW using a mesh-free approach. Eng Comput. 2020;36(1):13–27.

Chapter 7
Surface Modification via Friction Stir Processing

7.1 Surface Composite Fabricated by Direct Friction Stir Processing

7.1.1 Introduction

Friction stir processing (FSP), a new surface modification method has been introduced to produce surface composite [1, 2]. SiC particles [1], nickel particles [3], Al_3Ti and Al_2O_3 nanoparticles [4] were separately embedded into Al matrix by FSP to produce composite. Morisada et al. fabricated AZ31 Mg matrix composite with SiC [5] and carbon nanotubes [6]. The processing of surface composite by FSP was carried out at temperature under the melting point of the substrate. Therefore, the interface reaction between reinforcements and metal matrix was very limited. Before FSP, composite particles needed to be placed on the base metal. Mishra et al. [1] added SiC powders into a small amount of methanol, and then applied them to the surface of the plate before FSP. The SiC powders could also be filled into a groove on the plate before FSP [5–7]. In order to prevent the particles from being displaced out of the groove during FSP, surface "repair" was accomplished by Lee et al. [8] with a modified FSP tool that only had a shoulder and no pin. The particles were confined in the groove after the surface "repair" process. For the process of fabricating surface composite by FSP, a rotating FSP tool was inserted into the base metal after the reinforcement particles were placed on the base metal. The FSP tool always had pin on it to fabricate composite, and the fabricated composite was in the stir zone which was produced by the pin [9]. These preplaced particles were usually prone to be driven along the surface or groove by the advancing tool. The remaining particles on the base metal were stirred into the substrate in a gathered way. Therefore, they had difficulty in dispersing well and tend to assemble especially when the size of reinforcement particles was under a certain value. Lee et al. [8] reported that the size of clustered reinforcements became smaller and smaller with increasing FSP passes.

Y. Huang et al., *Friction Stir Welding and Processing*, Materials Forming, Machining and Tribology, https://doi.org/10.1007/978-981-99-8688-0_7

Faraji et al. [10] revealed that the particle distribution pattern was mainly affected by the FSP parameters such as traverse speed, rotational speed, and FSP pass number. Better particle distribution can be achieved in higher rotational speed/welding speed ratios and subsequent FSP passes. The change of tool rotation direction between FSP passes would lead to significantly more homogenous dispersion of SiC particles in metal matrix composite [9]. Besides, the effect of probe profile on microstructure of AZ31/Al_2O_3 composites fabricated by FSP was also investigated [11]. The influence of FSP and reinforcement on the microstructure and wear performance of aluminum matrix composite was reported [12, 13].

In this section, a hollow direct friction stir processing (DFSP) tool consisting of a concave shoulder and no pin on it was designed, as shown in Fig. 7.1. The reinforcement particles were not preplaced on the base metal but in the through-hole of the DFSP tool. As the rotating tool advanced along the workpiece surface, the reinforcement particles directly and dispersedly flowed into the enclosed space between the workpiece and the rotating shoulder through the through-hole in the tool. Therefore, instead of being driven along the workpiece surface by the advancing tool, these particles were confined in the space between the concave shoulder and workpiece. And they were stirred and pressed into the substrate dispersedly. Thus, they dispersed in a scattered and uniform way in the stir zone (SZ) after only one pass. Magnesium alloys, as the promising materials in the aerospace and automobile industries, were chosed. The surface composite was fabricated on the surface of the magnesium via the designed DFSP method to strengthen its surface mechanical properties, whose use-life was thus extended.

7.1.2 Materials and Experimental Procedure

Commercially available SiC powder (mean diameter: 1 μm, 99.9% pure) and AZ31 plate (5 mm thickness) of rolling state were used in this section. And the mechanical properties and chemical compositions of the AZ31 plate are listed in the Table 7.1. The DFSP to fabricate surface composite was performed by the FSW-3LM-003 machine. The DFSP tool made of high speed steel has a columnar shape (D_1: ϕ24 mm) with a through-hole (D_2: ϕ8 mm) in the midpoint of it. The shoulder of the tool is concave of which the inclination angle (θ') is 10°, as shown in Fig. 7.2. During the DFSP, the tool was titled back [lead angle (θ)] at 0.5°. A constant tool rotational speed of 400 r/min counterclockwise and a constant welding speed of 30 mm/min were adopted. As for the plunge depth (d), it should be less than the height difference (d_1) between the bottom of through-hole and the back edge of the tilted shoulder so that the particles exit will not be blocked. And the plunge depth should also be more than the height difference (d_2) between the front edge and the back edge of the tilted shoulder, which guarantees that the shoulder front can touch workpiece and that the concave shoulder and workpiece would form an enclosed space, as shown in Fig. 7.2. The value of d_1 and d_2 can be calculated by the equations below:

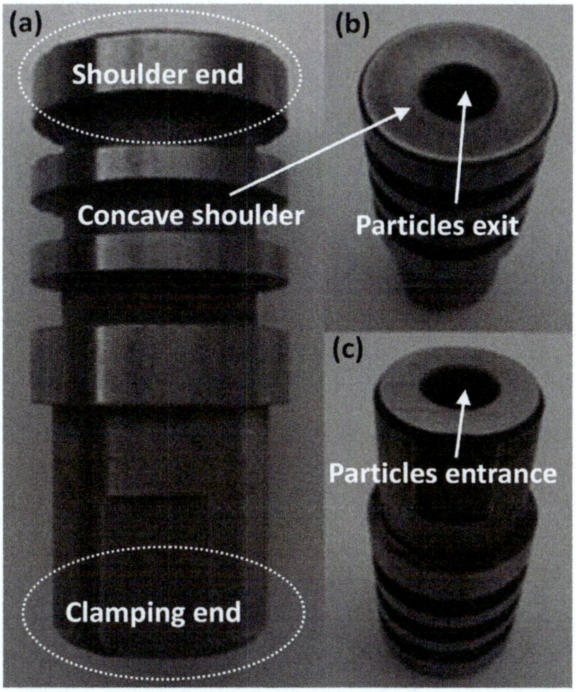

Fig. 7.1 DFSP tool: **a** front view of DFSP tool, **b** view of shoulder end, and **c** view of clamping end

$$(d_1 = (D_1 - D_2)\sin(\theta + \theta\prime))/(2\cos\theta\prime) \tag{7.1}$$

$$d_2 = D_1 \sin\theta \tag{7.2}$$

The calculated value of d_1 and d_2 are 1.48 and 0.21 mm, respectively. Thus, the plunge depth should be between 0.21 and 1.48 mm. The plunge depth of 0.3 mm was adopted during the DFSP.

Before DFSP, SiC particles were loaded in the hollow DFSP tool. The particles exit on the shoulder end of the DFSP tool was sealed with a piece of aluminum foil, so that the reinforcements were held in the hollow tool. The moment the rotating tool touched the workpiece, the foil was rubbed away and the SiC powders which were preloaded in the hollow tool flowed into the enclosed space between the rotating

Table 7.1 Mechanical properties and chemical compositions of AZ31 magnesium alloy

Mechanical properties			Chemical compositions/wt.%						
Tensile strength/ MPa	Elongation/%	Hardness/ HV	Si	Fe	Cu	Mn	Al	Zn	Ni
240	12	55	0.007	0.003	0.007	0.42	3.12	1.12	0.001

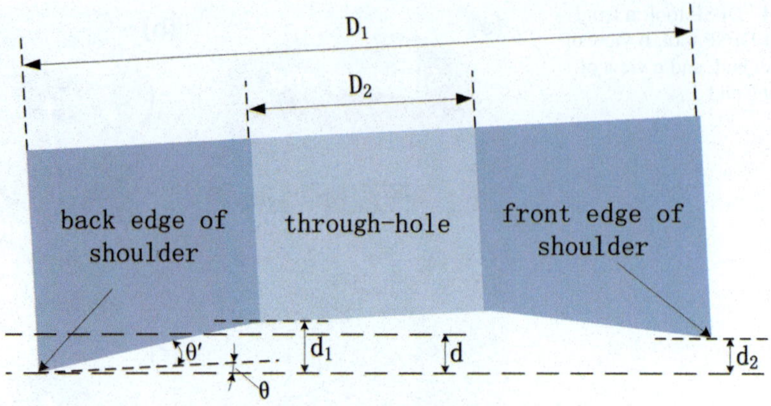

Fig. 7.2 Schematic illustration of scoping the plunge depth of DFSP tool

concave shoulder and the workpiece, and were immediately pressed in the workpiece by the shoulder, as shown in Fig. 7.3. Thus, the particles were stirred into the base metal dispersedly by the rotating DFSP tool just like a "planter".

To contrast the surface composite fabrication by FSP and DFSP, the conventional process to fabricate surface composite by FSP was carried out. The reinforcement particles were preplaced on the workpiece surface, and then the rotating FSP tool was inserted into the workpiece and traveled along its surface, as shown in Fig. 7.4. The size and shape of FSP tool and DFSP tool were the same except that the FSP tool did not have a through-hole in it. During the process to fabricate surface composite by FSP, the lead angle of the tool was 1° and the other process parameters of FSP were completely the same with those of the DFSP. The lead angle was increased to 1° so that the front edge of the shoulder could not contact with the plate when the plunge

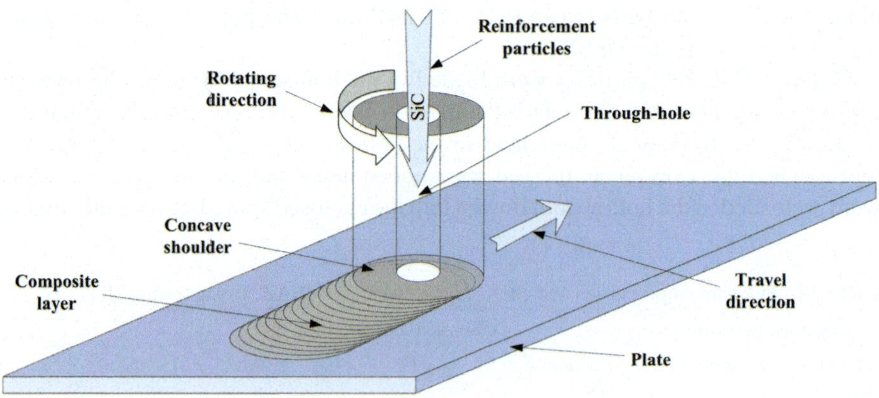

Fig. 7.3 Schematic illustration of the DFSP to fabricate surface composite

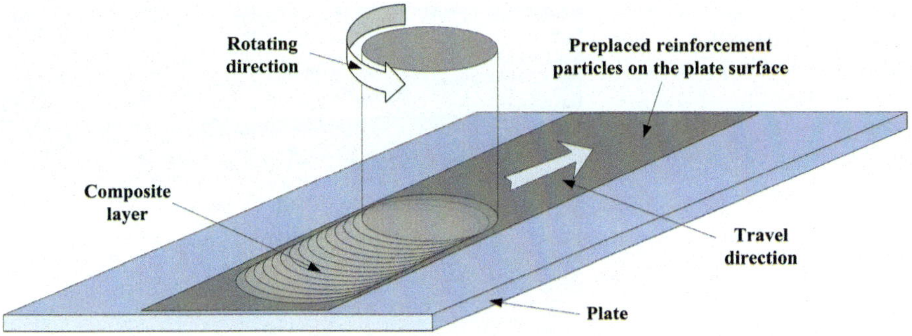

Fig. 7.4 Schematic illustration of the conventional FSP to fabricate surface composite

depth remained to be 0.3 mm. The preplaced particles on the substrate surface were not completely pushed away by the FSP tool.

The pure FSP can be used to refine the grain size in matrix with a relatively uniform microstructure in aluminum alloy [14, 15] and magnesium alloy [16]. To investigate the effect of the particles on the microstructure and properties of the AZ31, another experiment of FSP without SiC particles loaded in the hollow tool (pure FSP) was carried out using the same DFSP tool, and the process parameters of the experiment were completely the same with those of the DFSP with SiC particles.

The distribution of SiC particles in the SZ of the AZ31 were observed by the optical microscope (OM, Olympus-MPG3) and scanning electron microscopy (SEM, Hitachi-S4700). And the grain size was measured using the mean linear intercept method. The micro-hardness was measured by a micro-Vickers hardness tester (HVS-1000Z) using a load of 50 g for 10 s according to China national standard GB/T 4340.1–2009. And the dry sliding wear tests were carried out on a CJSIIIA ball-disk wear instrument with a SiN ball (ϕ8 mm), using a load 500 g for 6 min, and the rotational speed was 300 r/min. The wear scratches were examined by OM.

7.1.3 Microstructure

Figure 7.5 shows OM of the transverse section of AZ31 Mg/SiC$_p$ surface composite fabricated by the conventional FSP. In the Fig. 7.5, the dark band-regions are agglomerated SiC particles. And the image of the assembled particles was further enlarged by the SEM, as shown in Fig. 7.6. Figure 7.6 shows that the SiC particles gathered into clusters of which the diameter reached dozens of micrometer. During the conventional process to fabricate surface composite by FSP, the lead angle was increased to 1° and the front edge of the shoulder did not touch the plate. A certain amount of the preplaced particles were driven away along the workpiece surface by the advancing and rotating tool. And the remaining particles on the substrate surface were stirred

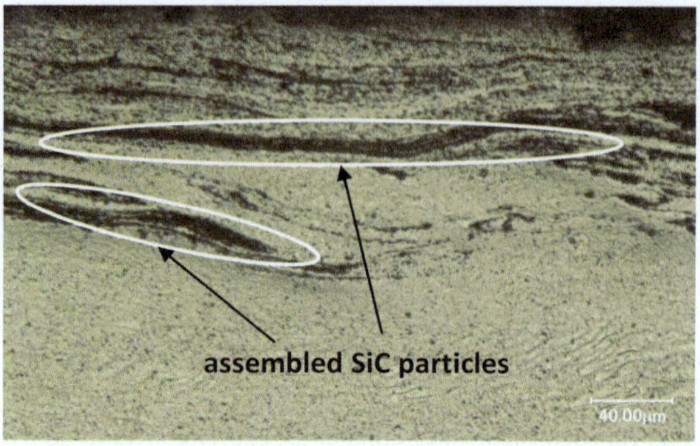

Fig. 7.5 OM image of SZ of the conventional FSPed AZ31 with the SiC particles

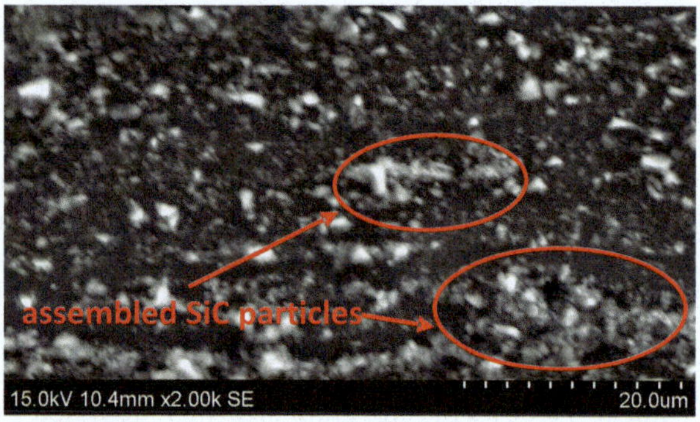

Fig. 7.6 SEM image of composite layer of conventional FSPed AZ31 with the SiC particles

into the base metal in clusters. Thus, they dispersed not well and assembled seriously in the SZ.

Figure 7.7a shows the microstructure of the as-received AZ31. During the pure FSP, the grains in the SZ experienced severe plastic deformation and recrystallization which produced a SZ consisting of quite fine grains, as shown in Figure 7.7b. Figure 7.7c shows the microstructure in the transverse section of DFSPed AZ31. The composite layer was approximately 150 μm thick. During DFSP, the particles were pressed and stirred into the base metal as they dispersedly flowed out from the through-hole to the enclosed space between the concave shoulder and workpiece. Therefore, the particles dispersed in a homogeneous and scattered way in the SZ after only one pass process. In the previous fabrication of composite by FSP, the reinforcements would always gather seriously in the SZ after one pass process. The method of increasing FSP passes was always adopted to decrease the size of clustered reinforcements. Besides, the previous FSP tool always had a pin on it. And the produced composite layer was only in the stir zone produced by the FSP pin. In this section, a pinless DFSP tool was used to produce the composite. And the composite layer was in the stir zone produced by the DFSP shoulder. And the dispersed particles led the grains to be further refined by the DFSP through the recrystallization. Measured by the mean linear intercept method, the average grain size were 16.57,

Fig. 7.7 OM images of **a** as-received AZ31, **b** SZ of the one pass pure FSPed AZ31, and **c** SZ of the DFSPed AZ31 with the SiC particles

5.21 and 1.24 μm for the as-received AZ31, one pass pure FSPed AZ31 and DFSPed AZ31, respectively.

Besides, the SEM of the composite layer in the DFSPed AZ31 is shown in Fig. 7.8a and b. The reinforcement particles dispersed in a scattered and homogenous way in the SZ, as shown in Fig. 7.8a. And the grain boundary was pinned by the SiC particles, as shown in Fig. 7.8b. Contrasting the SEM of the composite layer produced by the conventional FSP (Fig. 7.6) with that by the DFSP (Fig. 7.8a), the reinforcement particles were more scattered and homogenous in the composite layer produced by the DFSP than the conventional FSP.

7.1.4 Micro-hardness and Surface Wear Properties

Figure 7.9 shows the calculated and measured micro-hardness of transverse section of the samples. The micro-hardness of the samples can be estimated using the Hall-Patch equation [17]:

$$HV = 40 + 72d^{-1/2} \tag{7.3}$$

where d is the average grain size of the AZ31. And based on Eq. (7.3), the calculated micro-hardness of the as-received AZ31, the one pass pure FSPed AZ31 and DFSPed AZ31 with SiC were calculated to be 57.69, 62.10 and 104.66 HV.

The measured average micro-hardness of the as-received AZ31 is 57.77 HV. The pure FSP increases the micro-hardness of surface section by refining the grains in the SZ. The measured micro-hardness of the one pass pure FSPed AZ31 is 62.36 HV. The DFSP increases the micro-hardness further to 115.51 HV in the SZ, which depends on the further refining by the presence of SiC particles and the high hardness of SiC particles.

The measured micro-hardness value shows a good agreement with the calculated results except the DFSPed sample's, as shown in Fig. 7.9. For the DFSPed sample, the calculated value indicated the micro-hardness contributed only by the grain refinement via the Hall-Patch equation, yet the measured value indicated the

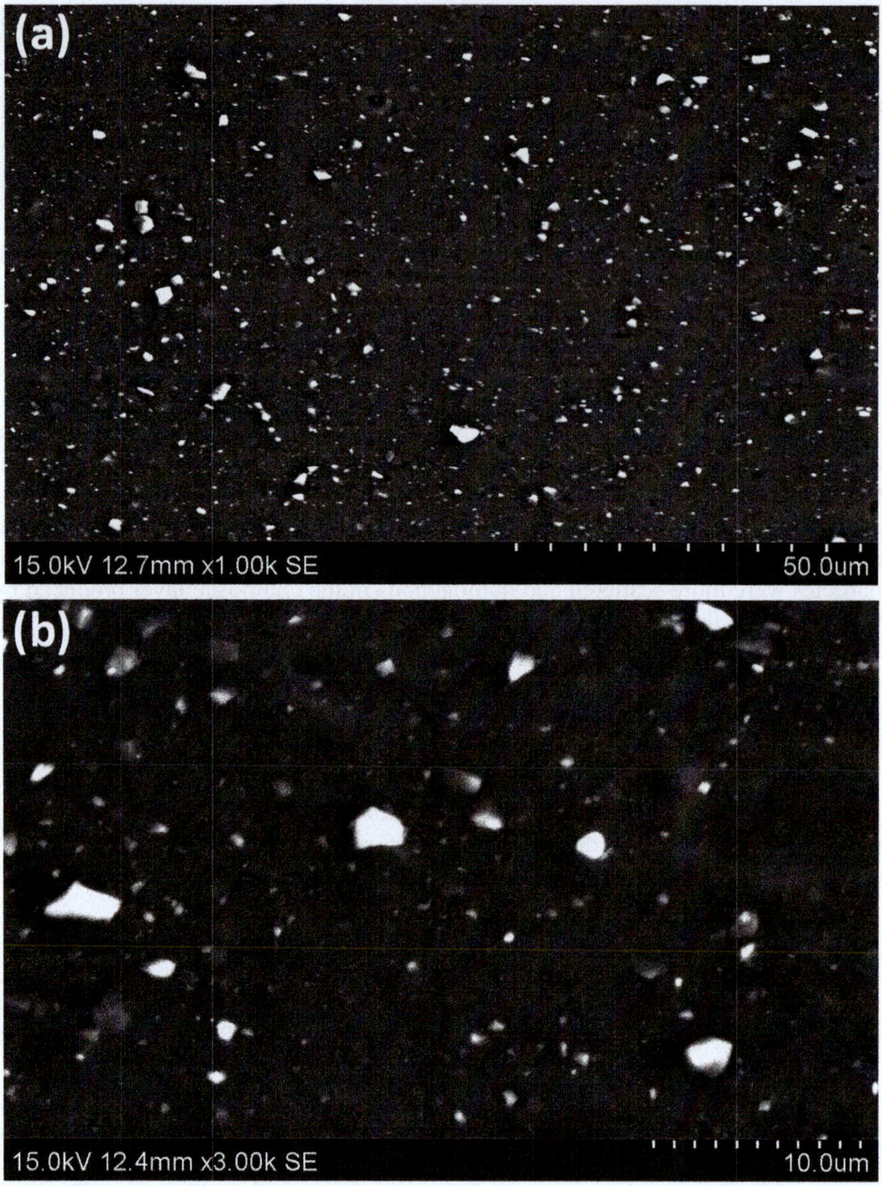

Fig. 7.8 SEM images of the composite layer of DFSPed AZ31 with the SiC particles: **a** 1000× and **b** 3000× magnification

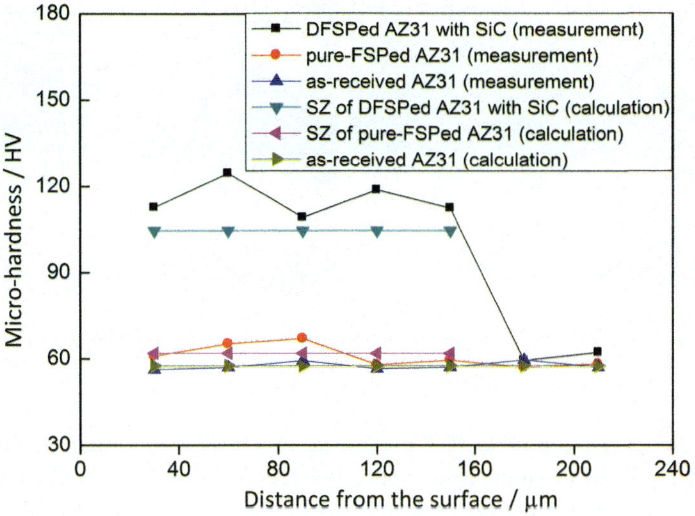

Fig. 7.9 Micro-hardness profile of the transverse section in as-received AZ31, pure-FSPed AZ31, and DFSPed AZ31 with the SiC particles

micro-hardness contributed by both the grain refinement and the reinforcement particles. Therefore, its measured value was higher than the calculated value. Furthermore, the DFSPed sample's micro-hardness had a higher fluctuation than those of other samples', as shown in Fig. 7.9. The main reason for this is the presence of the SiC particles, the micro-hardness value was higher when the indention was close to or just on the reinforcement particles, and decreased when the indention was on the matrix around the SiC particles.

The surface of base metal was worn by the rotating ball for 6 min, and the OM of the surface wear plows were shown in the Fig. 7.10a. And the plow width (PW) is approximately 620 μm. After the DFSP with SiC, a composite layer with dispersed and homogeneous SiC particles and quite fine grains was fabricated. The SiC particles have high wear resistance and the fine grains have higher wear property than those in base metal. Therefore, the surface wear property of the DFSPed AZ31 improved a lot, and the PW was reduced to about 410 μm, as shown in Fig. 7.10b.

7.1.5 Summary

The DFSP to produce surface composite was proposed, and the SiC particle dispersed AZ31 was successfully produced by the DFSP. The microstructure (the grain size and the dispersion of the SiC particles) and micro-hardness of the SZ and the surface wear property of the DSPed AZ31 with SiC were observed and measured. The obtained results can be summarized as follow:

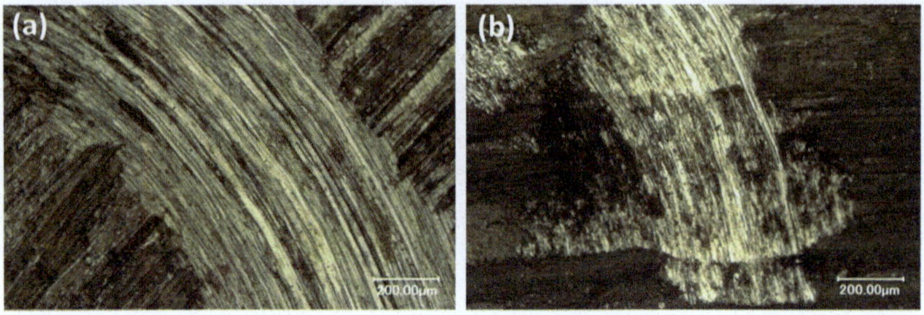

Fig. 7.10 OM images of worn surfaces in the **a** as-received AZ31 and **b** DFSPed AZ31 with the SiC particles

(1) For the fabrication of surface composite by DFSP, the reinforcement particles were confined in an enclosed space between the concave shoulder and workpiece. And they were pressed and stirred into the base metal as they dispersedly flowed out from the through-hole to the enclosed space. The SiC particles were dispersed and homogenous in SZ fabricated by DFSP compared with the conventional FSP by only one pass.
(2) A stable composite layer up to 150 μm thick was produced under the optimal process parameters of the DFSP: rotational speed = 400 r/min, welding speed = 30 mm/min, $\theta = 0.5°$, $d = 0.3$ mm, and the average grain size in the composite layer is 1.24 μm. The grains in the SZ are further refined by the DFSP with SiC through the recrystallization compared with the pure FSP.
(3) The micro-hardness of the SZ with SiC fabricated by DFSP increases to 115.51 HV. And in the wear property test, the plow width on the DFSPed AZ31 surface with SiC decreases about 210 μm.

7.2 Cryogenic Surface-Grinding Assisted Friction Stir Processing

7.2.1 Introduction

Despite wide application in the fabrication of structural materials with high strength, alloying needs to improve in alloy element recycling [18]. However, ductility and toughness frequently drop sharply as the grain size fall to 1 μm. Lu et al. [19] suggested that combining grains with different sizes could maintain the strength advantage of refined grains and reduce the loss of ductility. Fang et al. [20] fabricated a gradient nano-grained structure of copper via surface mechanical grinding treatment (SMGT). The processed sample exhibited high strength. Meanwhile, the tensile strain could exceed 100% without fractures. Gradient distribution of dislocation densities in high-purity zirconium sheets was obtained via asymmetrical rolling and subsequently partial annealing, which resulted in an extraordinary synergy of high strength and uniform tensile ductility [21].

Severe plastic deformation methods were used to fabricate gradient structures such as FSP [8, 9], SMAT, and SMGT [22]. The mechanism of grain refinement in SMAT was identified in terms of the formation of dislocation cells and twin-matrix lamellae [23]. Zhang et al. [24] suggested that grain refinement in FSP was achieved through dynamic recovery and dynamic recrystallization (DRX). In this section, we adopted in-situ surficial cryogenic grinding as a simple method to get charming gradient ultra-fine grained structure. The formation and evolution of ultra-fine grained microstructure were revealed. Two modes of grain refinement were identified in detail, corresponding to the different rotational velocities.

7.2.2 Materials and Experimental Procedure

Copper sheets (100 mm × 100 mm × 3 mm) of 99.9% purity were used. The tool was illustrated in Fig. 7.11a, whose major parts were 4 grinding ZrO_2 balls (spherical centers of the balls were 14 mm from the tool's axis and the balls' diameter was 8 mm). More details of the tool were illustrated [25]. When the balls plunged into the surface, severe deformation on the topmost layer was induced. Strain, strain rate, and temperature decreased with increasing depth, which was conducive to the formation of gradient structure. The processing speed and the plunge depth of the balls were respectively 30 mm/min and 0.3 mm. During processing, the copper sheets were immersed in liquid nitrogen to keep the material in cryogenic state, as shown in Fig. 7.11b. The rotational velocities were 50 r/min and 500 r/min, denoted as sample-50 and sample-500, respectively.

After processing, cross-section specimens were cut from the samples. Macrostructures of the matrix and the deformed region were characterized by OM. The detailed microstructures were characterized by SEM at an operating voltage of 20 kV. Microhardness of each samples was measured using a Vickers microhardness tester with a load of 50 g and a dwelling time of 10 s. Transmission electron microscope (TEM, FEI TALOS F200x) operated at a voltage of 200 kV was carried to further characterize the microstructural evolution and reveal the modes of grain refinement.

7.2.3 Microstructure Evolution and Properties

Typical macrostructure of stir zone was shown in Fig. 7.11d, which was characterized by a depth-dependent gradient structure. The stir zone was divided into three regions with different features: ultra-fine grained zone (UGZ), thermo-mechanically affected zone (TMAZ) and base metal (BM). Thickness of the UGZ was approximately 200 μm, while thickness of the TMAZ was 200–300 μm. Grains in TMAZ

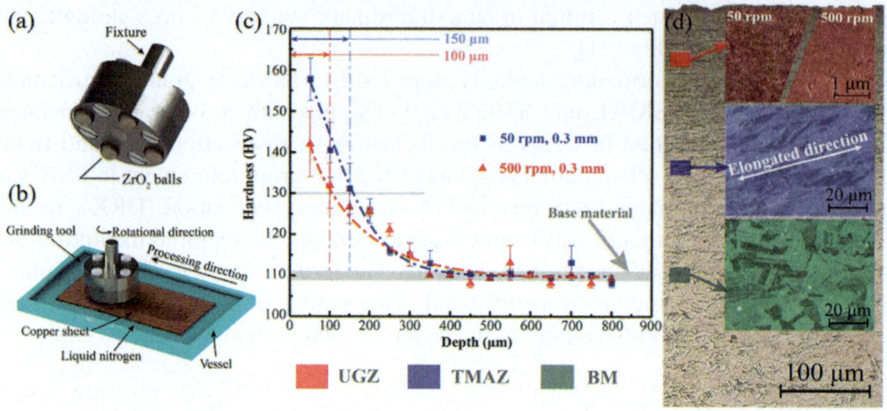

Fig. 7.11 Experimental method, microhardness distribution and macrostructure: **a** Illustration of the in-situ grinding tool, **b** illustration of the in-situ surficial cryogenic grinding, **c** microhardness variation at different rotational velocities, and **d** macrostructure of the gradient structure and the corresponding SEM images (inlens mode)

were elongated along the travel direction. The deepest zone of the sheet kept the as-received status, composed of coarse grains. Microhardness test results were shown in Fig. 7.11c. The microhardness decreased gradually from the surficial materials to the undeformed matrix at two velocities alike while some difference still existed yet. Peak value of hardness at low rotational velocity was higher than that at high rotational velocity. The regions where microhardness was over 130 HV were defined as strengthened layers. Strengthened layers' thicknesses of sample-50 and sample-500 were respectively 150 μm and 100 μm. In consideration of purity of the materials, there was no precipitates, which meant that the dominating strengthening mechanism was refinement strengthening. The microhardness of sample-50 maintained was higher than that of sample-500 at the same depth, which indicated that strengthening at low rotational velocity was more effective. Such result was attributed to two routes of grain refinement and was confirmed by following TEM observations then.

Difference between the two hardness curves attested difference of the microstructures corresponding to two rotational velocities. Figure 7.12 showed TEM bright field images of the ultra-fine grains. At the lower strain rate, elongated lamellar structures with small misorientation boundaries were observed (Fig. 7.12a). Grain size in transverse direction was less than 250 nm. Grains were composed of many wavy extended dislocation walls. Dense dislocation walls roughly perpendicular to the grain boundaries were substantially observed inside these elongated fine grains (Fig. 7.12c). Such features indicated that the ultra-fine grains evolved through grain splitting. In the beginning, strain-induced dislocations formed randomly inside original grains, and dislocation accumulated as strain increased. Then those activated dislocations were easier to gather around grain boundaries to minimize the total energy and form dislocation walls (Fig. 7.12b and c). Geometric necessary dislocations proliferated persistently as plastic strain further increased. Such dislocation activities led to

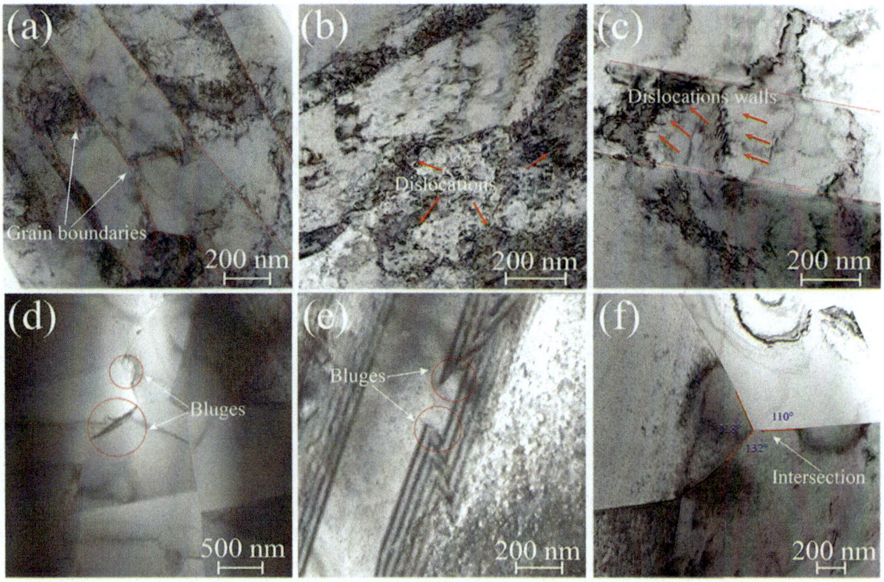

Fig. 7.12 TEM images of processed materials at different strain rates: **a** Ultra-fine lamellar structure at lower strain rate, **b** dislocation clusters at lower strain rate, and **c** dislocations walls

the increasing misorientation along the dislocation walls, and then the bonds intragranularly were weakened. These behaviors were reproduced and enhanced in the continuous processing, which finally led to the formation of elongated grains with low angle grain boundaries (LAGBs).

The TEM bright field images of the sample-500 were highly distinct from the sample-50 images. The majority of grains were roughly equiaxed or irregularly shaped, and dislocation density inside the grains was low (Fig. 7.12d), indicating that the grains evolved via DRX. Although the copper sheet was at liquid nitrogen temperature, sufficient plastic deformation occurred in a very small local area during a short period of time due to the enough strain rate. Adequate deformation energy was released, providing conditions for DRX. Different dislocation densities appeared because of the incompatibilities among grains. The slide of grain boundaries was enhanced, and partial LAGBs of original grains bulged towards localities with dense dislocations under the effect of boundary migration (Fig. 7.12d and e). Nucleation of the new grains occurred with further dislocation activities. Afterwards, recrystallized grains gradually grew via boundary migration. The migration of grain boundaries always pointed to the curvature center of grain boundaries, which drove the grain boundaries to be flat. The angle of triple grain boundaries tended to 120°, which made the grain boundaries in an equilibrium state.

7.2.4 Grain Refinement Modes

A numerical simulation was carried out to verify the temperature fields during the severe cryogenic plastic deformation. A computational domain with dimension of 100 mm × 100 mm × 3 mm was used in the mathematical evaluation model.

The constitutive model [26] was used to determine the relationship between temperature and flow stress:

$$\sigma = 30.914 - 31.2203T_y + 2.9946T_y^2 + 4.8141T_y^3 - 1.0736T_y^4 \tag{7.4}$$

where σ was the flow stress, T_y was defined as:

$$T_y = \frac{T - 690.875}{293.4607} \tag{7.5}$$

where T was the transient temperature.

Frictional heat and plastic deformation heat were synchronously considered. The frictional heat term Q_1 was a surficial heat source and was governed as follows:

$$Q_1 = 4 \cdot \beta_1 \cdot \int_0^{2\pi}\int_0^{r2} \frac{\mu F}{4\pi r_2^2}\sqrt{(r_1 - r_2\cos\theta)^2 + (r_2\sin\theta)^2} \cdot \omega r dr d\theta \tag{7.6}$$

The plastic deformation heat term Q_2 was a volumetric heat source, which was given as:

$$Q_2 = \beta_2 \sigma_e \dot{\varepsilon}_e \tag{7.7}$$

where β_1 and β_2 were empirical coefficients which represented thermal conversion rate, r_1 was the distance from center of the ball to axis of the tool, and r_2 was diameter of the ball, σ_e was the effective stress, and $\dot{\varepsilon}_e$ was effective strain rate.

The boundary heat exchange including convection and radiation of the surface was given as follows:

$$k\left.\frac{\partial T}{\partial z}\right|_{work\ face} = h(t - t_\infty) + \sigma\varepsilon\left(T^4 - T_\infty^4\right) \tag{7.8}$$

where σ was Stefan-Boltzmann constant, ε was an emissivity which was taken as 0.5, h was a heat transfer coefficient which was taken as 300 W/(m^2 K), 15 W/(m^2 K), 200 W/(m^2 K) on upper surface, lateral surface and lower surface, respectively.

The heat simulation result illustrated in Fig. 7.13 showed the difference between temperature fields at two strain rates. At the lower strain rate, the temperature of the whole materials kept below 307 K. At the higher strain rate, the peak value of the temperature located in the area underneath the grinding balls, reaching 618 K.

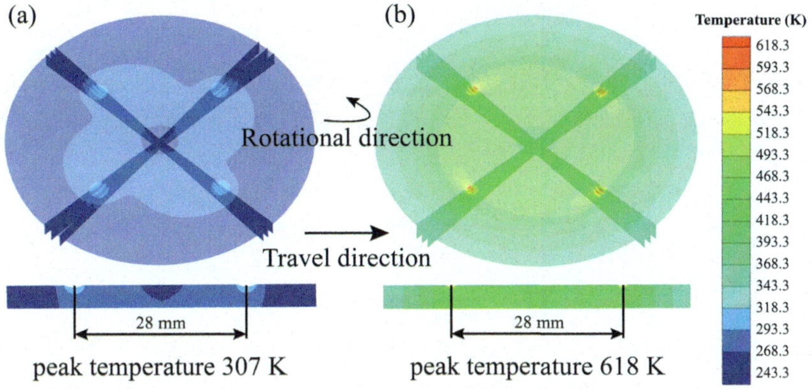

Fig. 7.13 Temperature fields at different strain rates: **a** lower strain rate and **b** higher strain rate

According to the research [27], the critical temperature where recrystallization took place was 523 K. Through the calculation of geometric dimensions and traveling speed of the tool, a fact that the duration of DRX was less than 0.93 s was found. Therefore, the hypothesis that DRX took place at a small area during a short time was confirmed.

In summary, the mode of grain refinement in the severe cryogenic plastic deformation was proposed. When the strain rate applied to the material was low, the dislocations rapidly proliferated, forming dislocation arrays and dislocation cells. Under a series of dislocation activities, the original grains were broken to form fine grains. When the strain rate applied to the material was high, in a very short period of time, the material in a small region underwent sufficient plastic deformation. Sufficient deformation energy was released to promote the occurrence of DRX. The grain boundaries bulged out to form nucleuses, and grains grew through the grain boundary migration. Eventually fine equiaxed grains formed (Fig. 7.14).

7.2.5 Summary

The gradient ultra-fine grained structure of pure copper was obtained via in-situ surficial cryogenic grinding. The microstructural evolution at different rotational velocities was investigated, and proved by simulation results. The main conclusions are as follows:

(1) The gradient structure was divided into three regions: 200-μm-thick UGZ, TMAZ and BM, corresponding to different microstructure features.
(2) Microhardness decreased smoothly as depth increased, which indicated that grain sizes were positively correlated with depth. The peak value of microhardness was 43.6% higher than that of the BM in sample-50, while the promotion was 29.1% in sample-500. Furthermore, the promotion at the same depth was

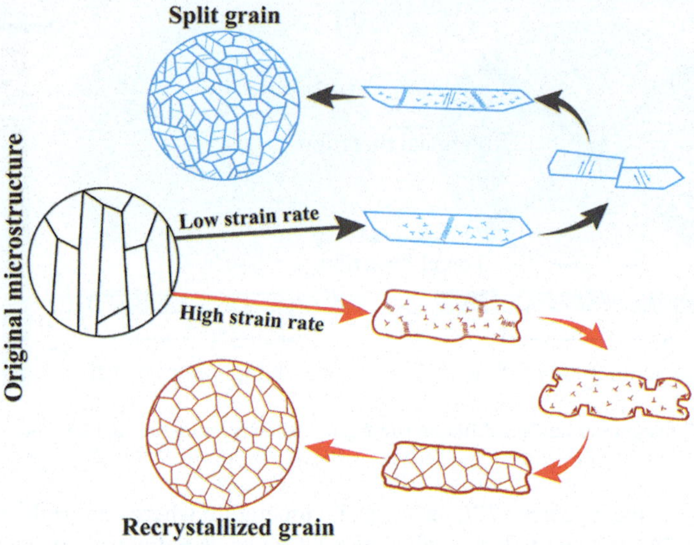

Fig. 7.14 Schematics of the grain refinement modes

also more significant in sample-50, indicating that the degree of grain refinement at low rotational velocity was higher.

(3) The modes of grain refinement in the severe cold plastic deformation in terms of dislocation behaviors and grain boundary activities were proposed: Accumulation of geometric necessary dislocations resulted in the split of the grains, dominating the refining process at low strain rate. Dynamic recovery and recrystallization took control of the refining process due to the sufficient energy of the local area at high strain rate.

7.3 Arc Surface-Nitriding Assisted Friction Stir Processing

7.3.1 Introduction

Functionally gradient coating is characterized by unique physical/mechanical properties, in which hard ceramic particles dispersedly distribute in a relatively ductile matrix and act as the super abrasives or gradient nanostructures on the surface [28, 29]. It has widespread applications in the fields of aerospace, automotive and so on [30, 31]. Aluminum alloys are extensively used as the weight-reducing materials [32]. However, low hardness and poor wear resistance restrict their application and development [33]. In order to solve these problems, many methods have been proposed to fabricate functionally gradient coating on the surface of aluminum alloys, such as spark plasma sintering [34], laser melting [35], electroforming [36], thermal/cold spray [37, 38] and severe plastic deformation [39, 40]. As a kind of improved methods, particle-reinforced coating can improve the hardness and the wear resistance of aluminum alloys, because the ceramics have been well integrated into these substrates [41]. Functionally gradient coating can increase the service life due to the combination of metal property (high toughness) and ceramics property (excellent wear resistance).

Nitride-reinforced coatings, such as AlN and TiN, have attracted extensive attentions due to high hardness and excellent wear resistance [42]. Arc nitriding technology has the advantage of low cost, simple operation, high efficiency and high-quality metallurgical bonding between coating and substrate. Arc nitriding has been successfully used to prepare TiN coating on the surfaces of copper and steel substrates, and then improved the hardness and wear resistance [43]. However, during arc nitriding, the segregation of the nitride forms on the surface of substrate, reducing the wear properties of the coating [44].

In this section, aluminum alloys, as the weight-reducing materials, were selected. An in-situ arc surface nitriding with subsequent FSP, characterized by low cost, simple operation, high efficiency, and good dispersion of the AlN, was proposed to fabricate the functionally gradient coating on the Al matrix, which explored the large area fabrication and extensive application in the field of aerospace or automotive. Microstructural evolution mechanism, hardness, and wear resistance of the functionally gradient coating were mainly investigated in detail.

7.3.2 Materials and Experimental Procedure

The substrate was AA6082-T6 sheet, whose the dimensions were 300 mm × 100 mm × 5 mm. The chemical compositions of the AA6082-T6 substrate consisted of 0.8% Si, 0.5% Fe, 0.1% Cu, 0.5% Mn, 0.6% Mg, 0.25% Cr, 0.1% Ti and Al in balance (mass fraction). A self-assembled TIG system containing a HK-5B-D welding trolley was used to perform in-situ arc surface nitriding. Protective gas was a mixture of argon (Ar) and nitrogen (N_2). A traverse speed of the TIG gun was 3.5 mm/s and an alternating current was 80 A. The distance between the tungsten electrode and the surface of the AA6082-T6 was 10 mm. The total flux of the gas was 10 L. The ratio between the N_2 and the Ar was 3:1. Schematic of the in-situ arc surface nitriding is shown in Fig. 7.15.

FSP was carried out by the commercial FSW machine (FSW-3LM-003) after completing in-situ arc surface nitriding. A processing tool was made of H13 steel, which was featured by a threaded pin and a concave shoulder. The diameters of the shoulder, the pin bottom and top were respectively 14 mm, 4 mm and 3 mm. The length of the rotational pin was 2.7 mm and the tilting angle with respect to Z-axis was 1.5°. A processing speed of 300 mm/min and a rotational speed of 800 r/min were employed to redisperse the nitride. The plunge depth of the processing tool was 0.1 mm. The processing direction was parallel to the traverse direction of the TIG gun.

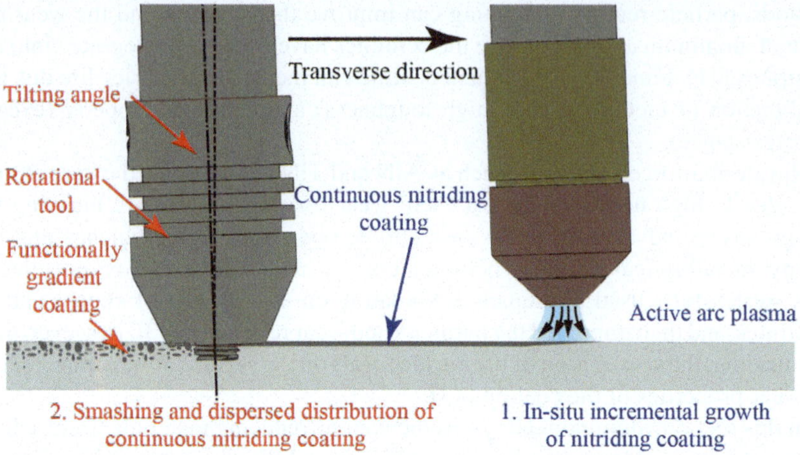

Fig. 7.15 Schematic of the fabrication for functionally gradient coating

The microstructural and mechanical specimens of the coatings and the substrate were cut perpendicular to the processing direction by an electrical discharge cutting machine. The metallographic specimens were polished by different grades of emery papers (from 400# to 7000#), while the final polishing was carried out using the diamond compounds (1.5 μm) in a disc polishing machine. The macro/micro structures were observed by an optical microscope and a SEM and an X-ray electron probe microanalyzer (EPMA). The phase composition of the surface coatings was determined by X-ray diffraction (XRD) analysis at 40 kV and 30 mA of the Cu-Kα radiation (0.1542 nm). The scan was conducted in 2θ mode and spanned across a range of 10°-120° at a step resolution of 0.033°.

Microhardness measurements were carried out using a Vickers hardness indenter with a load of 200 g and a dwell time of 15 s in the processing zone. Nanoindentation was performed to evaluate the local mechanical properties of the functionally gradient coating by the nano indenter (G200, USA) with a load capacity of 10 N. A Berkovich diamond indenter was applied. The maximum indentation depth of 1 μm was constant and the maximum indentation load corresponding to the maximum indentation depth was recorded. Five indentations were measured, while mean values of the three indentations were used after ignoring the indentations with extreme values (maximum and minimum values). The scratch tests were also performed by the nano indenter at a scratch load of 100 mN, a scratch distance of 5000 μm and a scratch rate of 30 μm/s.

Wear resistances of the AA6082-T6 substrate and the functionally gradient coating were conducted by a friction abrasion testing machine. The tests were conducted at room temperature with a fixed load of 20 g. The radius of friction, the rotational speed and the friction time were 3 mm, 360 r/min and 8 min, respectively. Coefficient of friction (COF) and sliding times were automatically recorded. The wear surface morphologies were characterized by SEM.

7.3.3 Microstructure of Nitriding Coating

Figure 7.16 shows the microstructures of the nitriding coating fabricated by the in-situ arc surface nitriding. A nitriding coating with a thickness of 600 μm is achieved. There are no defects appearing in the coating even if extremely high peak temperature induced by the arc, while good metallurgical bonding forms at the interface between the AA6082-T6 substrate and the coating, as displayed in Fig. 7.16a. The dendrites formed on the coating are marked by the square in Fig. 7.16a, as depicted in Fig. 7.16b. The elemental mapping views corresponding to the Fig. 7.16b are presented in Fig. 7.16c and d. The black region in Fig. 7.16b is composed of the Al and N elements, while the white region mainly contains the Al element. Therefore, it can be inferred that the white region is the AA6082-T6 substrate and the black region is the nitride. In order to further clarify the nitride, the XRD result of the nitriding coating is exhibited in Fig. 7.17. There are only main two phases of Al and AlN, and no other diffraction peaks are detected, which indicate that the AlN forms on the coating. Based on these results, it can be validated that the black regions character ized by the dendrites are mainly the AlN phase, and they present an inhomogeneous distribution.

The formation mechanism of the nitriding coating by the in-situ arc surface nitriding is displayed in Fig. 7.18. Under the thermal effect of the arc, the materials in the surface of AA6082-T6 substrate are locally heated up and melt beyond its melting point, forming a molten pool. The surface of the molten pool is covered with the nitrogen gas, as exhibited in Fig. 7.18a. The nitrogen gas plays a role in protecting the molten pool from the oxidation, and acts as the basis for the formation of the AlN. During the in-situ arc surface nitriding, the nitrogen in the arc zone is ionized into the atoms and ions, as exhibited in Fig. 7.18b. These highly active nitrogen ions (N^+ or N^{3-}) and nitrogen atoms (N) are adsorbed by the liquid Al, and then overcome the surface energy into the molten pool. In the molten pool, the convection of liquid metal and the diffusion of N improve the metallurgical reaction at high temperature, as Eq. (7.9).

$$\mathrm{Al} + \mathrm{N}\left(\mathrm{N}^{+}, \mathrm{N}^{3-}\right) \rightarrow \mathrm{AlN} \tag{7.9}$$

This reaction firstly happens at the interface between the molten pool bottom and the substrate. The AlN firstly separates out from the molten pool and nucleates, and grows along the opposite direction, forming the dendrites. At the same time, the formed AlN dendrite acts as the new nucleation of the new AlN phase, which further grows at the AlN dendrite. Because the reaction between the Al and the N is a severe exothermic reaction, the heat of reaction can still sustain the reaction and eventually form a continuous AlN coating after moving away the arc. The consequent AlN dendrites form due to the inconsistent direction, while the unreacted Al solidifies and reserves the original characteristics (Figs. 7.18c and 7.16a).

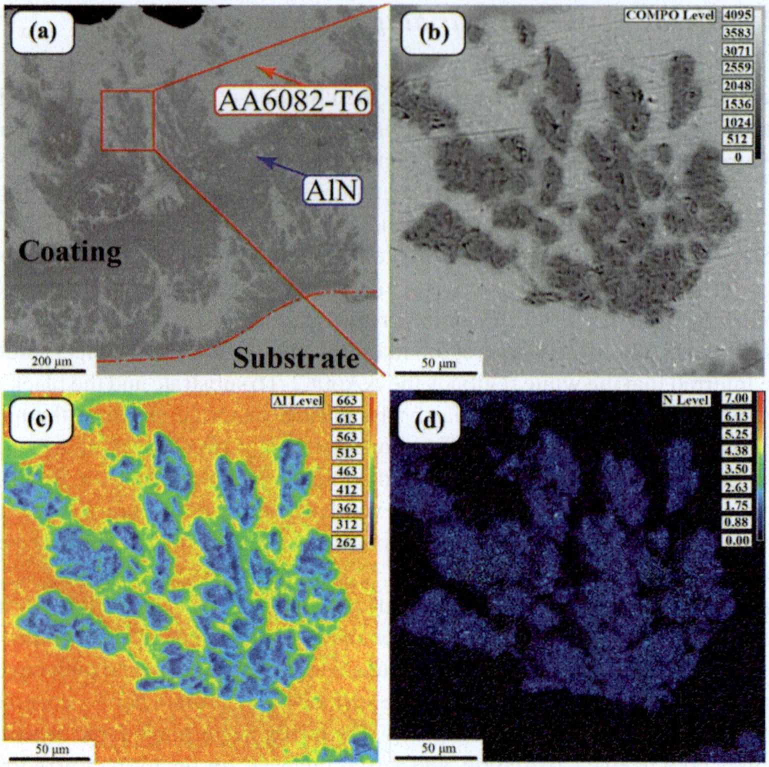

Fig. 7.16 Microstructures and element distributions in the nitriding coating fabricated by in-situ arc surface nitriding: **a** nitriding coating, **b** magnified view marked by the square in Fig. 16a, **c** distribution of Al element, and **d** distribution of N element

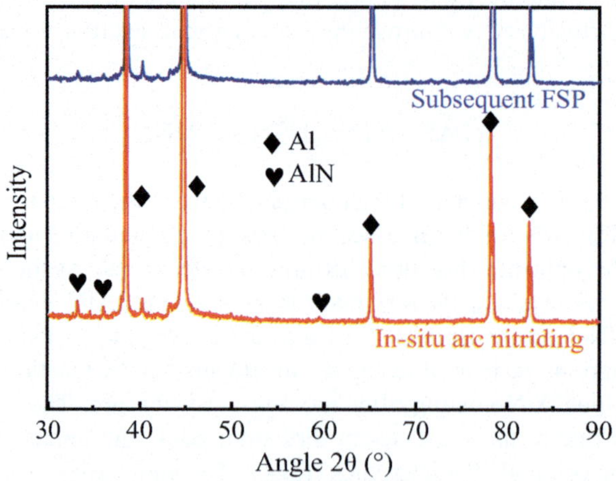

Fig. 7.17 XRD patterns of the nitriding and the funtionally gradient coatings

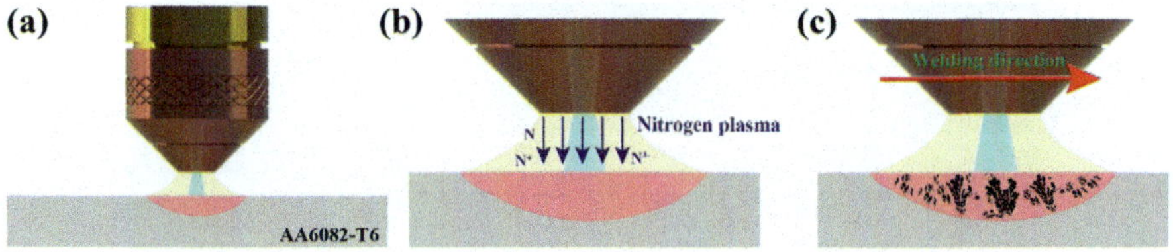

Fig. 7.18 Illustration of the in-situ arc surface nitriding: **a** arcing, **b** ionization of N_2, and **c** formation of dendrite structures characterized by the AlN

7.3.4 Microstructure of the Functionally Gradient Coating

Figure 7.19a shows the macrostruture of the FSP joint, which can be divided into three zones: SZ, TMAZ, and heat affected zone. The SZ is filled with the smashed AlN particles. The interface between TMAZ and SZ on the advancing side (AS) present a sharp morphology, while that on the retreating side (RS) shows an unclear interface due to the differences of shear stress and material flow on the two sides. A comparable thickness of the functionally gradient coating is 1000 μm, which is far higher than the nitriding coating with the thickness of 600 μm (Fig. 7.19). During FSP, the nitriding coating containing the AlN dendrites and AA6082-T6 substrate are completely softened and smashed. The material flow induced by FSP tool transforms the smashed AlN particles along the thickness direction (Fig. 7.19a), which attributes to the thicker functionally gradient coating compared with the nitriding coating. With distancing away from the top surface, the AlN particles gradually decrease (Figs. 7.19b–e), which realizes the uniform transition of the AlN and then achieves the gradient coating. This is because that the main frictional heat is produced by the shoulder of processing tool [45], which can sufficiently soften the nitriding coating and mix the AlN particles. For Fig. 7.19e, low heat input and high processing speed are detrimental to the transfer of the AlN particles from the top to the bottom of the functionally gradient coating.

XRD pattern of the functionally gradient coating after FSP is exhibited in Fig. 7.17, which mainly consists of Al and AlN phases, and no other diffraction peaks are detected. No reaction occurs between the AlN particles and AA6082-T6 substrate after FSP. Compared to the in-situ arc surface nitriding, some peaks corresponding to the AlN phase of the functionally gradient coating after FSP become relatively lower, which mainly attribute to the high level of the separation and disperse distribution of the AlN particles [46]. This also indicates that the FSP after the arc surface nitriding is superior to the other direct conventional fusion surfacing techniques, such as laser beam and thermal spraying [2]. The good interfacial condition between the AlN particles and the AA6082-T6 substrate during FSP is more positive, potentially attributing to high interfacial bonding strength.

As a matter of fact, when fabricating metal matrix composites (MMCs) with ceramic particles via single FSP, the wear of processing tool is severe, which is detrimental to the redistribution of the strengthening particles or results in the processing

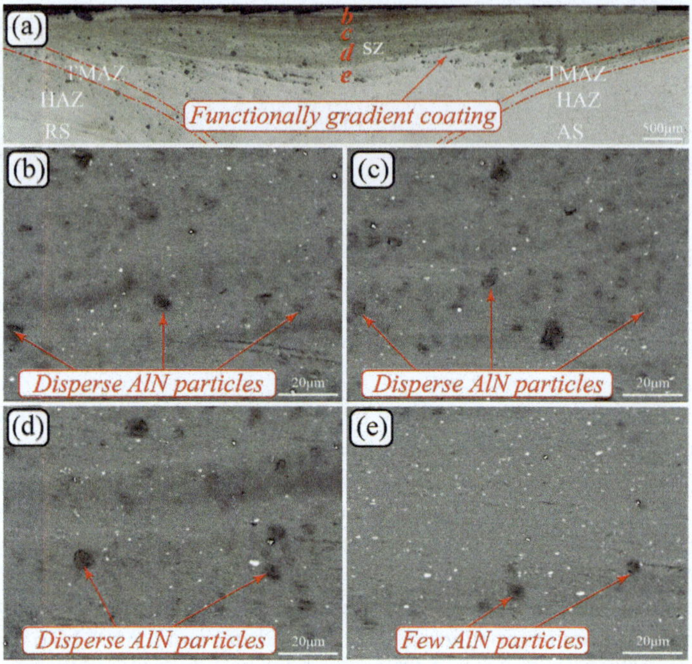

Fig. 7.19 SEM views of the functionally gradient coating: **a** macro-morphology in cross-section; b–**e** micro-morphologies of **b**, **c**, **d**, and **e** marked in Figure **a**

defects. In this section, the in-situ arc surface nitriding with subsequent FSP has the remarkable advantage compared with conventional single FSP: ① The nitriding process is low cost, simple operation and high efficiency. Meanwhile, no filler materials are added and the nitrogen is extremely economical compared with other ceramic particles. The good interfacial bonding can be achieved between nitriding coating and substrate, which reduces the process procedures for subsequent FSP. ② The nitriding coating with much more Al materials and the AlN dendrites can avoid the tool wear and then improve the coating formation due to low shear strength and good flowability of Al materials at high temperature. The severe stirring effects can break the AlN dendrites into small particles and refine them into nanoscale, playing the dispersion strengthening effects on the functionally gradient coating. Under severe thermo-mechanical effects induced by FSP, the good interfacial joining of Al/AlN in the functionally gradient coating is realized and no kissing bond defect appears, which effectively avoid the stress concentration. In a word, the in-situ arc surface nitriding with subsequent FSP has the feasibility and potential to fabricate the functionally gradient coating with low wear.

7.3.5 Microhardness

Figure 7.20 depicts the microhardness distribution of the functionally gradient coating. The microhardness of the AA6082-T6 substrate is approximately 90 HV. The in-situ arc surface nitriding with subsequent FSP leads to a dramatic increase in microhardness from 90 to 140 HV. This is due to the uniform redistribution and refinement of the AlN particles along the horizontal direction. This result has been reported by Hodder et al. [47]. Patel et al. [48] also stated that the microhardness of the FSPed Al–Zn–Mg–Cu alloy depended on the strengthening of phase particles, but not the grain size. FSP generates intense plastic deformation, and then refines the AlN particles in the processing zone. The AlN dendrites have been fragmented into different sizes. The uniform and disperse AlN particles attribute to the dispersion strengthening by restricting its deformation. Meanwhile, the microhardness distribution of the functionally gradient coating through the thickness direction presents a decreasing tendency, which is beneficial to the compatibility of deformation. This results indicate that FSP after the in-situ arc surface nitriding can availably improve the microhardness of the functionally gradient coating through the dispersion strengthening effect of the AlN particles.

In order to characterize the micromechanical properties of the functionally gradient coating, nano-indentation was performed, as exhibited in Fig. 7.21. The average hardness and the young's modules of the functionally gradient coating are respectively 1.77 GPa and 128.56 GPa. These results are consistent with Vickers hardness. Meanwhile, the young's modulus of the AA6082-T6 substrate is only about 70 GPa. The similar result has been reported by Hodder et al. [47]. This is because that the severe plastic deformation attributes to the extreme refinement and disperse distribution of the AlN particles in the grain boundaries of the Al matrix, which further decreases the mean free path distance and then improves the load sharing. Meanwhile, the pinning effect of the disperse AlN particles can inhibit the deformation of the materials and then enhance the hardness and the young's modules.

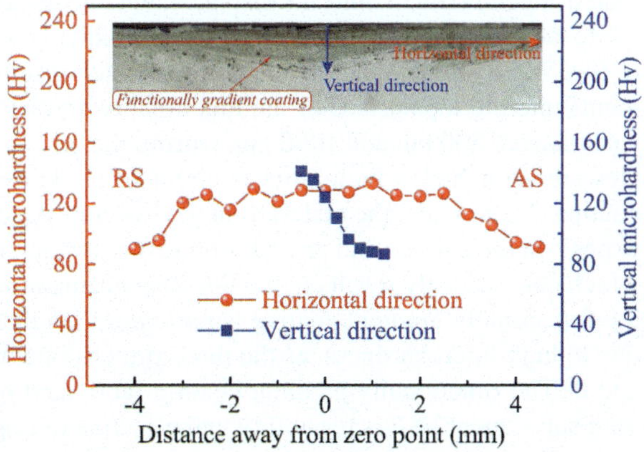

Fig. 7.20 Microhardness profile of the functionally gradient layer

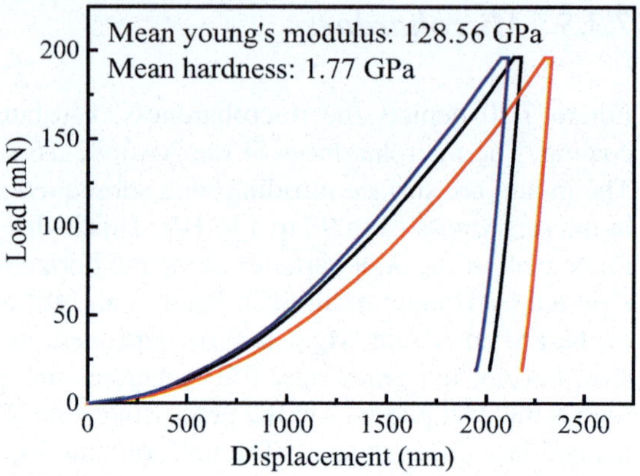

Fig. 7.21 Load versus depth curves from nano-indentation of the functionally gradient coating

These results further validate that the in-situ arc surface nitriding with subsequent FSP is an effective method to improve the micromechanical properties.

7.3.6 Scratch Property

Figure 7.22 shows the evolutions of the displacement into the top surface versus the scratch distance, and scratch surface morphologies for the AA6082-T6 substrate and the functionally gradient coating. For the AA6082-T6 substrate, the scratch depth for each time gradually increases with the increase of scratch distance due to low microhardness (Fig. 7.22a). As exhibited in Fig. 7.22c, there are no obvious variations of the displacement into the top surface for the functionally gradient coating, which are more superior compared with the AA6082-T6 substrate. This is because that the refined AlN particles in the functionally gradient coating can strengthen the processing zone and impede the material deformation. However, with further increasing the scratch time, the displacement into the top surface gradually decreases, while the displacement of the functionally gradient coating is still lower, as exhibited in Fig. 7.22a and c. When reaching the scratch distance of 3500 μm, the displacements into the top surface of the functionally gradient coating and the AA6082-T6 substrate are 500 nm and 1000 nm. During the first and second times, the AlN particles are scratched, which reduces effective thickness of the functionally gradient coating. Meanwhile, the materials at the lower of the coating with few AlN particles expose on the surface and provides relatively low strengthening effects (Fig. 7.19e), which consequently results in the low displacement. Moreover, the scratch width of the functionally gradient coating is narrower than that of the AA6082-T6 substrate due to high microhardness, as the three-dimensional topographies in the Fig. 7.22b and d. The functionally gradient coating fabricated by in-situ arc surface nitriding with subsequent FSP is feasibility and effective to improve the surface properties.

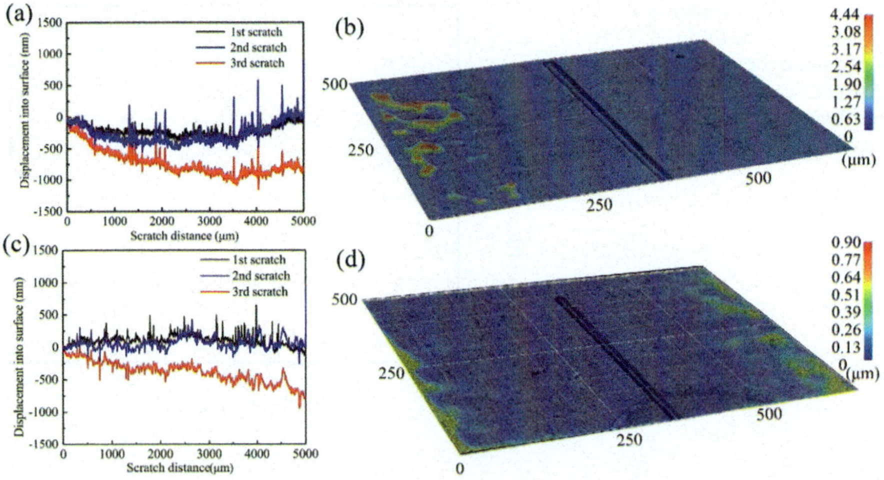

Fig. 7.22 Scratch tests of the AA6082-T6 substrate and the functionally gradient coating: **a** relation between scratch and plunge depth, **b** scratch morphology for the AA6082-T6 substrate, **c** relation between scratch and plunge depth, and **d** scratch morphology for the functionally gradient coating

7.3.7 Wear Property

Figure 7.23 shows the representative evolutions of the COFs versus time at 20 g for the functionally gradient coating and the AA6082-T6 substrate. The COF of the AA6082-T6 substrate is 0.5. The COF of the functionally gradient coating is only 0.15 until to the sliding time of 250 s, which is markedly lower. The COF of the AA6082-T6 substrate rises with the increase of sliding time, while the COF of the functionally gradient coating rises a steady low value and then increases with further increasing sliding time higher than 250 s. This originates from the disperse distribution of the refined AlN particles on the surface of the AA6082-T6 substrate. It can be concluded that the functionally gradient coating with the AlN particles plays a significant influence on the wear behavior.

Figure 7.24 shows the wear surface morphologies of the AA6082-T6 substrate and the functionally gradient coating. The wear track of the AA6082-T6 substrate is wider than that of the functionally gradient coating, which is attributed to the high wear resistance of the disperse AlN particles, as displayed in Fig. 7.24a and b. The main wear mechanism in the AA6082-T6 substrate is adhesive wear due to low hardness and shear strength (Fig. 7.24a and e). The worn out surface of AA6082-T6 substrate shows the deeper and wider grooves, which reveals severely crushed and dislodged delamination. In comparison, the functionally gradient coating indicates higher wear resistance. For the functionally gradient coating, the abrasive wear is the main mechanism, as shown in Fig. 7.24d and f. During the wear process, the micro interaction between the functionally gradient coating and the steel counterface is the first step. The Al materials are easily released from the functionally gradient coating

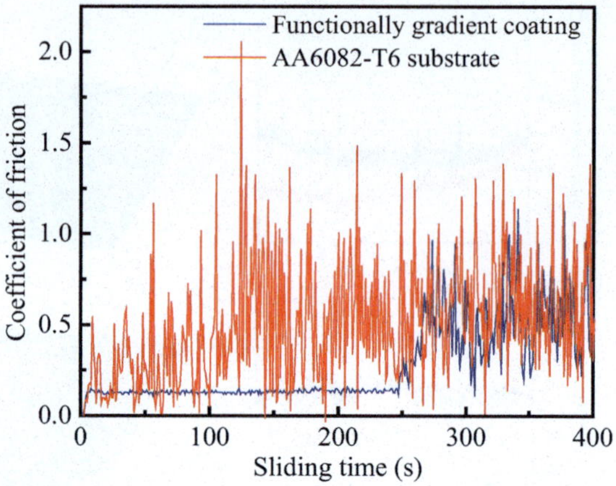

Fig. 7.23 Representative evolutions of the COFs versus time for the AA6082-T6 substrate and the functionally gradient coating

and then accumulate on the surface due to the repetitive press and low shear strength of the AA6082-T6 substrate, forming a shear layer with much more AlN particles. Moreover, the shear layer is continuously replenished by the AlN particles on the functionally gradient coating. Accordingly, the gradient shear layer can improve the load-bearing and lubricating capabilities, and then attributes to the low friction and wear properties.

In generally, the pinning effect of the disperse distribution of the refined AlN particles increases the lubricating effect and the shear force, thus increasing the wear resistance. The density and size of the disperse AlN particles are responsible for wear performance of the functionally gradient coating. Therefore, increasing the density and reducing the size of the AlN particles can further improve the wear resistance, which will be investigated in the future work.

Additionally, the potentiodynamic polarization behaviors of the AA6082-T6 substrate and the functionally gradient coating were performed. The addition of the AlN particles markedly reduces the corrosion potential and then weakens the susceptibility of corrosion.

7.3.8 Summary

(1) The nitriding coating with the thickness of about 600 μm was successfully deposited onto the surface of AA6082-T6 substrate by the in-situ arc surface nitriding. Under the thermal effect of the arc, the AA6082-T6 materials melted, while the nitrogen was ionized into highly active nitrogen ions and atoms, which easily reacted with the liquid Al, forming the AlN coating.

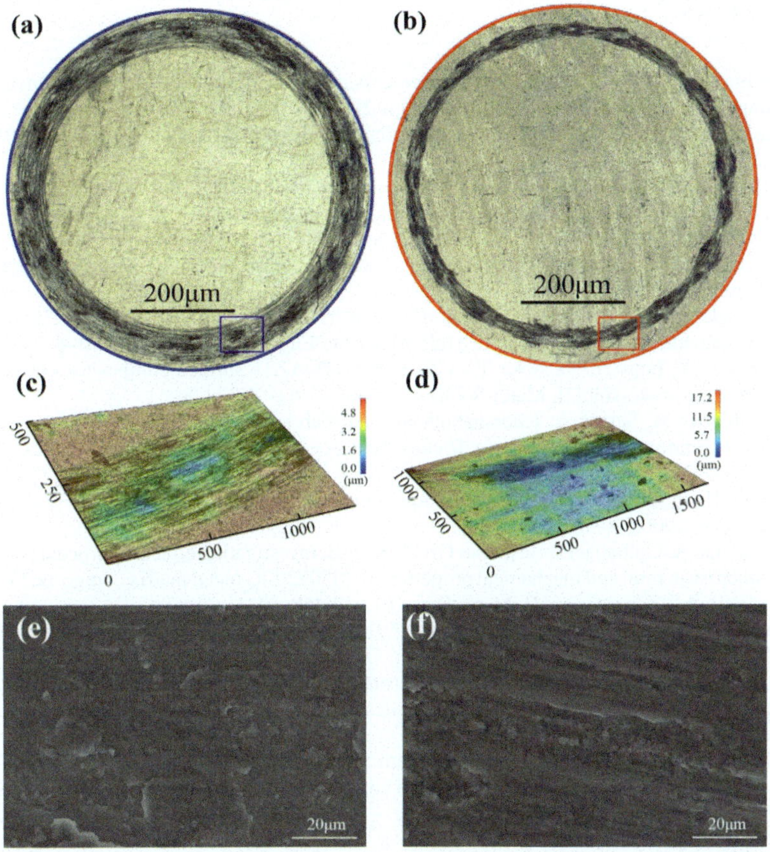

Fig. 7.24 Wear surface morphologies: **a** macro wear surface, **c** three-dimensional topography and **e** SEM morphology of the AA6082-T6 substrate; **b** macro wear surface, **d** three-dimensional topography, and **f** SEM morphology of the functionally gradient coating

(2) FSP broke the dendrites into small particles and made them dispersedly distribute via severe plastic flow. These synthesis effects attributed to the functionally gradient coating with the microhardness distributions from 140 to 90 HV along the thickness directions.

(3) The functionally gradient coating owned the better wear resistance and the lower COF. The introduction of the AlN transformed the adhesive wear of the AA6082-T6 substrate to the abrasive wear for the functionally gradient coating. The pinning effect of the dispersedly refined AlN particles increased the lubricating effect, significantly improving the wear resistance.

(4) The strategy to construct the functionally gradient coating based on the in-situ arc surface nitriding with subsequent FSP can achieve the remarkable wear resistance, which may satisfy the applications in the aerospace and automotive.

References

1. Mishra RS, Ma ZY, Charit I. Friction stir processing: a novel technique for fabrication of surface composite[J]. Mater Sci Eng A. 2003;341(1–2):307–10.
2. Mishra RS, Ma ZY. Friction stir welding and processing[J]. Mater Sci Eng R Rep. 2005;50(1–2):1–78.
3. Yadav D, Bauri R. Processing, microstructure and mechanical properties of nickel particles embedded aluminium matrix composite[J]. Mater Sci Eng A. 2011;528(3):1326–33.
4. Zhang Q, Xiao BL, Wang QZ, et al. In situ Al_3Ti and Al_2O_3 nanoparticles reinforced Al composites produced by friction stir processing in an Al–TiO_2 system[J]. Mater Lett. 2011;65(13):2070–2.
5. Morisada Y, Fujii H, Nagaoka T, et al. Effect of friction stir processing with SiC particles on microstructure and hardness of AZ31[J]. Mater Sci Eng A. 2006;433(1–2):50–4.
6. Morisada Y, Fujii H, Nagaoka T, et al. MWCNTs/AZ31 surface composites fabricated by friction stir processing[J]. Mater Sci Eng A. 2006;419(1–2):344–8.
7. Sharifitabar M, Sarani A, Khorshahian S, et al. Fabrication of 5052Al/Al_2O_3 nanoceramic particle reinforced composite via friction stir processing route[J]. Mater Des. 2011;32(8–9):4164–72.
8. Lee C, Huang J, Hsieh P. Mg based nano-composites fabricated by friction stir processing[J]. Scr Mater. 2006;54(7):1415–20.
9. Dolatkhah A, Golbabaei P, Besharati Givi MK, et al. Investigating effects of process parameters on microstructural and mechanical properties of Al5052/SiC metal matrix composite fabricated via friction stir processing[J]. Mater Des. 2012;37:458–64.
10. Faraji G, Asadi P. Characterization of AZ91/alumina nanocomposite produced by FSP[J]. Mater Sci Eng A. 2011;528(6):2431–40.
11. Azizieh M, Kokabi AH, Abachi P. Effect of rotational speed and probe profile on microstructure and hardness of AZ31/Al_2O_3 nanocomposites fabricated by friction stir processing[J]. Mater Des. 2011;32(4):2034–41.
12. Alidokht SA, Abdollah-zadeh A, Soleymani S, et al. Microstructure and tribological performance of an aluminium alloy based hybrid composite produced by friction stir processing[J]. Mater Des. 2011;32(5):2727–33.
13. Zahmatkesh B, Enayati MH, Karimzadeh F. Tribological and microstructural evaluation of friction stir processed Al2024 alloy[J]. Mater Des. 2010;31(10):4891–6.
14. Su J-Q, Nelson TW, Sterling CJ. Friction stir processing of large-area bulk UFG aluminum alloys[J]. Scr Mater. 2005;52(2):135–40.
15. El-danaf EA, El-rayes MM, Soliman MS. Friction stir processing: An effective technique to refine grain structure and enhance ductility[J]. Mater Des. 2010;31(3):1231–6.
16. Darras BM, Khraisheh MK, Abu-farha FK, et al. Friction stir processing of commercial AZ31 magnesium alloy[J]. J Mater Process Technol. 2007;191(1–3):77–81.
17. Chang CI, Lee CJ, Huang JC. Relationship between grain size and Zener-Holloman parameter during friction stir processing in AZ31 Mg alloys[J]. Scr Mater. 2004;51(6):509–14.
18. Reck BK, Graedel TE. Challenges in metal recycling[J]. Science. 2012;337(6095):690–5.
19. Lu K, Lu L, Suresh S. Strengthening materials by engineering coherent internal boundaries at the nanoscale[J]. Science. 2009;324(5925):349–52.
20. Fang TH, Li WL, Tao NR, et al. Revealing extraordinary intrinsic tensile plasticity in gradient nano-grained copper[J]. Science. 2011;331(6024):1587–90.
21. Wang L, Shi Y, Zhang Y, et al. High tensile ductility and strength in a gradient structured Zr[J]. Mater Lett. 2018;228:500–3.
22. Liu XC, Zhang HW, Lu K. Strain-induced ultrahard and ultrastable nanolaminated structure in nickel[J]. Science. 2013;342(6156):337–40.
23. Sun HQ, Shi Y-N, Zhang MX, et al. Plastic strain-induced grain refinement in the nanometer scale in a Mg alloy[J]. Acta Mater. 2007;55(3):975–82.
24. Zhang W, Ding H, Cai M, et al. Ultra-grain refinement and enhanced low-temperature superplasticity in a friction stir-processed Ti-6Al-4V alloy[J]. Mater Sci Eng A. 2018;727:90–6.

25. Huang Y, Wan L, Lv S, et al. In situ rolling friction stir welding for joining AA2219[J]. Mater Des. 2013;50:810–6.
26. Jin LZ, Sandström R. Numerical simulation of residual stresses for friction stir welds in copper canisters[J]. J Manuf Process. 2012;14(1):71–81.
27. Andrade U, Meyers MA, Vecchio KS, et al. Dynamic recrystallization in high-strain, high-strain-rate plastic deformation of copper[J]. Acta Metall Mater. 1994;42(9):3183–95.
28. Chen X, Han Z, Li X, et al. Lowering coefficient of friction in cu alloys with stable gradient nanostructures[J]. Sci Adv. 2016;2(12):1–8.
29. Lu L, Shen Y, Chen X, et al. Ultrahigh strength and high electrical conductivity in copper[J]. Science. 2004;304(5669):422–6.
30. Hao X, Wang H, Hua L, et al. Study on the law of ultrasonic vibration-assisted adhesive bonding of CFRP-to-Aluminum joints[J]. Int Conf Mech Mechatranics Eng. 2017;239–245.
31. Anwer A, Bagheri ZS, Fernie G, et al. Evolution of the coefficient of friction with surface wear for advanced surface textured composites[J]. Adv Mater Interfaces. 2017;4(6):1–11.
32. Chen S, Li X, Jiang X, et al. The effect of microstructure on the mechanical properties of friction stir welded 5A06 Al Alloy[J]. Mater Sci Eng A. 2018;735:382–93.
33. Miller WS, Zhuang L, Bottema J, et al. Recent development in aluminium alloys for the automotive industry[J]. Mater Sci Eng A. 2000;280(1):37–49.
34. Gong Y, Ping Y, Li D, et al. Preparation of high-quality graphene via electrochemical exfoliation & spark plasma sintering and its applications[J]. Appl Surf Sci. 2017;397:213–9.
35. Park J, Han HS, Park J, et al. Corrosion behavior of biodegradable Mg-based alloys via femtosecond laser surface melting[J]. Appl Surf Sci. 2018;448:424–34.
36. Zhao H, Liu L, Wu Y, et al. Investigation on wear and corrosion behavior of Cu-graphite composites prepared by electroforming[J]. Compos Sci Technol. 2007;67(6):1210–7.
37. Huang CJ, Yan XC, Li WY, et al. Post-spray modification of cold-sprayed Ni-Ti coatings by high-temperature vacuum annealing and friction stir processing[J]. Appl Surf Sci. 2018;451:56–66.
38. Li N, Li WY, Yang XW, et al. An investigation into the mechanism for enhanced mechanical properties in friction stir welded AA2024-T3 joints coated with cold spraying[J]. Appl Surf Sci. 2018;439:623–31.
39. Huang Y, Wang T, Guo W, et al. Microstructure and surface mechanical property of AZ31 Mg/SiC_p surface composite fabricated by direct friction stir processing[J]. Mater Des. 2014;59:274–8.
40. Li X, Lu K. Playing with defects in metals[J]. Nat Mater. 2017;16(7):700–1.
41. Torres B, Garrido MA, Rico A, et al. Wear behaviour of thermal spray Al/SiCp coatings[J]. Wear. 2010;268(5–6):828–36.
42. Yu S, Li D, Sun H, et al. Microanalysis of single-phase AlN nanocrystals and AlN-Al nanocomposites prepared by DC arc-discharge[J]. J Cryst Growth. 1998;183(3):284–8.
43. Li Y, Liu X, Zhou Z, et al. The microstructure and wear resistance of a copper matrix composite layer on copper via nitrogen-shielded arc cladding[J]. Coatings. 2016;6(4):67.
44. Huashun Y, Kim JD, Kang SB. The formation of AlN and TiN particles during nitrogen bearing gas injection into Al–Mg–Ti melt[J]. Mater Sci Eng A. 2004;386(1–2):318–25.
45. Ji SD, Meng XC, Liu JG, et al. Formation and mechanical properties of stationary shoulder friction stir welded 6005A–T6 aluminum alloy[J]. Mater Des. 2014;62:113–7.
46. Yang K, Li W, Niu P, et al. Cold sprayed AA2024/Al_2O_3 metal matrix composites improved by friction stir processing: microstructure characterization, mechanical performance and strengthening mechanisms[J]. J Alloys Compd. 2018;736:115–23.
47. Hodder KJ, Izadi H, Mcdonald AG, et al. Fabrication of aluminum–alumina metal matrix composites via cold gas dynamic spraying at low pressure followed by friction stir processing[J]. Mater Sci Eng A. 2012;556:114–21.

48. Patel VV, Badheka V, Kumar A. Influence of friction stir processed parameters on superplasticity of Al–Zn–Mg–Cu alloy[J]. Mater Manuf Process. 2016;31(12):1573–82.

Chapter 8
Friction Stir Processed Bulk Materials

8.1 Microstructural Evolution and Mechanical Properties of Mg–Zn–Y–Zr Alloy During Friction Stir Processing

8.1.1 Introduction

As the lightest structural metallic materials, magnesium alloys are potential in transportation industries because of their high specific strength [1]. Rare-earth (RE) elements are added into magnesium alloys in order to achieve excellent high-temperature strength and creep resistance, and thus to broaden their elevated temperature application [2]. It should be noticed that currently, magnesium alloys containing RE elements are fabricated by casting processing. The obtained cast components often have coarse grains and intergranular eutectic networks, resulting in their poor mechanical performance. Therefore, it is necessary to modify the microstructures of cast Mg alloys to enhance their strength and ductility.

Friction stir processing (FSP) is a novel severe plastic deformation technology derived from friction stir welding (FSW) [3]. During FSP, coupling of thermal and mechanical effects generated by the rotating tool can efficiently homogenize and refine microstructure. Not only grain refinement, intergranular eutectic phase can also be broken up into tiny particles and some can dissolve into Mg matrix. As a result, FSP can refine coarse casting microstructure and strengthen Mg alloys with a combination of fine-grain strengthening, second phase strengthening (including dispersion strengthening or/and precipitation strengthening) and solution strengthening [4–7]. Combined with heat treatment, FSP is an effective method to control microstructure evolution such as grain size, morphology and amount of the precipitates and solution degree [8, 9]. In addition, FSP develops a special texture characteristic of Mg alloys that basal planes aligned with the pin surface at the stir zone (SZ), which exerts a great role on mechanical properties [10, 11].

Y. Huang et al., *Friction Stir Welding and Processing*, Materials Forming, Machining and Tribology, https://doi.org/10.1007/978-981-99-8688-0_8

Mg–Zn–RE alloys are attractive because of different types of second phase related to various Zn/RE ratios [12, 13]. FSP of Mg–Zn–RE alloys has been widely studied to investigate the influence of severe plastic strain on second phase modification and achieve high mechanical properties. Yang et al. [14] intensively studied the influence of temperature, plastic strain and cooling rate on the distribution of long-period stacking ordered (LPSO) phase in FSPed Mg–Gd–Y–Zn–Zr alloy, and successfully developed a special microstructure of fine LPSO lamellae only existed within fine grains, which greatly improved strength and ductility [15]. Xie et al. [16, 17] investigated FSW and FSP of Mg–Zn–Y–Zr alloy and found that bulky I-phase were broken up and dispersed with some of them transformed to W-phase. Fine equiaxed recrystallized grains and dispersed W-phase particles were beneficial to the enhanced tensile properties and the excellent superplasticity of Mg–Zn–Y–Zr alloy. However, transformation mechanism from I-phase to W-phase still remains unanswered. The intrinsic relationship between processing parameters, microstructure modification and properties improvement need to be further explored as well. In this paper, different processing parameters of FSP were applied to Mg–Zn–Y–Zr castings to study the influence of thermal and mechanical effects on the microstructure modification and tensile properties improvement.

8.1.2 Experimental Procedures

Mg–6Zn–1Y–0.5Zr (wt.%) cast billet was used as base metal (BM). A tool with a shoulder 16 mm in diameter and a threaded cylinder pin 7 mm in diameter and 2.5 mm in length was used. 4 mm thick plates were cut from the as-received BM and then subjected to one-pass FSP at a constant rotation rate of 800 r/min and varied travel speeds of 20, 80 and 200 mm/min (hereafter denoted as 800/20, 800/80 and 800/200 for short). The principle directions of FSP samples were marked as PD (process direction, along the FSP direction), TD (transverse direction, transverse to the FSP direction), and ND (normal direction, normal to the plate surface).

The specimens used for microstructural examinations were cross sectioned perpendicular to the PD. Microstructural observations were conducted by optical microscopy (OM; Olympus-MPG3), scanning electron microscopy (SEM; FEI Quanta 200FEG) equipped with energy dispersive spectroscopy (EDS) and electron backscatter diffraction (EBSD), X-ray diffraction (XRD; D/max-RB), differential scanning calorimetry (DSC; STA449F3), and transmission electron microscopy (TEM; Talos F200x). The specimens for OM were prepared by mechanical polishing and etching using a solution of 6 g picric acid + 10 mL acetic acid + 70 mL ethanol + 10 mL water. The SEM specimens were only prepared by mechanical polishing. The grain sizes and particle volume fractions were counted by Image-pro plus 6.0 software. The specimens for EBSD were prepared by mechanical polishing followed by electropolishing (40 vol.% H_3PO_4 and 60 vol.% ethanol at 20 °C) using voltage of 2–3 V for 10–20 s. Thin foils for TEM were ion-milled by a PIPS691 miller.

Tensile specimens with a gauge length of 2.5 mm, a width of 1.4 mm, and a thickness of 0.8 mm were machined from BM and FSP samples. The FSP specimens were machined along TD with the gauge being completely at SZ. The tensile tests were performed using an INSTRON 5965 mini tester at room temperature at an initial strain rate of 1×10^{-3} s^{-1}. The fracture surfaces were examined by SEM.

8.1.3 Grain Refinement

Figure 8.1 shows the microstructures of BM and center of SZ. The BM is characterized as typical as-cast microstructure with coarse grains and intergranular eutectic networks (Fig. 8.1a). The SZs are composed of equiaxed and fine grains. The average grain size decreases with the increase of travel speed. Figure 8.2 presents the grain size distributions of FSPed samples at various parameters. The average grain sizes of 800/20, 800/80 and 800/200 samples are determined to be 3.20 ± 1.42, 2.37 ± 1.48, 1.65 ± 1.12 μm, respectively.

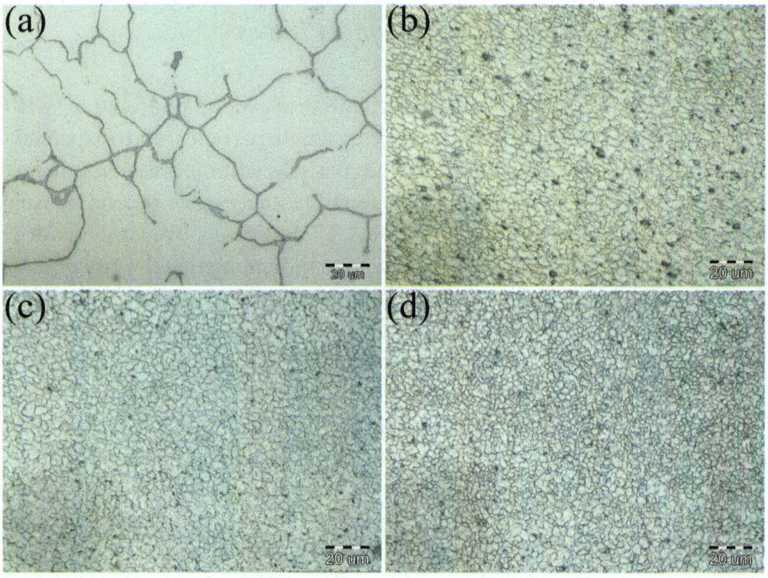

Fig. 8.1 Microstructures of Mg–Zn–Y–Zr samples: **a** BM, **b** 800 r/min–20 mm/min, **c** 800 r/min–80 mm/min and **d** 800 r/min–200 mm/min

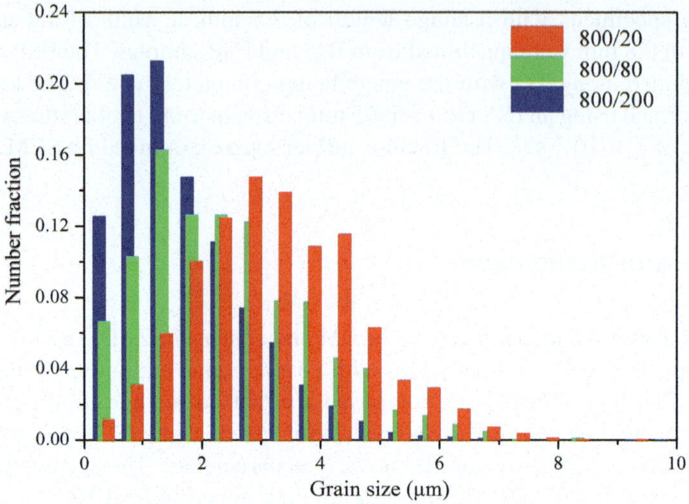

Fig. 8.2 Grain size distributions of FSP samples at various parameters

8.1.4 Transformation of the Second Phases

Figure 8.3 shows the morphology transformation of second phase after FSP. Coarse eutectic networks have been broken up into small particles and dispersed in SZ. Figure 8.4 shows the variation tendency that particle volume fraction increases with the increase of travel speed. Table 8.1 lists the EDS analyses of second phases of BM and SZ of 800/200 sample in Fig. 8.3. In BM, EDS results of several locations show that the Zn/Y ratio of eutectic phase is close to 6, indicating that the eutectic networks mainly consist of I-phase. However, Zn/Y ratio of large particles in SZ was close to 2.5, which implied that I-phase might transform to other phases. Figure 8.5a shows the XRD results of BM and SZ of three FSPed samples. In BM, α-Mg and I-phase are detected, whereas α-Mg and W-phase are detected in SZ. It indicates that I-phase transforms to W-phase during FSP. In Fig. 8.5b, DSC analyses show that there are two endothermic peaks locating at 454 and 520 °C in BM and one endothermic peak locating at 513 °C in SZ of 800/200 sample. The endothermic peak of 454 and 520 °C corresponds to the dissolution of I-phase and the eutectic temperature of W-phase, respectively [16, 18]. The DSC results reveal that all I-phase transforms to W-phase after FSP. In addition, W-phase detected by DSC in BM is few, and cannot be figured out by EDS and XRD.

Figure 8.6a is a high angle annular dark field (HAADF) image of SZ of 800/200 sample, in which large and tiny particles distribute discretely. Figure 8.6b–f shows the element distribution of HAADF image. Despite of Mg matrix, it is obvious that all particles contain Zn element. There is a distinct difference between distributions of Y and Zr elements that nearly all Y element exists in large particles while Zr element mainly exists in tiny particles. Figure 8.7a and b are the typical EDS spectrums of

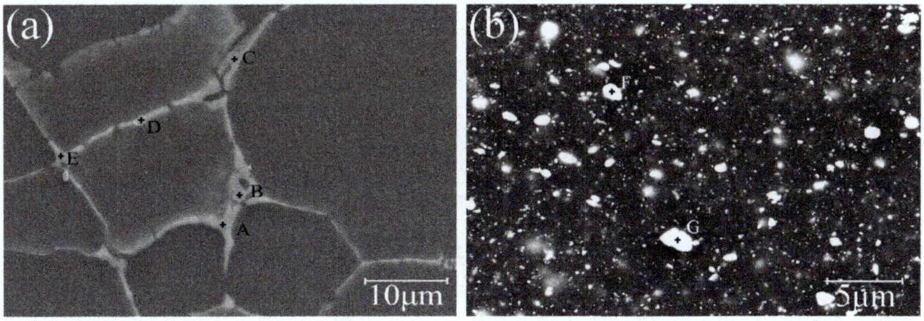

Fig. 8.3 SEM images of second phase in **a** BM and **b** SZ

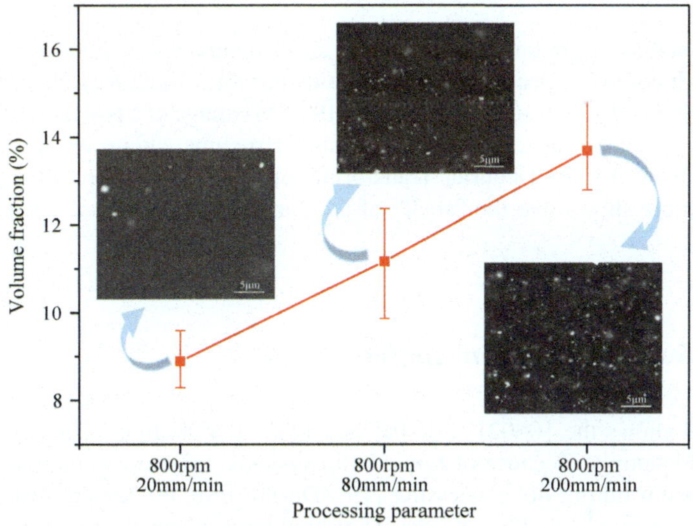

Fig. 8.4 Volume fraction of particles with different FSP parameters

Table 8.1 Results of EDS analysis of locations marked in Fig. 8.3 (at.%)

	A	B	C	D	E	F	G
Mg	55.87	56.74	63.60	65.27	67.41	80.71	38.32
Zn	36.79	35.95	30.97	29.60	26.97	13.69	44.60
Y	7.34	7.05	5.43	5.13	5.61	5.60	17.08
Zn/Y	5.01	5.10	5.70	5.77	4.81	2.44	2.61

the representative particles marked in Fig. 8.6a. The corresponding EDS result of point 1 shows moderate amount of Mg, Zn and Y elements, and a small quantity of Zr elements. The Zn/Y ratio is 2.6, corresponding to EDS results in Table 8.1. Combined with the XRD and DSC results, large particles are W-phase. Point 2

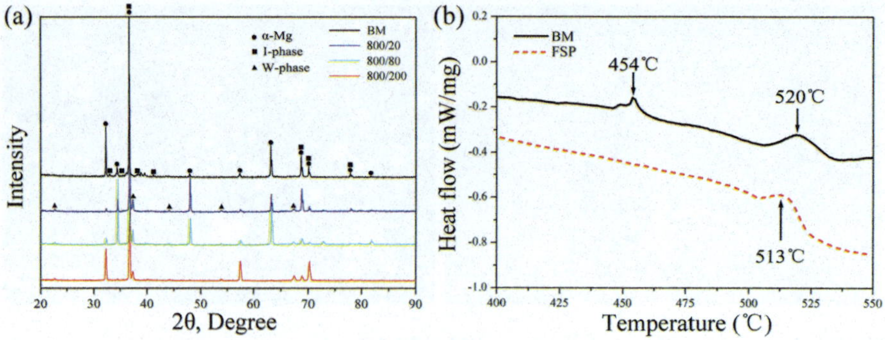

Fig. 8.5 Phase analysis results of **a** XRD and **b** DSC

contains moderate amount of Mg, Zn and Zr elements, and a small quantity of Y elements. It could be speculated that these tiny particles contain Zr elements.

Figure 8.8 is the high-resolution TEM (HRTEM) image of a typical large particle. Obviously, the large particles show a core–shell structure, where core part is pointed by large arrow. The Fast Fourier Transformation (FFT) pattern from the periphery of the particle shows that the "shell" of the particle is typical face-centered cubic W-phases.

8.1.5 Texture Evolution Analysis

Figure 8.9 shows the {0002}, {10-10} and {11-20} pole figures in the SZ center of 800/200 sample. Because of limited slip systems, texture can be easily formed in magnesium during hot processing [19]. During FSP, the texture distribution is affected by the tool shoulder and pin. The shoulder provides the compressive stress but only affects the top part of the processing zone [20]. The rotating pin generates a shear deformation and results in the typical texture in the SZ with the (0002) planes surrounding the pin surface [21]. The present EBSD result displays strong basal texture with the c-axis nearly parallel to PD, inferring the formation of typical FSP-Mg texture in SZ. In addition, Y element can weaken and randomize the texture of SZ and result in the deviation of c-axis of some grains from PD (shown in {0002} pole figure).

8.1.6 Tensile Properties and Fractography

The room temperature uniaxial tensile properties of BM and FSPed samples are summarized in Fig. 8.10. The as-cast BM exhibits a yield strength (YS) of 88.6 MPa, an ultimate tensile strength (UTS) of 196 MPa and an elongation (El) of 10.9%.

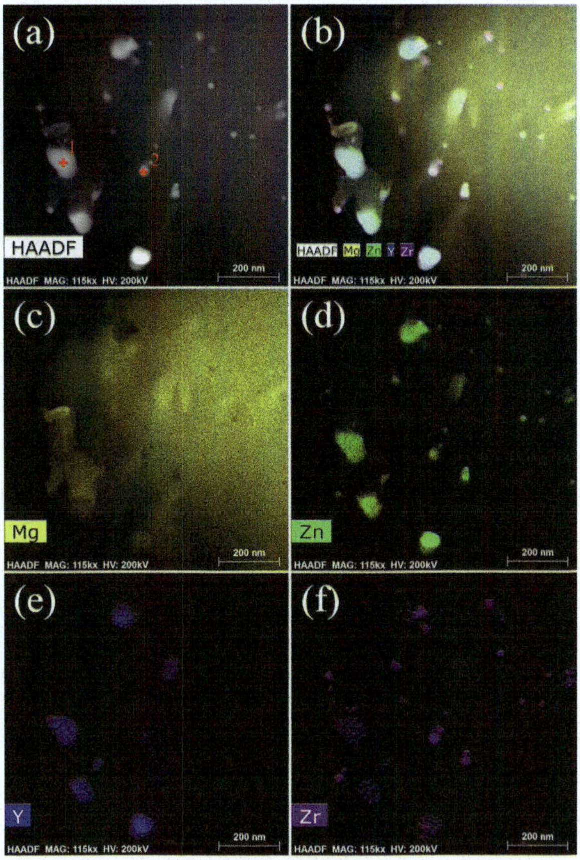

Fig. 8.6 **a** HAADF image of SZ, **b** corresponding mixed EDX mapping, **c**–**f** EDS element mappings for Mg, Zn, Y and Zr elements

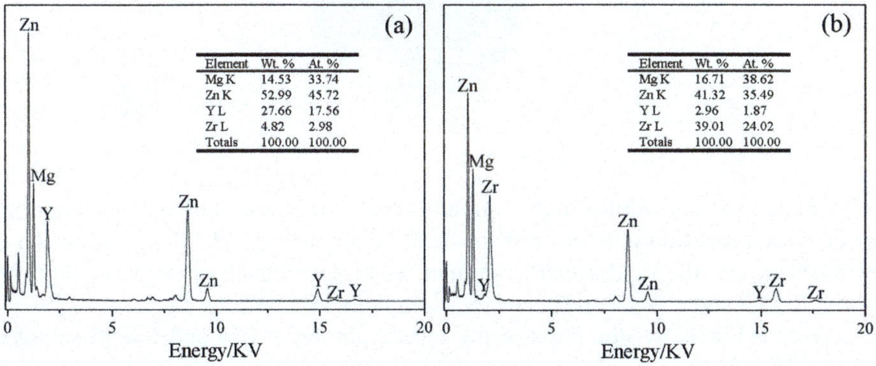

Fig. 8.7 EDS spectrums of particles in Fig. 8.6a

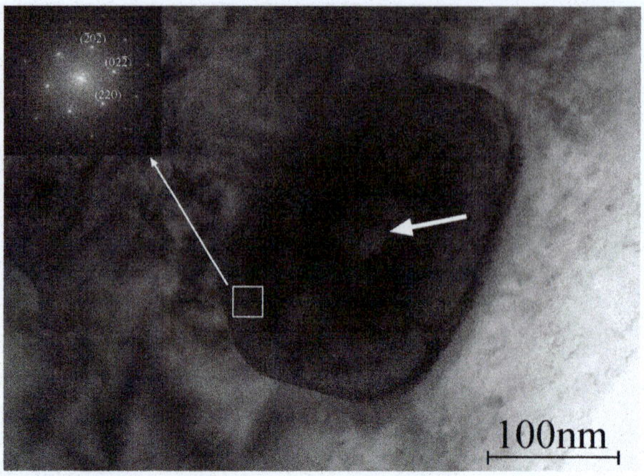

Fig. 8.8 HRTEM image of large particle at SZ. The inset is FFT pattern recorded from the framed region

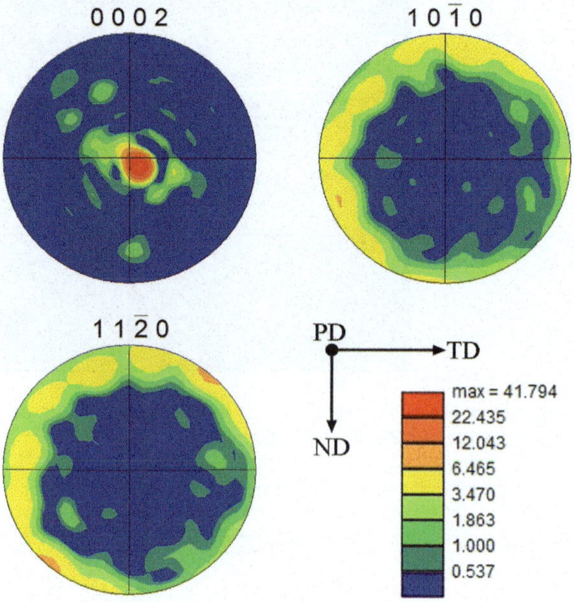

Fig. 8.9 Pole figures of the center part at SZ

All FSPed samples exhibit higher strength and elongation than BM and the total mechanical properties improve with the increase of travel speed. For the parameter of 800/200 sample, the yield strength, ultimate tensile strength and elongation improves by 93.1, 53.0 and 151.4%, respectively, compared with BM.

Figure 8.11a shows the engineering tensile curves of BM and FSPed samples. Clearly FSP favors the enhancement of the mechanical properties, and the mechanical properties gradually increases as the travel speed increases. All samples show a distinct strain hardening stage during tensile tests, which exerts a great role on

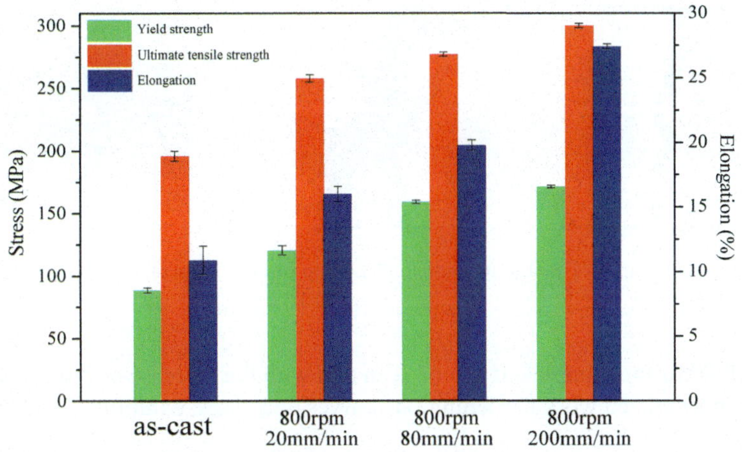

Fig. 8.10 Histogram showing yield strength, ultimate tensile strength and elongation of BM and FSP samples

mechanical properties. Figure 8.11b is normalized work hardening curves of BM and FSPed samples. Normalized work hardening rate is defined by the following Eq. (8.1) [22]:

$$\Theta = \frac{1}{\sigma}\left(\frac{\partial \sigma}{\partial \varepsilon}\right)_{\dot{\varepsilon}} \tag{8.1}$$

where σ, ε and $\dot{\varepsilon}$, , are the true stress, true strain and strain rate ($1 \times 10^{-3}\ s^{-1}$ in this work), respectively. In Fig. 8.11b, all curves exhibit an initial rapid drop due to the short period of elastoplastic transition. After then, all curves enter a steady stage III hardening range [23]. The inset of Fig. 8.11b shows the evident comparison of the

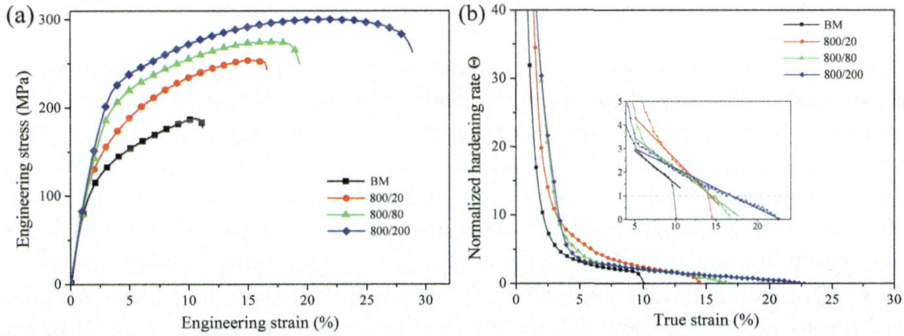

Fig. 8.11 Tensile properties of BM and FSP samples. **a** Engineering stress–strain curves and **b** work hardening rates as a function of ture strain characteristics such as cleavage steps (pointed by black arrow) and pits whose sizes are parallel to coarse cast grains (pointed by white arrow)

Fig. 8.12 Fractography of as BM: **a** macrography of top surface, **b** micrography of top surface and **c** fracture surface

stage III of various samples. It can be seen that with the increase of travel speed, the slope of normalized work hardening rate curves decreases, exhibiting more uniform strain.

Figure 8.12a and b show the top surface of fractured tensile sample of BM. The crack propagation path is roughly perpendicular to the tensile axis (Fig. 8.12a) and is mainly propagated along grain boundaries (Fig. 8.12b). A detail observation of the fracture as shown in Fig. 8.12c reveals the intergranular fracture.

Figure 8.13a shows the top surface of the fractured tensile sample of FSPed 800/200 sample. The crack propagation path is 45° deviated from tensile axis. Figure 8.13b shows that the morphology of fracture is not uniform, and can be roughly divided into two parts. Region I is close to the edge of SZ and it is full of dimples (Fig. 8.13c, d), inferring the ductile fracture behavior and crack initiation. Region II is close to the center of SZ with relatively smooth fracture surface. It can be speculated that the crack initiates near the edge of SZ and propagates along the path of 45° deviated from tensile axis to the center of SZ.

8.1.7 Relationship Between Microstructure Evolution and FSP Parameters

Rotation rate and travel speed are crucial parameters of FSP, which directly influence the thermal cycles and strain rate in SZ and lead to the final microstructure. In the present study, only travel speed increased from 20 to 200 mm/min with a constant rotation rate, and both heat input and strain rate decreased in this trend [24]. The final microstructure including grain size and fraction of dispersion phase are closely affected by these thermal–mechanical parameters. The final grain size is the competition result of grain refinement and grain coarsening. Increasing strain rate promotes the grain refinement while decreasing heat input helps to reduce the grain coarsening. The results in Fig. 8.2 show that the effect of heat input is larger than strain rate in controlling grain size.

During FSP process, drastic stirring effect breaks up the eutectic networks into small particles and part of the second phase dissolves into matrix. Decreasing travel

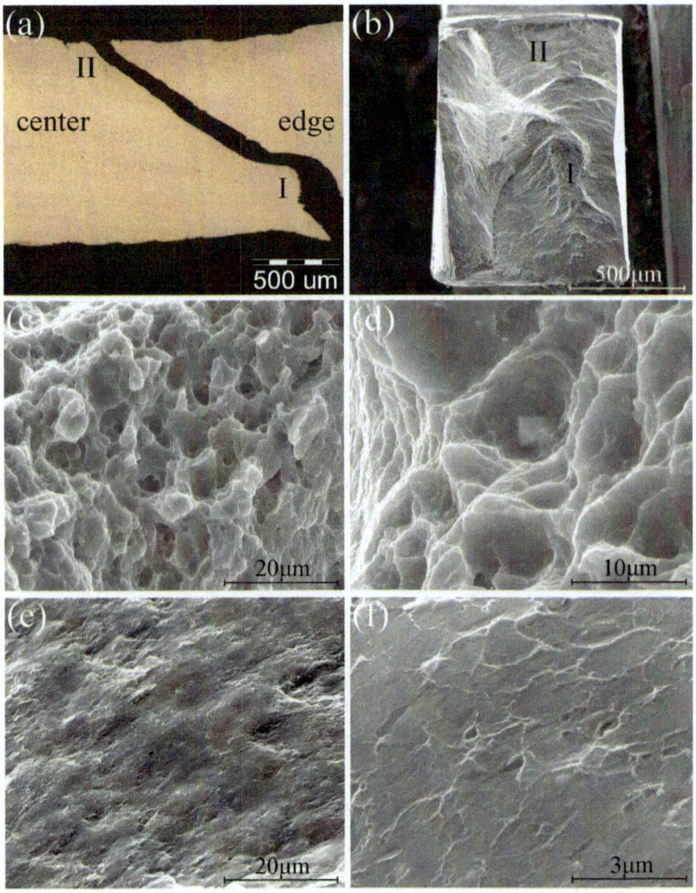

Fig. 8.13 Fractography of FSP sample: **a** macrography of top surface, **b** fracture surface, **c**, **d** Region I, **e**, **f** Region II

speed will improve both heat input and strain rate, which strengthens the solid solution of second phase. It corresponds to the particle volume fraction variation with travel speed shown in Fig. 8.4. In addition, the dispersion particles help to hinder grain coarsening and larger fraction of dispersed particles contributes to the smaller grain size at the travel speed of 200 mm/min. As a result, the microstructure of higher travel speed is characterized by finer grain size and larger fraction of dispersion particles.

8.1.8 Transformation of the Second Phase in Mg–Zn–Y–Zr System

Researches have reported the transformation of I-phase to W-phase during heat treatment process [25, 26]. Liu et al. [26] took a systematical investigation on the transform of I-phase and found that I-phase began to transform to W-phase at 400 °C in TEM samples and 447 °C in bulk samples and the growth of W-phase was controlled by diffusion. During the FSP of cast rare-earth Mg, the peak temperature of SZ can reach 520 °C [24] corresponding to the eutectic temperature of W-phase, which guarantees the temperature requirement of phase transformation during FSP. Zhang et al. [27] compared the reaction of Al and Ti during annealing and FSP, and found that mechanically activated effect of FSP substantially accelerated the reaction. They ascribed the reaction promotion to the pipe diffusion by dislocations, lower activation energy caused by more reaction interface of finer particles and activated atoms of reactants during severe plastic deformation process. The core–shell structure of large particles in SZ (shown in Fig. 8.8) demonstrates the element diffusion during phase transformation. During FSP, eutectic I-phase is broken up into small particles and more reaction interfaces are generated. High density of dislocations would be introduced near small particles and diffusion would be substantially accelerated through pipe diffusion. The activated atoms in FSP decrease activation energy and promote the phase transformation from I-phase to W-phase. Enough high temperature and mechanically activated effect of FSP substantially promote the phase transformation and result in the transformation of all I-phase to W-phase in SZ. The core part of the large particle is likely to be the remnant of I-phase with high Zn containment, which causes the Zn/Y ratio of EDS and EDX results higher than 1.5. Although some researchers have reported the existence of Zr-containing particles in Mg–Zn–Y–Zr alloy systems [28–30], the characterization and understanding of their effects still need further research.

8.1.9 Strengthening Mechanism of Mg–Zn–Y–Zr Material

The BM sample exhibits the lowest yield strength and ultimate tensile strength. It is mainly caused by its coarse grains and intergranular eutectic networks where cracks nucleation and debonding from matrix are preferred. FSP improves total mechanical properties in both strength and ductility, which is strongly related to the microstructural modification. During FSP, coarse grains are refined to several micrometers through dynamic recrystallization, which introduces great fine-grain strengthening. Eutectic networks are broken up into small particles and disperses. The dispersed particles could pin the mobile dislocations and thus hinder their movement, which introduces dispersion strengthening. In addition, the combination of thermal and mechanical effects dissolves a portion of the second phase into matrix, which introduces solution strengthening. The strong basal texture also contributes to the strengthening of FSPed materials. The Schmid factor on basal slip system with respect to TD is relatively low because of strong basal texture [31], which impedes the operation of basal slip and results in higher yield strength. All above factors strengthen the FSPed materials compared with as cast material.

The higher travel speed possesses higher yield and ultimate tensile strength. With the increase of travel speed, grain size becomes smaller and the fraction of dispersed particles increases. It improves the fine-grain strengthening and dispersion strengthening. However, the increase of travel speed decreases the dissolution of the second phase and results in a lower degree of solution strengthening. According to the tensile results, the increase of fine-grain strengthening and dispersion strengthening are obviously greater than the decrease of solution strengthening. Based on the discussion of strengthening related to microstructure, it could be concluded that FSP parameter of high travel speed with low heat input contributes to the fine-grain strengthening and dispersion strengthening, which is beneficial to the total mechanical properties.

8.1.10 Toughening Mechanism of Mg–Zn–Y–Zr Material

The BM sample is characterized by coarse grains and eutectic networks which are detrimental to ductility. However, the elongation of BM sample in the present study is pretty high compared with other cast rare earth magnesium such as GW103 [5] and Mg–Nd–Y alloy [32]. The excellent ductility of BM is mainly ascribed to the existence of I-phase. I-phase is quasicrystal phase and has a coherent interface structure with Mg matrix [33]. The atomic bonding between I-phase and α-Mg is rigid which would delay the debonding between I-phase and matrix during tensile test.

By FSP, dispersed particles pin the mobile dislocations and weaken the dynamic recovery of grains. With the accumulation and propagation of dislocations, the work hardening ability improves. It results in the reduction of slop in the stage III, which could enhance the ductility of materials [7]. Furthermore, Xin et al. [31] studied the influence of the strong basal texture around the pin surface on ductility and found that such texture combined with fine grains could help to activate non-basal slip systems during tensile tests, which could promote the deformation compatibility and enhance ductility.

As compared with curves in Fig. 8.11b, the stage III of 800/20 sample finishes before normalized hardening rate reaches 1 which is regarded as necking demarcation while the stage III of 800/200 sample lasts further longer. It can be concluded that with the increase of travel speed, the slope of the stage III decreases and leads to the increase of uniform stain, which is mainly due to the increased fraction of dispersed particles. The increase of dispersed particles helps to reduce low strength and defect-rich regions which provides stress concentration and degraded ductility. Yuan and Mishra [34] investigated the influence of grain size on work hardening behavior and the results showed that work hardening behavior is almost identical for grain sizes ranging 2.8–5.4 μm when tested in TD. Hence the influence of grain size in the present FSPed samples on the work hardening behavior doesn't differ a lot. Work hardening behavior is mainly improved by the dispersed particles and results in the enhancement of ductility.

8.1.11 Fracture Behavior

Fracture behavior can reflect the mechanical performance. The intergranular fracture behavior (Fig. 8.12) is caused by as cast microstructure of coarse grains and eutectic networks which would damage the mechanical properties especially ductility. This fracture behavior suggests a brittle fracture even though the elongation can reach 12%. FSPed samples exhibit a heterogeneous fracture surface which suggests the fracture initiates near the edge of SZ and propagates to the center of SZ. This fracture behavior is associated with the texture distribution of SZ. The present study shows the strong basal planes surrounding the pin surface in SZ. The basal slip is easy to activate at the edge of SZ but difficult at the center of SZ [31]. During tensile testing, basal slip is firstly operated at the edge of SZ and dislocations interacting with particles results in the dimples on the fracture surface (Fig. 8.13c, d). The propagation path of cracks is roughly along the onion ring layers at initiating position, as shown in Fig. 8.13a. It might be ascribed to heterogeneous dissolution extent and particle fraction in different position of a layer [6]. After crack propagates, the propagation direction remains unchanged and the propagation path is not along the onion layer in the latter process. The basal slip operation is difficult in the center of SZ and dimples are less in fracture surface. Additionally, the rapid crack propagation is also a reason for a smooth fracture surface as shown in Fig. 8.13e and f.

8.1.12 Summary

In this section, microstructural evolution and its effect on mechanical properties of Mg–Zn–Y–Zr alloy after FSP are investigated in the present study. FSP causes great grain refinement, dissolution and dispersion of eutectic networks and strong basal texture around pin surface. The dynamic recrystallized grain size decreases and volume fraction of dispersed particles increases with the increase of travel speed, mainly resulting from the decrease of heat input. The I-phase in BM transforms to W-phase during FSP and forms large particles with a core–shell structure. Mechanically activated effect induced by FSP promotes the transformation. FSPed Mg–6Zn–1Y–0.5Zr alloy samples exhibit better mechanical properties including yield strength (93.1%), ultimate tensile strength (53.0%) and elongation (151.4%) than the cast one, due to the fine-grain refinement strengthening, solution strengthening, dispersion strengthening and texture strengthening. The greater grain refinement and higher fraction of dispersed particles are beneficial for the pronounced ductility.

8.2 Dynamic Recrystallization and Mechanical Properties of Friction Stir Processed Mg–Zn–Y–Zr Alloys

8.2.1 Introduction

As an effective severe plastic deformation technique, FSP can greatly refine the microstructure of Mg alloys with coarse grains [35]. Xiao et al. [5] reported that FSP can significantly improve the mechanical properties of Mg–Gd–Y–Zr casting via microstructural modification, such as fundamental dissolution of eutectic networks and remarkable grain refinement. Yang et al. [15] employed FSP to modify the morphology and distribution of LPSO phase in a Mg–Gd–Y–Zn–Zr alloy, and successfully fabricated high-performance Mg alloy with a special microstructure that LPSO lamellae only existed inside fine grains. Ammouri et al. [36] established the relation between grain size and FSP parameters using an experimentally verified finite element model, shedding more insights in fine tuning the microstructure and mechanical performance of Mg alloys.

Dynamic recrystallization (DRX) is an important mechanism of microstructure refinement during FSP. Three main DRX mechanisms in Mg alloys, i.e. discontinuous DRX (DDRX), continuous DRX (CDRX) and twin DRX (TDRX) during hot deformation have been studied in details [37]. Mironov et al. [38] investigated grain evolution of AZ31 alloy during friction stir welding and found out the sequence of twinning, TDRX, CDRX and DDRX. Feng and Ma [39] studied DRX mechanisms in different FSP passes of AZ80 alloy, and found that TDRX and CDRX dominated the first pass, while CDRX and DDRX dominated the second pass. Until now, most studies of DRX behavior in FSP mainly focus on traditional AZ alloys. However, there are few literatures regarding the influence of RE elements on DRX behavior during FSP. Xin et al. [40] explored the FSP of Mg–Gd–Y–Nd–Zr alloy of solid solution and aging, and found that solute or precipitates of RE elements exerted a crucial role on grain homogenization during FSP. Although CDRX and DDRX were characterized, detailed information regarding the influence of RE on grain refinement mechanisms and the corresponding mechanical properties is still needed. In this study, as-cast and solid solution Mg–Zn–Y–Zr alloys were subjected to FSP with attention being paid to the following: ① the influence of Y form on DRX behavior and final microstructure and ② mechanical property variations corresponding to initial alloy states.

8.2.2 Experimental Procedure

Mg–6Zn–1Y–0.5Zr (wt.%) alloy was used as base metal (BM) with two initial conditions, i.e., as-cast condition (AC-BM) and solid solution condition (SS-BM, solutionized at 525 °C for 8 h, and then hot water quenched). The processing tool had a 16 mm shoulder, a cylinder threaded pin of 7 mm in diameter and 2.5 mm in length.

4 mm thick plates were cut from AC-BM and SS-BM, and then subjected to FSP at the rotation rate of 800 r/min and travel speed of 200 mm/min. After FSP, the materials of as-cast and solid solution conditions were designated as AC-FSP and SS-FSP for sake of convenience.

Microstructures of all samples were characterized on the cross-section, perpendicular to the processing direction, by OM, SEM, XRD, and TEM. The specimens for OM were prepared by mechanical polishing and etched using a solution of 6 g picric acid + 10 mL acetic acid + 70 mL ethanol + 10 mL water. Grain sizes were counted by Image-pro plus 6.0 software.

The microhardness tests were conducted along the centerline of SZ on the cross-section perpendicular to the processing direction, using a load of 200 g and an indentation time of 10 s. Tensile specimens with a gauge length of 2.5 mm, a width of 1.4 mm, and a thickness of 0.8 mm were machined from AC-BM, SS-BM, AC-FSP and SS-FSP. The FSP specimens were machined along transverse direction with the gauge being completely at the center of SZ. The room temperature tensile tests were performed using an INSTRON 5965 mini tester with an initial strain rate of $1 \times 10^{-3}\ s^{-1}$. Tensile fracture surfaces were examined by SEM.

8.2.3 *Grain Refinement Process*

Figure 8.14a shows the schematic illustration of thermal mechanical affected zone (TMAZ). Position A locates at the interface between TMAZ and heat affected zone (HAZ) while position B locates at the interface between SZ and TMAZ. As shown in Fig. 8.14b and c, AC-BM is characterized by coarse grains and intergranular eutectic networks. After solid solution treatment, a lot of eutectic networks dissolve into matrix and a few particles remain near grain boundaries.

Figure 8.14d and f show the microstructures of position A in AC-FSP and SS-FSP, respectively. Twins found in position A are circled, which are not found in BM. These twins are generated during FSP and AC-FSP exhibits more twins. In SS-FSP, the solution of Y promotes the activation of non-basal slips and reduces the influence of twinning on deformation, which decreases the amounts of twins [41]. Figure 8.14e and g exhibit twin morphologies in AC-FSP and SS-FSP, indicating the occurrence of TDRX at position A. Microstructures in circles indicate DRXed grains inside twins, which can be attributed to low-angle boundaries inside twins gradually absorbing dislocations and transforming to high-angle boundaries. Figure 8.14e shows that an intact twin is divided into two parts along specific orientation (indicated by arrow), which is caused by the secondary twinning inside primary twins. Figure 8.14g displays the bulges of twin boundaries (circled by rectangles) in SS-FSP, which is not clear in AC-FSP. It can be regarded as a sort of nucleation mechanism similar to DDRX.

In Fig. 8.14h and j, the original grain boundary can be roughly recognized by the streamlines containing black particles. Both images show "necklace-like structure" that a large number of DRXed grains exist at the vicinity of initial coarse grain

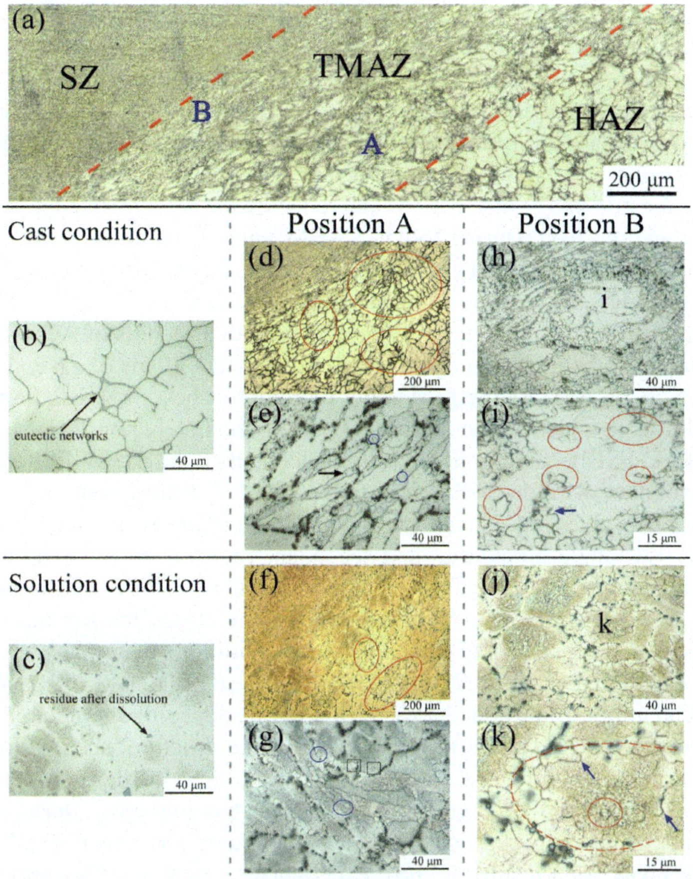

Fig. 8.14 **a** Schematic illustration of TMAZ; microstructures of **b** AC-BM and **c** SS-BM; microstructures of position A in **d** AC-FSP and **f** SS-FSP, **e** and **g** are microstructures of twins in **d** and **f**; microstructures of position B in **h** AC-FSP and **j** SS-FSP, **i** and **k** magnified marked areas in **h** and **j**

boundaries. DRXed grains in AC-FSP are much more than those in SS-FSP, which suggests a more pronounced DRX in AC-FSP. Figure 8.14i and k are magnified images of marked areas in Fig. 8.14h and j, which show relatively intact grains with less deformation. The bulges of grain boundary are marked by arrows, suggesting the nucleation of DDRX. Large particles of more than 1 μm close to initial grain boundaries act as nucleation sites for DRX, known as particle-simulated nucleation (PSN). CDRX also contributes to the "necklace-like structure" owing to the rapid development of strain gradients near grain boundaries [42]. Eutectic I-phase can promote the CDRX in Mg–Zn–Y–Zr alloy as well [43]. Consequently, DRXed grains generated near initial grain boundaries can be ascribed to DDRX, PSN, and probably CDRX. Circles indicate CDRXed grains inside intact initial grains. The similar phenomenon was also observed in the TMAZ of FSPed Mg–3Gd–1Zn cast alloy [44]. DRXed grains inside the initial coarse grains are much more than those in AC-FSP, indicating the easier occurrence of CDRX. In SS-FSP, the solution of Y plays an important role on DRX behavior. Abaspour and Cáceres [45, 46] found that the solution of Y could facilitate the formation of short-range order in Mg alloys. Short-range order has strong effects on dislocation behaviors, such as inhibiting the cross-slip, promoting the pile-ups of dislocations and reducing the velocity of edge segments of paired dislocations [47]. During FSP in SS-BM, short-range order could impede the movement of dislocations and grain boundaries decreasing both CDRX and DDRX, which leads to less DRXed grains inside and around initial grains of SS-FSP.

Figure 8.15 illustrates the microstructural evolution models of AC-FSP and SS-FSP. Apart from the microstructural differences, solute atoms of Zn and Y diffuse into matrix in SS-BM. As the processing tool approaches, materials first enter the cold condition, in which temperature and strain are relatively low. Pronounced twins emerge in AC-FSP because of limited slip systems. By contrast, SS-FSP experiences a small amount of twins due to the activation of non-basal slip systems by the solution of Y. Then DRX takes place inside twins which infers TDRX. The nucleus of TDRX are mainly low-angle boundaries and secondary twinning in AC-FSP. In SS-FSP, however, twin boundaries fluctuate and bulges of twin boundaries may transform into subgrains as nucleus of DRX. With the tool coming closer, temperature and strain further increase and materials enter the warm condition. Few twins are observed because of the large extent of TDRX. The occurrence of DDRX, PSN and possible CDRX near initial grain boundaries results in "necklace-like structure". AC-FSP consists of more DRXed grains in both "necklace-like structure" and interior of initial grains, and more second phase particles promoting PSN. In addition, nanoparticles inhibit the growth of DRXed grains through Zener-pinning effect during FSP [48]. AC-FSP possesses more dispersed particles and exerts a stronger inhibitory effect on grain growth. As a result, AC-FSP experiences a greater DRX during FSP and possesses finer grains of the Mg alloy.

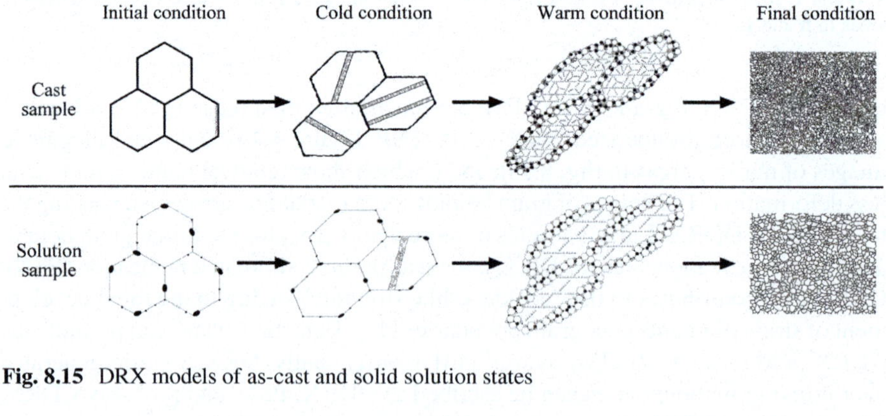

Fig. 8.15 DRX models of as-cast and solid solution states

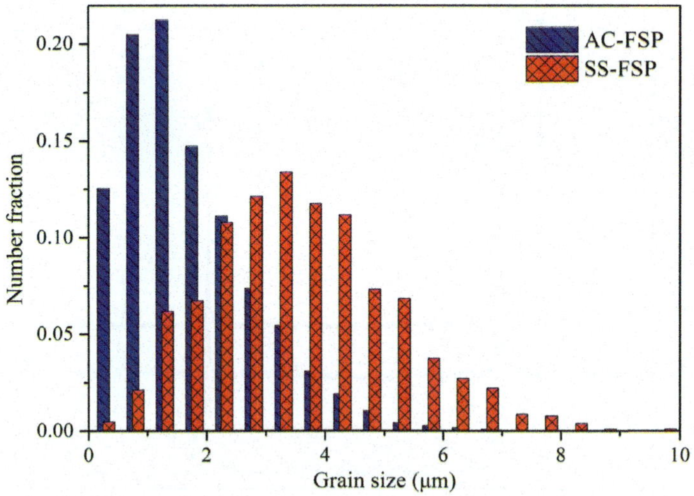

Fig. 8.16 Grain size distributions of AC-FSP and SS-FSP in the center of SZ

Figure 8.16 presents the grain size distributions in SZs of two FSP samples. The average grain sizes of AC-FSP and SS-FSP are (1.65 ± 1.12) μm and (3.58 ± 1.59) μm, respectively. Both average value and variance of grain sizes are smaller in AC-FSP than those in SS-FSP. The smaller average grain size is caused by greater DRX and PSN mechanism introduced by more particles. Less grain size variance is related to homogeneity of DRX. DRX is easier to occur near initial grain boundaries because of PSN and stress concentration. Larger difference of DRX around and inside initial grains leads to higher variance of grain sizes in SS-FSP.

8.2.4 Characterization of the Second Phase

Figure 8.17 shows the XRD patterns of two BM samples and SZs of two FSP samples. In AC-BM, only α-Mg and I-phase (Mg_3Zn_6Y) are detected. In other three samples, detected phase containing Mg, Zn and Y is W-phase ($Mg_3Zn_3Y_2$) rather than I-phase, which infers that I-phase has transformed to W-phase during FSP and solid solution treatment. Such transformation was also reported by previous reports using heat treatment [29] and FSW [16]. Liu et al. [26] adopted in situ TEM technique to investigate the evolution of I-phase in elevating temperature, and found that the transformation from I-phase to W-phase began at 447 °C in bulk samples. The transformation was dominated by element diffusion. During FSP, friction heat guarantees the temperature and severe plastic deformation accelerates the element diffusion by mechanically activated effect [27]. The combination of thermal and mechanical effects during FSP results in the transformation from I-phase to W-phase in a short time. In addition, Zn_2Zr phase is detected in SS-BM and SS-FSP. The formation

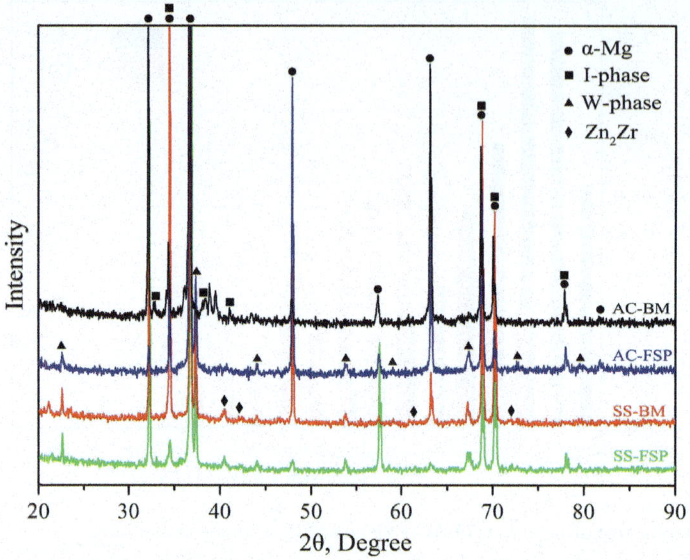

Fig. 8.17 XRD patterns of AC-BM, AC-FSP, SS-BM and SS-FSP samples. (XRD data of AC-BM and AC-FSP samples are cited from previous work [49])

of Zn_2Zr phase may be caused by the precipitation of solute elements during solid solution treatment.

Figure 8.18 exhibits the back scattered electron (BSE) images of SZs in AC-FSP and SS-FSP. All the tiny dispersed second phase particles in AC-FSP are spherical while both spherical and rod-shaped second phases appear in SS-FSP. Compared with Fig. 8.18a and b, the amount of tiny dispersed second phase in AC-FSP is greater than that in SS-FSP (at least five areas were observed in each sample to check the reproducibility of the results).

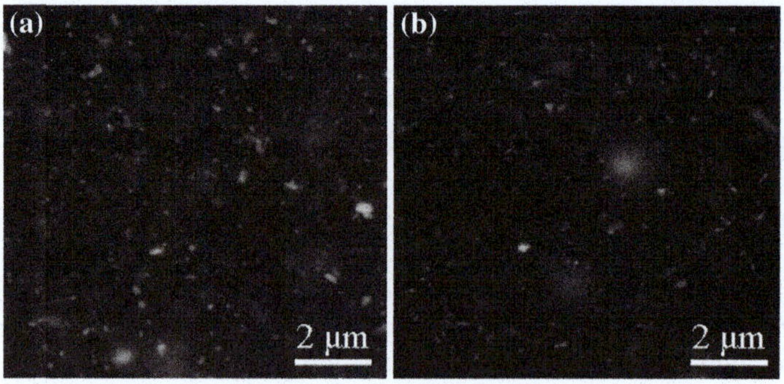

Fig. 8.18 BSE images of SZs in **a** AC-FSP and **b** SS-FSP samples

Figures 8.19 and 8.20 show the HAADF images and corresponding element distributions of Zn, Y and Zr in the SZs of AC-FSP and SS-FSP, respectively. A large irregular particle and a lot of tiny spherical particles are presented in Fig. 8.18a. The enrichment of Y (Fig. 8.19c) suggests that the irregular particle is W-phase. The element distributions of Zn and Zr (Fig. 8.20b and d) suggest that tiny spherical particles are Zn–Zr phase. Figure 8.20a presents spherical and rod-shaped tiny particles in the SZ of SS-FSP. The corresponding element distributions (Fig. 8.20b–d) suggest that these tiny particles are Zn–Zr phase (only one large spherical particle rich in Y is W-phase). Table 8.2 lists the EDS results in Figs. 8.19a and 8.20a. Testing points 1 and 2 correspond to tiny spherical particles and testing points 3–5 correspond to rod-shaped phases. EDS results show that the Zn/Zr ratios of testing points 1 and 2 are close to 1, suggesting that tiny spherical particles are ZnZr phase. The Zn/Zr ratios of testing points 3–5 are close to 2, suggesting that rod-shaped phase is Zn_2Zr precipitate. The precipitation of Zn_2Zr in solid solution samples may be derived from the interaction between large amount of Zn and Zr solutes during solid solution treatment.

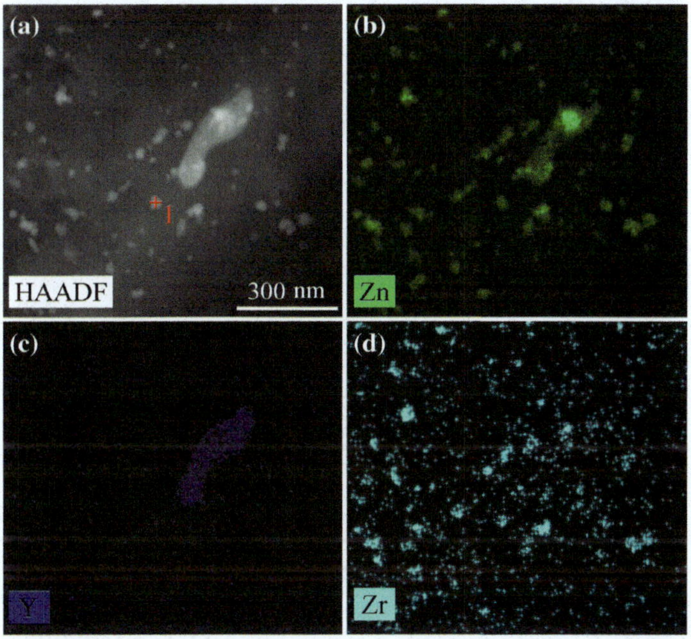

Fig. 8.19 **a** HAADF image of SZ of AC-FSP sample, and **b–d** EDS element mappings for Zn, Y and Zr

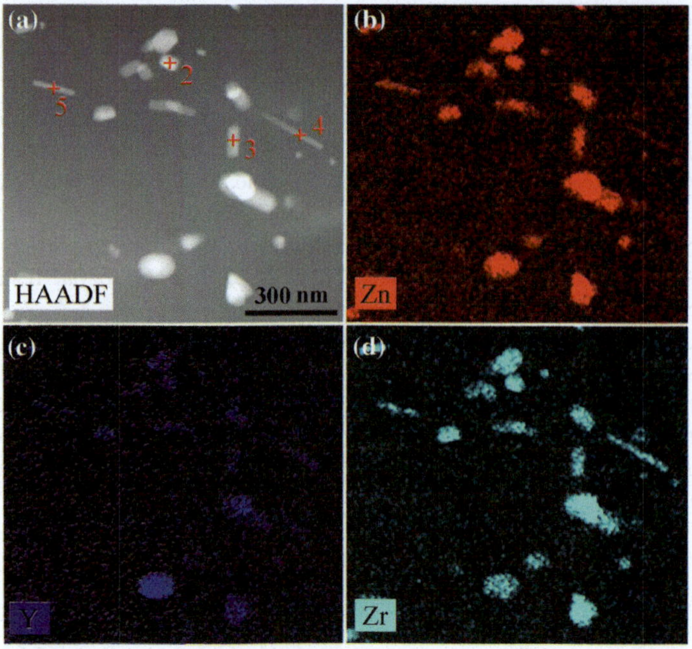

Fig. 8.20 **a** HAADF image of SZ of SS-FSP sample, and **b–d** EDS element mappings for Zn, Y and Zr

Table 8.2 Results of EDS analyses of Figs. 8.6a and 8.7a (at.%)

Position	Mg	Zn	Y	Zr
1	80.73	10.04	0.32	8.91
2	78.33	11.77	0.42	9.48
3	74.34	16.67	0.4	8.59
4	77.39	14.47	0.15	7.79
5	69.02	19.76	0.38	10.84

8.2.5 Mechanical Properties

Figure 8.21 shows the microhardness distributions of AC-BM, AC-FSP, SS-BM and SS-FSP. Zero on x-coordinate represents the center of SZ in AC-FSP and SS-FSP. For AC-BM, the great fluctuation of microhardness distribution is ascribed to the hardness difference between eutectic I-phase and α-Mg matrix. After solid solution treatment, microhardness distribution of SS-BM tends to be homogeneous, which can be attributed to the less hard second phase, and solution strengthening of the matrix due to the solid solution treatment. Although AC-BM and SS-BM exhibit large difference in microhardness distributions, the average values are almost identical.

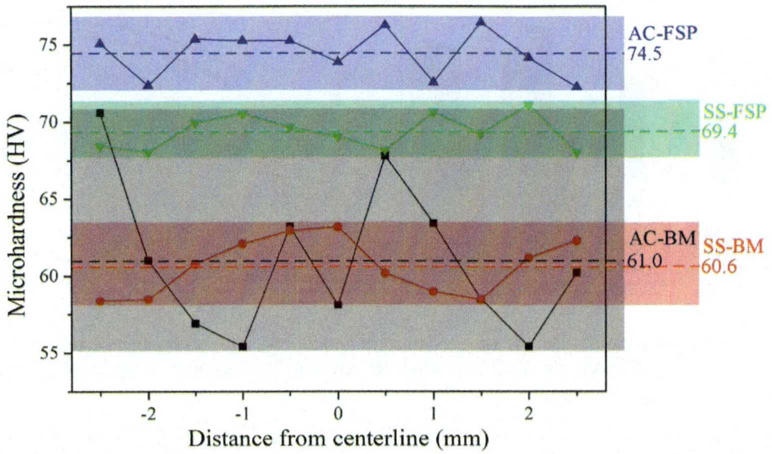

Fig. 8.21 Microhardness distributions of AC-BM, AC-FSP, SS-BM and SS-FSP samples

After FSP, average microhardness values are improved by – 10 HV compared with corresponding BM samples, resulting from grain refinement and dispersion of the second phase particles. The microhardness values of AC-FSP are higher than that of SS-FSP, which is caused by finer grain size and more dispersed particles.

Figure 8.22a shows the YS, UTS and El of all these samples. The YS of AC-BM, SS-BM, AC-FSP and SS-FSP are 90, 90, 170 and 140 MPa, respectively. The comparison of YS among these four samples corresponds well with their average microhardness values due to the similar strengthening effect. Figure 8.22b shows engineering stress–strain curves of four samples in which the comparisons of UTS and El. are clearly present. AC-BM exhibits the lowest UTS of 200 MPa and El. of 11%. After solid solution treatment, UTS and El. of SS-BM increase to 240 MPa and 18%. The improvement of ductility after solid solution is ascribed to dissolution of eutectic networks, delaying nucleation and expansion of cracks. However, both BM samples do not exhibit ultimate uniform elongation according to the Considère criterion. It is ascribed to coarse grains, eutectic networks and large intergranular particles.

Contrary to BM samples, FSP samples exhibit ultimate uniform elongation. After FSP, grain size decreases to several microns and eutectic networks and large particles are broken into uniformly dispersed small particles. Such homogeneous and fine microstructure remarkably strengthens and toughens FSP samples by two mechanisms. Firstly, grains with micron size and dispersed particles not only enhance strength through fine-grain and dispersion strengthening effect, but also contribute to the stress dispersion, delaying stress concentration to a large extent. Numerous dispersed W-phase particles contribute to high strength [50] and their submicron sizes decrease the possibility of cavitation and micro cracks at the interface of the W-phase/α-Mg [51]. Secondly, non-basal slips can be activated in such fine microstructure

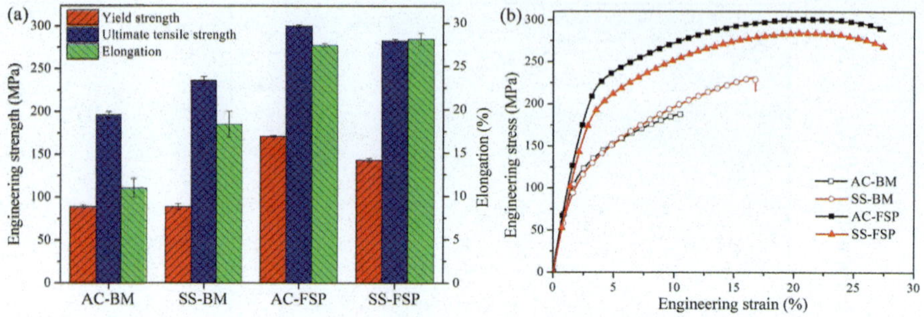

Fig. 8.22 **a** Yield strength, ultimate tensile strength and elongation and **b** engineering stress–strain curves of AC-BM, SS-BM, AC-FSP and SS-FSP samples Tensile data of AC-BM and AC-FSP samples are cited from previous work [26]

during deformation, favoring the deformation harmonization and ductility enhancement. The activation of non-basal slips is ascribed to fine grains, addition of Y and special texture. Substantial dislocation activities on non-basal slips in fine-grained AZ31B Mg alloys during room temperature tensile tests were characterized and attributed to grain-boundary compatibility stress [52]. The addition of Y could decrease the stacking fault energy and promote pyramidal slip in Mg alloys [53]. Non-basal slips, especially pyramidal slip, could promote ductility effectively [54]. Texture is an important factor influencing mechanical behavior of Mg alloys. Previous research reported that rotating pin introduced strong semicircular basal texture in SZ of Mg–Zn–Y–Zr alloy during FSP [49]. Xin et al. [31] studied Schmid factor distributions of various regions and found that non-basal slips were inclined to operate inside SZ. Fine grains, the addition of Y and semicircular basal texture can promote non-basal slips in FSPed Mg–Zn–Y–Zr alloys, which are beneficial to the enhancement of ductility. AC-FSP and SS-FSP samples possess similar plastic behavior during tensile tests and have nearly the identical El. (27% for AC-FSP and 28% for SS-FSP). Because of higher YS and similar plastic behavior, UTS of AC-FSP (300 MPa) is higher than that of SS-FSP (280 MPa). As a result, a simple FSP on AC-BM can achieve the best mechanical properties with YS of 170 MPa (90% improvement), UTS of 300 MPa (50% improvement) and El. of 27% (150% improvement).

8.2.6 Fracture Behavior Analysis

Figure 8.23 shows the fracture surfaces of BM and FSP samples, which reflect their mechanical performances. Figure 8.23a shows the fracture surface of AC-BM with cleavage steps (indicated by black arrow) and large dents caused by separation of coarse grains (indicated by white arrow). These features indicate that cracks mainly propagate along eutectic I-phase networks and some traverse through coarse grains,

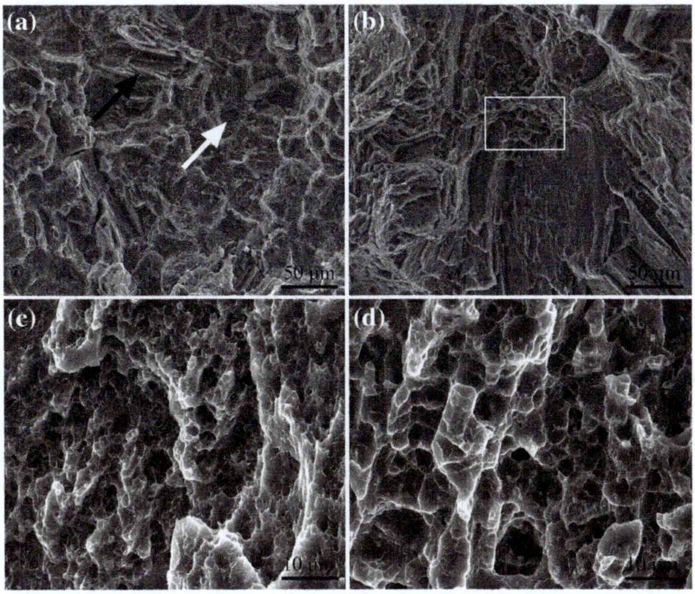

Fig. 8.23 SEM images of fracture surfaces of **a** AC-BM, **b** AC-FSP, **c** SS-BM and **d** SS-FSP samples

which are typical intergranular fracture. I-phase is a kind of icosahedral phase and possesses strong bonding with Mg matrix, which contributes to high El. (> 10%) in spite of the features of intergranular fracture. Compared with AC-BM, the fracture surface of SS-BM (Fig. 8.23b) shows some dimples (circled by white frame) caused by large intergranular particles. The structure of discontinuous particles in SS-BM is conducive to inhibit the propagation and expansion of cracks, which improves tensile ductility effectively. Figure 8.23c and d show the fracture surfaces of AC-FSP and SS-FSP with similar fracture features containing a large number of dimples, which demonstrates that the fracture feature transforms from brittle to ductile due to the formation of fine grains and uniformly dispersed particles.

8.2.7 Summary

To sum up, the influence of Y form in as-cast and solid solution Mg–Zn–Y–Zr alloys on DRX behavior during FSP and mechanical property variations has been studied. For the samples of both as-cast and solid solution states, twinning firstly occurs in cold condition, and then TDRX takes place in twins. With the increase of thermal–mechanical effect, PSN, DDRX and CDRX occur. Greater DRX and more particles contribute to finer and more uniform grains during FSP of as-cast Mg–Zn–Y–Zr alloy. Great grain refinement and dispersed small particles caused by FSP lead to pronounced strength and ductility in both samples of as-cast and solid solution states. FSP of as-cast Mg–Zn–Y–Zr alloy can favor the formation of fine microstructure and large volume of dispersed second phase particles, causing the excellent combination of mechanical performance, i.e., yield strength of 170 MPa, ultimate tensile strength of 300 MPa and elongation of 27%, improved by 90, 50 and 150%, respectively.

8.3 Ultrafine-Grained Mg–Zn–Y–Zr Alloy with Remarkable Improvement in Superplasticity

8.3.1 Introduction

Mg alloys are the lightest structural metal among all structure metal materials, which exhibit high specific strength, high specific stiffness, good casting performance and high thermal conductivity [55, 56]. However, Mg alloys, the hexagonal close packing crystal structure have few slip systems, resulting in low strength and poor plasticity. Fabricating the ultrafine-grained structure is an effective choice because grain refinement usually improves the mechanical properties and superplastic properties [57, 58]. Severe plastic deformation (SPD) can be considered as an effective method to achieve the ultrafine-grains with superplastic properties [59–61]. SPD techniques provide abnormal grain refinement in the metal materials, and ultrafine-grained structures can be achived in metal materials via imposing high strains [62]. FSP as a novel SPD technique for materials solid state processing, is originated from friction stir welding [63], with the advantages of microstructure refinement [64, 65], dispersion distribution and DRX [66–69].

In this letter, the ultrafine-grained structure in Mg–Zn–Y–Zr alloy with remarkable improvement in superplasticity was fabricated by FSP. The evolution microstructure and deformation of ultrafine-grained Mg–Zn–Y–Zr alloy structures were explored during the SPD. The main purpose of this letter is to utilize FSP to fabricate the ultrafine-grained Mg–Zn–Y–Zr alloy with remarkable improvement in superplasticity from the as-casted state, and make clear the superplastic deformation mechanism after FSP.

8.3.2 Experimental Procedures

The BM was Mg–6Zn–1Y–0.5Zr (wt.%) as-casted sheet with a thickness of 4.0 mm. FSP was applied to these materials to enhance the superplasticity. The processing tool was composed of a shoulder 16 mm in diameter, a threaded cylinder pin 6.8 mm in diameter and 3.4 mm in length. Selected processing speed was 100 mm/min, while rotational velocities were 400, 600, and 900 r/min, respectively.

Specimens for OM (Olympus-MPG3), micro-area XRD (D/max-RB) and SEM (ZEISS-SUPRA55) were used to characterize the evolution of microstructure in processed zones (PZs). EDS was used to examine chemical composition of precipitates phases in the PZs. The Vickers microhardness tests (HXD-1000TM) were performed with a load of 200 g and a dwelling time of 10 s. Mechanical properties were assessed at a range of high temperatures 350–450 °C and strain rate of $8.8 \times 10^{-4}\ s^{-1}$–$1 \times 10^{-2}\ s^{-1}$, which tested on SHIMADZU AG–X PLUS testing machine.

8.3.3 Microstructures Characteristics Analysis of the FSPed Mg–RE Alloys

The microstructure characteristics and average grain sizes of as-casted and FSPed Mg–Zn–Y–Zr alloys are shown in Fig. 8.24 The microstructure of the as-casted Mg–Zn–Y–Zr alloys is characterized as the coarser intergranular network eutectic phase with the average grain size is about (170.4 ± 19.3) μm (Fig. 8.24a). The microstructures of PZs with rotational velocities of 400, 600 and 900 r/min are the typical fined, which the grain sizes are (3.6 ± 0.8) μm at 400 r/min, (1.9 ± 0.4) μm at 600 r/min and (2.7 ± 0.9) μm at 900 r/min, respectively (Fig. 8.24b–d). FSP has a prominent effect on grain refinement. During FSP process, the SPD promotes original coarser grains to be broken and dispersed, and high strain accelerates grain refinement. The coarser intergranular network eutectic phases in BM are transformed into equiaxed crystal structures, and extremely DRX occurs in PZs.

SEM images and EDS analysis in the BM and PZs with different rotational velocities are as depicted in Fig. 8.25. The phases precipitate from the crystal boundary of in the as-casted Mg–Zn–Y–Zr alloy, which is the eutectic phase characterized by coarse-grained eutectic networks. For the coupled thermo-mechanical effect, the primary precipitates are broken and refined to 0.1–0.5 μm with uniform distribution in the PZs. With the increase of rotational velocity, the crushing effect of the processing pin is strengthened. The precipitates in the PZs by the effect of SPD are more refined and uniform. The EDS results of the eutectic phases in the BM are the proportion of Zn and Y elements. In Fig. 8.26, the atomic ratio is about 4.6, closed to 6. Referring to the XRD pattern, it is detected the α-Mg and Mg_3Zn_6Y eutectic phases. In PZs, the element proportion of Zn and Y in the PZs ranges from 2.5 to 3.2. The precipitates were only containing Mg, Zn, and Y elements, which is closing to the phase of $Mg_3Zn_3Y_2$. The phases detected by XRD contain α-Mg, Mg_3Zn_6Y and $Mg_3Zn_3Y_2$ phases. Coarser Mg_3Zn_6Y phases are transformed into the refined and dispersed $Mg_3Zn_3Y_2$ phases, which infer that Mg_3Zn_6Y has been transformed to $Mg_3Zn_3Y_2$ via FSP and solid solution treatment. The reference has shown that the precipitates of Mg–Zn–Y alloys transformed from Mg_3Zn_6Y to $Mg_3Zn_3Y_2$ during annealing at 720 K [26]. Element diffusion is the main mechanism of phase transformation. SPD provides effective strain to mechanically activate the element diffusion, leading to the transition from Mg_3Zn_6Y to $Mg_3Zn_3Y_2$ in a short time. During FSP, the

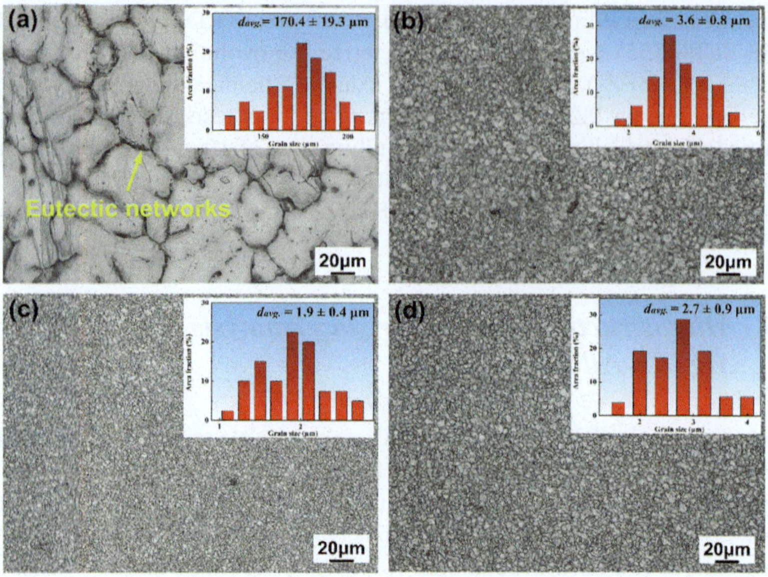

Fig. 8.24 Microstructures of Mg–RE alloy specimens: **a** BM, **b** 400, **c** 600, and **d** 900 r/min

core–shell structure of particles in PZs were found by Wang et al. that demonstrated the element diffusion during phase transformation [49]. Eutectic Mg_3Zn_6Y broken up into some minor particles and generated more reaction interfaces. Enough mechanically activated effect of FSP significantly accelerated the phase transformation and result in the transformation of Mg_3Zn_6Y to $Mg_3Zn_3Y_2$ in PZs.

Figure 8.27 shows the average Vickers microhardness of 77.9 HV at 400 r/min, 82.9 HV at 600 r/min, and 81.4 HV at 900 r/min. The microhardness of 600 r/min is exceeding 27.7% of BM (64.9 HV). After FSP, the grains in the PZs are further refined, and dispersion distribution of the precipitates are uniform. Fine grains and gradually strengthened dispersion distribution are the main reason of increasing the microhardness in the PZs.

8.3.4 Superplasticity Behavior Analysis of the FSPed Mg–RE Alloys

Fig. 8.28a and b show the failure tensile specimens with different temperature and the variation in elongation with ductility and flow stress. Mg–Zn–Y–Zr alloy specimens show high elongation with the range of 337–642%. With the increase of tensile temperature, more slip systems and grain boundary slip phenomena can be occurred during plastic deformation, why the materials show superplasticity behavior. The maximum elongation of 642% is obtained at the tensile temperature of 450 °C,

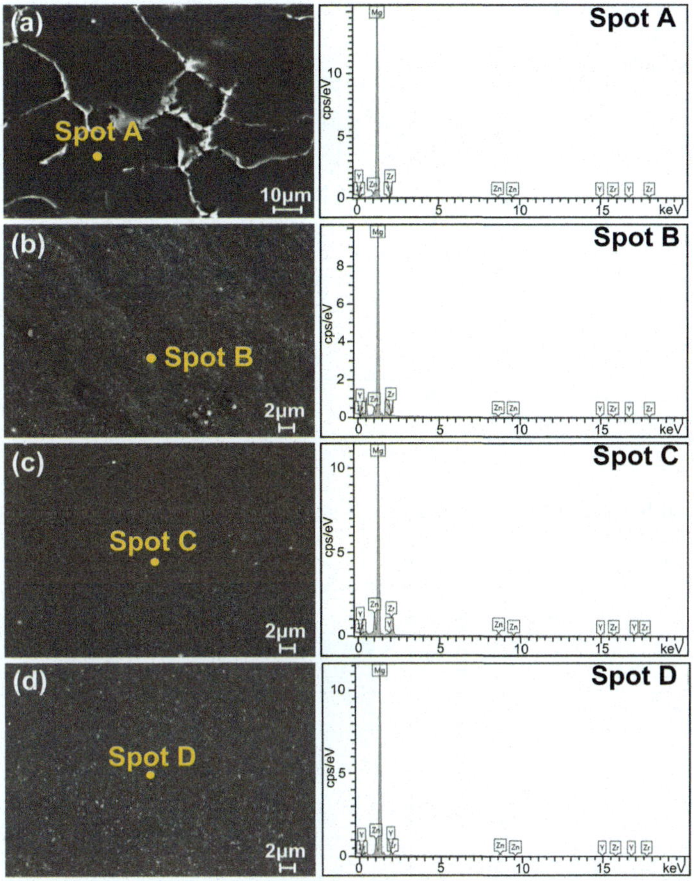

Fig. 8.25 Microstructure characteristics analysis: SEM images of **a** BM, **b–d** the FSPed Mg–Zn–Y–Zr alloys

which is the exceeding more than 58 times of the as-casted Mg–Zn–Y–Zr alloy. Figure 8.28c and d exhibit the failure tensile specimens with different strain rate and variation in elongation with ductility and flow stress. The ultrafine-grained structures exhibit the superplasticity at 400 °C with the strain rate of $5.6 \times 10^{-3}\ s^{-1}$, which 540% elongation is achieved. The variation of the flow stress in the alloys is characterized by linking the microstructural evolution during deformation at high temperatures [56]. The strain rate sensitivity index m value of 0.43 is observed at strain rates of 8.8×10^{-4}–1.1×10^{-2} with a temperature of 400 °C. As strain rate sensitivity index m value is close to 0.5, it is related to grain boundary sliding mechanism [70].

Fine and dispersed precipitates $Mg_3Zn_3Y_2$ and fine, equiaxed and uniform recrystallized grains are obtained in Mg–Zn–Y–Zr alloy after FSP, which provide the basic guarantee for the superplasticity of Mg–Zn–Y–Zr alloy. However, precipitates and recrystallized grains, the grain boundary characteristics are also very important. The

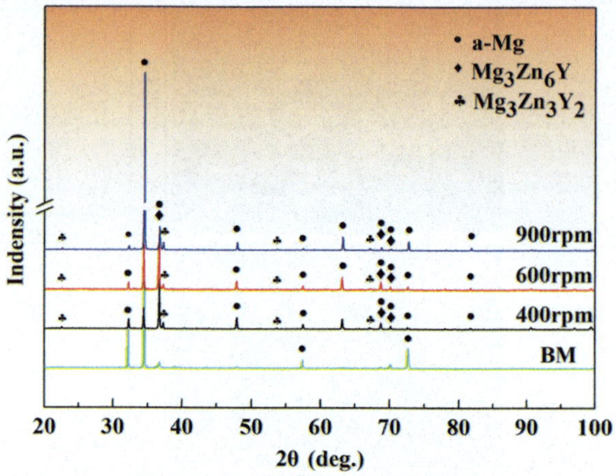

Fig. 8.26 XRD patterns of BM and the FSPed Mg–Zn–Y–Zr alloys

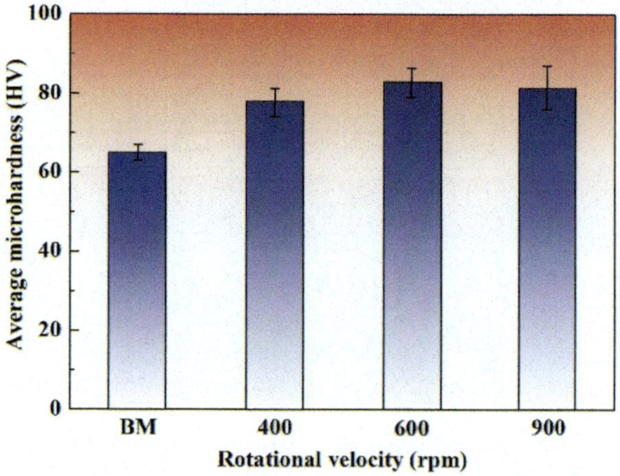

Fig. 8.27 Hardness analysis of BM and the FSPed Mg–Zn–Y–Zr alloys

high angle grain boundary is conducive to the occurrence of grain sliding and rotational, which is beneficial to the superplastic deformation. Grain boundary sliding is considered as the predominant superplastic deformation mechanism in ultrafine-grained materials. The microstructural evolution of CDRX is shown in Fig. 8.29. Mg–Zn–Y–Zr alloys are the coarser eutectic microstructures, which undergo the primary deformation in the severe lattice distortion area near the grain boundary and forming the ultrafine-grains. The recrystallized grains tend to surround the primary grain boundaries and have different grain orientation distribution of the primary grains. Based on FSP, CDRX prevents excessive grain growth caused by excess frictional heat during high strain deformation. CDRX promotes grain boundary sliding and enhance ductility since both CDRX and grain boundary sliding are accommodated by the grain boundary diffusion and migration [71]. With the increase of

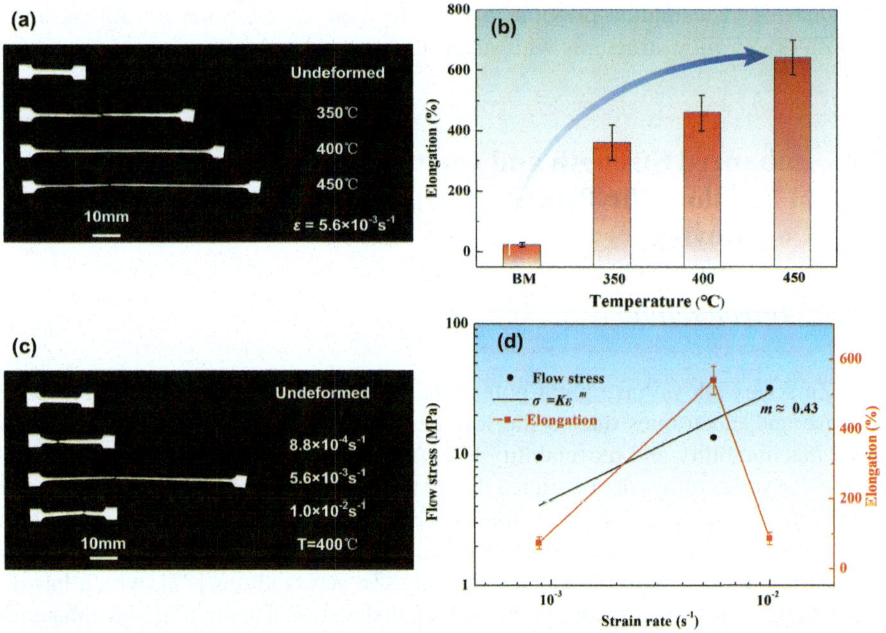

Fig. 8.28 Failure tensile specimen of **a** tensile temperatures of 350–450 °C, **c** strain rates of 8.8×10^{-4} s^{-1}–1×10^{-2} s^{-1} **b** variation in elongation, **d** variation in elongation with ductility and flow stress

SPD degree, more and more CDRX grains are obtained, and finally the grains are ultrafine-grained structure that with superplasticity behavior.

In summary, ultrafine-grained Mg–Zn–Y–Zr alloy structures were constructed. The thermo-mechanical coupled effect promoted the grain refinement and the precipitates dispersion homogeneously. SPD promoted element diffusion and accelerated the transition of precipitates phase from Mg_3Zn_6Y to $Mg_3Zn_3Y_2$. The ultrafine-grained particles in Mg–Zn–Y–Zr alloys with uniform distribution, which led to maximum superplastic reached 642%. Grain boundary sliding is considered as the predominant superplastic deformation mechanism in ultrafine-grained materials.

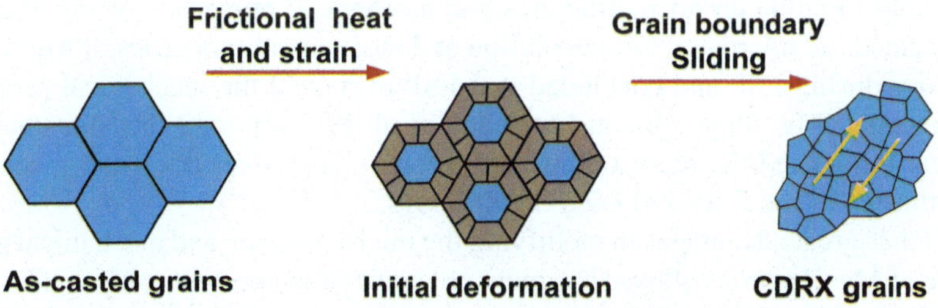

Fig. 8.29 Microstructural evolution of rotational dynamic recrystallization

The developed technique provides a feasible route to construct ultrafine-grained Mg–Zn–Y–Zr alloy structures with superplasticity behavior.

8.4 Enhanced Strength and Ductility of Friction-Stir-Processed Mg–6Zn Alloys via Y and Zr Co-Alloying

8.4.1 Introduction

Mg–Zn series alloys have great potentials in the fields of automobile, aerospace, defense and electronics due to the low density, suitable strength-to-weight ratio, good machinability and recyclability [72–74]. However, the industrial applications of Mg–Zn series alloys are restricted due to the low strength at ambient and elevated temperatures. Alloying is the feasible strategy to reduce the grain size, modify the microstructure and enhance the mechanical properties in Mg–Zn series alloys [75, 76]. Zr is the main alloying element for Mg–Zn series alloys [77], which has the crucial effect on grain refinement, increasing dislocation density [78] and enhancing mechanical properties [79]. Alloying of RE elements in Mg–Zn series alloys, such as elements of Gd, Y and Ce achieves promising high-temperature strength, creep and corrosion resistance [80–82]. Y has the smallest atomic mass compared with other RE elements [83] and the micro-alloying of Y can improve the structural stability [84] and modify the basal texture of Mg alloys [85], which contributing to the enhancement of strength, ductility and stretch formability [86]. Generally, co-alloying of Zr and RE elements in Mg–Zn series alloys acts the multiple effects on microstructure and mechanical properties compared with that of the single micro-alloyed alloys.

FSP, a newly developed severe plastic deformation method, has been proved to be effective on tailoring the microstructure and mechanical properties of Mg alloys [69, 87, 88]. FSP is a short-route, green and energy efficient solid-state processing technique [89–93], in which the grain refinement of Mg matrix and the uniform distribution of second phase are achieved under the severe thermo-mechanical effect [94], thus leading to the enhancement of mechanical properties [14]. At present, FSP of Mg–Zn–RE alloys has been researched to investigate the influence of severe plastic deformation on microstructures and mechanical properties. Wang et al. [49] investigated the microstructural evolution and mechanical properties of Mg–Zn–Y–Zr alloys during FSP and concluded that FSP improved the mechanical properties of the casted Mg alloys obviously. Huang et al. [95] reported that the state of Y element influenced the recrystallization, second phases dispersion and mechanical performance of the FSPed Mg–Zn–Y–Zr alloys.

FSP has great advantages in modifying the microstructure and mechanical performance of Mg–Zn series alloys. The main objective of the present study is to evaluate the effect of Y and Zr co-alloying on microstructural evolution and the mechanical properties of the as-casted and the FSPed Mg–6Zn alloys. The grain refinement and evolution of the eutectic phases can be expected, and the mechanical properties as well as the strengthening mechanism are revealed.

8.4.2 Materials and Experimental Procedure

The alloys with nominal compositions of Mg–6Zn and Mg–6Zn–1Y–0.5Zr (wt.%) were casted. The self-made casting test rig is composed of vacuum electric resistance furnace, protective atmosphere and temperature controlling systems. The appropriate amounts of raw ingots including commercially pure Mg, pure Zn, Mg–30Y and Mg–30Zr (wt.%) master alloys were melted at 720 °C for 20 min in the inert atmosphere of CO_2 and SF_6. Then the molten ingots were poured into the pre-heated steel mould under the pressure of 100 MPa for solidification. The actual chemical compositions of the casted alloys were analyzed and listed in Table 8.3. The billets were cut into plates with the thickness of 4 mm and processed by subsequent FSP. The FSP test rig is the gantry-moving type numerical control machine (FSW-3LM-003) with the welding speed of – 1400 mm/min and the rotational speed of – 3000 rpm. The FSP tool was made of H13 steel, of which the shoulder diameter, threaded cylinder pin diameter and length were 16, 7 and 3.4 mm respectively. The tilting angle of the FSP tool with respect to Z-axis was 2° and the plunge depth was 0.2 mm. The rotational velocity of 600 r/min and the processing speed of 100 mm/min were employed after the optimization of processing parameters.

The microstructural analysis of samples were along the transverse direction of FSP, which were characterized by SEM equipped with EDS and electron backscattered diffraction. The size and volume fraction of the second phases were measured and calculated by the Image-Pro Plus 6.0 software. Vickers micro-hardness was conducted with the load of 200 g and the dwell time of 10 s. The tensile samples were machined along the longitudinal direction of FSP in which all the gauge zone were in the fine grain zone. Three dog-bone-shaped specimens with the dimensions of 20 mm in length, 4 mm in width and 2 mm in thickness were employed. The tensile tests were conducted on the electronic universal tester with the constant strain rate of 1 mm/min at ambient temperature.

Table 8.3 Chemical compositions of the casted Mg alloys (wt.%)

Elements	Zn	Y	Zr	Mg
Mg–6Zn	5.95	–	–	Bal.
Mg–6Zn–1Y–0.5Zr	5.86	0.77	0.52	Bal.

8.4.3 Microstructures of the Mg–6Zn–(1Y–0.5Zr) Alloys

Figure 8.30a and b exhibit the microstructures of the as-casted Mg–6Zn and Mg–6Zn–1Y–0.5Zr alloys, respectively. The as-casted Mg alloys are both characterized by the coarser α-Mg dendrites with networks of intermetallic compounds distributing along grain boundaries. The co-alloying of Y and Zr refined the size of the as-casted α-Mg dendrites, which decreased from 54.4 μm for Mg–6Zn alloys to 33.5 μm for Mg–6Zn–1Y–0.5Zr alloys. The co-alloying of Y and Zr also have an obvious influence on the content of the intermetallic compounds for the as-casted Mg alloys, which increased from 2.7 to 11.4 vol.%.

Figure 8.30c and d show the SEM images of the FSPed Mg–6Zn and Mg–6Zn–1Y–0.5Zr alloys respectively, and Fig. 8.30e exhibits the schematics of FSP. The eutectic networks were broken and crushed under the thermal and mechanical effect of FSP, leading to the formation of uniformly-distributed second-phase particles. The volume fraction of the second-phase particles in the FSPed Mg–6Zn–1Y–0.5Zr alloys is much higher than that of the Mg–6Zn alloys, of which the calculated values are 4.7 and 1.2 vol.% respectively. Solid dissolution occurred during FSP and the contents of the second phases decreased to about 40% than that of the as-casted Mg alloys, owing to the effect of thermal exposure and severe plastic deformation of FSP [96].

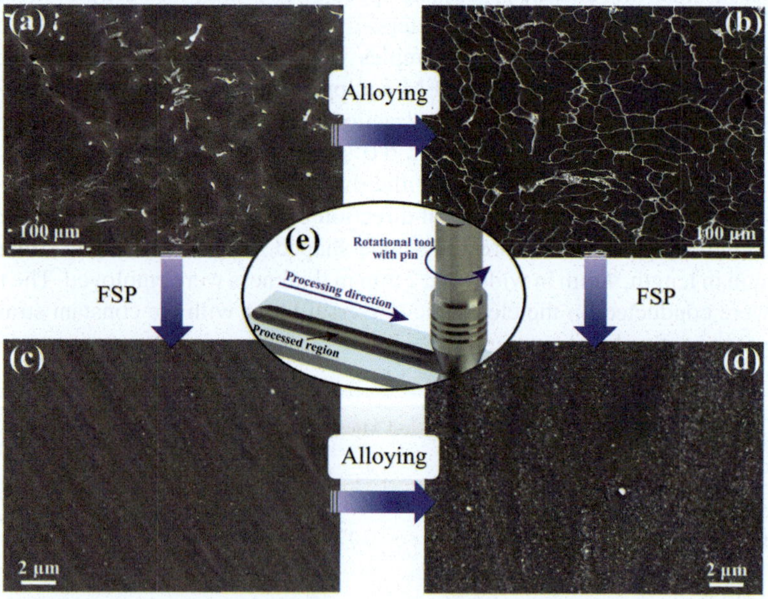

Fig. 8.30 Morphology of **a**, **b** as-casted and **c**, **d** FSPed Mg alloys: **a**, **c** Mg–Zn and **b**, **d** Mg–6Zn–1Y–0.5Zr, **e** represents the schematics of FSP

Figure 8.31 shows the morphology of the as-casted and FSPed Mg alloys with higher magnification. As shown in Fig. 8.31a and b, the morphology of the intermetallic compounds changed from semi-continuous, granular or rod-like into continuous networks and skeleton-like via co-alloying of Y and Zr in the as-casted Mg alloys. As exhibited in Fig. 8.31c and d, the fine particles are prone to distribute uniformly without preferred orientation in the FSPed Mg alloys. The detailed values of the size distribution for the second phases are calculated and displayed in Table 8.4. The particles are ranging 0.042–0.423 μm and the average particle size is about 0.146 μm in the FSPed Mg–6Zn alloys. In the FSPed Mg–6Zn–1Y–0.5Zr alloys, as shown in Fig. 8.31d, the second phases are ranging 0.048–0.641 μm and the average particle size is about 0.159 μm, larger than that in the FSPed Mg–6Zn alloys. The corresponding EDS results of the point A–H are displayed in Table 8.5. The eutectic phases in the as-casted Mg–6Zn alloys are Mg–Zn phase, and the Zn content in the eutectic phases of the as-casted Mg–6Zn–1Y–0.5Zr alloys increased greatly, which is mainly due to the formation of Mg–Zn–Y and Zn–Zr phases. Compared with the EDS results of the as-casted Mg alloys, the composition of the second phases remained unchanged after FSP. The Mg–Zn–Y ternary phases and the Zn–Zr binary phases are detected in the FSPed Mg–6Zn–1Y–0.5Zr alloys, together with the Mg–Zn phases in the FSPed Mg–6Zn alloys.

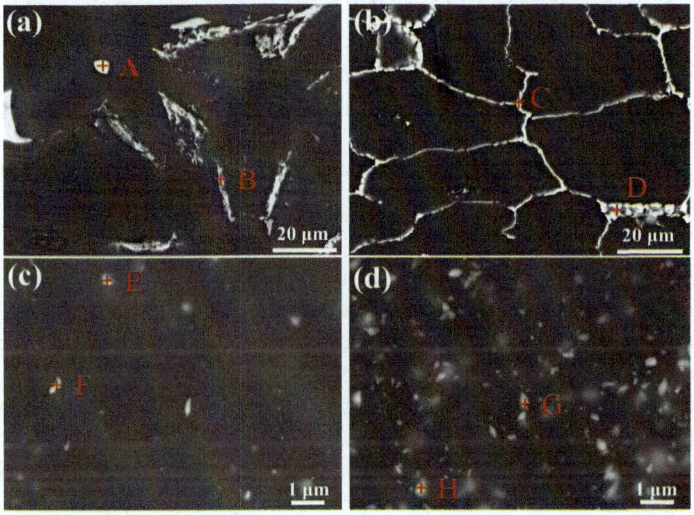

Fig. 8.31 Detailed morphology of **a**, **b** as-casted and **c**, **d** FSPed Mg alloys: **a**, **c** Mg–6Zn and **b**, **d** Mg–6Zn–1Y–0.5Zr with high magnification

Table 8.4 Size distribution of the second phases in the FSPed Mg alloys (μm)

Composition	d_{max}	d_{min}	d_{avg}
Mg–6Zn	0.423	0.042	0.146
Mg–6Zn–1Y–0.5Zr	0.641	0.048	0.159

Table 8.5 EDS results of the point A–H indicated in Fig. 8.32 (at.%)

Point	A	B	C	D	E	F	G	H
Mg	93.85	92.85	53.65	64.78	94.55	91.75	39.96	91.66
Zn	6.15	7.15	41.14	32.50	5.45	8.25	53.42	4.79
Y	–	–	4.77	2.72	–	–	6.13	0.08
Zr	–	–	0.44	0.00	–	–	0.48	3.48

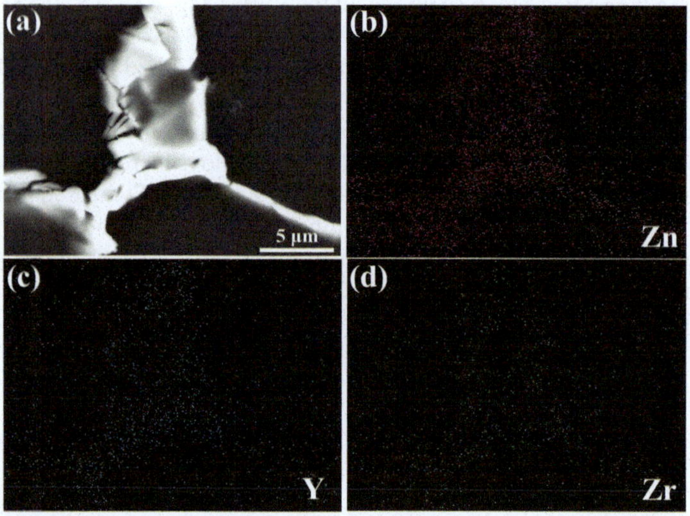

Fig. 8.32 As-casted Mg–6Zn–1Y–0.5Zr alloys: **a** morphology of eutectic phase and **b–d** the corresponding EDS mapping of Zn, Y and Zr

Figure 8.32 illustrates the EDS element mapping showing the distribution of the Zn, Y and Zr elements in the as-casted Mg–6Zn–1Y–0.5Zr alloys. The Zn, Y and Zr elements are densely enriched in the skeleton-like eutectic phases, indicating that these eutectic phases are mainly composed of Mg–Zn–Y ternary phases and Zn–Zr binary phases [97–99]. The EDS mapping results in Fig. 8.32 are consistent with the EDS results in Table 8.5.

Figure 8.33a and b show the inverse pole figure (IPF) maps of the FSPed Mg–6Zn and Mg–6Zn–1Y–0.5Zr alloys respectively. Dynamic recrystallization (DRX) occurred and the coarser α-Mg dendrites transformed into fine and equiaxed grains under the thermal and mechanical effect of FSP. Meanwhile, it is distinguished that the DRXed grain size of the FSPed Mg–6Zn–1Y–0.5Zr alloys is much smaller than that of the FSPed Mg–6Zn alloys. The grain size distribution histograms of the FSPed Mg–6Zn and Mg–6Zn–1Y–0.5Zr alloys are shown in Fig. 8.34a and b. The DRXed grain size is ranging – 10 and – 7 μm for the FSPed Mg–6Zn and Mg–6Zn–1Y–0.5Zr alloys, and the average grain size of the two alloys are 3.6 and 2.1 μm respectively. Besides the decrease of the average grain size, the grain orientation of the FSPed

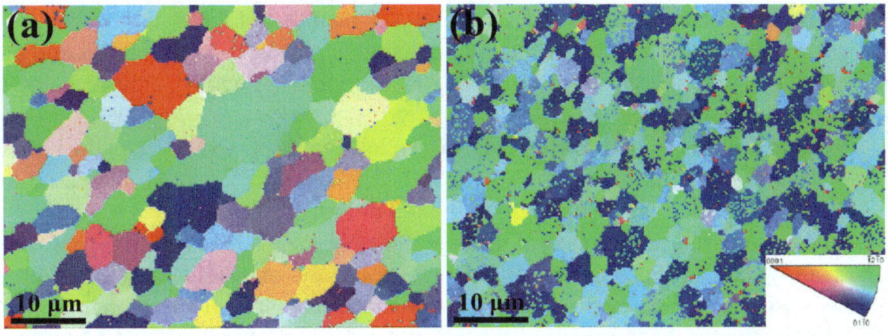

Fig. 8.33 IPF maps of the FSPed Mg alloys: **a** Mg–6Zn and **b** Mg–6Zn–1Y–0.5Zr, the legend shows the IPF coloring of Mg

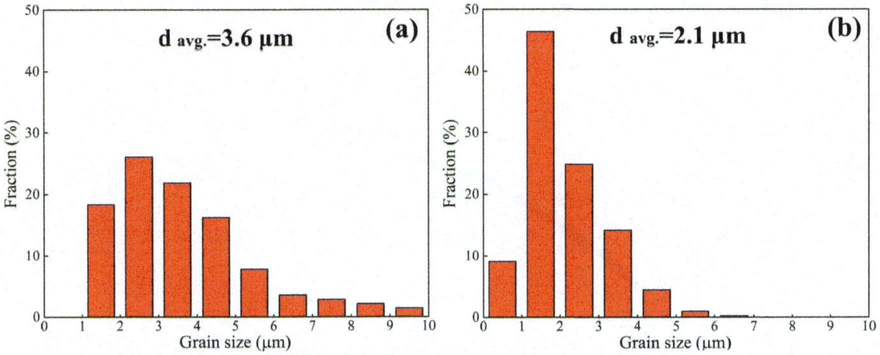

Fig. 8.34 Grain size distribution of the FSPed Mg alloys: **a** Mg–6Zn and **b** Mg–6Zn–1Y–0.5Zr

Mg–6Zn–1Y–0.5Zr alloys is different from that of the FSPed Mg–6Zn alloys, as exhibited in Fig. 8.33. Obviously, the proportion of the basal texture decreased with the increasing of the non-basal textures by Y and Zr co-alloying in the FSPed Mg alloys.

8.4.4 Mechanical Properties of the Mg–6Zn–(1Y–0.5Zr) Alloys

The typical tensile stress–strain curves at ambient temperature of the as-casted and FSPed Mg–6Zn versus Mg–6Zn–1Y–0.5Zr alloys are illustrated in Fig. 8.35, and Table 8.6 exhibits the detailed values of the mechanical properties, including YS, UTS, El. and the average micro-hardness. Co-alloying of Y and Zr enhanced the mechanical properties of the as-casted and FSPed Mg–6Zn alloys simultaneously. The YS, UTS and El. of the as-casted Mg–6Zn–1Y–0.5Zr alloys are 97, 200 MPa and

12.5%, which enhanced by 32.9, 53.8 and 40.4% than that of the as-casted Mg–6Zn alloys, respectively. The refinement and homogeneity of the α-Mg dendrites and the increased eutectic phases are responsible for strength and ductility enhancement in the as-casted Mg alloys.

Compared with that of the FSPed Mg–6Zn alloy, the YS, UTS and El. of the FSPed Mg–6Zn–1Y–0.5Zr alloys attained 170, 310 MPa and 27.7%, which increased by 27.8, 10.7 and 48.9% respectively, as exhibited in Fig. 8.35 and Table 8.6. As shown in Table 8.6, the micro-hardness of the as-casted Mg–6Zn alloys and Mg–6Zn–1Y–0.5Zr alloys are 55 $HV_{0.2}$ and 65 $HV_{0.2}$ respectively, of which the increment is mainly owing to the refined α-Mg dendrites and the comparatively higher content of eutectic phases. FSP enhanced the micro-hardness of the as-casted Mg alloys apparently. The average micro-hardness in the stir zone of the FSPed Mg–6Zn–1Y–0.5Zr alloys reached 80 $HV_{0.2}$, which enhanced by 17.6% than that of the FSPed Mg–6Zn alloys (68 $HV_{0.2}$).

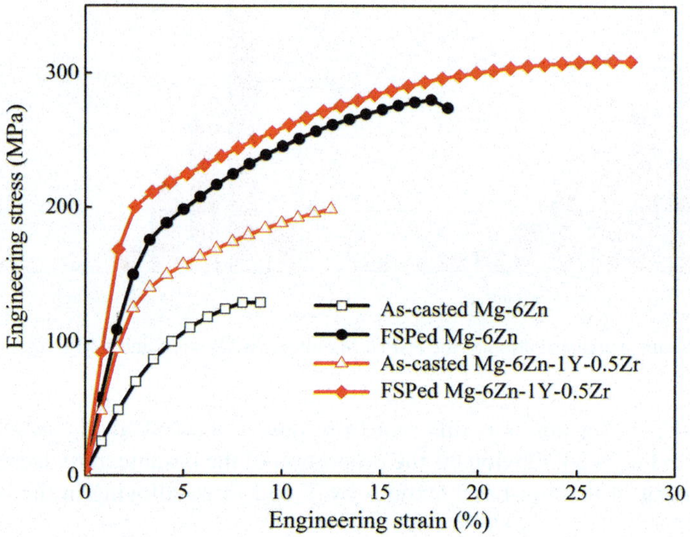

Fig. 8.35 Typical tensile stress–strain curves for the Mg–6Zn and Mg–6Zn–1Y–0.5Zr alloys

Table 8.6 Mechanical properties of the Mg–6Zn and Mg–6Zn–1Y–0.5Zr alloys

Composition	Condition	YS (MPa)	UTS (MPa)	El. (%)	Hardness ($HV_{0.2}$)
Mg–6Zn	As-cast	73 ± 3	130 ± 4	8.9 ± 1.8	55 ± 6
Mg–6Zn–1Y–0.5Zr	As-cast	97 ± 2	200 ± 6	12.5 ± 1.4	65 ± 3
Mg–6Zn	FSP	133 ± 2	280 ± 2	18.6 ± 0.8	68 ± 4
Mg–6Zn–1Y–0.5Zr	FSP	170 ± 3	310 ± 5	27.7 ± 1.2	80 ± 2

8.4.5 *Effect of Co-alloying on Microstructural Evolution*

The microstructural evolution of the as-casted Mg–6Zn alloys after Y and Zr co-alloying was characterized (Fig. 8.30a, b). The co-addition of Y and Zr elements in Mg–6Zn alloys acted the multiple effects on microstructural modification during casting. The addition of Zr results in the significant grain refinement of the as-casted Mg alloys by generating heterogeneous Zr-rich cores during solidification [100]. Meanwhile, the RE elements have been reported to refine the α-Mg dendrites in the as-casted Mg alloys, in which the segregated RE elements at the front of the advancing liquid–solid interface confine the grain growth effectively [97, 101, 102]. Y is confirmed to reduce the surface energy of the liquid Mg and decrease the critical nucleation energy to stimulate more nucleation. Besides the refinement of the α-Mg dendrites during solidification, the co-alloying of Y and Zr in the as-casted Mg–6Zn alloys caused the increase in the content of intermetallic compounds (from 2.7 to 11.4 vol.%), mainly attributed to the enhanced thermal stability of the compounds.

The eutectic networks were crushed into uniformly distributed small particles and DRX occurred during FSP. The initial grain size in the as-casted condition has great influence on the DRX behavior during FSP. The finer grain size, larger grain boundaries and increased intermetallic compounds resulted from Y and Zr co-alloying in the as-casted Mg–6Zn–1Y–0.5Zr alloys increased nucleation sites for the DRXed grains, thus leading to the higher volume fraction of DRXed grains during FSP. The content and size of the fragmented second-phases are responsible for the DRX process, in which the high-density dislocation accumulated around the dispersed particles during FSP. The size of the second phases was reported to determine the DRX behaviors with different ways, in which the particles with the size larger than 0.5 μm promoted the DRX through the mechanisms of particles-stimulated nucleation (PSN) while the particles with the size smaller than 0.5 μm pinned the grain boundaries to prevent the DRXed grain boundaries migration and grain growth. Compared with the Mg–6Zn alloys, higher volume fraction of the dispersed second phases in the FSPed Mg–6Zn–1Y–0.5Zr alloys led to the increased nucleation sites and higher quantities of DRXed grains through the mechanisms of PSN. Besides, the uniformly-distributed second-phases in the FSPed Mg alloys can pin the grain boundaries to prevent the DRXed grain growth [103, 104]. For example, the Zr alloying in Zn-containing Mg alloys has been proved to be effective in pinning grain boundaries and hindering the grain growth during DRX process due to the fine Mg–(Zn, Zr) precipitates. As a result, the increased volume fraction of the dispersed second-phases is the key point for the grain refinement of the FSPed Mg–6Zn–1Y–0.5Zr, compared with that of the FSPed Mg–6Zn alloys.

The variations in the grain orientations were determined by the DRX behaviors during FSP. The nucleation of the DRXed grains in Mg alloys was facilitated by the fragmented eutectic-phase particles via PSN mechanism and the nucleation sites were located in vicinity of the particles. The PSN mechanism is confirmed to randomize the orientation of DRXed grains and thus weaken the basal texture in Mg alloys [105, 106]. The fragmented second-phases with the size larger than 0.5 μm in the FSPed

Mg–6Zn–1Y–0.5Zr alloys provide the sites for the nucleation of the DRXed grains. The content of the second-phases increased by co-alloying of Y and Zr, thus leading to the weakening of the basal texture as a result of the enhanced DRX process.

8.4.6 Evaluation and Modelling of Mechanical Properties Enhancement

The significant enhancement of mechanical properties for the FSPed Mg–6Zn alloys after co-alloying of Y and Zr elements can be attributed to the two following aspects: dispersion strengthening of the homogeneously-dispersed second-phases [107] and grain-refinement strengthening induced by DRX process. The fine second-phase particles with the maximum size of 641 nm are uniformly distributed in the Mg matrix under the severe plastic deformation of FSP, thus a significant dispersion-strengthening effect is achieved. The sub-micron scaled second-phases play a role in interaction with dislocations, thus the resistance of plastic deformation is enhanced by hindering the motion of dislocations. Grain refinement is effective in both enhance the strength and ductility of the FSPed Mg alloys, in which the increased area of grain boundaries can impede dislocations movement and the grain boundaries tend to coordinate mutually during plastic deformation. As discussed above, the dispersion strengthening and grain-refinement strengthening are responsible for the increased YS value of the FSPed Mg–6Zn–1Y–0.5Zr, which could be predicted and modelled by the following equations:

According to the classical Ashby-Orowan strengthening theory [108, 109], the undissolved intermetallic compounds which are homogenously dispersed into the Mg matrix can resist the dislocation motion during plastic deformation. The YS enhancement due to the Ashby-Orowan strengthening can be described by the Eq. (8.2) [110, 111]:

$$\Delta\sigma_{OR} = \frac{0.538Gb\sqrt{f}}{d_p}\ln\frac{d_p}{2b} \tag{8.2}$$

where $\Delta\sigma_{OR}$ is the YS increment of the FSPed Mg alloys via dispersion strengthening, G is the shear modulus (G=1.66 $*$ 10^4 MPa for Mg), b is the Burgers vector (b=3.2 $*$ 10^{-10} m), f is the volume fraction and d_p is the effective particle-diameter of the second phases. The YS increment by dispersion strengthening calculated from the Ashby-Orowan equation for the FSPed Mg–6Zn and the FSPed Mg–6Zn–1Y–0.5Zr alloys are 11.6 and 21.5 MPa, respectively. The co-alloying of Y and Zr into the Mg–6Zn alloys lead to the YS increment of 9.9 MPa by second-phase strengthening after FSP, compared with that of the FSPed Mg–6Zn alloys.

YS of Mg alloys and the grain size accord with the Hall–Petch equation ($\sigma_y = \sigma_0 + kd^{-\frac{1}{2}}$), thus the increased YS due to grain refinement of the FSPed Mg alloys by Y and Zr co-alloying can be described by the Eq. (8.3) [112–114]:

$$\Delta\sigma_{\mathrm{HP}} = k\left(d_{\mathrm{Mg-6Zn-1Y-0.5Zr}}^{-\frac{1}{2}} - d_{\mathrm{Mg-6Zn}}^{-\frac{1}{2}}\right) \tag{8.3}$$

where $\Delta\sigma_{\mathrm{HP}}$ is the increment of YS between the FSPed Mg–6Zn–1Y–0.5Zr and Mg–6Zn alloys, k is the Hall–Petch proportional constant of Mg alloys (k=0.13 MPa·m$^{1/2}$), $d_{\mathrm{Mg-6Zn-1Y-0.5Zr}}$ and $d_{\mathrm{Mg-6Zn}}$ are the average grain size of the FSPed Mg–6Zn–1Y–0.5Zr and Mg–6Zn alloys, respectively. The calculated $\Delta\sigma_{\mathrm{HP}}$ for the FSPed Mg–6Zn–1Y–0.5Zr alloys is 21.2 MPa. Accordingly, the calculated YS increment of the FSPed Mg–6Zn–1Y–0.5Zr alloys reaches 31.1 MPa, which is consistent with the experimental value (about 37 MPa). About two-thirds of the calculated YS enhancement results from the strengthening effect of grain refinement, thus grain refinement is dominated for the strengthening contributions of the FSPed Mg–6Zn–1Y–0.5Zr alloys.

Besides grain refinement of coordinating the mobility of grain boundary, texture is another vitally important factor concerning about the ductility of Mg alloys. The co-alloying of Y and Zr in Mg–6Zn alloys increased the quantity of second phases, which modified the texture components via stimulating the DRX during FSP. The increased proportion of the non-basal texture component is confirmed to promote the sliding of the non-basal slip systems and enhance the ductility of the FSPed Mg alloys. Overall, the decreased average grain size and the weakened basal texture are responsible for the enhanced ductility of the FSPed Mg–6Zn–1Y–0.5Zr alloys. Meanwhile, the alloying of Y in Mg alloys was also confirmed to be effective in dismissing the mobility of grain boundaries and improve the ability of homogeneous deformation [115]. The co-alloying of Y and Zr elements has great potentials in simultaneously improving the strength and ductility of the FSPed Mg–Zn series alloys.

8.4.7 Summary

The effect of Y and Zr co-alloying on the evolutions of microstructures and mechanical properties for the as-casted and FSPed Mg–6Zn alloys were evaluated, and the strengthening mechanisms were discussed. The co-alloying of Y and Zr changed the type, morphology and content of the eutectic phases in the as-casted Mg–6Zn alloys. Grain refinement was obtained, of which the average dendrite size decreased from 54.4 to 33.5 μm. Grain refinement and uniform distribution of the sub-micron second-phases were obtained after FSP. The dispersed second-phases promoted dynamic recrystallization and hindered grain growth, thus the average grain size of the FSPed Mg–6Zn–1Y–0.5Zr alloys refined to 2.1 μm. Besides grain refinement, the intensity of non-basal texture also increased due to particle-stimulated nucleation of the fragmented second-phases. The yield strength, ultimate tensile strength and elongation of the FSPed Mg–6Zn–1Y–0.5Zr alloys increased simultaneously compared with that of the FSPed Mg–6Zn alloys. Grain refinement and dispersion strengthening are mainly responsible for the strength enhancement, according to the modelled equations and the calculated yield-strength values. The grain refinement and the weakened basal texture are responsible for the enhanced ductility.

8.5 Strengthening and Toughening Mechanisms of CNTs/Mg–6Zn Composites via Friction Stir Processing

8.5.1 Introduction

The applications of lightweight structural materials are expanding under the demand of lower fuel consumption and CO_2 emission. As the lightest structural metals, Mg alloys are emerging as a new candidate for aerospace, automobile and military applications due to the high strength-to-weight ratio [94, 95, 116, 117]. Nevertheless, the applications of Mg alloys are limited by lower elastic modulus, strength and toughness [118–120]. Developing high-strength and light-weight metal matrix composites is feasible to overcome the limitations. Carbon nanotubes (CNTs) are considered as the potential reinforcements to strengthen metal matrix composites due to the superior mechanical properties such as high strength (100 GPa) and elastic modulus (1 TPa), favorable geometrical properties such as high aspect ratio, high specific surface area, and good physical properties. Accordingly, CNTs reinforced Mg [121–124], aluminum [125–127] and copper [128, 129] matrix composites are synthesized experimentally and exhibit considerable mechanical properties, which are effective in coping with the urgent demands of light-weight and high-strength metal materials.

However, there still remains critical issues in manufacturing CNTs reinforced metal matrix composites: ① obtaining the uniform distribution of CNTs due to the strong Van Der Walls force, ② maintaining the structural integrity of CNTs during fabrication procedure, ③ controlling sound bonding and compact interfaces due to the weak wettability between CNTs and metal matrix [121–123]. Conventional processing techniques, such as stirring casting [122] and powder metallurgy [130] are not effective to solve these technical issues.

FSP is a novel solid state processing technique to actualize metal matrix modification and mechanical properties enhancement, via its severe plastic deformation and low heat input [118, 131, 132]. In-situ FSP [124, 125, 133] is reported to fabricate CNTs reinforced composites, but CNTs agglomerations and weak interfacial bonding [124, 125] are not solved efficiently due to the poor compatibility between CNTs and matrix. Multi-pass FSP is applied to achieve the further dispersion of CNTs in matrix, while the severe shortening of CNTs and in-situ solid-state chemical reaction deteriorate the survivability of CNTs, due to the thermo-mechanical conditions during multi-pass FSP [134, 135]. Recently two-step approaches are also applied to achieve the uniform-distribution and the stability of CNTs, such as powder metallurgy with subsequent FSP [136] or subsequent hot extrusion [137], and stirring casting with subsequent hot extrusion [123]. The strengthening improvements achieved by these two-step approaches have fulfilled the expectations, while relative work about stirring casting with subsequent FSP has been rarely studied to date. Therefore, it is necessary to develop a new, effective route for fabricating CNTs/Mg composites with exceptional properties.

In this section, an efficient two-step approach via the combination of stirring casting and subsequent FSP was designed to fabricate 1.0 vol.% CNTs/Mg–6Zn composites. The pre-treatment of CNTs and stirring casting were applied to disperse the agglomerated CNTs and obtain the sound interfacial bonding with Mg–6Zn matrix. The subsequent FSP achieved the synergetic effects: the re-dispersion of the incorporated CNTs and the grain refinement of the matrix induced by severe plastic deformation, thus leading to the enhancement of mechanical properties. The main objective is to investigate the microstructure evolution and the enhancement mechanical properties, especially the strengthening mechanisms.

8.5.2 Experimental Procedures

The whole fabrication process includes the following steps: the pre-dispersion of CNTs, the stirring casting and the subsequent FSP, as illustrated in Fig. 8.36. The as-received CNTs, pure Mg and pure Zn were used as the starting materials. The external diameter and length of the as-received CNTs (provided by Chengdu Organic Chemistry Co Ltd., Chengdu, China) are 40–60 nm and − 2 μm, which were synthesized by chemical vapor deposition method. The as-fabricated Mg–6Zn alloys composition (in wt.%) was 6.0 Zn and balanced Mg, and the incorporated CNTs in CNTs/Mg–6Zn composites were 1.0 vol.% of the metal matrix.

According to the desired compositions of the tested alloys and the composites, appropriate amounts of Mg, Zn and CNTs were calculated in advance. The pre-dispersion of CNTs includes the ultrasonic treatment in the chemical dispersant and the subsequent ball-milling, as illustrated in Fig. 8.36a and b. The mixture of the as-received CNTs and the pure Mg chips was firstly stirred at 250 r/min in chemical dispersant solution under ultrasonic to pre-disperse the CNTs clusters and clean the CNTs surface. Then the mechanical ball-milling was carried out to further disperse the mixture in a planetary ball milling apparatus under the rotational speed of 400 r/min and the whole time of 4 h. The mixture of the CNTs and the pure Mg chips was put into the stainless steel container with alcohol as the medium during ball-milling.

The stirring casting test rig is self-made facility equipped with LGT-3-1 electric resistance furnace, temperature and protective atmosphere controlling systems. The appropriate amounts of Mg and Zn ingots were melted at 720 °C for 10 min in the protective atmosphere of CO_2 and SF_6 and then the melt was cooled to 600 °C at which the melt was in a liquid–solid two phase region. The ball-milled Mg chips coated with CNTs were added into the melt and then semi-solid stirring started under the stirring rate of 800–1200 r/min and the stirring time of 10 min, as illustrated in Fig. 8.36c. The stirrer is self-designed equipped with variable-frequency speed-controller and four stirring blades (30 mm × 20 mm × 3 mm).Then the melt was

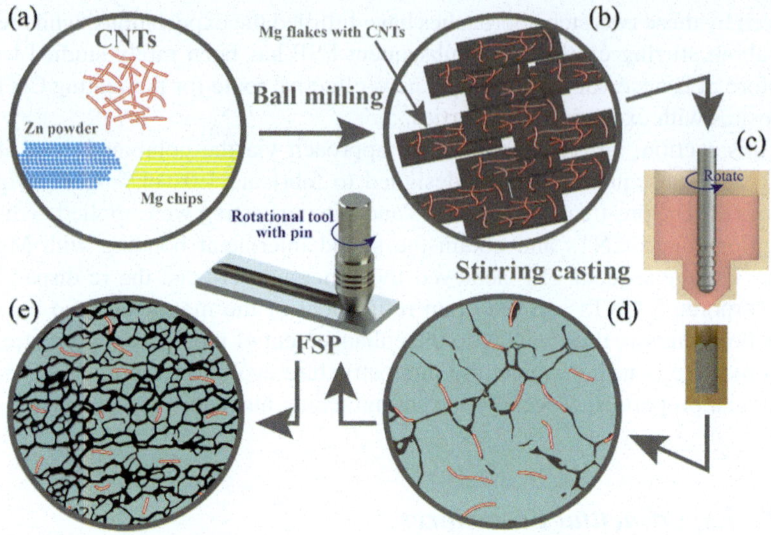

Fig. 8.36 Schematics for the fabrication procedures of the CNTs/Mg–6Zn composites: **a** the as-received materials, **b** the deposited CNTs with Mg chips by pre-treatments, **c** stirring casting, **d** microstructure morphology and CNTs distribution for the as-cast and **e** the FSPed composites

reheated to 690 °C to carry out ultrasonic vibration using the ultrasonic probe with the power level of 500 W and the vibration time of 20 min. Finally the melt was solidified into the preheated steel mold under 100 MPa pressure to obtain the ingots without porosity. The ingots were cut into samples with the size of 120 mm × 40 mm × 5.5 mm and homogenized at 350 °C for 12 h before FSP.

The FSP test rig is the gantry-moving type numerical control machine (FSW-3LM-003) with the welding speed of – 1400 mm/min and the rotational speed of – 3000 r/min. The shoulder with 15 mm in diameter as well as the threaded cylinder pin with 3 mm in diameter and 4.5 mm in length were employed during FSP, as illustrated between Fig. 8.36d and e. A constant rotation velocity of 1000 r/min and processing speed of 50 mm/min were employed under the tool tilting angle with respect to Z-axis of 1.5° and a plunge depth of 0.2 mm. Figure 8.36d and e show the illustrations of the grain morphology, the morphology and distribution of CNTs for the as-cast and FSPed composites, respectively. The subsequent FSP after stirring casting procedure is supposed to achieve the synergistic effects: the grain refinement of the matrix and the re-dispersion of CNTs. The Mg–6Zn alloys without CNTs addition were also fabricated under the same fabricating parameters.

The microstructural evolution was observed by OM, SEM, TEM. The composition was analyzed by EDS, XRD, selected area electron diffraction (SAED). All the samples for microstructural analysis were observed along the transverse direction of FSP and prepared by the standard procedures of mechanical grinding, polishing and etching. The XRD examination mode was continuous scan within the scan speed of 10(°)/min and the scan range of 20°–90°. Vickers micro-hardness tests were performed with a load of 200 g and a dwell time of 10 s. Five dog-bone shaped samples with a gauge length of 2.5 mm, width of 1.4 mm and thickness of 1 mm were used in ambient tensile test. The tensile specimens were machined along the transverse direction of FSP to ensure that all the gauge zone were in the fine grain nugget zone. The tensile tests were conducted at a strain rate of 1×10^{-3} s^{-1} using an INSTRON 5965 mini tester and the fractographs were observed by SEM.

8.5.3 Microstructure Characterization

Figure 8.37a and b exhibit the microstructure of the CNTs/Mg–6Zn composites after stirring casting and FSP, respectively. The as-cast samples in Fig. 8.37a consist of coarse primary α-Mg surrounded by the network-like discontinuous eutectics structures which mainly present along grain boundaries. The eutectics are mainly composed of the eutectic α-Mg and intermetallic compound phase. Grain refinement was achieved after FSP and the coarser inter-grain eutectics phases were broken into fine, dispersed second phases, as shown in Fig. 8.37b. Based on the XRD diagrams in Fig. 8.37c, only α-Mg and $Mg_{0.97}Zn_{0.03}$ phase are detected in all samples, including the as-cast, FSPed alloys and the FSPed composites, and no additional diffraction peaks arise after FSP.

Figure 8.37d–f are EDS results of the tested point D, E and F marked in Fig. 8.37a and b, respectively. EDS results in Fig. 8.37d and e confirm that the particles marked in Fig. 8.37b are Zn-containing second phases with approximately 5.84% atom fraction of Zn element, which is higher than that of the matrix. According to the XRD and EDS results, the second phases in Fig. 8.37a and b are identified to be $Mg_{0.97}Zn_{0.03}$ intermetallic compounds, and FSP caused the fragmentation and partially dissolution of the second phases [138]. Meanwhile, the dispersed CNTs (as circled) in Fig. 8.37b and the corresponding EDS in Fig. 8.37f prove that CNTs dispersed homogeneously without oxidation in the FSPed composites, and the incorporated CNTs survived through undergoing the severe thermal and mechanical effects, such as ultrasonic dispersion, ball-milling, stirring casting and subsequent FSP. Therefore, the novel process adopted in this research can fabricate metal matrix composites with uniform distribution and integrity of the CNTs.

Figure 8.38a–d exhibit the OM and TEM micrographs of the FSPed Mg–6Zn alloys and CNTs/Mg–6Zn composites, respectively. Dynamic recrystallization occurred during FSP, causing significant grain refinement. Whilst, the grains are much finer and the distribution of grain size is much more homogeneous for the FSPed composites than those of the FSPed alloys. The average grain size calculated by using the linear intercept method for the FSPed alloys and the FSPed composites are 5.6 and 4.4 μm, respectively. According to the dynamic recrystallization theory, the incorporated CNTs were proved to accelerate grain refinement through increasing the nucleation sites and inhibiting the grain boundaries migration [121, 135]. Meanwhile, the uniform distribution of CNTs is responsible for the homogeneity of the recrystallized grain size.

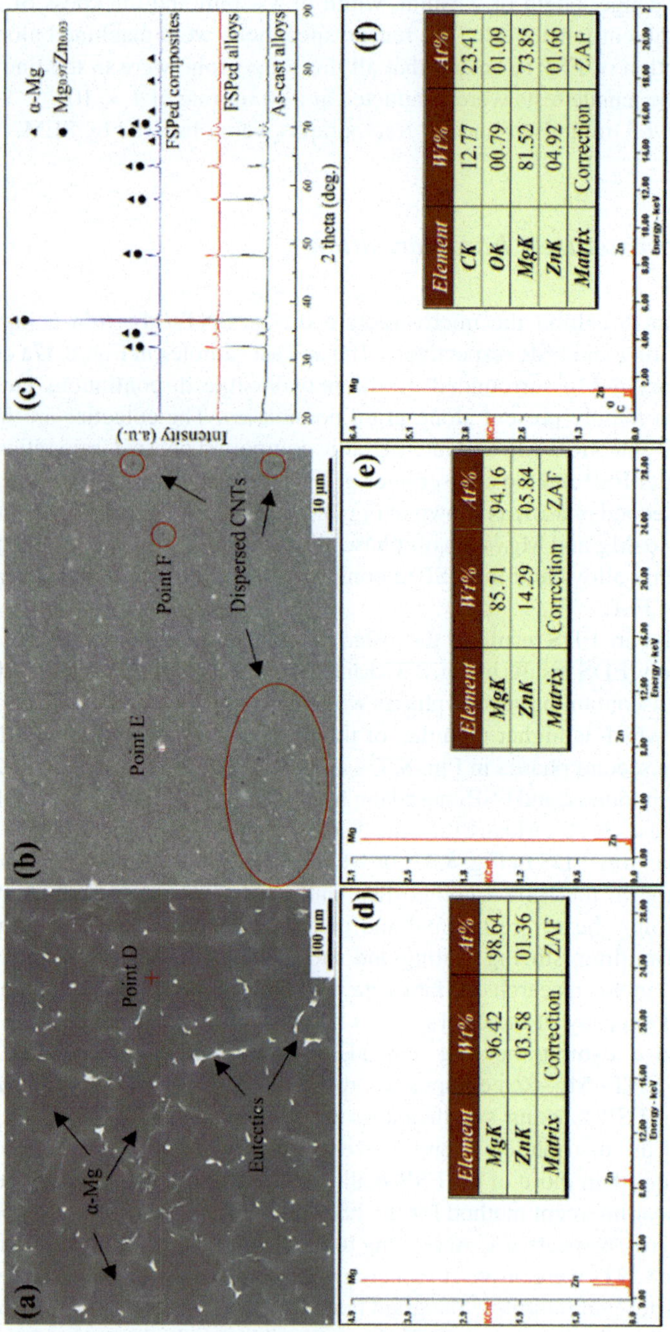

Fig. 8.37 SEM images of the CNTs/Mg–6Zn composites and the corresponding XRD and EDS results: **a** as-cast composites, **b** FSPed composites, **c** XRD diagrams, **d–f** corresponding EDS of the point D, E and F, respectively

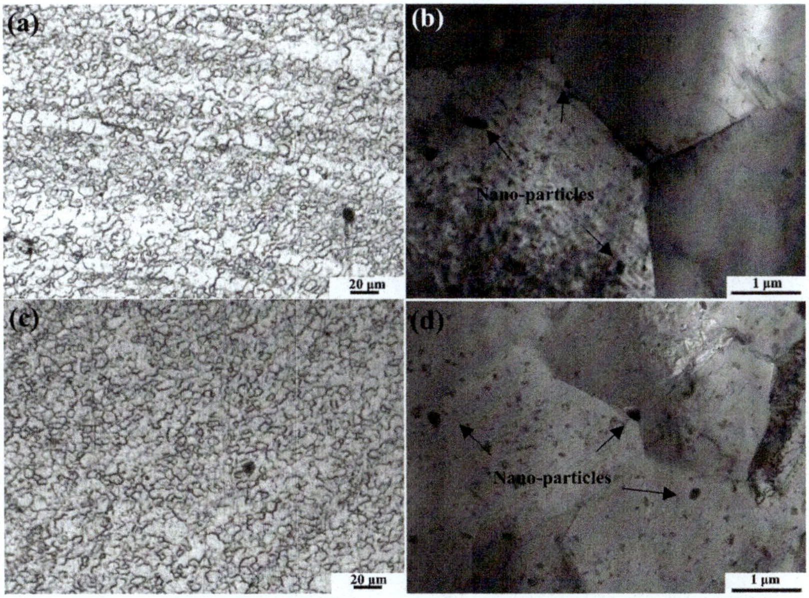

Fig. 8.38 OM and TEM images of the recrystallized grains and nano-scaled phases for **a**, **b** the FSPed Mg–6Zn alloys and **c**, **d** the FSPed CNTs/Mg–6Zn composites

Spherical particles were mainly induced by the coupled thermo-mechanical effect of FSP, as illustrated in Fig. 8.38b and d. The eutectic networks of the as-cast samples were crushed into the nano-sized particles and partially dissolved into the Mg matrix. These re-distributed constituent phases in nanoscale contribute to the grain refinement and thermal stability, thus leading to the enhancement of mechanical properties [139].

8.5.4 Morphology, Distribution and Integrity of CNTs

The distribution and morphology of the incorporated CNTs in FSPed CNTs/Mg–6Zn composites are shown in Fig. 8.39. Figure 8.39a–c reveal that CNTs are singly dispersed in matrix without any clusters after FSP. The FSP process is proved to contribute to the further promotion of CNTs re-distribution effectively. Figure 8.39d shows the morphology of a singly dispersed CNT and the interfaces between the CNT and matrix. The specific features of the CNT, such as the tubular structure, the CNT walls and the axial hollow channel as well as the interfaces with matrix can be identified clearly. No defects exist in the CNTs/matrix interface and the interface is continuous and compact. Meanwhile, the stirring casting process with the CNTs pretreatment modifies the surface of CNTs and promotes the compact interfacial bonding with the matrix, which benefits to the enhancement of mechanical properties.

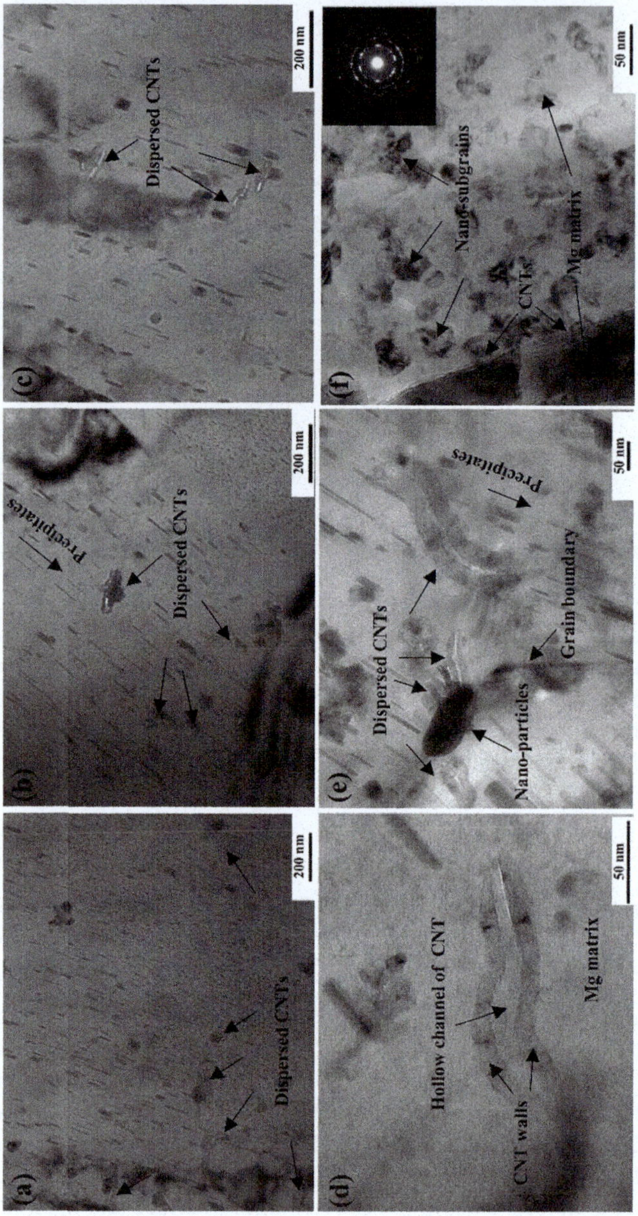

Fig. 8.39 TEM images for the FSPed CNTs/Mg–6Zn composites: **a**, **b** and **c** singly-distributed CNTs, **d**, **e** the morphology of CNTs and **f** the nano-sized sub-grains next to CNTs, the inset is the SAED pattern

CNTs in Fig. 8.39e locate both on the grain boundary and in the grain interior, and interestingly one of them locates next to a nano-sized particle. It can be inferred that the incorporated CNTs and the nano-sized phases can facilitate sites for the nucleation of dynamic recrystallization. Meanwhile, the CNTs locate on the grain boundaries hinder grain growth effectively by blocking grain boundaries migration. Rod-shaped precipitates in Fig. 8.39b and e (length of about 20–100 nm and width of about 5–10 nm, arranging along a particular direction) are mainly dynamically re-precipitated under the thermal–mechanical effect of FSP. It is reported that the incorporated CNTs in the Mg matrix have assisted in the precipitates nucleation and growth, which is attributed to the activation effect of the dislocations [140]. But the precipitates cannot be detected within the sensitivity limits of XRD due to the low content.

Figure 8.39f shows the TEM image of the nano-scaled sub-structures formed around CNTs. The SAED pattern presented in the insert is composed of diffraction rings with superimposed scattered spots and arcs, which confirms the formation of nano-crystalline grains [121]. Stress concentration and dislocations were induced near the incorporated CNTs, forming a high density of dislocations and orientation gradient gap. Dislocations are prone to pile up around the CNTs and this region is propitious to the nucleation of dynamic recrystallization, leading to the formation of the massive nano-scaled sub-grains near CNTs.

As shown in Fig. 8.39, the shortening of CNTs occurred because of the shearing effect during fabrication, as also described in previous studies [134–136]. However, the structural integrity of these open ended CNTs is still remained, which is propitious to realize the reinforcing potentials of CNTs.

8.5.5 Mechanical Properties

The tensile behaviors of the tested samples are compared in Fig. 8.40. Table 8.7 exhibits the detailed values of the mechanical properties, including YS, UTS and El. It is clear that both the CNTs addition and the FSP procedure enhanced the mechanical properties of the as-cast Mg–6Zn matrix. The YS, UTS, El. of the FSPed composites reach 171, 330 MPa and 15.2%, with an increase by 144, 156 and 87% respectively, compared with that of the as-cast Mg–6Zn alloy. The grain refinement and the incorporated CNTs are responsible for the enhancement of mechanical properties. While the El. of the FSPed composites decreased from 18.9 to 15.2%, compared with that of the FSPed Mg–6Zn alloy. Ductility loss is inevitable for the CNTs reinforced metal matrix composites after the secondary thermo-mechanical processing [123, 135]. The mismatch in interfaces between the incorporated CNTs and the matrix can result in stress concentration, which could act as the nucleation sites for micro-cracks.

The micro-hardness of the alloys and the composites are also listed in Table 8.7. The indentation hardness of the FSPed composites increased 20%, compared with

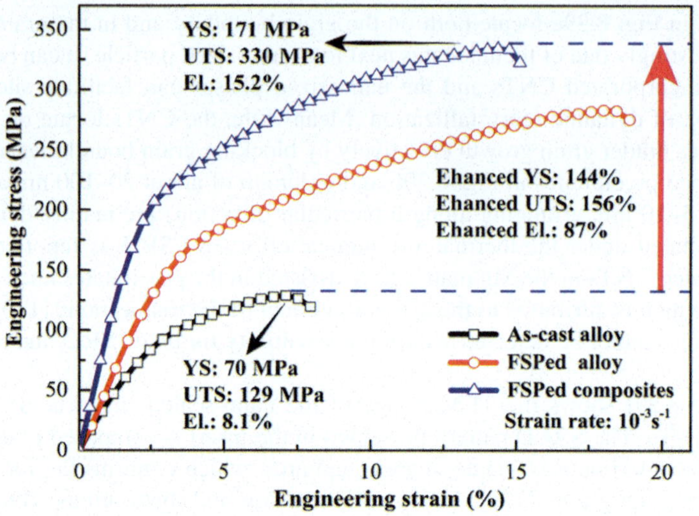

Fig. 8.40 The tensile stress–strain curves of the as-cast Mg–6Zn alloys, FSPed Mg–6Zn alloys and FSPed CNTs/Mg–6Zn composites

Table 8.7 Tensile properties and micro-hardness of the Mg–6Zn alloys and the CNTs/Mg–6Zn composites for the as-cast and the FSPed samples

Material	Condition	YS/MPa	UTS/MPa	El./%	Hardness/HV
Mg–6Zn	As-cast	70 ± 3.3	129 ± 2.4	8.1 ± 2.1	55 ± 5.8
Mg–6Zn	FSP	134 ± 4.8	281 ± 4.3	18.9 ± 1.1	69 ± 3.8
CNTs/ Mg–6Zn	FSP	171 ± 2.0	330 ± 5.5	15.2 ± 1.4	83 ± 7.2

that of the FSPed Mg alloys. The homogeneous dispersion of CNTs is predominant in micro-hardness increasing in spite of grain refinement.

8.5.6 *Fracture Behaviors*

Figure 8.41 exhibits the typical tensile fractographs of the FSPed Mg–6Zn alloys and the FSPed CNTs/Mg–6Zn composites. The detailed morphology in Fig. 8.41a and b both exhibits the features of ductile fracture, which is characterized by a large number of round, deep dimples and high tearing ridges. The micro-voids initiated under the effect of shearing stress and then propagated across the fracture surface, leading the formation of deep dimples. Hardly any fractographic differences existed in the size and homogeneity of the dimples between Fig. 8.41a and b, which indicated that the FSPed CNTs/Mg–6Zn composites could maintain the high ductility. It can

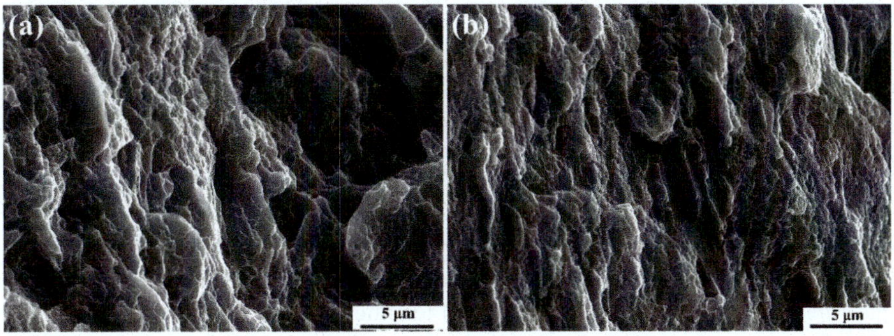

Fig. 8.41 Fractographs for **a** the FSPed alloys and **b** the FSPed composites

also be extracted from Fig. 8.41b that no CNTs entanglements or broken CNTs were detected on the fractographs, due to the desirable dispersion and the limited volume fraction of CNTs. The incorporated CNTs dispersed singly and randomly in matrix, and the compact bonding together with the high strain-hardening capacity of CNTs all contribute to avoiding the premature fracture initiation of the FSPed composites.

8.5.7 Strengthening Mechanisms

The mechanical properties enhancement of the FSPed CNTs/Mg–6Zn composites is mainly attributed to the grain refinement, the uniform dispersion of CNTs and the sound interfacial bonding. The strengthening mechanisms can be analyzed as the following.

(1) Grain refinement strengthening mechanism.

Grain refinement strengthening is equivalent to grain boundary or Hall–Petch strengthening. Grain boundary is occupied with disordered impurity atoms and there are grain misorientations between the neighboring grains. Dislocation migrations are retarded by the grain boundaries thus the multiplication of dislocations occurs. Eventually, the pilling-up of dislocations lead to the formation of clusters and gradually the flow stress is reaching to a peak value. According to the Hall–Petch relationship ($\sigma_y = \sigma_0 + kd^{-\frac{1}{2}}$), the proportional constant k of Mg alloys [141] is proved to be as 4 times higher as that of the aluminum alloys, thus grain refinement strengthening is considered to be of vital importance for Mg alloy and its composites.

(2) Load transfer strengthening mechanism.

For composites reinforced with the nano-scaled fibers or whiskers, the reinforcements can carry a certain part of load. The main concept of the load transfer theory is initially introduced by Kelly and Tyson [139], in which the shearing stress can be transferred from the matrix to the interfaces, the reinforcement tips, and then to the whole subject of the reinforcements. The strengthening effect mainly depends on the bonding interface, the volume fraction and the dispersion condition of the reinforcements, and also the properties of the reinforcement itself, including strength, length and aspect ratio, chemical stability and so on.

(3) Orowan looping strengthening mechanism.

Precipitation strengthening makes significant contributions to the strength enhancement, based on the interaction between the precipitates and the dislocations. Similarly, the incorporated CNTs can act as the impenetrable particles that interrupting the dislocation migration and leading to the dislocation bending. Thus the dislocations migration is hindered, leading to the increase of the matrix strength. The contribution of the Orowan strengthening increases with the decreasing size and increasing the volume fraction of the nano-scaled reinforcements [142].

Therefore, the YS of the FSPed CNTs/Mg–6Zn composites can be predicted by the following combined Eq. (8.4), on account of the grain refinement, load-transfer and Orowan looping mechanisms [123, 135, 139, 143, 144]:

$$\sigma_{yc} = \left[\sigma_{ym} + k\left(d_{com}^{-1/2} - d_{matrix}^{-1/2}\right)\right][V_{CNT}(S+4)/4 + (1 - V_{CNT})] + 0.8G_m Mb/L_p \quad (8.4)$$

where σ_{yc} and σ_{ym} are the YS of the composites and the matrix respectively; k is given 0.13 MPa·m$^{1/2}$; d_{com} and d_{matrix} are the mean grain size of the composites and the matrix, respectively; V_{CNT} is the volume fraction of the incorporated CNTs, and S is the aspect ratio of the CNTs; G_m is the shear modulus of matrix (G_m=1.66 × 10^4 MPa); M is the Taylor factor, which is chosen to be 3 in this case [145]; b is the Burgers vector, which is given 3.21 × 10^{-10} m; L_p is the inter-particle distance of the composites. The average length and diameter of CNTs are about 2 μm and 50 nm respectively, which can be considered as rod-shaped reinforcements, thus L_p can be calculated by Eq. (8.5):

$$L_p = \sqrt{\frac{\pi d_{CNT}^{2}}{2V_{CNT}}} \quad (8.5)$$

The UTS of the FSPed CNTs/Mg composites can be predicted by the following combined Eq. (8.6) [122, 130]:

$$\sigma_c = V_{CNT}\sigma_{CNT}\left(\frac{l}{2l_c}\right) + V_m\sigma_m \quad (8.6)$$

$$l_c = \sigma_{CNT} d_{CNT}/\sigma_m \quad (8.7)$$

where σ_c and σ_m are the UTS of the composites and the matrix respectively; σ_{CNT} is the strength of the CNTs (30 GPa); V_{CNT} and V_m are the volume fraction of the CNTs and matrix respectively; d_{CNT} and l are the mean diameter and length of CNTs; the calculated l_c in Eq. (8.7) is a critical value to evaluate whether the CNTs will be pulled-out or pulled-off from the matrix during load applying.

The calculated YS and UTS of the FSPed composites based on the Eq. (8.4)–(8.7) are 174 and 335 MPa respectively, which are in good agreement with the experimental values. The deviations between the calculated and the experimental values probably result from the shortening and deterioration of the CNTs. The calculated l_c of the composites is about 5.3 μm while the length of the incorporated CNTs is approximately 2 μm, indicating that the incorporated CNTs could not reach its fracture strength and be pulled out from the Mg matrix. In conclusion, despite the inevitable degradation and shortening of CNTs, the strengthening effect can be obtained through grain refinement, load-transfer and Orowan looping mechanisms. Thus, the fabrication route is proved to be effective to develop innovative CNTs-reinforced metal matrix composites with the exceptional mechanical properties.

8.5.8 Summary

High strength and toughness CNTs/Mg–6Zn composites were fabricated by stirring casting integrated with FSP. Grain refinement was achieved and the CNTs were singly and uniformly distributed in matrix. The interfaces between the CNTs and Mg–6Zn matrix were compact without pores. The structural integrity of the CNTs was remained though CNTs shortening occurred partially. The YS, UTS and El. of the FSPed composites reached 171 MPa, 330 MPa and 15%, with a significant enhancement than that of the as-cast Mg–6Zn alloy. Thus, stirring casting and subsequent FSP are proved to be an effective and feasible method to fabricate high strength and toughness CNTs/Mg composites. According to the calculated and the experimental strength values, the grain refinement mechanism, load-transfer mechanism and Orowan looping mechanism are mainly responsible for the strengthening contributions simultaneously.

References

1. Mordike B, Ebert T. Magnesium. Mater Sci Eng A. 2001;302(1):37–45.
2. Mordike BL. Creep-resistant magnesium alloys. Mater Sci Eng A. 2002;324(1–2):103–12.
3. Mishra R, Mahoney M, Mcfadden S, et al. High strain rate superplasticity in a friction stir processed 7075 Al alloy. Scr Mater. 1999;42(2):163–8.
4. Feng AH, Xiao BL, Ma ZY, et al. Effect of friction Stir processing procedures on microstructure and mechanical properties of Mg–Al–Zn casting. Metall Mater Trans A. 2009;40(10):2447–56.
5. Xiao BL, Yang Q, Yang J, et al. Enhanced mechanical properties of Mg–Gd–Y–Zr casting via friction stir processing. J Alloys Compd. 2011;509(6):2879–84.
6. Yang Q, Xiao BL, Ma ZY. Influence of process parameters on microstructure and mechanical properties of friction-stir-processed Mg–Gd–Y–Zr casting. Metall Mater Trans A. 2012;43(6):2094–109.

7. Yuan W, Panigrahi SK, Mishra RS. Achieving high strength and high ductility in friction stir-processed cast magnesium alloy. Metall Mater Trans A. 2013;44(8):3675–84.
8. Freeney TA, Mishra RS. Effect of friction stir processing on microstructure and mechanical properties of a cast-magnesium-rare earth alloy. Metall Mater Trans A. 2010;41(1):73–84.
9. Kumar N, Choudhuri D, Banerjee R, et al. Strength and ductility optimization of Mg–Y–Nd–Zr alloy by microstructural design. Int J Plast. 2015;68:77–97.
10. Park SHC, Sato YS, Kokawa H. Basal plane texture and flow pattern in friction stir weld of a magnesium alloy. Metall Mater Trans A. 2003;34(4):987–94.
11. Yuan W, Mishra RS, Carlson B, et al. Effect of texture on the mechanical behavior of ultrafine grained magnesium alloy. Scr Mater. 2011;64(6):580–3.
12. Lee JY, Kim DH, Lim HK, et al. Effects of Zn/Y ratio on microstructure and mechanical properties of Mg–Zn–Y alloys. Mater Lett. 2005;59(29–30):3801–5.
13. Huang ZH, Liang SM, Chen RS, et al. Solidification pathways and constituent phases of Mg–Zn–Y–Zr alloys. J Alloys Compd. 2009;468(1–2):170–8.
14. Yang Q, Xiao BL, Wang D, et al. Study on distribution of long-period stacking ordered phase in Mg–Gd–Y–Zn–Zr alloy using friction stir processing. Mater Sci Eng A. 2015;626:275–85.
15. Yang Q, Xiao BL, Wang D, et al. Formation of long-period stacking ordered phase only within grains in Mg–Gd–Y–Zn–Zr casting by friction stir processing. J Alloys Compd. 2013;581:585–9.
16. Xie GM, Ma ZY, Geng L, et al. Microstructural evolution and mechanical properties of friction stir welded Mg–Zn–Y–Zr alloy. Mater Sci Eng A. 2007;471(1–2):63–8.
17. Xie GM, Ma ZY, Geng L, et al. Microstructural evolution and enhanced superplasticity in friction stir processed Mg–Zn–Y–Zr alloy. J Mater Res. 2008;23(5):1207–13.
18. Zeng X, Zhang Y, Lu C, et al. Precipitation behavior and mechanical properties of a Mg–Zn–Y–Zr alloy processed by thermo-mechanical treatment. J Alloys Compd. 2005;395(1–2):213–9.
19. Wang J, Guo Y, Li J, et al. Microstructure, texture and mechanical properties of extruded Mg–5Al–2Nd–0.2Mn alloy. J Alloys Compd. 2015;653:100–7.
20. Sato YS, Kokawa H, Ikeda K, et al. Microtexture in the friction-stir weld of an aluminum alloy. Metall Mater Trans A. 2001;32(4):941–8.
21. Woo W, Choo H, Brown DW, et al. Texture variation and its influence on the tensile behavior of a friction-stir processed magnesium alloy. Scr Mater. 2006;54(11):1859–64.
22. Kai XZ, Li ZQ, Fan GL, et al. Enhanced strength and ductility in particulate-reinforced aluminum matrix composites fabricated by flake powder metallurgy. Mater Sci Eng A. 2013;587:46–53.
23. Kocks UF, Mecking H. Physics and phenomenology of strain hardening: the FCC case. Prog Mater Sci. 2003;48(3):171–273.
24. Han J, Chen J, Peng L, et al. Influence of processing parameters on thermal field in Mg–Nd–Zn–Zr alloy during friction stir processing. Mater Des. 2016;94:186–94.
25. Farzadfar SA, Sanjari M, Jung I-H, et al. Experimental and calculated phases in two as-cast and annealed Mg–Zn–Y alloys. Mater Charact. 2012;63:9–16.
26. Liu JF, Yang ZQ, Ye HQ. In situ transmission electron microscopy investigation of quasicrystal-crystal transformations in Mg–Zn–Y alloys. J Alloys Compd. 2015;621:179–88.
27. Zhang Q, Xiao BL, Ma ZY. Mechanically activated effect of friction stir processing in Al–Ti reaction. Mater Chem Phys. 2013;139(2–3):596–602.
28. Li J, Zhang Y, Wang J, et al. Microstructure evolution of Mg–4.3Zn–0.7Y–0.6Zr Alloy during solution heat treatment. Mater Trans. 2014;55(2):264–9.
29. Yan B, Dong X, Ma R, et al. Effects of heat treatment on microstructure, mechanical properties and damping capacity of Mg–Zn–Y–Zr alloy. Mater Sci Eng A. 2014;594:168–77.
30. Chen X, Liu L, Liu J, et al. Microstructure, electromagnetic shielding effectiveness and mechanical properties of Mg–Zn–Y–Zr alloys. Mater Des. 2015;65:360–9.
31. Xin R, Li B, Liao A, et al. Correlation between texture variation and transverse tensile behavior of friction-stir-processed AZ31 Mg alloy. Metall Mater Trans A. 2012;43(7):2500–8.

32. Cao G, Zhang D, Zhang W, et al. Microstructure evolution and mechanical properties of Mg–Nd–Y alloy in different friction stir processing conditions. J Alloys Compd. 2015;636:12–9.
33. Xu DK, Han EH. Effects of icosahedral phase formation on the microstructure and mechanical improvement of Mg alloys: a review. Prog Nat Sci Mater Int. 2012;22(5):364–85.
34. Yuan W, Mishra RS. Grain size and texture effects on deformation behavior of AZ31 magnesium alloy. Mater Sci Eng A. 2012;558:716–24.
35. Ma ZY, Pilchak AL, Juhas MC, et al. Microstructural refinement and property enhancement of cast light alloys via friction stir processing. Scr Mater. 2008;58(5):361–6.
36. Ammouri AH, Kridli G, Ayoub G, et al. Relating grain size to the Zener-Hollomon parameter for twin-roll-cast AZ31B alloy refined by friction stir processing. J Mater Process Technol. 2015;222:301–6.
37. Sitdikov O, Kaibyshev R. Dynamic recrystallization in pure magnesium. Mater Trans. 2001;42(9):1928–37.
38. Mironov S, Onuma T, Sato YS, et al. Microstructure evolution during friction-stir welding of AZ31 magnesium alloy. Acta Mater. 2015;100:301–12.
39. Feng AH, Ma ZY. Microstructural evolution of cast Mg–Al–Zn during friction stir processing and subsequent aging. Acta Mater. 2009;57(14):4248–60.
40. Xin R, Zheng X, Liu Z, et al. Microstructure and texture evolution of an Mg–Gd–Y–Nd–Zr alloy during friction stir processing. J Alloys Compd. 2016;659:51–9.
41. Cottam R, Robson J, Lorimer G, et al. Dynamic recrystallization of Mg and Mg–Y alloys: crystallographic texture development. Mater Sci Eng A. 2008;485(1–2):375–82.
42. Sakai T, Belyakov A, Kaibyshev R, et al. Dynamic and post-dynamic recrystallization under hot, cold and severe plastic deformation conditions. Prog Mater Sci. 2014;60(1):130–207.
43. Xu SW, Zheng MY, Kamado S, et al. Dynamic microstructural changes during hot extrusion and mechanical properties of a Mg–5.0 Zn–0.9 Y–0.16 Zr (wt.%) alloy. Mater Sci Eng A. 2011;528(12):4055–67.
44. Sabbaghian M, Mahmudi R. Microstructural evolution and local mechanical properties of friction stir processed Mg–3Gd–1Zn cast alloy. J Mater Eng Perform. 2016;25(5):1856–63.
45. Abaspour S, Cáceres CH. Thermodynamics-based selection and design of creep-resistant cast Mg alloys. Metall Mater Trans A. 2015;46(12):5972–88.
46. Abaspour S, Cáceres CH. High temperature strength and stress relaxation behavior of dilute binary Mg alloys. Metall Mater Trans A. 2016;47(3):1313–21.
47. Castany P, Pettinari-Sturmel F, Douin J, et al. TEM quantitative characterization of short-range order and its effects on the deformation micromechanims in a Ti–6Al–4V alloy. Mater Sci Eng A. 2017;680:85–91.
48. Navazani M, Dehghani K. Fabrication of Mg–ZrO_2 surface layer composites by friction stir processing. J Mater Process Technol. 2016;229:439–49.
49. Wang Y, Huang Y, Meng X, et al. Microstructural evolution and mechanical properties of Mg Zn Y Zr alloy during friction stir processing. J Alloys Compd. 2017;696:875–83.
50. Wang Q, Liu K, Wang Z, et al. Microstructure, texture and mechanical properties of as-extruded Mg–Zn–Er alloys containing W-phase. J Alloys Compd. 2014;602:32–9.
51. Yang Q, Xiao BL, Ma ZY, et al. Achieving high strain rate superplasticity in Mg–Zn–Y–Zr alloy produced by friction stir processing. Scr Mater. 2011;65(4):335–8.
52. Koike J, Kobayashi T, Mukai T, et al. The activity of non-basal slip systems and dynamic recovery at room temperature in fine-grained AZ31B magnesium alloys. Acta Mater. 2003;51(7):2055–65.
53. Sandlöbes S, Zaefferer S, Schestakow I, et al. On the role of non-basal deformation mechanisms for the ductility of Mg and Mg–Y alloys. Acta Mater. 2011;59(2):429–39.
54. Cáceres CH, Lukáč P. Strain hardening behaviour and the Taylor factor of pure magnesium. Philos Mag. 2008;88(7):977–89.
55. Shao Y, Zeng RC, Li SQ, et al. Advance in antibacterial magnesium alloys and surface coatings on magnesium alloys: a review. Acta Metall Sin. 2020;33(5):615–29.
56. Bae DH, Kim SH, Kim DH, et al. Deformation behavior of Mg–Zn–Y alloys reinforced by icosahedral quasicrystalline particles. Acta Mater. 2002;50(9):2343–56.

57. Karakulak E. A review: past, present and future of grain refining of magnesium castings. J Magnes Alloy. 2019;7(3):355–69.
58. Liu C, Chen X, Yuan Y, et al. Altered age-hardening behavior in the ultrafine-grained surface layer of Mg–Zn–Y–Ce–Zr alloy processed by sliding friction treatment. J Mater Sci Technol. 2021;78:20–9.
59. Li J, Meng X, Li Y, et al. Friction stir extrusion for fabricating Mg–RE alloys with high strength and ductility. Mater Lett. 2021;289: 129414.
60. Xie Y, Meng X, Mao D, et al. Homogeneously dispersed graphene nanoplatelets as long-term corrosion inhibitors for aluminum matrix composites. ACS Appl Mater Interfaces. 2021;13(27):32161–74.
61. Xie Y, Meng X, Li Y, et al. Insight into ultra-refined grains of aluminum matrix composites via deformation-driven metallurgy. Compos Commun. 2021;26:100776.
62. Demirtas M, Purcek G. Room temperature superplaticity in fine/ultrafine grained materials subjected to severe plastic deformation. Mater Trans. 2019;60(7):1159–67.
63. Aktarer SM, Sekban DM, Saray O, et al. Effect of two-pass friction stir processing on the microstructure and mechanical properties of as-cast binary Al–12Si alloy. Mater Sci Eng A. 2015;636:311–9.
64. Sekban DM, Akterer SM, Saray O, et al. Formability of friction stir processed low carbon steels used in shipbuilding. J Mater Sci Technol. 2018;34(1):237–44.
65. Sekban DM, Aktarer SM, Purcek G. Friction stir welding of low-carbon shipbuilding steel plates: microstructure, mechanical properties, and corrosion behavior. Metall Mater Trans A Phys Metall Mater Sci. 2019;50(9):4127–40.
66. Heidarzadeh A, Mironov S, Kaibyshev R, et al. Progress in materials science friction stir welding/processing of metals and alloys: a comprehensive review on microstructural evolution. Prog Mater Sci. 2020;117:100752.
67. Meng X, Huang Y, Cao J, et al. Recent progress on control strategies for inherent issues in friction stir welding. Prog Mater Sci. 2021;115:100706.
68. Li J, Li Y, Wang F, et al. Friction stir processing of high-entropy alloy reinforced aluminum matrix composites for mechanical properties enhancement. Mater Sci Eng A. 2020;792:139755.
69. Huang Y, Li J, Wan L, et al. Strengthening and toughening mechanisms of CNTs/Mg–6Zn composites via friction stir processing. Mater Sci Eng A. 2018;732:205–11.
70. Yang Q, Xiao BL, Ma ZY. Enhanced superplasticity in friction stir processed Mg–Gd–Y–Zr alloy. J Alloys Compd. 2013;551:61–6.
71. Demirtas M, Kawasaki M, Yanar H, et al. High temperature superplasticity and deformation behavior of naturally aged Zn–Al alloys with different phase compositions. Mater Sci Eng A. 2018;730:73–83.
72. Sun WT, Qiao XG, Zheng MY, et al. Evolution of long-period stacking ordered structure and hardness of Mg–8.2Gd–3.8Y–1.0Zn–0.4Zr alloy during processing by high pressure torsion. Mater Sci Eng A. 2018;738:238–52.
73. Homma T, Kunito N, Kamado S. Fabrication of extraordinary high-strength magnesium alloy by hot extrusion. Scr Mater. 2009;61(6):644–7.
74. Rashad M, Pan F, Asif M. Exploring mechanical behavior of Mg–6Zn alloy reinforced with graphene nanoplatelets. Mater Sci Eng A. 2016;649:263–9.
75. Zhang BB, Yan FK, Zhao MJ, et al. Combined strengthening from nanotwins and nanoprecipitates in an iron-based superalloy. Acta Mater. 2018;151:310–20.
76. Sun WT, Qiao XG, Zheng MY, et al. Achieving ultra-high hardness of nanostructured Mg–8.2Gd–3.2Y–1.0Zn–0.4Zr alloy produced by a combination of high pressure torsion and ageing treatment. Scr Mater. 2018;155:21–5.
77. Li B, Hou L, Wu R, et al. Microstructure and thermal conductivity of Mg–2Zn–Zr alloy. J Alloys Compd. 2017;722:772–7.
78. Niu R, Yan F, Wang Y, et al. Effect of Zr content on damping property of Mg–Zr binary alloys. Mater Sci Eng A. 2018;718:418–26.

79. Shahzad M, Waqas H, Rafi-ud-din AH, et al. The roles of Zn distribution and eutectic particles on microstructure development during extrusion and anisotropic mechanical properties in a Mg–Zn–Zr alloy. Mater Sci Eng A. 2015;620:50–7.
80. Solomon ELS, Natarajan AR, Roy AM, et al. Stability and strain-driven evolution of β' precipitate in Mg–Y alloys. Acta Mater. 2019;166:148–57.
81. Gao S, Liu Y, Zhao D, et al. Effect of different Co contents on the microstructure and tensile strength of Mg–Co–Y alloys. Mater Sci Eng A. 2019;750:91–7.
82. Du Y, Zheng M, Qiao X, et al. Improving microstructure and mechanical properties in Mg-6 mass% Zn alloys by combined addition of Ca and Ce. Mater Sci Eng A. 2016;656:67–74.
83. Yong H, Guo S, Yuan Z, et al. Improved hydrogen storage kinetics and thermodynamics of RE–Mg-based alloy by co-doping Ce–Y. Int J Hydrogen Energy. 2019;44(31):16765–76.
84. Chen Y, Wang Y, Gao J. Microstructure and mechanical properties of as-cast Mg–Sn–Zn–Y alloys. J Alloys Compd. 2018;740:727–34.
85. Chino Y, Sassa K, Mabuchi M. Texture and stretch formability of a rolled Mg–Zn alloy containing dilute content of Y. Mater Sci Eng A. 2009;513–514:394–400.
86. Wang Q, Shen Y, Jiang B, et al. A micro-alloyed Mg–Sn–Y alloy with high ductility at room temperature. Mater Sci Eng A. 2018;735:131–44.
87. Huang R, Ji S, Meng X, et al. Drilling-filling friction stir repairing of AZ31B magnesium alloy. J Mater Process Technol. 2018;255:765–72.
88. Yan K, Sun J, Bai J, et al. Preparation of a high strength and high ductility Mg–6Zn alloy wire by combination of ECAP and hot drawing. Mater Sci Eng A. 2019;739:513–8.
89. Ji SD, Meng XC, Huang RF, et al. Microstructures and mechanical properties of 7N01-T4 aluminum alloy joints by active-passive filling friction stir repairing. Mater Sci Eng A. 2016;664:94–102.
90. Zhang W, Ding H, Cai M, et al. Low-temperature superplastic deformation mechanism in Ti–6Al–4V alloy processed by friction stir processing. Mater Sci Eng A. 2019;764:138261.
91. Meng X, Huang Y, Liu S, et al. Functionally gradient coating of aluminum alloy via in situ arc surface nitriding with subsequent friction stir processing. Adv Eng Mater. 2019;21(1):1800841.
92. Liu CY, Zhang B, Ma ZY, et al. Effect of Sc addition, friction stir processing, and T6 treatment on the damping and mechanical properties of 7055 Al alloy. J Alloys Compd. 2019;772:775–81.
93. Xue P, Li WD, Wang D, et al. Enhanced mechanical properties of medium carbon steel casting via friction stir processing and subsequent annealing. Mater Sci Eng A. 2016;670:153–8.
94. Huang Y, Wang T, Guo W, et al. Microstructure and surface mechanical property of AZ31 Mg/SiCp surface composite fabricated by direct friction stir processing. Mater Des. 2014;59:274–8.
95. Huang Y, Wang Y, Meng X, et al. Dynamic recrystallization and mechanical properties of friction stir processed Mg–Zn–Y–Zr alloys. J Mater Process Technol. 2017;249:331–8.
96. Fang XY, Yi DQ, Nie JF, et al. Effect of Zr, Mn and Sc additions on the grain size of Mg–Gd alloy. J Alloys Compd. 2009;470(1–2):311–6.
97. Zhang JS, Zhang WB, Ruan XQ, et al. Effect of zirconium on the microstructure and mechanical properties of long period stacking ordered Mg96Gd3Zn1 alloy. Mater Sci Eng A. 2013;560:847–50.
98. Chia TL, Easton MA, Zhu SM, et al. The effect of alloy composition on the microstructure and tensile properties of binary Mg-rare earth alloys. Intermetallics. 2009;17(7):481–90.
99. Ali Y, Qiu D, Jiang B, et al. Current research progress in grain refinement of cast magnesium alloys: a review article. J Alloys Compd. 2015;619:639–51.
100. Zhang W, Ding H, Cai M, et al. Ultra-grain refinement and enhanced low-temperature superplasticity in a friction stir-processed Ti–6Al–4V alloy. Mater Sci Eng A. 2018;727:90–6.
101. Fu W, Wang RH, Xue H, et al. Effects of Zr addition on the multi-scale second-phase particles and fracture behavior for Mg–3Gd–1Zn alloy. J Alloys Compd. 2018;747:197–210.
102. Robson JD, Paa-Rai C. The interaction of grain refinement and ageing in magnesium–zinc–zirconium (ZK) alloys. Acta Mater. 2015;95:10–9.

103. Yazdipour A, Shafiei MA, Dehghani K. Modeling the microstructural evolution and effect of cooling rate on the nanograins formed during the friction stir processing of Al5083. Mater Sci Eng A. 2009;527(1–2):192–7.
104. Jeong HY, Kim B, Kim SG, et al. Effect of Ce addition on the microstructure and tensile properties of extruded Mg–Zn–Zr alloys. Mater Sci Eng A. 2014;612:217–22.
105. Li ZT, Qiao XG, Xu C, et al. Ultrahigh strength Mg–Al–Ca–Mn extrusion alloys with various aluminum contents. J Alloys Compd. 2019;792:130–41.
106. Robson JD, Henry DT, Davis B. Particle effects on recrystallization in magnesium–manganese alloys: particle-stimulated nucleation. Acta Mater. 2009;57(9):2739–47.
107. Ji S, Niu S, Liu J, et al. Friction stir lap welding of Al to Mg assisted by ultrasound and a Zn interlayer. J Mater Process Technol. 2019;267:141–51.
108. Yu H, Kim YM, You BS, et al. Effects of cerium addition on the microstructure, mechanical properties and hot workability of ZK60 alloy. Mater Sci Eng A. 2013;559:798–807.
109. Liu L, Chen X, Pan F, et al. Effect of Y and Ce additions on microstructure and mechanical properties of Mg–Zn–Zr alloys. Mater Sci Eng A. 2015;644:247–53.
110. Wang X, Wang Y, Su Y, et al. Synergetic strengthening effects on copper matrix induced by Al_2O_3 particle revealed from micro-scale mechanical deformation and microstructure evolutions. Ceram Int. 2019;45(12):14889–95.
111. Peng J, Li Z, Fu L, et al. Carbide precipitation strengthening in fine-grained carbon-doped FeCoCrNiMn high entropy alloy. J Alloys Compd. 2019;803:491–8.
112. He ML, Luo TJ, Liu YT, et al. Effects of Cu and Ce co-addition on the microstructure and mechanical properties of Mg–6Zn–0.5Zr alloy. J Alloys Compd. 2018;767:1216–24.
113. Sun WT, Qiao XG, Zheng MY, et al. Exceptional grain refinement in a Mg alloy during high pressure torsion due to rare earth containing nanosized precipitates. Mater Sci Eng A. 2018;728:115–23.
114. Zhang JF, Andrä H, Zhang XX, et al. An enhanced finite element model considering multi strengthening and damage mechanisms in particle reinforced metal matrix composites. Compos Struct. 2019;226:111281.
115. Ji BQ, Ma Y, Wu R, et al. Effect of Y and Ce addition on microstructures and mechanical properties of LZ91 alloys. J Alloys Compd. 2019;800:72–80.
116. Faraji G, Asadi P. Characterization of AZ91/alumina nanocomposite produced by FSP. Mater Sci Eng A. 2011;528(6):2431–40.
117. Xu N, Bao Y. Enhanced mechanical properties of tungsten inert gas welded AZ31 magnesium alloy joint using two-pass friction stir processing with rapid cooling. Mater Sci Eng A. 2016;655:292–9.
118. Raja A, Pancholi V. Effect of friction stir processing on tensile and fracture behaviour of AZ91 alloy. J Mater Process Technol. 2017;248:8–17.
119. Liao H, Chen J, Peng L, et al. Fabrication and characterization of magnesium matrix composite processed by combination of friction stir processing and high-energy ball milling. Mater Sci Eng A. 2017;683:207–14.
120. Cao F, Song GL, Atrens A. Corrosion and passivation of magnesium alloys. Corros Sci. 2016;111:835–45.
121. Han G, Wang Z, Liu K, et al. Synthesis of CNT-reinforced AZ31 magnesium alloy composites with uniformly distributed CNTs. Mater Sci Eng A. 2015;628:350–7.
122. Li CD, Wang XJ, Liu WQ, et al. Effect of solidification on microstructures and mechanical properties of carbon nanotubes reinforced magnesium matrix composite. Mater Des. 2014;58:204–8.
123. Li CD, Wang XJ, Wu K, et al. Distribution and integrity of carbon nanotubes in carbon nanotube/magnesium composites. J Alloys Compd. 2014;612:330–6.
124. Morisada Y, Fujii H, Nagaoka T, et al. MWCNTs/AZ31 surface composites fabricated by friction stir processing. Mater Sci Eng A. 2006;419(1–2):344–8.
125. Hosseini SA, Ranjbar K, Dehmolaei R, et al. Fabrication of Al5083 surface composites reinforced by CNTs and cerium oxide nano particles via friction stir processing. J Alloys Compd. 2015;622:725–33.

126. Chen M, Fan G, Tan Z, et al. Design of an efficient flake powder metallurgy route to fabricate CNT/6061Al composites. Mater Des. 2018;142:288–96.
127. Zare H, Jahedi M, Toroghinejad MR, et al. Compressive, shear, and fracture behavior of CNT reinforced Al matrix composites manufactured by severe plastic deformation. Mater Des. 2016;106:112–9.
128. Wang H, Zhang Z-H, Zhang H-M, et al. Novel synthesizing and characterization of copper matrix composites reinforced with carbon nanotubes. Mater Sci Eng A. 2017;696:80–9.
129. Huang Z, Zheng Z, Zhao S, et al. Copper matrix composites reinforced by aligned carbon nanotubes: mechanical and tribological properties. Mater Des. 2017;133:570–8.
130. Liu ZY, Zhao K, Xiao BL, et al. Fabrication of CNT/Al composites with low damage to CNTs by a novel solution-assisted wet mixing combined with powder metallurgy processing. Mater Des. 2016;97:424–30.
131. Khodabakhshi F, Gerlich AP, Simchi A, et al. Hot deformation behavior of an aluminum-matrix hybrid nanocomposite fabricated by friction stir processing. Mater Sci Eng A. 2015;626:458–66.
132. Liu ZY, Xiao BL, Wang WG, et al. Developing high-performance aluminum matrix composites with directionally aligned carbon nanotubes by combining friction stir processing and subsequent rolling. Carbon N Y. 2013;62:35–42.
133. Liu Q, Ke L, Liu F, et al. Microstructure and mechanical property of multi-walled carbon nanotubes reinforced aluminum matrix composites fabricated by friction stir processing. Mater Des. 2013;45:343–8.
134. Izadi H, Gerlich AP. Distribution and stability of carbon nanotubes during multi-pass friction stir processing of carbon nanotube/aluminum composites. Carbon N Y. 2012;50(12):4744–9.
135. Khodabakhshi F, Gerlich AP, Švec P. Reactive friction-stir processing of an Al–Mg alloy with introducing multi-walled carbon nano-tubes (MW-CNTs): microstructural characteristics and mechanical properties. Mater Charact. 2017;131:359–73.
136. Liu ZY, Xiao BL, Wang WG, et al. Analysis of carbon nanotube shortening and composite strengthening in carbon nanotube/aluminum composites fabricated by multi-pass friction stir processing. Carbon N Y. 2014;69:264–74.
137. Han GQ, Shen JH, Ye XX, et al. The influence of CNTs on the microstructure and ductility of CNT/Mg composites. Mater Lett. 2016;181:300–4.
138. Ma R, Dong X, Yan B, et al. Mechanical and damping properties of thermal treated Mg–Zn–Y–Zr alloys reinforced with quasicrystal phase. Mater Sci Eng A. 2014;602:11–8.
139. Mokdad F, Chen DL, Liu ZY, et al. Deformation and strengthening mechanisms of a carbon nanotube reinforced aluminum composite. Carbon N Y. 2016;104:64–77.
140. Goh CS, Wei J, Lee LC, et al. Ductility improvement and fatigue studies in Mg–CNT nanocomposites. Compos Sci Technol. 2008;68(6):1432–9.
141. Mallikarjuna C, Shashidhara SM, Mallik US. Evaluation of grain refinement and variation in mechanical properties of equal-channel angular pressed 2014 aluminum alloy. Mater Des. 2009;30(5):1638–42.
142. Zhang Z, Chen DL. Contribution of Orowan strengthening effect in particulate-reinforced metal matrix nanocomposites. Mater Sci Eng A. 2008;483–484:148–52.
143. Liu ZY, Xiao BL, Wang WG, et al. Elevated temperature tensile properties and thermal expansion of CNT/2009Al composites. Compos Sci Technol. 2012;72(15):1826–33.
144. Liu ZY, Xiao BL, Wang WG, et al. Singly dispersed carbon nanotube/aluminum composites fabricated by powder metallurgy combined with friction stir processing. Carbon N Y. 2012;50(5):1843–52.
145. Li Q, Viereckl A, Rottmair CA, et al. Improved processing of carbon nanotube/magnesium alloy composites. Compos Sci Technol. 2009;69(7–8):1193–9.

Chapter 9
Graphene Nanoplatelet-Reinforced Aluminum Matrix Composites

9.1 Feasibility Verification of Deformation-Driven Metallurgy

9.1.1 Introduction

With high Young's modulus (about 1 TPa), super strength (about 130 GPa) and extraordinary-large specific surface area (about 2630 m^2/g), graphene nanoplatelets (GNPs) has the potential to achieve great strength metal matrix composites with very few additions [1, 2]. Besides, the relative-low cost (about$70 kg^{-1}) of industrial-produced GNPs nowadays, dispersing about 1–2 wt.% GNPs into aluminum matrix would only raise the costs of composites by less than half, which is considered to be an acceptable approach [3]. These composites exhibit great promise and can in principle provide solutions to a wide range of these industrial applications [4–7]. It is highly appreciated that comprehensive mechanical properties but not just one single property can be achieved without dramatic increased costs. However, most traditional and newly developed technologies are frequently accompanied by inherent issues due to limitations such as fatal ductility loss, high price and time costs, reinforcement agglomeration, poor interfacial compatibility or processing difficulties [7–10].

Generally, the fatal ductility loss is mainly attributed to the agglomeration [11] and weak interfacial bonding [12–14]. A critical step in powder metallurgy route is to apply high-energy ball milling (HEBM) before sintering [15]. Although the inhibition of agglomeration tendency and the promotion of uniform dispersion of GNPs can be obtained via HEBM, the ductility loss of most results is unsatisfactory, because HEBM cannot provide sufficient temperature conditions to facilitate interface bonding. As such, the gist of suppress ductility loss for a balance between strength and ductility is to propose a novel method combining the advantages of solid state sintering and HEBM synchronously.

Y. Huang et al., *Friction Stir Welding and Processing*, Materials Forming, Machining and Tribology, https://doi.org/10.1007/978-981-99-8688-0_9

Starting from these considerations, we present a high-performance, low-cost strategy for GNPs reinforced aluminum composites driven by severe plastic deformation, namely deformation-driven metallurgy (DDM). The extremely high plastic strain rate of this one-step approach induced by friction and stir of processing tools conducts ultrafast fabrication (only several seconds). Experimental results, corroborated by analytical calculation, showed great results in detail.

9.1.2 Experimental Procedures

The matrix was industrial 1060 pure aluminum powders (10 μm in mean diameter, purity $\geq$99.6%, atomized by Tianjin Hekenasi, produced by Southwest Aluminum Group), as depicted in Fig. 9.1a. The as-received GNPs with an average thickness of 5–15 nm were supplied by Nanjing Ji Cang Nano Tech, as illustrated in Fig. 9.1b. The 99.50 wt.% aluminum powders and the 1.50 wt.% GNPs were blended by ball milling in a planetary ball mill machine (QM-3SP2, Nanjing University Instrument Plant) at a rotational velocity of 200 r/min for 10 h with a ball-to-powder ratio of 5:1. The diameters of the ZrO_2 milling balls were 6 mm, and the blended powders were shown in Fig. 9.1c and d.

The milled powders were placed in an aluminum tubular mold and then cold compacted at about 4 MPa by a processing tool to densify the materials. Then, the

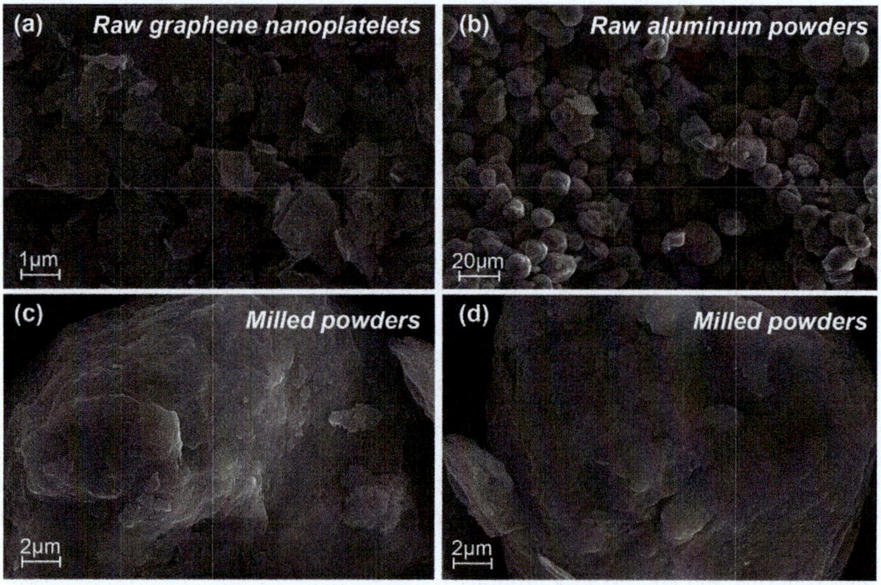

Fig. 9.1 SEM images of **a** raw GNP powders, **b** raw aluminum powders, **c** and **d** powders after 10 h ball milling

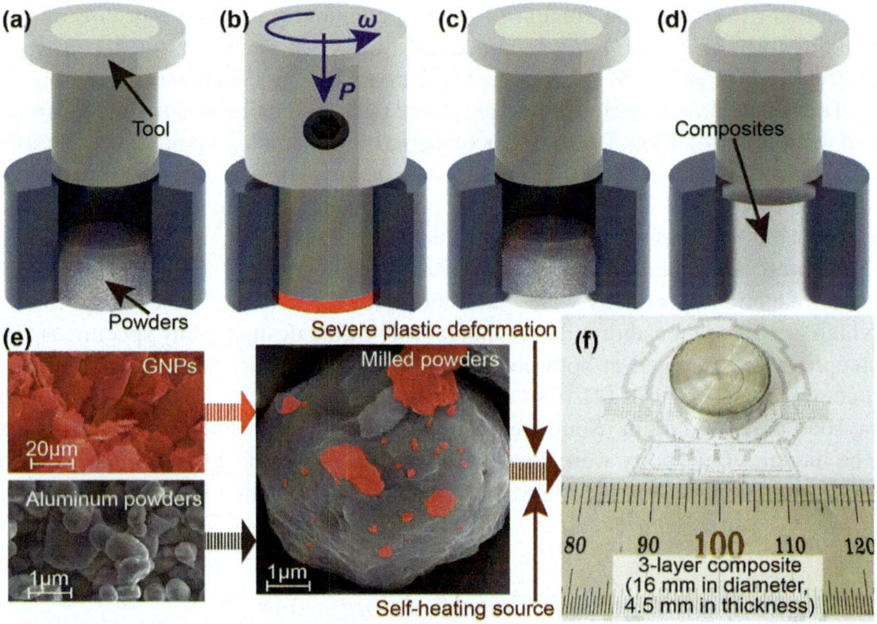

Fig. 9.2 Schematic diagram and products of the one-step DDM method: **a** Preparation of the first layer, **b** deformation-driven process, **c** preparation of the second layer, and **d** finished product. **e** SEM images of raw GNPs, raw aluminum powders and milled powders. **f** A 3-layer composite specimen with a diameter of 16 mm and thickness of 4.5 mm

processing tool rotated at a rotational velocity of 800 r/min for several seconds, as shown in Fig. 9.2a–d. The densified composites were obtained by the frictional heat input and severe plastic deformation of the deformation-driven processing. The final diameter of the composite disc was 16 mm. By adjusting the rotational time (10–60 s), bulk composites with different degrees of mechanical performances can be obtained.

Electron backscattered diffraction (EBSD) was performed using a scanning electron microscopy equipment (SEM, Zeiss Merlin Compact) with an HKL EBSD detector at a working voltage of 20 kV. The specimen for EBSD was polished by a JEOL IB-09020CP cross-section polisher. The data files obtained from EBSD were processed using HKL Channel 5 software. The sizes of the grains d are represented by the equivalent diameter calculated from the mapped grain area A by the approximate equation $d = \sqrt{\frac{4A}{\pi}}$. The average grain size is calculated by analyzing all grains in a single EBSD map. Transmission electron microscopy (TEM, FEI Talos F200X) was performed at a working voltage of 200 kV. Tip-enhanced Raman spectroscopy (TERS, Renishaw inVia-Reflex) was conducted to investigate the structural integrity of the GNPs using 532 nm line of an argon-ion laser under ambient conditions. Five test points of each sample were conducted to make sure the results credibly. A TERS mapping was carried out to the distribution of the GNPs with a spatial resolution of 20 nm. The content of oxygen-containing functional groups and the bonding form of Al, C, O were characterized by X-ray photoelectron spectroscopy (XPS, ThermoFisher ESCALAB 250Xi) with an Al X-ray source as excitation.

The tensile properties of the samples were measured using an Instron 5569 tester at a work mode of 1 kN. Due to the actual size of the samples, standard tensile specimens were unable to obtain. Therefore, tensile specimens with equal reduction in accordance with GB/T 229.1–2010 were prepared, whose dimensions were approximately 14 mm × 6 mm × 1 mm. Through the comparison test for 1060 pure aluminum (Southwest Aluminum Group), the ultimate tensile strength and elongation of the equal reduction specimens were consistent with those of the standard specimens, which proved the feasibility and comparability of the designed tiny specimens. The samples were clamped at both ends and stretched along the length direction with a constant test speed of 0.1 mm/min at room temperature. A tensile sample made of 1060 pure aluminum (Southwest Aluminum Group) was also tested as control. The moduli of elasticity were measured using an Agilent G200 nano indenter. The penetration depths were measured with the maximum applied force of 10 mN.

9.1.3 Coupled Thermal-Flow Modeling

A computational fluid dynamic model was established to calculate the coupled thermal-flow field during the DDM process. A computational domain with a diameter and height of 16 mm and 1.5 mm was used. The sample was treated as an incompressible, continuous fluid and was dominated by the Navier–Stokes equations. The viscosity μ of the materials is dependent of temperature and effective strain rate, as written blow [16]:

$$\mu = \frac{\sigma}{3\dot{\varepsilon}} \tag{9.1}$$

where σ is equivalent flow stress and $\dot{\varepsilon}$ is equivalent strain rate. The flow stress equals to the yield stress approximately and it is highly temperature and strain rate dependent according to the constitutive equation introduced by the Sellars-Tegart model [17]:

$$\sigma_{\mathrm{e}} = \left(1 - \sqrt{\frac{T - 273.15}{T_{\mathrm{melt}} - 273.15}}\right)\frac{1}{\alpha} arsinh\left[\left(\frac{Z}{A}\right)^{\frac{1}{n}}\right] + \sigma_{\mathrm{melt}} \tag{9.2}$$

$$Z = \dot{\varepsilon}_e \, e^{\frac{Q}{RT}} \tag{9.3}$$

where α, A and n are material constants, T is transient temperature during FSM process, T_{melt} is melt temperature of the pure Al, σ_{melt} is yield stress above the melt temperature of 1060 pure aluminum, Z is Zener-Hollomon parameter, Q is activation energy, and R is gas constant.

Two parts of the self-heating behavior were taken into consideration. The surficial heat source term F represents the friction heat between the DDM tool and the materials:

$$F = R\omega \cdot \left[\frac{\delta\sigma_e}{\sqrt{3}} + (1-\delta)\mu P\right] \tag{9.4}$$

where R is the distance between the element and the axis, ω is rotational velocity, μ is friction coefficient, and P is pressure. δ is slip coefficient. The other part is plastic deformation heat source, which is a volumetric heat source [18]:

$$\Phi = 2\left(\left(\frac{\partial u_1}{\partial x_1}\right)^2 + \left(\frac{\partial u_2}{\partial x_2}\right)^2 + \left(\frac{\partial u_3}{\partial x_3}\right)^2\right) + \left(\frac{\partial u_1}{\partial x_2} + \frac{\partial u_2}{\partial x_1}\right)^2 + \left(\frac{\partial u_1}{\partial x_3} + \frac{\partial u_3}{\partial x_1}\right)^2 + \left(\frac{\partial u_3}{\partial x_2} + \frac{\partial u_2}{\partial x_3}\right)^2 \tag{9.5}$$

9.1.4 Microstructural Characteristics

The milled powders were sintered by the processing tool directly, and the metallurgical behavior is illustrated in Fig. 9.2. A sample was obtained with a diameter of 16 mm and thickness of 4.5 mm, as shown in Fig. 9.2f. The temperature condition for DDM process, that is, the self-heating source is induced by the deformation-driven effects, which can be divided into two parts: ① the frictional heat at the interface of tools and powders, ② the deformation heat generated from the plastic deformation of powders. According to the Sellars-Tegart constitutive equation [19], the flow stress of plasticized materials drops sharply with the increase of transient temperature. High flow stress at low temperatures brings about sharp increase of temperature, that is, the temperature condition for sintering is achieved in just several seconds. Subsequently, the self-heating behavior dependent on the flow stress reaches to a balance between heat generation and dissipation, which maintains a thermal equilibrium with relatively low temperature. The entire process requires very little energy, which contributes to green manufacturing. The temperature is significantly lower than the temperature required for conventional sintering method, inhibiting the coarsening of microstructures and the growth of Al_4C_3 brittle intermetallic phase.

To verify this proposal, we calculated the evolution of temperature and strain rate during DDM process via computational fluid dynamics. The simulated results including the distributions of the temperature and strain rate versus time are shown in Fig. 9.3a, taking the rotating velocity of 800 r/min as an example. The overall temperature gradually increases with time, and reaches the quasi-steady state 716 K in the 7.0 s at the fabricating process of GNPs reinforced aluminum matrix composites. The temperature remains almost unchanged in the following time (Fig. 9.3), while the large strain rate results in the severe plastic deformation in the composites. The TEM images of the composite are shown in Figs. 9.4, 9.5 and 9.6. Rod-shape Al_4C_3 cannot be observed in these figures, indicating that the major brittle phase detrimental to the mechanical performance is inhibited by the extremely low temperature condition.

The introduction of severe plastic deformation promoted the interfacial reaction and the formation of the multiscale nanostructure in the composites. As shown in Fig. 9.3a, the minimum equivalent strain rate is greater than 10 s^{-1} starting from 5.5 s, implying that the plasticized materials experienced severe plastic deformation.

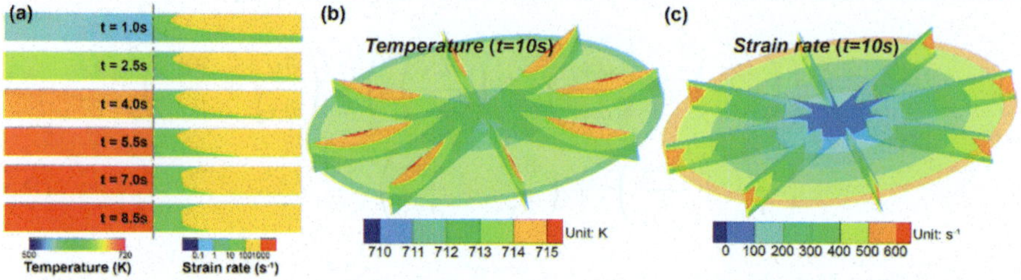

Fig. 9.3 Self-heating behaviors: **a** Simulated distributions of temperature and strain rate versus processing time, **b** distribution of quasi-steady temperature at 10 s, and **c** distribution of quasi-steady strain rate at 10 s

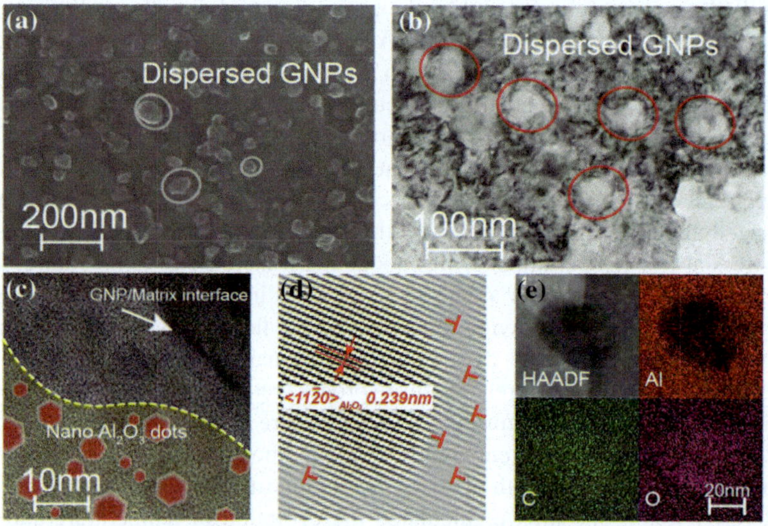

Fig. 9.4 Characteristics of the multiscale nanostructure: **a** SEM image of the dispersed GNPs with 30 s sintering time, **b** TEM observation of the intragranular dispersed GNPs with 30 s sintering time, **c** Distribution of nano Al_2O_3 dots in GNP particles and the interface between the GNP and the matrix with 30 s sintering time, **d** Inverse fast Fourier transform (IFFT) image of the nano Al_2O_3 dots, and **e** Super energy dispersive spectroscopy (EDS) image of the composition distribution in the GNP particle

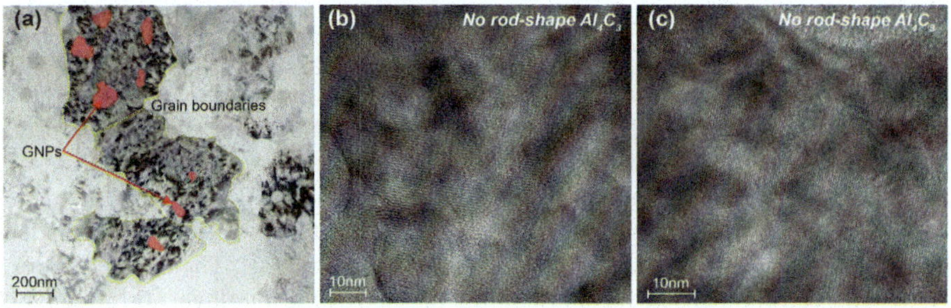

Fig. 9.5 HR-TEM observation of the GNPs and the intermetallic compounds. **a** Uniform distribution of GNPs inside the grains, **b** and **c** Bright field images of the aluminum matrix and the interface between GNPs and aluminum grains. It can be seen that there are no rod-like Al_4C_3 phase dispersed in the composites

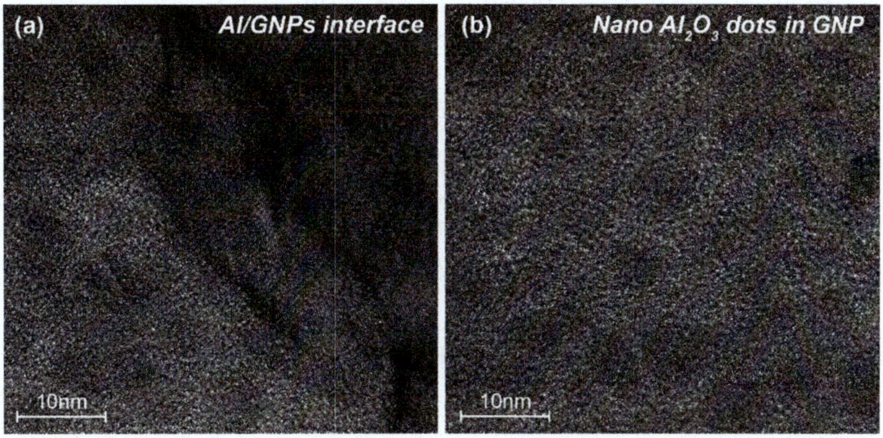

Fig. 9.6 HR-TEM observation of the multiscale nanostructure. **a** Bright field image of Al/GNPs interface and **b** Bright field image of the nano Al_2O_3 dots inside GNP

The oxide film on the surface of the powders is broken and redistributed under the deformation-driven effects of the tools. The bare aluminum particles contact directly with each other and GNPs. Subsequently, the interfaces between the bare particles and the GNPs form metallurgical bonding, and the GNP particles are coated by the exposed aluminum surface. In the case of samples with 30 s sintering time, the GNPs are distributed uniformly and most of them are situated inside the grains, as shown in Figs. 9.4b and 9.5a. It is widely accepted that the strengthening phase would be better located in the grain interior rather than on the grain boundary to avoid negative impact on the ductility of metals [20, 21]. Besides, no Al_4C_3 was observed in the HR-TEM images of the composites as shown in Fig. 9.5. As such, the vast majority of GNPs dispersed uniformly in the grain interior improves the strength of composite without reducing ductility.

On a smaller scale, the massive nano Al_2O_3 dots are uniformly dispersed inside the GNP particles, as shown in Fig. 9.4c, which was identified in Fig. 9.4d. The high resolution bright field images are presented in Fig. 9.6. This indicates that the Al_2O_3 dots are concentrated into the GNP particles, which is driven by severe plastic deformation. This concentration phenomenon was also observed from the super EDS results as shown in Fig. 9.4e. Oxygen is concentrated inside the GNP particles. Since GNP has the structure of several graphene layers, it is prone to slip between the graphene layers when receiving a force parallel to graphene layers [22]. The embedded nano Al_2O_3 dots can compensate for shortcomings caused by the interlayer slip effectively. This phenomenon will be described in detail in the discussion section. Multiscale nanostructure is achieved with the structure of nano Al_2O_3 dots (several nanometers)-GNPs (tens of nanometers)-aluminum matrix (hundreds of nanometers), improving the comprehensive mechanical properties of the composites.

Besides, the Al-C and Al-O-C bonds form by means of temperature conditions and defect energy at the bare Al/GNP and Al_2O_3/GNP interface, and convert during the ball milling and the subsequent DDM process. To analyze the evolution during the entire process, XPS performed for C1s peaks is shown in Fig. 9.7a. In this figure, the spectrum consists of several components: sp^2 (about 284.5 eV), sp^3 (about 285.1 eV), C–O (about 286.4 eV), C = O (about 289.3 eV), C–O–Al (about 283.5 eV), and C-Al (about 282.9 eV). The intensity ratio of sp^2 component to sp^3 component characterizes the graphitization degree and the defects of the GNPs. One can clearly find that this intensity ratio of the milled powder is much higher than that of raw powder, which indicates that the GNPs are damaged by introducing open edges and vacancies in the graphene plane [23]. The subsequent DDM process in the first 10 s further fragmented GNPs, which provides reactional site for reaction between matrix and the GNPs [24]. As time goes on, GNPs are no longer broken but only dispersed and redistributed. A certain amount of C–O–Al bonds are produced, demonstrating that the Al/GNP is primarily bonded by the reaction between oxygen-containing functional groups on GNPs and aluminum powders or between graphene layers and oxide films on the surface of aluminum powders, rather than directly forming C-Al bonds. In the subsequent DDM process, there are three phenomena should be emphasized. First, cold welding dominates the behavior of powders in the first 10 s, because the overall temperature has not reached the quasi-steady state in most of the time (0–7 s). Consequently, the XPS spectrum shows an evolution behavior similar to the ball milling process. The C–O–Al bonds increase and occupy the major component. The C–Al bonds are not produced due to the low temperature; Second, as the entire materials reach quasi-steady state and the oxide film is sheared off on the aluminum surface, the C–Al bonds form rapidly between GNPs and bare aluminum, and replace the original C–O–Al bonds. Nevertheless, there are still a large amount of C–O–Al bonds existing due to the nano Al_2O_3 doped in GNPs. Third, with the time increasing further, the C–Al bonds continue to generate, which may lead to the nucleation and growth of the brittle Al_4C_3 phase [25].

To explain the reacting behavior further, we applied Raman spectroscopy to studying the properties of the materials, as shown in Fig. 9.7b. There are three typical regions: D-band (about 1350 cm^{-1}), G-band (about 1580 cm^{-1}), and Al_4C_3 region

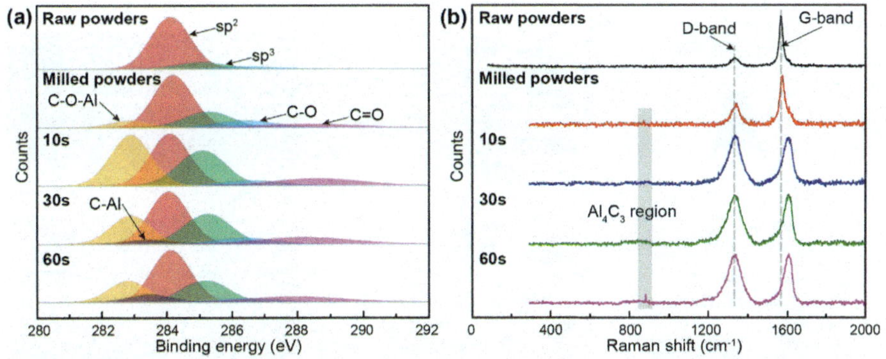

Fig. 9.7 Characteristics of the reacting behavior: **a** XPS spectrum of raw powders, milled powders, and DDMed composites with different processing time of 10 s, 30 s, and 60 s, and **b** Raman spectrum of raw powders, milled powders, and DDMed composites with different processing time of 10 s, 30 s, and 60 s

(about 860 cm^{-1}). The D-band and the G-band represent the breathing mode of A_{1g} symmetry and the stretching mode of sp^2 pairs, respectively [26]. The intensity ratio of D-band to G-band characterizes the defects of GNPs. The damage degree of GNPs increased first and then stabilized during ball milling and subsequent DDM process, which is consistent with the results obtained by XPS spectrum. After 30 s DDM process, there is no significant bulge in Al_4C_3 region. It proves that the bare aluminum bonds with GNPs without nucleation and growth of Al_4C_3 phase. Figure 9.8b is the TERS mapping of GNPs and Al_4C_3. One can see that the GNPs are evenly dispersed in the composites, whereas Al_4C_3 is not produced. However, it is not difficult to find that a tiny peak arose in Al_4C_3 region when the processing reaches 60 s, indicating the harmful brittle phase begins to be produced in the composites.

9.1.5 Mechanical Properties

The mechanical performances of the composites were investigated. The nano indentation results are shown in Fig. 9.9. The Young's modulus of our composites increases to 76.9 GPa, compared to that of the control 1060 pure aluminum (69.1 GPa), which is capable of large weight savings in airplane structures and improve the fuel efficiency [27]. Furthermore, it is of vital importance to strike a balance between high strength and low ductility loss for GNPs reinforced aluminum matrix composites [28]. Our method achieves an enhancement of tensile strength by 317% with only 27.0% ductility loss when DDM time is 30 s. The best result was obtained with 30 s processing time as shown in Fig. 9.10. Figure 9.11 summarizes the relationship of these two properties among other reported literature. Both ultrahigh tensile strength (497 MPa) and tenable ductility (15.2%) have been obtained by our strategy. From the EBSD results shown in Fig. 9.8, one can see that the materials undergo significant

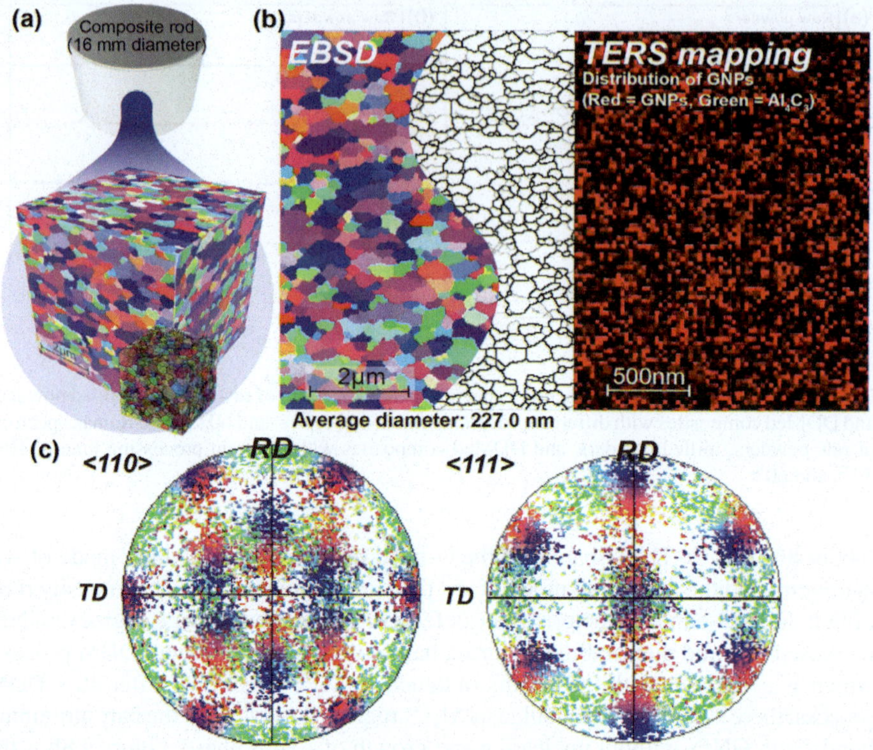

Fig. 9.8 Microstructure of the specimens: **a** Microstructure of the composites with 30 s sintering time, **b** EBSD and TERS mapping of the composites with 30 s sintering time (Red refers to GNPs and green refers to Al_4C_3 in the TERS mapping), and **c** Pole figures <110> and <111>

dynamic recrystallization at a slightly higher recrystallization temperature induced by self-heating. Simultaneously, the migration of the grain boundaries is effectively inhibited due to the GNPs dispersed in the composites [29]. Finally, a nanostructure with an average grain diameter of 227 nm is obtained. To our best knowledge, this is the smallest microstructure reported for carbon nanomaterials reinforced aluminum matrix composites so far [5–7]. Besides, high angle grain boundaries accounts for a major proportion in this microstructure, indicating that the materials have excellent thermal stability.

9.1.6 Strengthening Modes

The experimental characterizations suggest that our method can effectively fabricate GNPs reinforced aluminum matrix composites with a great balance of strength and

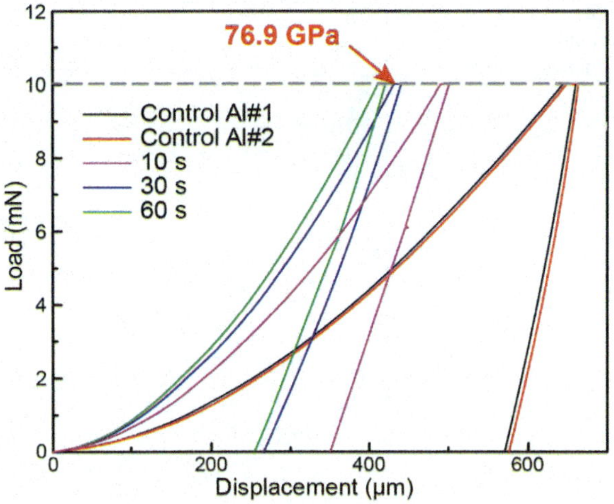

Fig. 9.9 Nano indentation of the control pure aluminum and the DDMed composites with different processing time of 10 s, 30 s, and 60 s

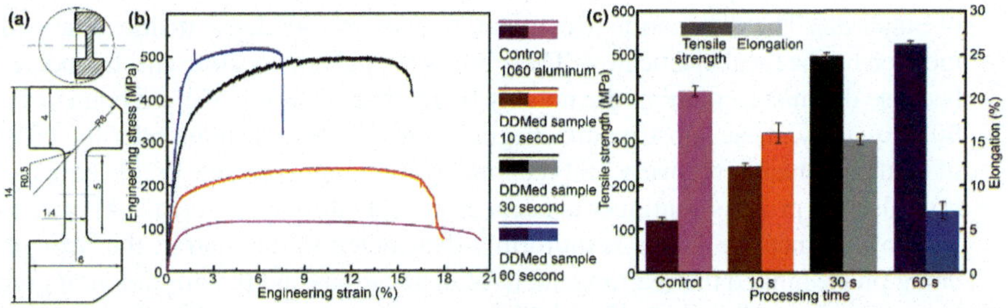

Fig. 9.10 Mechanical performances of the DDMed samples: **a** Sizes of the tensile specimens, **b** typical engineering stress–strain curves, and **c** their tensile strength / elongation of control 1060 aluminum and DDMed samples

ductility. This strategy is mainly embodied in the following three points for the strengthening and toughening modification:

1. **Nano crystalline strengthening**

   Size-affected plasticity is crucial in a wide range of alloys and composites. According to Hall–Petch relationship, higher strength can be achieved by nano crystalline structure [30]. Our strategy provides perfect conditions for grain refinement. In terms of temperature condition, the heat input is based on self-heating induced by deformation, and can be regulated in real time with flow stress. This "negative feedback" mechanism ensures that the working temperature is maintained at the nearly lowest temperature for dynamic recovery and DRX throughout the entire process [31]. In terms of critical strain energy for DRX,

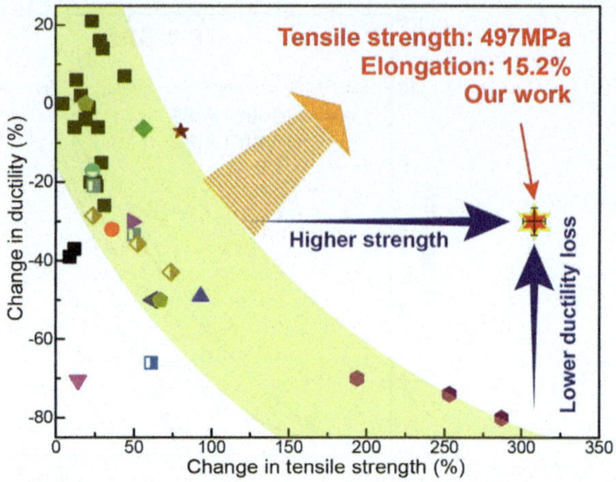

Fig. 9.11 Plot of the changes in comprehensive mechanical performances

the severe plastic deformation of composites produces a large amount of dislocation substructures [32]. The increase of plastic strain introduces high dislocation densities. Then, uneven strain hardening leads to localization of plastic flow of materials. The dislocation accumulation evolves into dense dislocation walls induced by the local plastic flow. These are essentially dislocation sub-boundaries because the misorientations are notably larger than the original grain boundaries. Subsequently, these substructures begin to grow, turn and slide under the effect of further dynamic recovery and recrystallization, which result in fine equiaxed crystals. The process continues to occur repeatedly throughout the DDM process. Moreover, since the GNPs is uniformly distributed in the matrix, the strengthening phase pinned the rearrangement of dislocations and the migration of grain boundary. According to the Smith-Zener pinning principle in Eq. (9.6) [33]:

$$P_{\mathrm{z}} = \frac{3 f_{\mathrm{v}} \gamma_{\mathrm{AB}}}{4r} \tag{9.6}$$

where P_z is Zener pinning pressure, f_v is volume fraction of GNPs, r is radius of GNPs, and γ_{AB} is grain boundary tension. Whether grain boundaries can migrate or not depends on their dislocation driving force, Smith-Zener pinning force and curvature force. The key to suppressing grain boundary migration is to achieve the balance of the Smith-Zener pinning force and the driving force. In this experiment, the Smith-Zener pinning force becomes great due to large addition of nanoscale GNPs (1.50 wt.%), which is high enough to stop the grain boundary migration. As such, refinement of microstructure induced by DRX continues to occur, while grain growth is significantly inhibited, which resulted in a nano crystalline structure of the composites.

2. **Suppression of grain boundary weakening**

   The grain boundaries in materials naturally presents as a continuous network. However, there are high defect energies near the network of grain boundaries, which are easy to become the nucleation of cracks under the external pressure [20]. Especially when GNPs are located at the grain boundary, there are serious geometric incompatibility between the one-dimensional edge of GNPs and the two-dimensional nature of grain boundaries [34]. It is prone to introduce stress concentration at the triple-boundary line of the two grains and the GNP sandwiched in the middle, as shown in Fig. 9.12a. Therefore, the GNPs would be better located in the grain interior rather than at the grain boundary to increase strength without preserving the ductility. In our strategy, we continuously apply severe plastic deformation on the materials through the stirring and friction effects of the processing tool. The aluminum powders are in contact with each other and deformed, while the oxide film on the surface is sheared off. Then, the GNPs are joined with the bare aluminum surface and coated by the aluminum powders on both sides, which result in a sandwich structure. This structure continues to deform and recrystallize under subsequent effects of strain and elevated temperature, and gradually incorporates the GNPs in the grain interior. Furthermore, although GNPs has a significant pinning effect on the grain boundaries, not all GNPs eventually pin the migration of grain boundaries. They just slow the migration velocity, which are eventually distributed inside the grains. At the same time, since the recrystallization of the composites is a dynamic process, the GNPs pinning the grain boundaries were redistributed inside the grains during the subsequent recrystallization. That is to say, in the final composites, only small amounts of GNPs are still distributed at the grain boundaries, and vast majority of GNPs are distributed inside the grains. Consequently, most GNPs are situated intragranularly rather than intergranularly, which suppress the grain boundary weakening effectively.

Since GNP is naturally a multilayer structure, it is easy to cause interlayer slip when subjected to a force parallel to the graphene layer [35, 36], and thus causes the materials to be spatially discontinuous. This provides nucleation sites for crack initiation. Therefore, it is a great challenge for conventional GNPs reinforced aluminum matrix composites to get a balance between ultrahigh strength and tenable ductility. Nevertheless, a significantly different deformation mode is also achieved by our

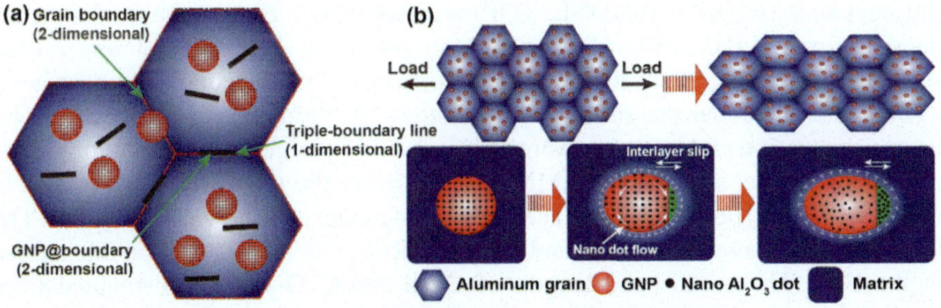

Fig. 9.12 Schematic of the strengthening mechanisms: **a** Intergranular and intragranular distribution of the GNP particles and **b** Multiscale deformation mode of the composites

strategy, owing to the multiscale nanostructure. As illustrated in Fig. 9.12b, during the tensile of the composites, large amounts of dislocations accumulate around the GNP particle, resulting in the interlayer slip inside the GNP particle and shearing along the tensile direction. At this time, the nano Al_2O_3 dots inside the GNP particle begin to move along the direction of deformation and "flow" into the vacancy induced by the interlayer slip. This behavior ensures the continuity inside the GNP particle on the one hand, and relieves the residual stress around the GNP particle on the other hand, which improves the strengthening and toughening effect of the GNPs during further deformation.

To explain the diffusion slip behaviors of the nano Al_2O_3 dots, a first principle calculation was performed using Quantum ESPRESSO [37]. The exchange–correlation energy was calculated by Perdew-Burke-Erzenh of revised for solid (PBEsol) approach [38] and Generalized gradient approximation (GGA) [39]. Ultrasoft pseudopotential was used to describe the real electronic functions. A k-point grid of 11 $\times$ 11 $\times$ 1 and an energy cutoff of 400 eV were used in this evaluation. Three highly symmetrical combination sites were studied: the TOP is located directly above the carbon atoms of GNPs, the BRIDGE is located at the middle of the C–C bonds, and the HEART is located at the center of the honeycomb lattice. The combination density was controlled by the lattice parameters, which was set as 9.51 Å $\times$ 9.51 Å $\times$ 20.00 Å. A three-layer GNP was placed at the lowest plane of Z-axis. The combination energy of each C–Al and C–O–Al group on GNPs can be defined as:

$$\Delta E = E^{\text{tot}} - \left(E^{\text{func}} + nE^{\text{c}}\right) \tag{9.7}$$

where E^{tot} is the total energy of the combination system, E^{func} is the energy of function groups, E^{c} is the energy of single carbon atoms in GNPs, and n is the number of carbon atoms in the calculated model. According to the definition, the stability of the system becomes higher with the decreasing combination energy. Table 9.1 shows the typical combination of the two groups calculated by ab initio method. The combination energies of C–Al groups are always higher than those of C–O–Al groups, which demonstrates that the C–O–Al tends to generate preferentially. Al_2O_3 dots have a great tendency to enter the interior of graphene, leading to the formation of multiscale nanostructure. Furthermore, the combination energies of C–O–Al at different sites (HEART, BRIDGE, TOP) are very close. The maximum difference does not exceed 0.06 eV/group. Al_2O_3 dots produces easy diffusion slip on the surfaces of GNPs, and are likely to fill automatically the discontinuous vacancies of the materials under the effect of tensile stress. Additionally, calculations of slip energy fluctuation of Al_2O_3 dots were performed using large-scale atomic/molecular massively parallel simulator (LAMMPS) [40] using the modified embedded atom method (MEAM) potentials [41]. The calculation model was shown in Fig. 9.13a. The layers of GNP were set as three and the nano dots grew along [001] on the surface of the GNP. The initial bond lengths of C–Al and C–O–Al were estimated from the ab initio calculation above. The energy fluctuation was evaluated by the energy change per Al atom. As shown in Fig. 9.13b, the energy amplitudes of Al and Al_2O_3 dots were 0.81 eV/Al atom and 0.42 eV/Al atom respectively. The energy barrier

Table 9.1 Combination energy of C–Al and C–O–Al groups at different sites

Combination position	Combination energy (eV/group)	
	C–Al	C–O–Al
TOP	−1.37	−2.33
BRIDGE	−1.54	−2.35
HEART	−1.40	−2.29

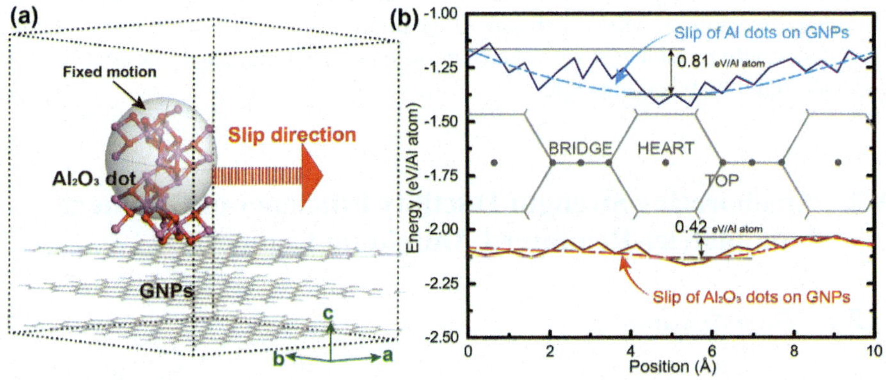

Fig. 9.13 Results of the molecular dynamics: **a** Calculation model for slip energy fluctuation and **b** Energy fluctuation during diffusion slip on GNPs of Al and Al_2O_3 dots

of diffusion slip of Al_2O_3 was lower significantly, which implied that the diffusion slips of Al_2O_3 dots occurred synchronously with applied external load. As such, enhanced ductility of the composites was achieved under the multiscale collaborative deformation mechanism. Eventually, a great balance is achieved between ultrahigh strength and inhibition of ductility loss.

9.1.7 Summary

We present an excellent strategy for achieving excellent balance between strength and ductility of GNPs reinforced aluminum composites, namely DDM. The above phenomena bring about triple strengthening and toughening effect of the composites:

Nano crystalline strengthening: "Negative feedback" mechanism of DDM ensured that the working temperature is maintained at the nearly lowest temperature for dynamic recovery and DRX. The isolation of the GNPs inhibited the growth of grains furtherly, which resulted in nano-sized microstructures with average grain diameter of 227.0 nm.

Suppression of grain boundary weakening: The vast majority of GNPs were incorporated in the grain interior under the severe deformation. The stress concentration at the three triple-boundary line of two grains and GNP sandwiched in the middle was avoided.

Multiscale deformation mode: Multiscale nanostructure was achieved with the structure of nano Al_2O_3 dots (several nanometers)-GNPs (tens of nanometers)-aluminum matrix (hundreds of nanometers). Surface diffusion slip of Al_2O_3 dots effectively reduced spatial discontinuity of materials during tensile test, which enhanced the ductility of the composites.

Through the synergistic effect of the above three mechanisms, great enhancement of tensile strength by 317% with only 27% ductility loss was achieved, indicating the DDM could provide a novel balance between high strength and low ductility loss for bulk metal matrix composites.

9.2 Ameliorating Strength-Ductility Efficiency of Graphene Nanoplatelet-Reinforced Aluminum Composites

9.2.1 Introduction

Carbonaceous nanomaterials, including carbon nanotube [42], graphene [43], graphene oxide [44], and fullerene [45], etc., have gained extensive attention for their distinctive physical and mechanical properties, which have been considered as ideal reinforcements for the next-generation aluminum matrix composites. GNPs become more favorable due to the better load transfer effect induced by two-dimensional morphology [43]. At present, its mass production price is lower than 500 USD per kilogram [46]. Although this is still much higher than the cost of the aluminum matrix, since it only requires a tiny amount of addition to obtaining high strengthening efficiency, the GNP-reinforced aluminum matrix composites have obtained huge industrial application potential, especially in high-value aerospace engineering structures [47].

At present, the applicable fabrication scheme for aluminum matrix composites can be generally divided into two categories: liquid-state method [48] and solid-state method [49]. For the fabrication scheme based on the liquid process, the reinforcements are added into the crucible together with the aluminum matrix before heating or added to the molten aluminum matrix later [50]. The major problem of the liquid-state method is the difficulty in obtaining a uniform dispersion of reinforcement [51]. The solid-state fabrication method is mainly powder metallurgy, such as hot pressing sintering [52] and spark plasma sintering [2]. Another state-of-the-art solid-state method is friction stir processing (FSP) [53]. Severe plastic deformation with a large strain rate was applied to promote uniform dispersion of GNPs into the matrix, contributing to high-quality composites [54]. However, there are still challenges to obtaining whole bulk ultra-fine-grained materials via the FSP process, restricting the shape and size of the desired structures.

Based on the composites fabricated by the liquid-state and solid-state methods, the main difference in the microstructures is the grain size, reinforcement distribution, and interfacial bonding between reinforcements and matrices [55]. For the aluminum matrix composites fabricated by the liquid-state method, the melt viscosity increases sharply when the amount of GNPs reaches a critical value, which causes the melt to no longer be thoroughly stirred and blended by external effects. The uneven distribution of the reinforcements may lead to the unsatisfactory mechanical performance of the composites. Also, aluminum and carbon tend to form hydrolyzable intermetallic compound Al_4C_3 under high temperature, which is detrimental to the corrosion resistance of the composites. For the solid-state methods, especially for the severe plastic deformation methods, since the deformation energy replaces partial heat input to provide sufficient driving force for the consolidation of the composites [56], high-performance composites can be obtained at a lower temperature, which avoids the nucleation of the detrimental intermetallic compounds. Additionally, uniform dispersion and dynamic recrystallization are significantly promoted, resulting in the refinement of homogenization of the composites [57]. It is mainly due to two points: ① Nano-sized particles tend to become the nucleation sites of the recrystallized grains [58]; ② The pinning effect of the nano-reinforcement on the grain boundary caused the suppression of grain growth [59]. These nano-sized particles hinder the rearrangement of dislocations from forming sub-grain boundaries and the migration of high-angle grain boundaries. When the reinforcement is uniformly dispersed into the matrix, the minimum theoretical grain size that the microstructure under the action of hot deformation can achieve through the Zener pinning effect is [29]:

$$D = \frac{4\alpha d_{\mathrm{p}}}{3 f_{\mathrm{p}}} \tag{9.8}$$

where d_{p}, f_{p}, and α is the average diameter of the reinforcements, the volume fraction of the reinforcements, and a constant, respectively. According to Eq. (9.8), reducing the average size and increasing the volume fraction of the reinforcements can promote the pinning effect on the growth of the matrix. However, Hassan et al. [60] found that greater grain refinement can be obtained in micro-sized particles instead of nano-sized particle-reinforced composites. It is because the nanoparticles are very prone to agglomerate [61], severely weakening their pinning effect and interfacial bonding areas. Therefore, good dispersion of GNPs is an essential prerequisite for achieving satisfactory grain refinement and interfacial bonding. Severe plastic deformation techniques are an ideal technical means, which is equivalent to introducing the HEBM process into the powder sintering process [62]. DDM was used to prepare GNP-reinforced aluminum matrix composites. This method aims to obtain bulk high-performance composites with ultra-fine microstructure and uniform dispersion of reinforcements, as well as appropriate interfacial bonding. These composites focused on ameliorating their strength-ductility efficiency. The microstructures and mechanical properties were investigated, and theoretical modeling was established to explain the strengthening performance in detail.

9.2.2 Experimental Procedures

The raw few-layer GNP (1 ~ 5 nm in thickness, 1 ~ 3 μm in diameter, Nanjing XFNANO Materials Tech Co., Ltd, China) was chosen, whose morphologies are shown in Figs. 9.14a–c. 1.5 wt.% graphene nanoplatelet and the balance aluminum powders were used as the powders for the preparation of the composites. DDM process (Fig. 9.15) was applied on the powders towards bulk composites with different rotational velocities from 600 r/minm to 1400 r/min.

The GNPs distribution on the milled aluminum powders and prepared bulk composites was examined using SEM (HITACHI SU5000). The microstructures were characterized by EBSD via a ZEISS SUPRA 55 FE-SEM equipment. The dispersion, structural evolution, and interface bonding state of the GNPs were observed by TEM (JEOL JEM-2100) equipped with a silicon drift detector EDS. Raman spectroscopic measurements were conducted using Renishaw inVia-Reflex to evaluate the disordering and defect density of graphene structures. XPS (ThermoFisher ESCALAB 250Xi) was utilized to obtain the chemical bonding evolution during the DDM process.

Tensile specimens with a gauge length of 14 mm and a width of 6 mm were cut from the prepared composites (Fig. 9.16). The tensile tests were conducted at a strain rate of $1 \times 10^{-3}\ s^{-1}$ at room temperature using an INSTRON 5569 tester. Tensile specimens

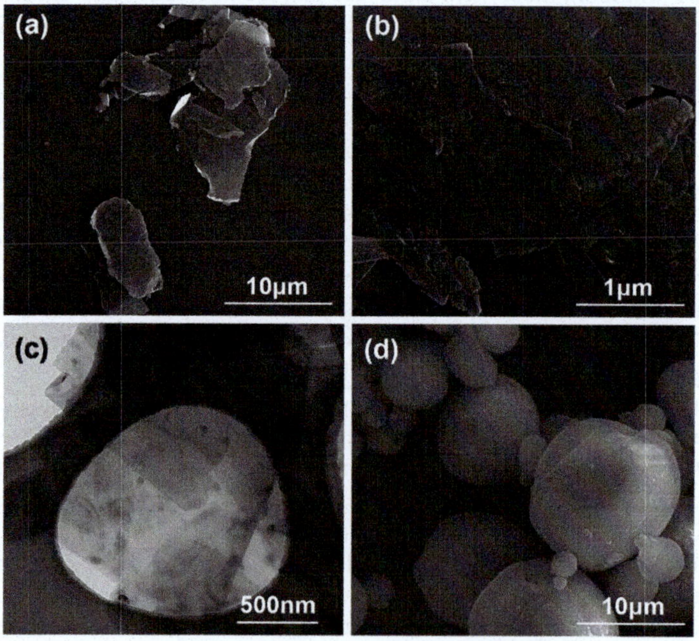

Fig. 9.14 Morphologies of raw materials: **a** GNPs in low magnification, **b** GNPs in high magnification, **c** dispersed GNPs, and **d** spherical pure aluminum powders

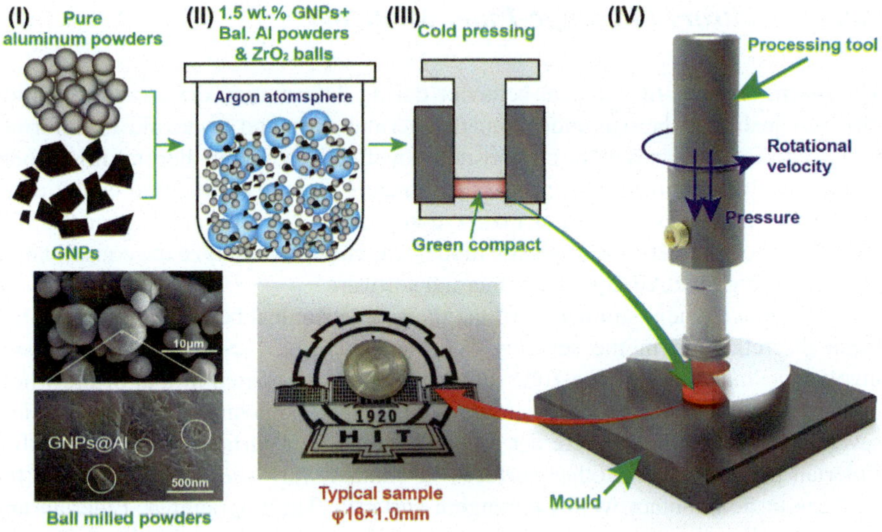

Fig. 9.15 Schematic of DDM preparation procedure and the typical sample

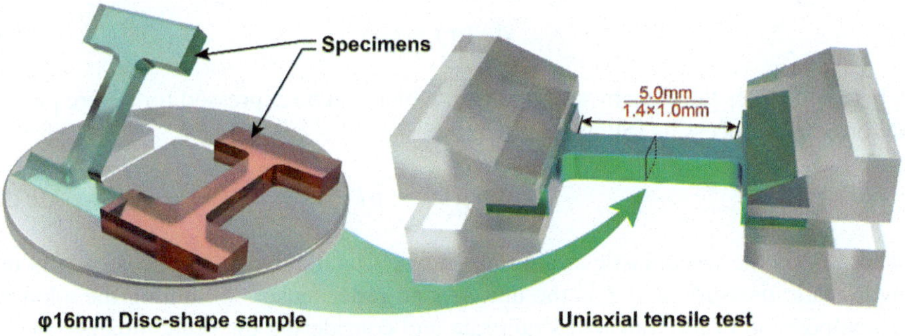

Fig. 9.16 Sampling position and dimensions for tensile testing

made of pure aluminum materials prepared by DDM were also tested as a control. At least three specimens were tested for each parameter. Vickers microhardness test was conducted to evaluate the homogeneity of mechanical properties with a load of 100 N and dwelling time of 10 s.

9.2.3 Arbitrary Lagrange-Euler Modeling

The macro thermal–mechanical behavior during deformation-driven consolidation process can directly help us understand the thermal effect and severe plastic deformation of each micro zones in the composites. An effective coupled thermal-flow model contributes to optimization of process parameters and microstructural evolution of the GNPs reinforced aluminum matrix composites. An arbitrary Lagrange-Euler (ALE) method based on elastoplastic model was chosen to analyze the consolidation process through the ABAQUS commercial software.

ALE finite element simulation of the thermomechanical behavior requires a fine mesh discretization in the severely deformed zone [63]. Since DDM processes involved a small size stirring zone (composite disc) compared to that of the whole model (mould, tool, and composite disc), only the composites disc elements and several vicinity elements were set as ALE zone. Their surfaces were regarded as Eulerian surfaces, while other surfaces in contact were sliding surfaces and the surfaces in non-contact were Lagrangian surfaces. The Lagrangian, Eulerian and ALE formalisms as s function of the part of the model were considered. For the composite zone including GNPs, assuming that viscous stresses are predominant, the momentum balance can be calculated by the following equation [64]:

$$\mathrm{div}(\sigma) = 0 \tag{9.9}$$

where σ denotes the Cauchy stress tensor, which can be expressed for severe plastic deformation process as following equation:

$$\sigma = 2\mu \mathrm{D}^{\upsilon} - pI \tag{9.10}$$

where D^{υ} is the viscoplastic strain rate tensor, μ is the dynamic viscosity, p is the hydrostatic pressure, and I is the unit second order tensor. Assuming the additive decomposition of the transient strain rate and considering the predominance of the viscoplastic strain rate in the composite micro zone, the viscoplastic strain rate tensor can be expressed as:

$$D^{\upsilon} = D - D^{\mathrm{e}} - D^{\mathrm{temp}} \tag{9.11}$$

where D^{e} and $\mathrm{D}^{\mathrm{temp}}$ is the strain rate tensor induced by elastic and thermal effect. However, since the plastic deformation is much higher than elastic deformation and the deformation caused by temperature change. D^{e} and $\mathrm{D}^{\mathrm{temp}}$ can be ignored, thus the viscoplastic strain rate tensor can be calculated by:

$$\mathrm{D}^{\upsilon} = \mathrm{D} = \frac{1}{2}\left(\mathrm{grad}^{t}(\upsilon) + \mathrm{grad}(\upsilon)\right) \tag{9.12}$$

where υ is nodal velocity in calculation domain. Besides, the non-Newtonian dynamic viscosity of the metallic materials μ is dependent of temperature and plastic strain rate attested by Zienkiewicz and Cormeau [65], which is governed by the following equation [66]:

$$\mu = \frac{\sigma^p}{3\dot{\varepsilon}^p} \tag{9.13}$$

where σ^p is the plastic flow stress scalar which is regarded as the driving force to sustain the non-Newtonian plastic deformation [67], and $\dot{\varepsilon}^p$ is the transient plastic strain rate scalar. The plastic flow stress can be calculated by Johnson–Cook constitutive equation [68]:

$$\sigma = \left(A + B\varepsilon^{pn}\right)\left[1 + C\ln\frac{\dot{\varepsilon}^p}{\dot{\varepsilon_0}}\right]\left[1 - \left(\frac{T - T_0}{T_m - T_0}\right)^m\right] \tag{9.14}$$

where A, B, C, m, and n are Johnson–Cook constants, ε^p is the equivalent plastic strain scalar, $\dot{\varepsilon_0}$ is the unit strain rate, and T_m is the melting point of pure aluminum matrix.

During DDM process, there are two part of thermal boundary conditions. The first part is the temperature rise induced by mechanical dissipation at the interface between processing tool and the composites, as well as the aluminum mould. Heat transfers can be calculated by the following equation:

$$\rho C\left(\frac{\partial T}{\partial t} + \mathrm{v} \cdot \mathrm{grad}(T)\right) = \mathrm{div}(\lambda \mathrm{grad}(T)) - Q \tag{9.15}$$

where ρ is the density of the composites, C is the specific heat, λ is the thermal conductivity, v is the convective velocity tensor, and Q is the thermal dissipation which can be expressed as:

$$Q = \alpha\eta \mathrm{S} : \mathrm{D}^v \tag{9.16}$$

where α is the Taylor-Quinney coefficient [69], which is set as 0.9, and η is the slip rate [70]. In addition, one should note that the tool-composite interface is considered as a perfect thermo-mechanical contact. The second part of boundary conditions is the heat exchange between the samples and ambient atmosphere, which is governed by:

$$\lambda\frac{\partial T}{\partial z} = h(T - T_\infty) + \sigma_s\varepsilon\left(T^4 - T_\infty^4\right) \tag{9.17}$$

where h, σ_s, ε, and T_∞ are respectively heat transfer coefficient, Stefan-Boltzmann constant, surficial emissivity and ambient temperature.

9.2.4 Characterization of the Powders and the Composites

Figure 9.17 shows the quasi-steady temperature and equivalent plastic strain distribution at different rotational velocities. Since the angular velocity is constant throughout the DDM process, the local linear velocity is proportional to the radial distance from the center of the composite disc [66]. However, because the mold is made of aluminum, whose thermal conductivity is high, massive thermal exchange occurs at the edge of the composite disc, making the temperature farther from the center lower than the center area, thus appearing the quasi-steady temperature distribution characteristics as shown in Fig. 9.17a. Figure 9.17c shows the distribution of the radial temperature at different rotational velocities. The overall temperature gradually rises as the rotational velocity increases. Even the curve at the lowest rotational velocity of 600 r/min, its edge temperature (649 K) is still higher than the minimum temperature of 0.5 T_m (0.5×933 K ≈ 467 K) required for recrystallization, implying that all the DDM composite discs undergo dynamic recrystallization.

Figure 9.17b shows the contour distribution of equivalent plastic strain. The cumulative plastic strain is gradually increased around a radial distance less than 6.5 mm. The maximum plastic strain can reach 60 ~ 100. This huge plastic strain can effectively mix materials, thereby inducing uniform redistribution of GNPs and dynamic recovery and recrystallization. The strain in the central area of the composite disc is relatively small, while its quasi-steady temperature is high. The outer region has a

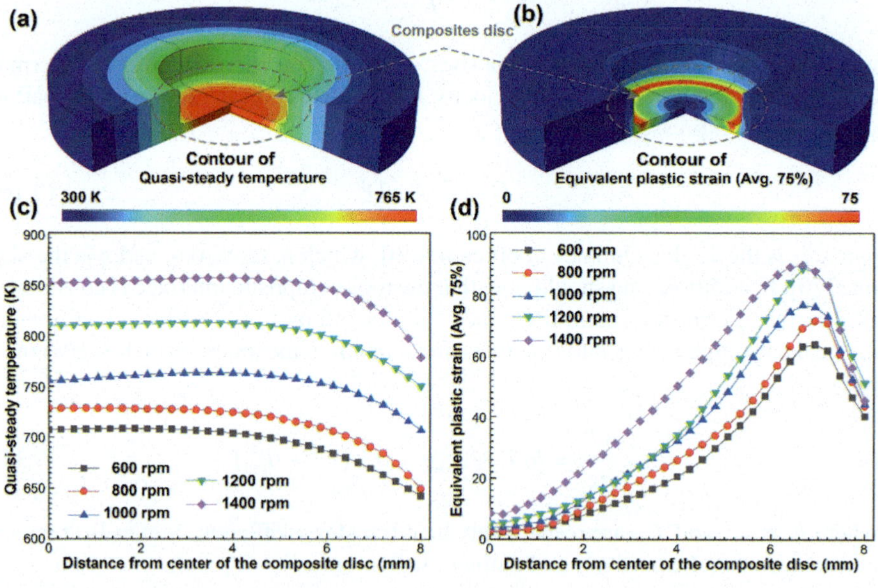

Fig. 9.17 Coupled thermal-flow behaviors during DDM process: **a** contour of quasi-steady temperature at 1000 r/min, **b** contour of equivalent plastic strain at 1000 r/min, **c** radial distribution of quasi-steady temperature, and **d** radial distribution of equivalent plastic strain

lower temperature, but there is a large stain rate and more cumulative plastic strain. We consider the plastic strain energy as compensation for the lack of sintering caused by insufficient temperature. The phenomenon of the negative correlation between temperature and plastic strain fits this conjecture. In outer areas with lower temperatures, a large amount of plastic deformation may make the mechanical performances of the composite more uniform.

Raman spectroscopy was performed to describe the evolution of GNPs, as shown in Fig. 9.19. The intensity ratio of the D-band to the G-band (I_D/I_G) characterizes the degree of GNP damage. This value of raw GNPs is 0.055 and reaches 1.121 after ball milling, indicating that the graphene has been significantly broken and redistributed, and large amounts of 2-dimensional edge on GNPs are exposed. These edges serve as active reactional sites and provide necessary conditions for the diffusion and reaction between GNPs and aluminum matrices in the subsequent DDM process. I_D/I_G is observed to further increase for the DDM composites, which is positively correlated with the rotational velocity. The high plastic strain significantly aggravated the damage and redistribution of GNPs. However, one should note that a more uniform and finer distribution of GNPs provides a guarantee for better strengthening efficiency, but this also results in loss of the load transfer effect of GNPs. Besides, higher defects density also potentially increased the driving force for the formation of Al_4C_3. Figure 9.18c shows the fitting curves of I_D/I_G versus rotational velocity. The R-squared value is 0.989, which proves the accuracy and reliability of the fitted curve. The damage degree of GNPs is exponentially related to the rotational velocity; that is, a high velocity will seriously aggravate the damage of graphene, implying that we need to strictly control the rotational velocity to avoid excessive damage of GNPs and the spontaneous formation of Al_4C_3 compounds. Also, we performed Raman mapping about GNPs damage and the characteristic signal of Al_4C_3 around 860 cm^{-1}. The damage distribution of GNPs at 1000 r/min and 1400 r/min is relatively uniform, which confirms our previous guess about the complementarity of heat input and plastic strain partially. Moreover, the emergence of the Al_4C_3 region is observed at 1400 r/min, which does not exist at lower rotational velocities. It proves our discussion that excessive GNP fragmentation and excessive heat input caused by high velocity contribute to the formation of Al_4C_3 brittle compounds.

Figure 9.19 shows the XPS spectra of the raw GNPs and the ball-milled powders. The micro-defects induced by severe plastic deformation serve as active reaction sites, which can react with the aluminum matrix to strengthen the interfacial bonding between GNPs and the aluminum matrix. Also, the ratio of sp^2 peak to sp^3 peak characterizes the graphitization and defects of GNPs. The raw GNPs have a high degree of graphitization and low defect density (Fig. 9.19b). The ratio of sp^2 peak to sp^3 peak of the ball-milled powders increases slightly (Fig. 9.20a), indicating that the GNPs undergo brokenness and damage. The plastic strain caused by ball milling introduces open edges and vacancies on the graphene plane of GNPs, which leads to an increase in the defect degree of GNPs. A small amount of Al–O–C bonds can also be observed, which indicates that two potential reactions occur between carbon element on GNPs and alumina films on undeformed aluminum powders, aluminum element on shear-broken aluminum powders, and oxygen-containing functional groups on GNPs. For

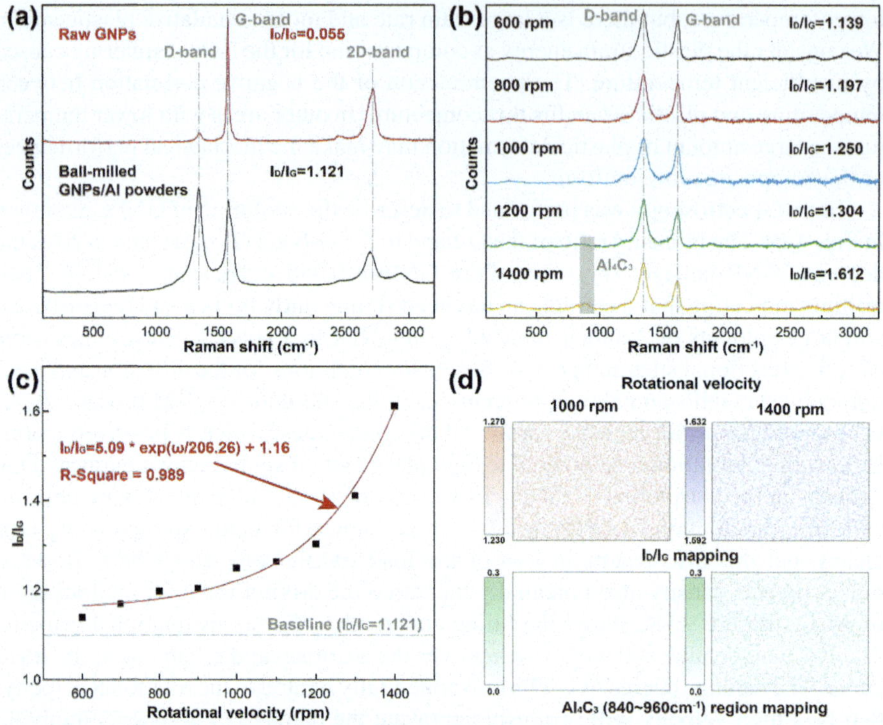

Fig. 9.18 Raman spectra of powders and DDM composites: **a** curves of powders, **b** curves of composites, **c** curve fitting of GNP damage degree versus rotational velocity, and **d** Raman mapping of damage and intermetallic compounds at different rotational velocities

the composite discs prepared at different rotational velocities, there is a process of Al–O–C bonds that first increases and decreases as the rotational velocity increases. Besides, the Al–C bond gradually appears and increases with the increase of the rotational velocity. It shows that the cold-welding behavior between GNPs and the aluminum matrix initially dominated the interfacial bonding behavior. During this process, part of the aluminum matrix powder that has not been subjected to severe plastic deformation maintains oxide films on the surface. It directly mechanically interlocks with the graphene to form the Al–O–C bond. The surficial oxide film on the other part of the aluminum matrix is broken under the shear-breaking effect of severe plastic deformation. The bare aluminum matrix is exposed and reacts with the oxygen-containing groups on the surface of the graphene, which also forms large amounts of Al–O–C bonds. With the further increase of the rotational velocity, the aluminum element can directly react with the carbon element due to more heat input and more defect energy to form Al–C bonds. These two bonding forms simultaneously constitute an efficient bonding between the aluminum matrices and GNPs. However, it is worth noting that the Al–C bond is also a characteristic of Al_4C_3. The excessive production of these hydrolyzable intermetallic compounds contributes to

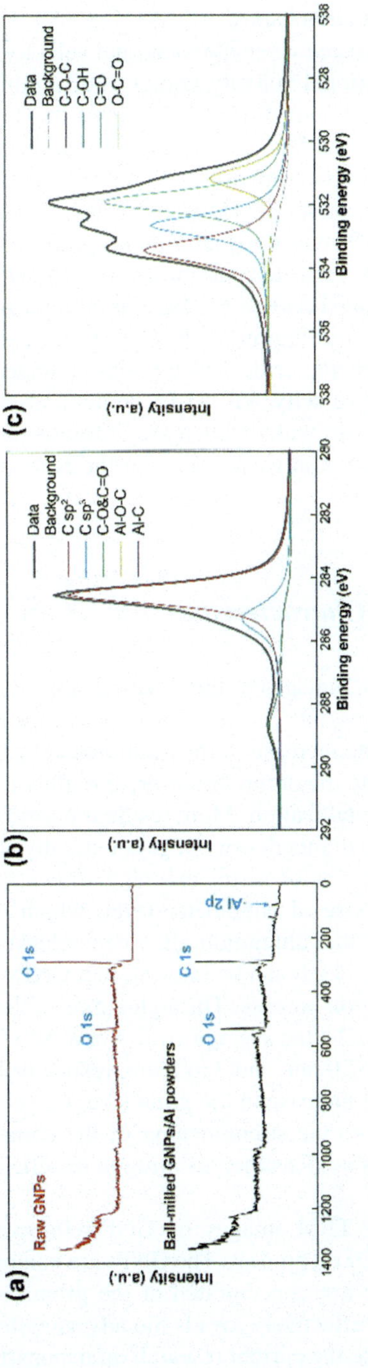

Fig. 9.19 XPS results of raw and ball-milled powders: **a** full spectra, **b** C 1 s spectrum of GNPs, and **c** O 1 s spectrum of GNPs

a severe loss of corrosion resistance and ductility of the composite discs. Therefore, the Al–C bond needs to be strictly controlled to meet the high-quality, comprehensive mechanical properties of the composites.

The Al-C bond can be detected when the rotational velocity reaches only 600 rpm in Fig. 9.20. When the rotational velocity comes 1000 r/min, its peak intensity is close to one-fifth of Al–O–C bonds. Figure 9.21 shows the deconvolution results of the Al 2p peak of the powder and the composite discs at different rotational velocities. The Al–C bond can be observed above 1000 r/min. This result is inconsistent with the result of Raman spectroscopy. The above Raman mapping shows no Al_4C_3 compounds exist in the GNP-reinforced aluminum matrix composites at the rotational velocity of 1000 r/min. This inconsistency implies that these Al–C bonds only exist as the chemical bonding between the bare aluminum matrix and the carbon element on GNPs rather than in the unsatisfactory Al_4C_3 compounds. Only when the Al–C bond reaches a specific critical value, that is, under the action of higher heat input brought by higher velocity, will Al_4C_3 compounds be produced. Also, the Al–Al bond gradually decreases with the increase of rotational velocity in Fig. 9.21, showing that more aluminum atoms participate in the reaction with carbon due to higher heat input.

9.2.5 Microstuctural Characteristics

Figure 9.22 depicts the EBSD images of the GNP-reinforced composites. In kernel average misorientation images, one can see that average misorientation is relatively high, and it increases with the increase of the rotational velocity, indicating that the high plastic strain induced by the deformation-driven effect is not entirely used for dynamic recovery and recrystallization. More residual plastic strain accumulates in the composites, resulting in a higher density of geometrically necessary dislocations, which can be proven by the Taylor factor mapping shown in Fig. 9.23. All three composites have the grain size of nanometer level, which is much smaller than the grain size (about 1 μm) that aluminum alloy can achieve under classic plastic deformation techniques [31, 71]. It can be inferred that GNPs play an essential role in the dynamic recrystallization process. Through a more extensive range of EBSD grain size statistics like Fig. 9.24, the average grain size at 800 r/min, 1000 r/min, and 1200 r/min are 216.3 nm, 267.0 nm, and 326.8 nm, respectively (Fig. 9.25). Higher heat input leads to a gradual increase in the grain size; that is to say, low rotational velocity is more conducive to the strengthening of the composites. However, the too low rotational velocity is not feasible, resulting in insufficient consolidation and porosity of the composite.

Figure 9.26 depicts the TEM images of GNP-reinforced aluminum matrix composites prepared by the DDM process. The GNPs are homogeneously distributed inside the matrix. They are not concentrated at the grain boundaries but mainly dispersed intragranularly. It effectively avoids the adverse effects of GNPs on two-dimensional boundaries. Like the current research on nanoparticle-reinforced metal

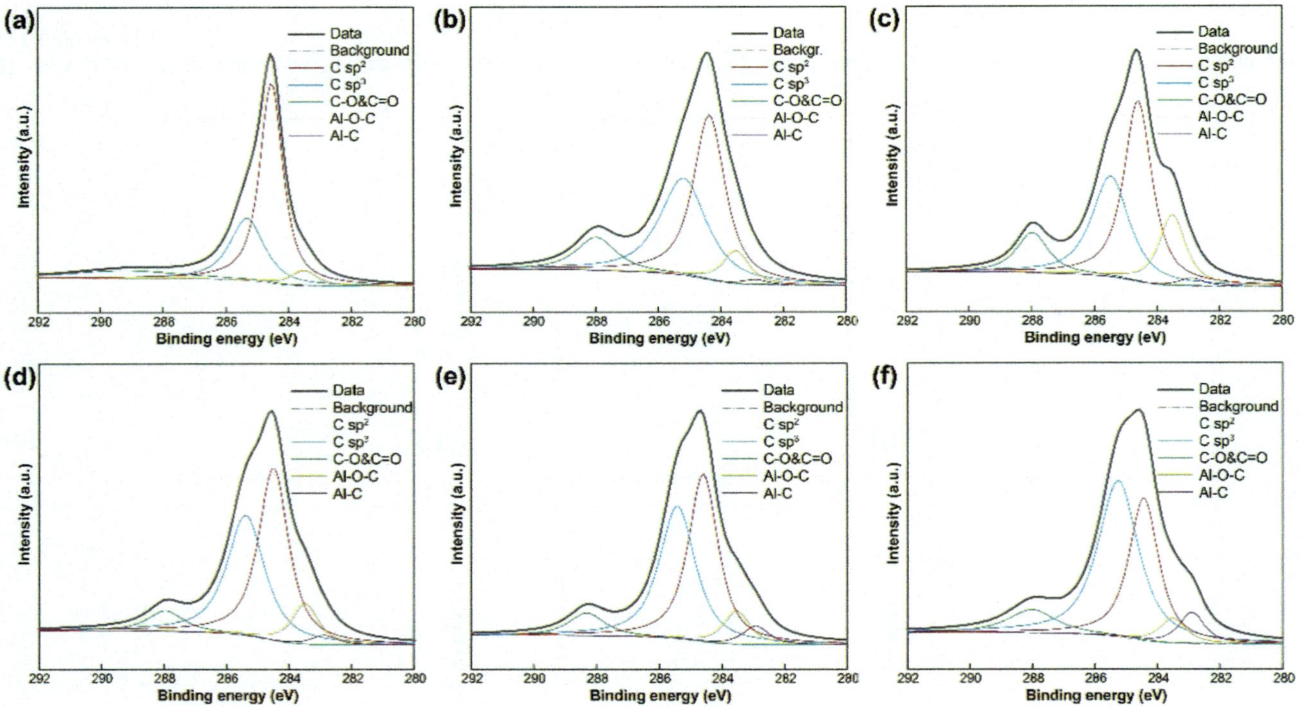

Fig. 9.20 XPS results of ball-milled powders and DDM composites at C 1 s peak: **a** ball-milled powders, **b** 600 r/min, **c** 800 r/min, **d** 1000 r/min, **e** 1200 r/min, and **f** 1400 r/min

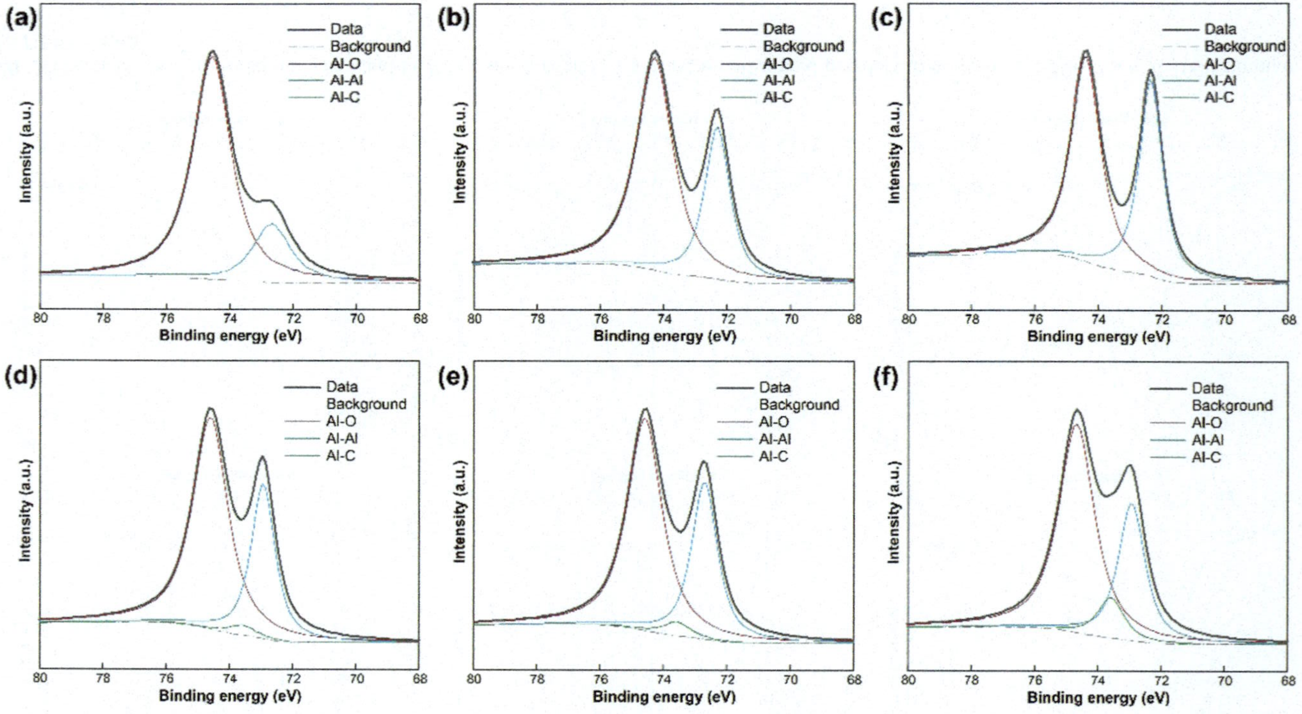

Fig. 9.21 XPS results of ball-milled powders and DDM composites et al. 2p peak: **a** ball-milled powders, **b** 600 r/min, **c** 800 r/min, **d** 1000 r/min, **e** 1200 r/min, and **f** 1400 r/min

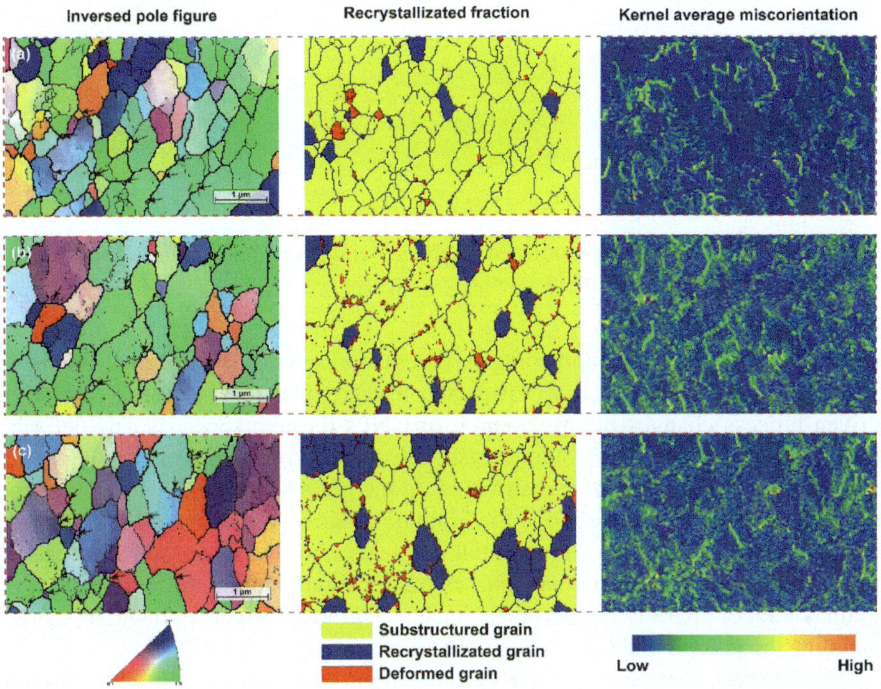

Fig. 9.22 EBSD images of DDM composites at different rotational velocities: **a** 800 r/min, **b** 1000 r/min, and **c** 1200 r/min

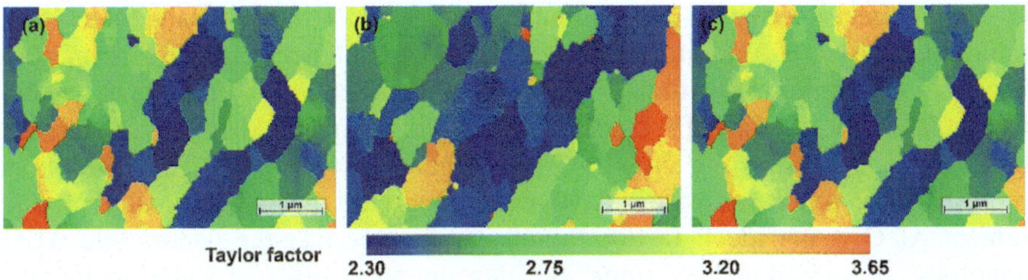

Fig. 9.23 Taylor factor mapping of DDM composites at different rotational velocities: **a** 800 r/min, **b** 1000 r/min, and **c** 1200 r/min

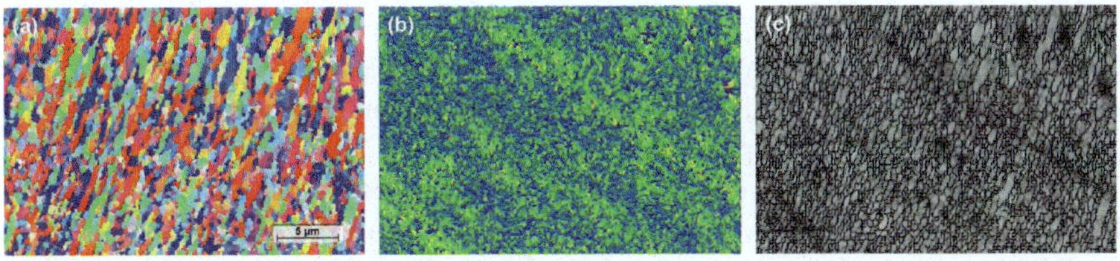

Fig. 9.24 EBSD images of DDM composites at 1000 r/min: **a** inversed pole figure, **b** kernel average misorientation, and **c** grain boundaries

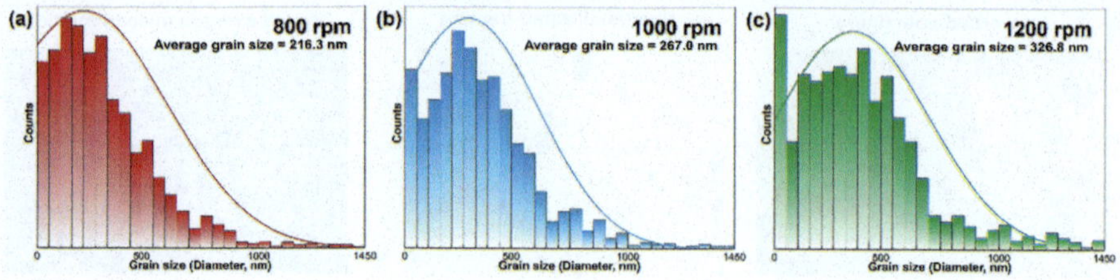

Fig. 9.25 Grain size distribution of DDM composites at different rotational velocities: **a** 800 r/min, **b** 1000 r/min, and **c** 1200 r/min

matrix composites, the strengthening particles dispersed inside the grains often better strengthen efficiency than those dispersed intergranularly [72]. The deformation-driven process changes the distribution of GNPs through severe plastic deformation, thereby improving the comprehensive mechanical properties of the composites. Additionally, partial graphene distributed on the grain boundary exhibits the Zener pinning effect on the migration of the grain boundaries, as shown in Fig. 9.26b and c. It implies that GNPs play an essential role in dynamic recovery and recrystallization. These two-dimensional nanophases have a significant inhibitory effect on the grain boundary migration of the newly nucleated grains via recrystallization, thereby avoiding the growth and coarsening of these grains.

The interfaces between the GNPs and the aluminum matrices also show different characteristics, as shown in Figs. 9.26d–f. At the lower velocity, these two phases are mainly bonded through the GNPs-amorphous Al_2O_3–Al indirect interface. This interface changes to the dual-modal interface, including the indirect interface and GNPs-Al direct interface at the rotational velocity of 1000 r/min, called the GNPs-(amorphous Al_2O_3) Al semi-direct interface. At the higher velocity, this interface begins to be dominated by the GNPs-Al direct interface. As the rotational velocity increases, the proportion of Al–O–C bonds in the composite discs firstly increases and then decreases, while the Al–C bonds gradually increase. The Al–C bond represents the formation of the Al–C direct interface, but the excessive production of Al–C is not ideal. In Fig. 9.26i, one can see that the undesirable Al_4C_3 nanorods are formed at 1400 r/min. Besides, due to insufficient sintering at 600 r/min, pores are observed in the composites for the too-low heat input, causing a severe loss of the mechanical properties of the composites. Therefore, it can be indicated that we can avoid the formation of pores and detrimental intermetallic compounds synchronously under appropriate heat input and plastic strain. Although part of the GNPs-amorphous Al_2O_3–Al indirect interface still exists on the interface between the aluminum matrices and GNPs, it is hopeful of obtaining the best mechanical performance due to the suppression of adverse factors.

There are large amounts of Al_2O_3 nanodots dispersed in the GNPs in Fig. 9.26h. It is the self-assembly behavior caused by the severe plastic deformation brought by the DDM process. Alumina nanodots fall off from the surface of aluminum powders by shear friction and enter the GNPs under the driving force of the Gibbs free energy change. Since the graphene layers of GNPs rely on van der Waals force to realize

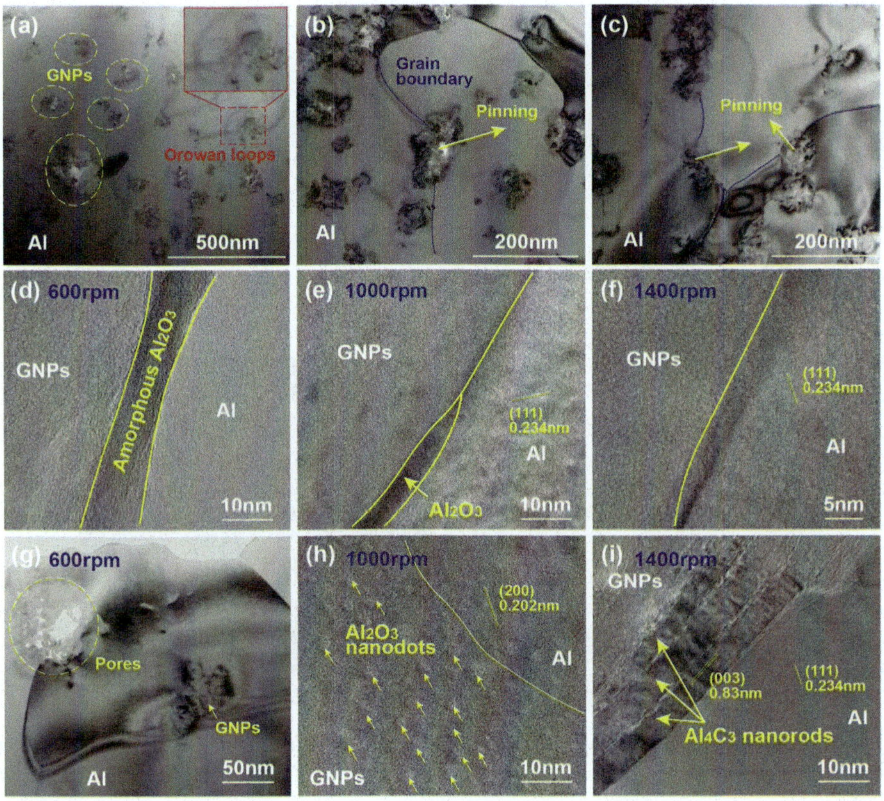

Fig. 9.26 TEM images of DDM composites: **a** GNP distribution, **b** and **c** grain boundaries and pinning effect, **d** GNPs-amorphous Al_2O_3–Al indirect interface, **e** GNPs-(amorphous Al_2O_3)–Al semi-direct interface, **f** GNPs-Al direct interface, **g** pores at 600 r/min, **h** Al_2O_3 nanodots in GNPs at 1000 r/min, and **i** Al_4C_3 nanorods at 1400 r/min

the bonding, interlayer slip is prone to occur due to external tangential load. These nano-alumina particles embedded in GNPs can effectively pin and compensate the interlayer slip, thereby avoiding the inherent issue of performance degradation caused by interlayer slip. Figure 9.27 shows the TEM images of the stretched region at the rotational velocity of 1000 r/min. GNPs are distributed on the fracture surface. An interlayer slip can be seen by observing the corresponding high-resolution image (Fig. 9.27c). According to the oxygen element distribution mapping by EDS in Fig. 9.27d, there are large amounts of Al_2O_3 nanodots in the slip gap, indicating that the nano-alumina has slipped into the space discontinuity caused by the interlayer slip under external load. It inhibits the nucleation and propagation of defects caused by the discontinuities in the two-dimensional phase-reinforced composites, thereby avoiding premature failure.

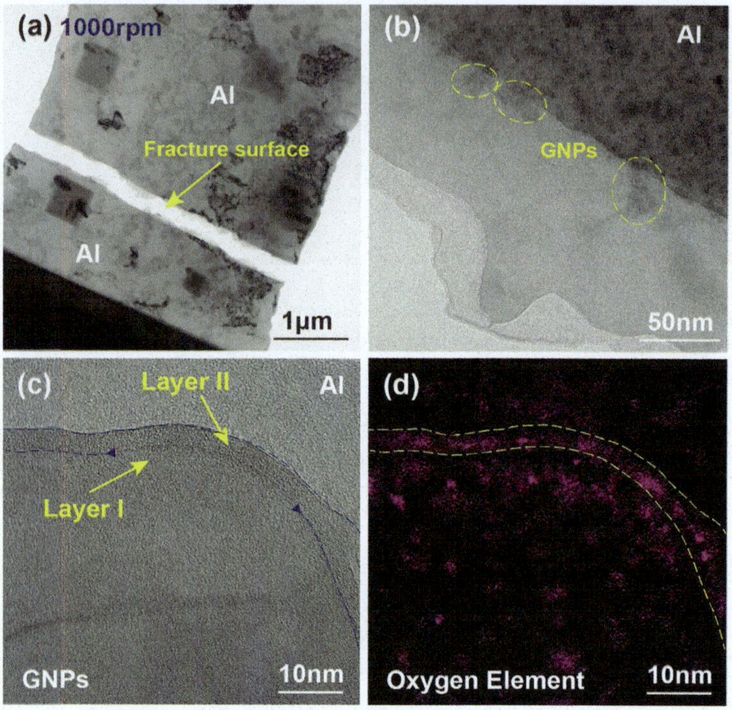

Fig. 9.27 TEM images of DDM composites at 1000 r/min: **a** fracture surface in FIB specimen, **b** GNPs near fracture surface, **c** interlayer slip in deformed GNP, and **d** oxygen element mapping of **c**

9.2.6 Mechanical Performances

Figure 9.28 shows the radial hardness distribution of the composite. There are weakened zones in the center of the composite at the rotational velocities of 600 r/min and 800 r/min, showing that the materials are not uniform enough at these parameters. When the rotational velocity reaches 1000 r/min, a uniform radial hardness distribution can be obtained. The low strain and high heat input zone in the center and the high strain and low heat input zone on edge result in similar mechanical properties. When the rotational velocity is 1400 r/min, the hardness of the center is even greater than that of the edge. It can be inferred that the excessive heat input in the center results in the production of Al_4C_3 brittle intermetallic compounds, which significantly increased the microhardness. Plastic strain and heat input can complement each other to achieve the homogenization of the composites. The morphology of the surface and indentation of the composites are observed in Figs. 9.28 and 9.29. The porosity of the composites is relatively high at the rotational velocity of 600 r/min. When the velocity reaches 1200 r/min and 1400 r/min, there are almost no pores on the surface of the materials. No cracks were observed near the indentation at all

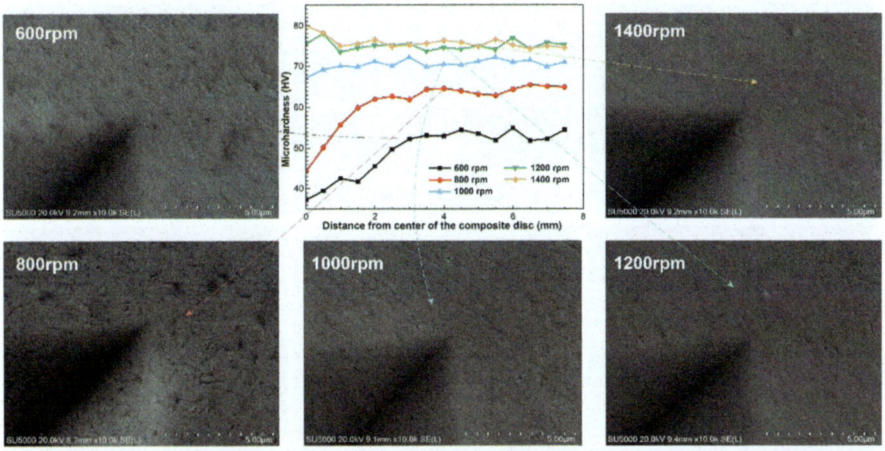

Fig. 9.28 Radial microhardness distribution of DDM composites at different rotational velocities and corresponding observation of the indentations

parameters, indicating that the interfacial bonding from by coupling Al–C and Al–O–C bonds did not fail under the plastic strain induced by microhardness indentation. It proves that the interfacial bonding of the materials is satisfactory.

Figure 9.30 shows the tensile properties of GNP-reinforced aluminum matrix composites. Both ultimate tensile strength and elongation firstly increase and then decrease with the increase of the rotational velocity. The highest values of the two are not reached at the same parameter. When the rotational velocity is 1200 r/min, the ultimate tensile strength reaches (485 $\pm$ 11) MPa. The higher rotational velocity contributes to the formation of Al_4C_3 intermetallic compounds, which weakens the load-bearing capacity of the phase interface. When the rotational velocity is 1000 r/min, the maximum elongation reaches 19.9 $\pm$ 0.6%. The high-quality interface bonding and GNPs dispersion brought by the high plastic strain were realized, along with no Al_4C_3 formation and ultra-fine grains brought by the lower heat input. The combination of the above features finally leads to the preparation of high-quality aluminum matrix composites. Moreover, we multiply these two values to obtain an evaluation index to evaluate the comprehensive mechanical properties simultaneously. Figure 9.30b shows the comparison between the composites prepared and the DDM pure aluminum disc. The highest index of the composites reaches 9.31 GPa%, which exceeds the index of matrix by 269.4% and also exceeds the index of the composites prepared by our previous work [73]. Through the comparison of the index shown in Fig. 9.31, this index is better than most available literature. The index 9.31 GPa% is also higher than the performance of 2xxx and 7xxx series aluminum alloys, including the third-generation 2195-T8 Al–Cu–Li alloys. As such, we prepared the GNP-reinforced aluminum matrix composites with the best strength-ductility efficiency via the DDM technique.

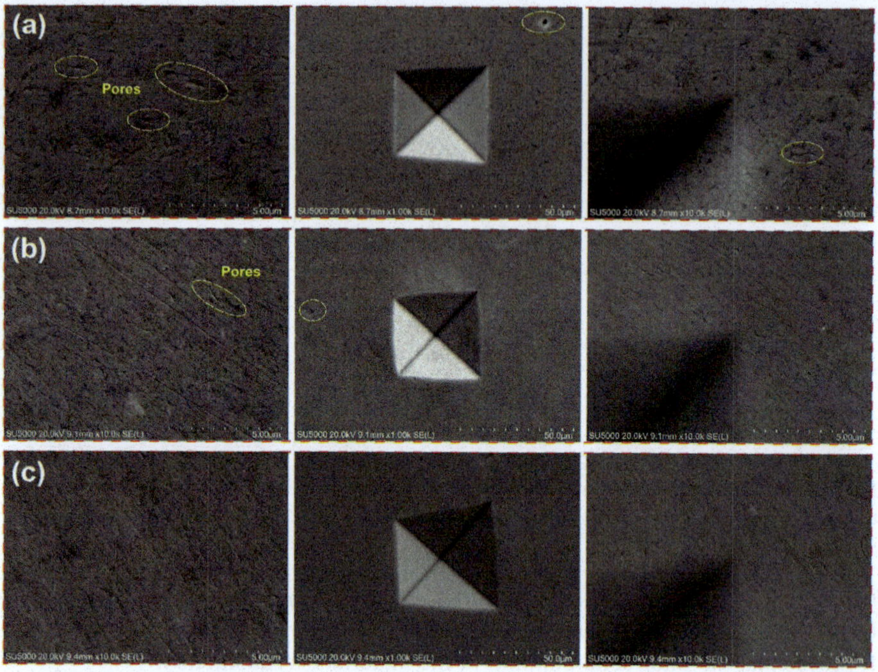

Fig. 9.29 SEM images of surface and indention of DDM composites at different rotational velocities: **a** 800 r/min, **b** 1000 r/min, and **c** 1200 r/min

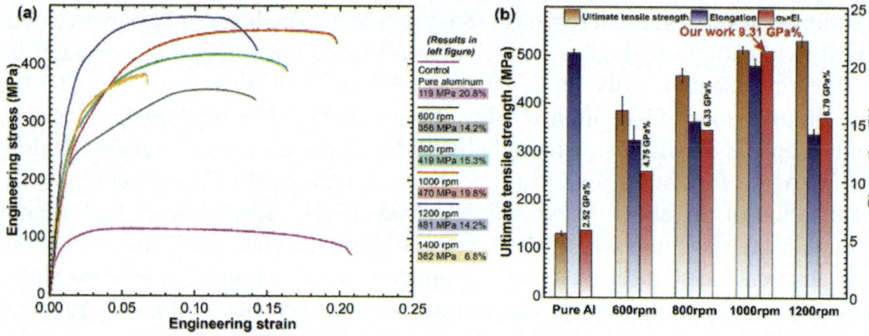

Fig. 9.30 Tensile properties of DDM composites: **a** typical engineering stress–strain curves and **b** comparison among pure aluminum and composites

9.2.7 Summary

A novel route named DDM from powders to bulk materials directly without fusion process was proposed to ameliorate the strength-ductility efficiency of GNP-reinforced aluminum matrix composites. The DDM takes only one minute to apply

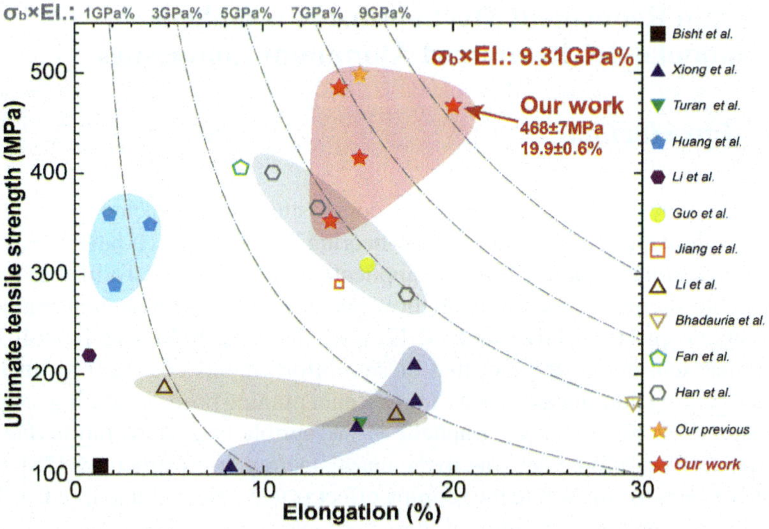

Fig. 9.31 Plot about product of ultimate tensile strength and elongation in relevant literature and our work

high plastic strain and frictional/deformation heat required for dynamic recovery and recrystallization, contributing to the sound composites. Good structural homogeneity and uniform dispersion of GNPs were successfully obtained by designing the thermo-mechanical parameters. With only 1.5 wt.% addition of GNPs into the pure aluminum matrix, an enhancement of strength by 293.3% with almost no loss in ductility was obtained. The product of strength and elongation was proposed as an index to evaluate the comprehensive mechanical performances of the composites, and our work reaches 9.31 GPa%, which is the highest result to our best knowledge. Uniform dispersion of graphene nanoplatelets and ultra-fine microstructures (267.0 nm) was achieved via severe plastic deformation and dynamic recrystallization. High-efficiency interfacial bonding was realized via graphene nanoplatelets-(amorphous Al_2O_3)-Al semi-direct interface without the formation of Al_4C_3. The automatic flow of Al_2O_3 nanodots to compensate for the spatial discontinuity caused by an interlayer slip of graphene was observed to achieve self-compensating spatial continuity. In conclusion, this novel fabrication route effectively develops innovative GNP-reinforced aluminum matrix composites with the exceptional amelioration of strength-ductility efficiency.

9.3 Grain Refinement Mechanisms of Graphene Nanoplatelet-Reinforced Aluminum Composites

9.3.1 Introduction

Aluminum alloys and aluminum matrix composites have attracted great attention for their light weight and excellent load-bearing capacity, which have been widely utilized in astronautic and aeronautic applications [71, 74, 75]. To further enhance their specific strength towards ultra-lightweight design, grain refinement engineering is one of the most effective method [76]. GNPs have been widely utilized for grain refinement due to the fact that these reinforcements could act as nucleation substrates during liquid-state processing or Zener pinning positions during solid-state processing [77]. Nevertheless, graphene is susceptible to agglomeration due to the great van der Waals force, leading to the unsatisfactory microstructures [78]. One of the major issues to ameliorate the refining effect is to conduct efficient redistribution and realize dispersion uniformity of these carbonaceous nanoparticles.

Severe plastic deformation techniques, including friction stir processing, high pressure torsion and equal channel angular pressing, etc., are ideal approaches towards the uniform distribution of reinforcements, as well as the great grain refinement [79, 80]. High-value plastic strain performed by external deformation contributes to dynamic recovery/recrystallization of the matrix and fragmentation/dispersion of the nanoparticles, thereby obtaining excellent mechanical performances of aluminum matrix composites [81]. DDM was utilized to introduce the ultra-fine microstructures into the aluminum matrix composites reinforced by few-layer GNPs. The solid-state preparation based on blended metallic powders with GNPs further enhances the grain refinement. The synergy mechanism for grain refinement was discussed via severe plastic deformation and evolution behavior of the reinforcements.

9.3.2 Experimental Procedures

Few-layer GNPs and pure aluminum powders with average diameter of respectively 1–3 μm and 10 μm were chosen as raw powders. The addition of GNPs in the composites was 1.5 wt.%. Ball milling in argon atmosphere was conducted to obtain the fully-blended powders with the milling time of 10 h and rotational velocity of 200 r/min. The blended powders were then placed in an aluminum mould and consolidated by DDM at a rotational velocity of 1000 r/min with severe plastic deformation and corresponding frictional/deformation heat synchronously, as shown in Fig. 9.32a.

Following the DDM process, tensile specimens with a gauge length of 5 mm and width of 1.4 mm was machined and tested were conducted at a strain rate of $1 \times 10^{-3}\ s^{-1}$ at room temperature. Three specimens were tested for each parameter. The

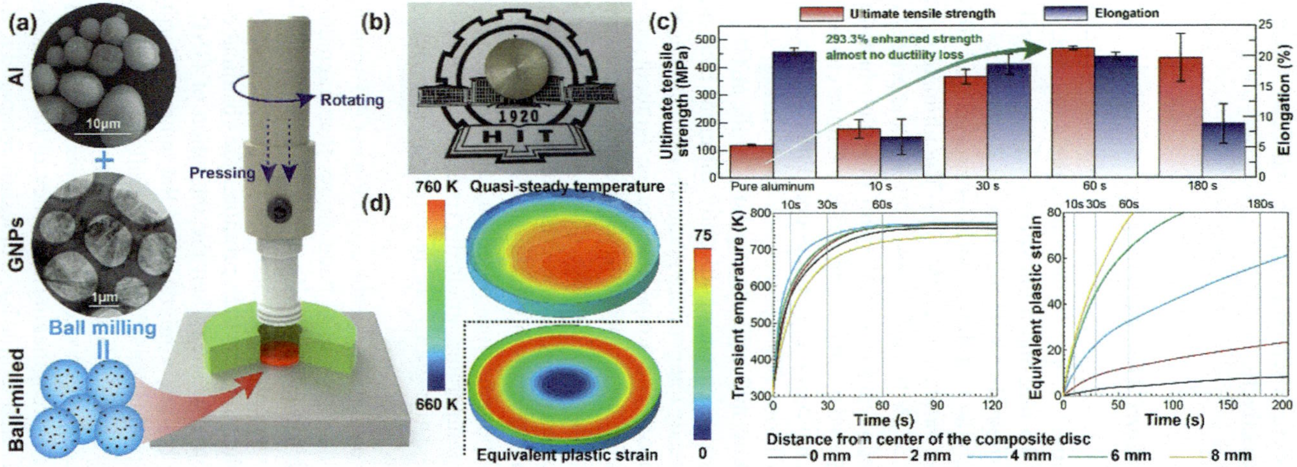

Fig. 9.32 Preparation process of GNP-reinforced aluminum matrix composites: **a** schematic of DDM procedure, **b** macrostructures of the typical composite disc, **c** tensile properties of the composites with different processing time and **d** evolution of the temperature and plastic deformation

samples were characterized by Raman spectroscopy with 532 nm laser excitation and XPS to obtain the chemical evolution. The microstructures including grain sizes and dispersion of GNPs were observed by EBSD and TEM with operating voltage of 20 kV and 200 kV, respectively.

9.3.3 Grain Refinement Mechanisms

Figure 9.32b shows the macroscopic formation of the composites. The composite discs have a diameter of 16 mm and a thickness of 1 mm. Sound composites are obtained and continuous streamline with the shape of Archimedes spiral is observed on the upper surface, implying that the composites experienced sliding friction and severe plastic deformation induced by DDM [73]. The ultimate tensile strength and elongation reached (468 $\pm$ 7) MPa and 19.9% $\pm$ 0.6% at the processing time of 60 s, showing a strengthening efficiency of 293.3% with almost no ductility loss compared to the pure aluminum prepared by DDM (Fig. 9.32c). A finite element analysis based on ABAQUS was conducted to quantify the coupled thermo-mechanical effect and microstructural evolution during DDM process, as shown in Fig. 9.32d. The contours of quasi-steady temperature and equivalent plastic strain at 60 s show that each micro zone of the disc experienced significantly different heat input and plastic deformation. The quasi-steady temperature gradually decreased as the distance from center of the disc increased. The temperature range is about 660–760 K, which are all higher than the minimum temperature required for dynamic recrystallization of aluminum ($0.5T_m \approx 467$ K) [82]. Besides, the whole composites undergo great accumulative plastic deformation. The synthesis combination implies that sufficient dynamic recrystallization occurs on the composites to refine the microstructures. In addition, the transient temperature does not reach a constant value at the beginning of the DDM process. The temperature gradually rises during the first 60 s, indicating that there is a stage where the grains are modified by cold deformation without dynamic recrystallization.

Raman spectroscopy can characterize the evolution of GNPs during DDM process (Fig. 9.33a). The ratio of D-band to G-band shows that fragmentation degree of graphene. Severe plastic deformation significantly enhances the smashing and dispersion of GNPs. This means that more exposed two-dimensional edges and active sites are produced on the graphene [83], which promotes the chemical bonding between GNPs and aluminum matrix to strengthen the inhibition of GNPs on dislocation rearrangement and grain boundary migration of the matrices during recrystallization and growth. The fragmentation of GNPs gradually increases with the increase of DDM time, indicating that the average diameter of graphene is continuously decreasing. However, excessively broken graphene weakens its load transfer effect, thereby impairing the high strengthening efficiency of GNPs themselves. In addition, the G-band of the DDMed sample has a blueshift compared to the raw powders. This shift of G-band was related to the average number of graphene layers due to the weakening of the bonding energy between graphene layers [84], which can be calculated by the following equation:

$$\omega_G = 1581.6 + \frac{11}{1 + n^{1.6}} \tag{9.18}$$

where ω_G is the wavenumber of the G-band and n is the number of graphene layers, which was calculated as 5.51, 2.63, 2.45, 2.17 and 1.65 for raw powders, 10 s, 30 s, 60 s and 180 s, respectively. The number of layers decreases with severe plastic deformation, indicating that the DDM process is capable of introducing interlayer slip in GNPs to further improve their dispersion. Similar results can be seen in the deconvoluted C 1 s peak of the XPS results in Fig. 9.33b. The gradual increase in the ratio of sp^3 to sp^2 indicates the fragmentation of GNPs. The emergence of C–O–Al bond and its gradual substitution by C–Al bond, showing that there is a good chemical bonding between GNPs and matrices [85], thus ensuring the inhibitory effect of graphene on the coarsening of the recrystallized grains. The average grain size of the composite at the preparation time of 60 s is 267.0 nm (Fig. 9.33c), which is much smaller than the grain size of conventional GNPs reinforced aluminum matrix composites, proving that GNPs are capable of realizing the nano-sized grain refinement via severe plastic deformation.

Figure 9.34 depicts the distribution of GNPs in the aluminum matrix composites. The GNPs are uniformly dispersed intergranularly and intragranularly. Those reinforcement located at the grain boundary exhibits obvious pinning effect on the migration of the grain boundary and the inhibitory effect on the rearrangement of dislocations [86]. The fragmentation of GNPs and dynamic recrystallization are proven to occur simultaneously during the DDM process. The severe plastic deformation promotes the rapid multiplication of the necessary geometric dislocations and realizes the nucleation of recrystallized grains synchronously. The GNPs suppresses the movement of dislocation, promoting the localized dislocation aggregation and accelerating the dynamic recrystallization [87]. In the subsequent growth process of the recrystallized grains, those GNPs adjacent to the grain boundaries realize the Zener pinning of the grain boundary due to the significant fragmentation and re-dispersion behaviors via severe plastic deformation. This pinning force is inversely proportional to the average size of GNPs [29]. The pinning force produced by the thinned GNPs rapidly balances with the curvature driving force and dislocation driving force, thereby effectively inhibiting the growth of recrystallized grains, and finally obtaining a nano-sized microstructure. We conducted a model to analysis the grain growth qualitatively. The stored dislocation energy during severe plastic deformation was used to trigger dynamic recrystallization and support grain growth. The stored energy for each dislocation can be calculated as:

$$U_{\text{single}} = \alpha G b \tag{9.19}$$

where α, G and b are a constant, shear modulus and Burgers vector of aluminum, respectively. The dynamic dislocation density induced by plastic deformation is written as [88]:

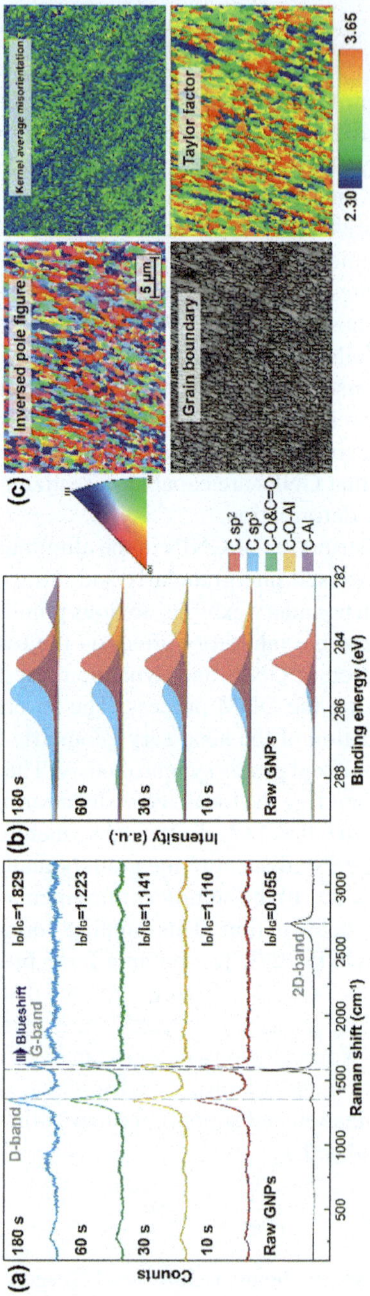

Fig. 9.33 Microstructural factors: **a** Raman spectra, **b** XPS results and **c** EBSD images of the DDMed composites

$$\frac{d\rho}{d\varepsilon_{\mathrm{p}}} = \frac{G\sqrt{\rho_0}}{100G_0Mb} \tag{9.20}$$

where ε_{p} is the plastic strain, ρ_0 is the initial density of dislocation, G_0 is the nondimensionalized constant and M is the average Taylor factor as shown in Fig. 9.33c. Since the strain are not constant during DDM process (Fig. 9.32d) and the GNPs are capable of accelerating the dislocation aggregation, Eq. (9.20) can be manipulated as:

$$\frac{d\rho}{dt} = \frac{\mu G f_v\sqrt{\rho_0}}{100rG_0Mb}\frac{d\varepsilon_p}{dt} = \frac{G\dot{\varepsilon}_p\sqrt{\rho_0}}{100G_0Mb} \tag{9.21}$$

where μ is a constant relevant to reinforcements, f_v is the volume fraction of GNPs, r is the equivalent radius of GNPs, t is the time and $\dot{\varepsilon}_p$ is the plastic strain rate. In fact, since the Consequently, the total stored energy is described via the combination of Eqs. (9.19) and (9.21):

$$U = \frac{\mu\alpha G^2 f_v\sqrt{\rho_0}}{100rG_0M}\int_t \dot{\varepsilon}_p \tag{9.22}$$

The critical flow stress determines the occurrence of the dynamic recrystallization [89], which is describe as:

$$\sigma_c = MGb\sqrt{\rho_c} \tag{9.23}$$

where ρ_c is the critical dislocation density. According to the combination of Eqs. (9.22) and (9.23), the energy consumed for dynamic recrystallization is written as:

$$U_{\mathrm{drx}} = \frac{\alpha\sigma_c^2}{M^2Gb} \tag{9.24}$$

Dynamic recrystallization will be triggered when U is larger than U_{drx}, and the diameter of the newly recrystallized grains can be calculated by $a + b\ln Z$. Z is the Zener-Hollomon parameter [16]. Moreover, considering the pinning energy of GNPs and the driving energy of dislocation and curvature, the growth equation of recrystallized grains is expressed as:

$$\frac{dD}{dt} = \begin{cases} \beta\left(\frac{dU}{dt} + \frac{2\gamma_b}{D} - \sum\frac{3\gamma_b f_v}{2r}\right)exp\left(-\frac{Q_G}{RT} - \dot{\varepsilon}\right), & \frac{dD}{dt} > 0 \\ 0, & \frac{dD}{dt} \le 0 \end{cases} \tag{9.25}$$

where β is the migration coefficient of grain boundaries, γ_b is the grain boundary energy, $\frac{2\gamma_b}{D}$ is the curvature driving energy, D is the diameter of grains, $\sum\frac{3\gamma_b f_v}{2r}$ is the pinning energy, and Q_G is the migration activation energy of grain boundaries.

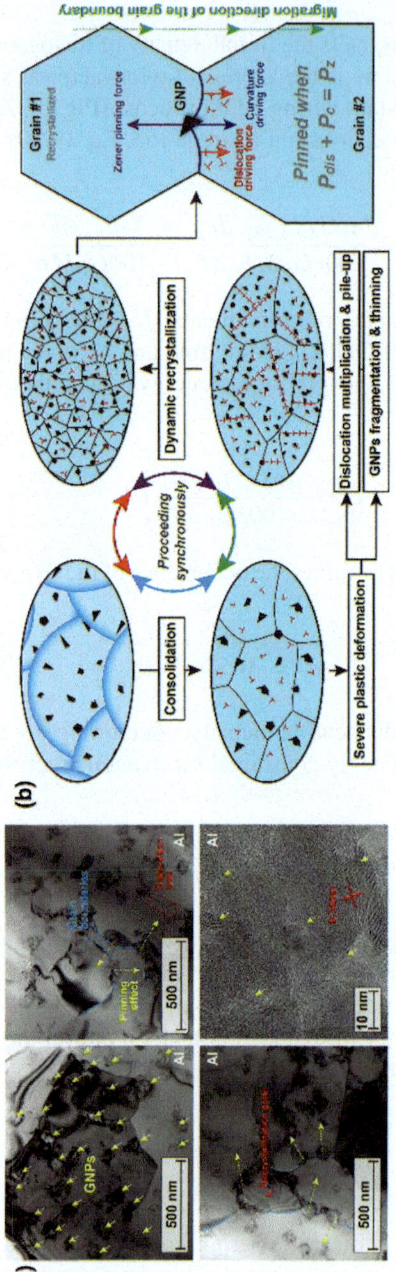

Fig. 9.34 GNPs and their pinning effect: **a** TEM images of the DDMed composites, and **b** synergy grain refinement mechanism during DDM process

Based on the above model, the evolution behaviors of the grain sizes with and without GNPs were obtained as shown in Fig. 9.35. The equivalent radius of GNPs is calculated by using the data from non-linear fitting of XPS and Raman results versus time. Nano-sized grain refinement is observed in the composites containing GNPs. The calculated grain diameter with the DDM time of 60 s is 289.9 nm, which is basically in agreement with the experimental results and is reduced by about 82% compared to that of the composites without GNPs. This proves the inhibitory effect of the fragmented and thinned graphene on the migration of recrystallized grain boundaries. Besides, the composites containing GNPs trigger dynamic recrystallization earlier, which is consistent with our previous discussion. The evolution curve includes 5 different stages: Incubation, recrystallization, growth, steady and 2nd growth stages. The first growth stage is attributed to the tiny newly-nucleated grains and insufficient pinning effect. The steady stage corresponds to the state where the pinning force and the dislocation/curvature driving force strike a balance. The 2nd growth stage is due to the limitation of fragmentation of the GNPs. It is difficult for GNPs to continue to be thinned and fragmented to provide greater pinning effect, which triggers the migration of the grain boundaries again. In addition, the excessive fragmentation of GNPs also damages the Orowan strengthening and load transfer effect. The better grain refinement needs to be realized by adjusting the temperature and strain input adequately towards higher strength and ductility. As such, enhanced grain refinement is achieved via the synergy grain refinement mechanism based on severe plastic deformation induced by DDM and the improved pinning effect of the dispersed GNPs towards nano-crystalline aluminum matrix composites.

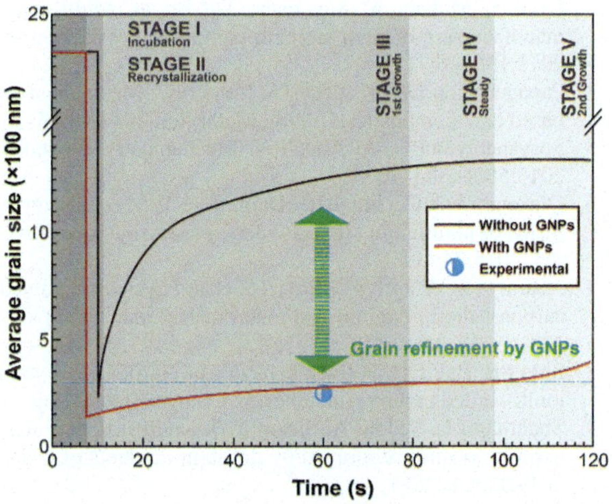

Fig. 9.35 Evolution behaviors of the grain sizes with and without GNPs

9.3.4 Summary

DDM was utilized as a feasible technique for realizing nano-sized grain refinement of few-layer GNPs reinforced aluminum matrix composites, in which severe plastic deformation and corresponding frictional/deformation heat contributed to solid-state sintering and microstructural modification. Fragmentation, thinning and re-dispersion of GNPs were obtained with the occurrence of dynamic recrystallization. These GNPs suppressed the movement of dislocation, promoting the localized dislocation aggregation and accelerating the dynamic recrystallization in the nucleation stage. They also introduced pinning force to balance with the curvature and dislocation driving force, inhibiting the growth of recrystallized grains effectively to obtain a nano-crystalline microstructure. The average grain size reached 267.0 nm, which was much smaller than the grain size of traditional aluminum matrix composites to realize high strengthening-toughening efficiency, indicating that DDM was capable of enhancing the grain refinement via severe plastic deformation.

References

1. Hekner B, Myalski J, Valle N, et al. Friction and wear behavior of Al–SiC(n) hybrid composites with carbon addition[J]. Compos Part B Eng. 2017;108:291–300.
2. Zhou W, Mikulova P, Fan Y, et al. Interfacial reaction induced efficient load transfer in few-layer graphene reinforced Al matrix composites for high-performance conductor[J]. Compos Part B Eng. 2019;167:93–9.
3. Dixit S, Mahata A, Mahapatra DR, et al. Multi-layer graphene reinforced aluminum—manufacturing of high strength composite by friction stir alloying[J]. Compos Part B Eng. 2018;136:63–71.
4. Papageorgiou DG, Kinloch IA, Young RJ. Mechanical properties of graphene and graphene-based nanocomposites[J]. Prog Mater Sci. 2017;90:75–127.
5. Suryanarayana C, Al-Aqeeli N. Mechanically alloyed nanocomposites[J]. Prog Mater Sci. 2013;58(4):383–402.
6. Olakanmi EO, Cochrane RF, Dalgarno KW. A review on selective laser sintering/melting (SLS/SLM) of aluminium alloy powders: processing, microstructure, and properties[J]. Prog Mater Sci. 2015;74:401–77.
7. Azarniya A, Azarniya A, Sovizi S, et al. Physicomechanical properties of spark plasma sintered carbon nanotube-reinforced metal matrix nanocomposites[J]. Prog Mater Sci. 2017;90:276–324.
8. Shin SE, Bae DH. Fatigue behavior of Al2024 alloy-matrix nanocomposites reinforced with multi-walled carbon nanotubes[J]. Compos Part B Eng. 2018;134:61–8.
9. Shahrdami L, Sedghi A, Shaeri MH. Microstructure and mechanical properties of Al matrix nanocomposites reinforced by different amounts of CNT and SiCW[J]. Compos Part B Eng. 2019;175: 107081.
10. Mistry JM, Gohil PP. Experimental investigations on wear and friction behaviour of Si_3N_4p reinforced heat-treated aluminium matrix composites produced using electromagnetic stir casting process[J]. Compos Part B Eng. 2019;161:190–204.
11. Uzun A. Production of aluminium foams reinforced with silicon carbide and carbon nanotubes prepared by powder metallurgy method[J]. Compos Part B Eng. 2019;172:206–17.
12. Chu K, Wang F, Li Y, et al. Interface structure and strengthening behavior of graphene/CuCr composites[J]. Carbon NY. 2018;133:127–39.

13. Shin SE, Ko YJ, Bae DH. Mechanical and thermal properties of nanocarbon-reinforced aluminum matrix composites at elevated temperatures[J]. Compos Part B Eng. 2016;106:66–73.
14. Şenel MC, Gürbüz M, Koç E. Fabrication and characterization of synergistic Al–SiC-GNPs hybrid composites[J]. Compos Part B Eng. 2018;154:1–9.
15. Kavimani V, Prakash KS, Thankachan T. Experimental investigations on wear and friction behaviour of SiC@r-GO reinforced Mg matrix composites produced through solvent-based powder metallurgy[J]. Compos Part B Eng. 2019;162:508–21.
16. Zienkiewicz OC, Cormeau IC. Visco-plasticity-plasticity and creep in elastic solids-A unified numerical solution approach[J]. Int J Numer Methods Eng. 1974;8(4):821–45.
17. Su H, Wu CS, Pittner A, et al. Thermal energy generation and distribution in friction stir welding of aluminum alloys[J]. Energy. 2014;77:720–31.
18. Nandan R, Roy GG, Lienert TJ, et al. Three-dimensional heat and material flow during friction stir welding of mild steel[J]. Acta Mater. 2007;55(3):883–95.
19. Sheppard T, Jackson A. Constitutive equations for use in prediction of flow stress during extrusion of aluminium alloys[J]. Mater Sci Technol. 1997;13(3):203–9.
20. Liu G, Zhang GJ, Jiang F, et al. Nanostructured high-strength molybdenum alloys with unprecedented tensile ductility[J]. Nat Mater. 2013;12(4):344–50.
21. Zhu T, Li J, Samanta A, et al. Interfacial plasticity governs strain rate sensitivity and ductility in nanostructured metals[J]. Proc Natl Acad Sci. 2007;104(9):3031–6.
22. Zhang Y, Pan C. Measurements of mechanical properties and number of layers of graphene from nano-indentation[J]. Diam Relat Mater. 2012;24:1–5.
23. Munir KS, Wen C. Deterioration of the strong sp2 carbon network in carbon nanotubes during the mechanical dispersion processing—a review[J]. Crit Rev Solid State Mater Sci. 2016;41(5):347–66.
24. Li Z, Fu X, Guo Q, et al. Graphene quality dominated interface deformation behavior of graphene-metal composite: the defective is better[J]. Int J Plast. 2018;111:253–65.
25. Dasari BL, Morshed M, Nouri JM, et al. Mechanical properties of graphene oxide reinforced aluminium matrix composites[J]. Compos Part B Eng. 2018;145:136–44.
26. Saadallah S, Cablé A, Hamamda S, et al. Structural and thermal characterization of multi-wall carbon nanotubes (MWCNTs)/aluminum (Al) nanocomposites[J]. Compos Part B Eng. 2018;151:232–6.
27. Lavernia EJ, Grant NJ. Aluminium-lithium alloys[J]. J Mater Sci. 1987;22(5):1521–9.
28. Ferreira F, Ferreira I, Camacho E, et al. Graphene oxide-reinforced aluminium-matrix nanostructured composites fabricated by accumulative roll bonding[J]. Compos Part B Eng. 2019;164:265–71.
29. Huang Y, Xie Y, Meng X, et al. Atypical grain coarsening of friction stir welded AA6082-T6: characterization and modeling[J]. Mater Sci Eng A. 2019;740–741:211–7.
30. El-Awady JA. Unravelling the physics of size-dependent dislocation-mediated plasticity[J]. Nat Commun. 2015;6:1–9.
31. Simar A, Bréchet Y, de Meester B, et al. Integrated modeling of friction stir welding of 6xxx series Al alloys: Process, microstructure and properties[J]. Prog Mater Sci. 2012;57(1):95–83.
32. Sevillano JG, van Houtte P, Aernoudt E. Large strain work hardening and textures[J]. Prog Mater Sci. 1980;25(2–4):69–34.
33. Agnoli A, Bozzolo N, Logé R, et al. Development of a level set methodology to simulate grain growth in the presence of real secondary phase particles and stored energy—application to a nickel-base superalloy[J]. Comput Mater Sci. 2014;89:233–41.
34. So KP, Kushima A, Park JG, et al. Intragranular dispersion of carbon nanotubes comprehensively improves aluminum alloys[J]. Adv Sci. 2018;5(7):1800115.
35. Lee C, Wei X, Li Q, et al. Elastic and frictional properties of graphene[J]. Phys Status Solidi-Basic Res. 2009;246(11–12):2562–7.
36. Zhang J, Yang S, Chen Z, et al. Graphene encapsulated SiC nanoparticles as tribology-favoured nanofillers in aluminium composite[J]. Compos Part B Eng. 2019;162:445–53.
37. Giannozzi P, Baroni S, Bonini N, et al. QUANTUM ESPRESSO: a modular and open-source software project for quantum simulations of materials[J]. J Phys Condens Matter. 2009;21(39): 395502.

38. Csonka GI, Perdew JP, Ruzsinszky A, et al. Assessing the performance of recent density functionals for bulk solids[J]. Phys Rev B—Condens Matter Mater Phys. 2009;79(15):1–14.
39. Perdew JP, Burke K, Ernzerhof M. Generalized gradient approximation made simple[J]. Phys Rev Lett. 1996;3:3865–8.
40. Plimpton S. Fast parallel algorithms for short-range molecular dynamics[J]. J Comput Phys. 1995;117(1):1–19.
41. Lee BJ, Baskes MI. Second nearest-neighbor modified embedded-atom-method potential[J]. Phys Rev B. 2000;62(13):8564–7.
42. Zhou W, Yamaguchi T, Kikuchi K, et al. Effectively enhanced load transfer by interfacial reactions in multi-walled carbon nanotube reinforced Al matrix composites[J]. Acta Mater. 2017;125:369–76.
43. Liu ZY, Wang LH, Zan YN, et al. Enhancing strengthening efficiency of graphene nanosheets in aluminum matrix composite by improving interface bonding[J]. Compos Part B Eng. 2020;199: 108268.
44. Zhang Y, Li X. Bioinspired, graphene/Al_2O_3 doubly reinforced aluminum composites with high strength and toughness[J]. Nano Lett. 2017;17(11):6907–15.
45. Turan ME. Investigation of mechanical properties of carbonaceous (MWCNT, GNPs and C60) reinforced hot-extruded aluminum matrix composites[J]. J Alloys Compd. 2019;788:352–60.
46. Peplow M. Graphene booms in factories but lacks a killer app[J]. Nature. 2015;522(7556):268–9.
47. Zan YN, Zhang Q, Zhou YT, et al. Introducing graphene (reduced graphene oxide) into Al matrix composites for enhanced high-temperature strength[J]. Compos Part B Eng. 2020;195:1–10.
48. Zhao H, Yue Y, Guo L, et al. Cloning nacre's 3D interlocking skeleton in engineering composites to achieve exceptional mechanical properties[J]. Adv Mater. 2016;28(25):5099–105.
49. Heim FM, Zhang Y, Li X. Uniting strength and toughness of Al matrix composites with coordinated Al_3Ni and Al_3Ti reinforcements[J]. Adv Eng Mater. 2018;20(1):1–11.
50. Ramanathan A, Krishnan PK, Muraliraja R. A review on the production of metal matrix composites through stir casting—furnace design, properties, challenges, and research opportunities[J]. J Manuf Process. 2019;42:213–45.
51. Ju JM, Wang G, Sim KH. Facile synthesis of graphene reinforced Al matrix composites with improved dispersion of graphene and enhanced mechanical properties[J]. J Alloys Compd. 2017;704:585–92.
52. Sun W, Zhan K, Yang Z, et al. Facile fabrication of GO/Al composites with improved dispersion of graphene and enhanced mechanical properties by Cu doping and powder metallurgy[J]. J Alloys Compd. 2020;815: 152465.
53. Izadi H, Gerlich AP. Distribution and stability of carbon nanotubes during multi-pass friction stir processing of carbon nanotube/aluminum composites[J]. Carbon NY. 2012;50(12):4744–9.
54. Khodabakhshi F, Arab SM, Švec P, et al. Fabrication of a new Al–Mg/graphene nanocomposite by multi-pass friction-stir processing: dispersion, microstructure, stability, and strengthening[J]. Mater Charact. 2017;132:92–107.
55. Feng S, Guo Q, Li Z, et al. Strengthening and toughening mechanisms in graphene-Al nanolaminated composite micro-pillars[J]. Acta Mater. 2017;125:98–108.
56. Aristizabal K, Katzensteiner A, Bachmaier A, et al. Study of the structural defects on carbon nanotubes in metal matrix composites processed by severe plastic deformation[J]. Carbon NY. 2017;125:156–61.
57. Zare H, Jahedi M, Toroghinejad MR, et al. Microstructure and mechanical properties of carbon nanotubes reinforced aluminum matrix composites synthesized via equal-channel angular pressing[J]. Mater Sci Eng A. 2016;670:205–16.
58. Khodabakhshi F, Nosko M, Gerlich AP. Effects of graphene nano-platelets (GNPs) on the microstructural characteristics and textural development of an Al–Mg alloy during friction-stir processing[J]. Surf Coatings Technol. 2018;335:288–305.
59. Mokdad F, Chen DL, Liu ZY, et al. Deformation and strengthening mechanisms of a carbon nanotube reinforced aluminum composite[J]. Carbon NY. 2016;104:64–77.

60. Hassan SF, Gupta M. Effect of particulate size of Al_2O_3 reinforcement on microstructure and mechanical behavior of solidification processed elemental Mg[J]. J Alloys Compd. 2006;419(1–2):84–90.
61. Bisht A, Srivastava M, Kumar RM, et al. Strengthening mechanism in graphene nanoplatelets reinforced aluminum composite fabricated through spark plasma sintering[J]. Mater Sci Eng A. 2017;695:20–8.
62. Mishra R, Ma Z, Charit I. Friction stir processing: a novel technique for fabrication of surface composite[J]. Mater Sci Eng A. 2003;341(1–2):307–10.
63. Feulvarch E, Roux JC, Bergheau JM. A simple and robust moving mesh technique for the finite element simulation of friction stir welding[J]. J Comput Appl Math. 2013;246:269–77.
64. Robe H, Claudin C, Bergheau J-M, et al. R-ALE simulation of heat transfer during friction stir welding of an AA2xxx/AA7xxx joint on a large process window[J]. Int J Mech Sci. 2019;155:31–40.
65. Zienkiewicz OC, Cormeau IC. Visco-plasticity—plasticity and creep in elastic solids—a unified numerical solution approach[J]. Int J Numer Methods Eng. 1974;8:821–45.
66. Huang Y, Xie Y, Meng X, et al. Numerical design of high depth-to-width ratio friction stir welding[J]. J Mater Process Technol. 2018;252:233–41.
67. Huang Y, Xie Y, Meng X, et al. Joint formation mechanism of high depth-to-width ratio friction stir welding[J]. J Mater Sci Technol. 2019;35(7):1261–9.
68. Yan S, Yang H, Li H, et al. A unified model for coupling constitutive behavior and micro-defects evolution of aluminum alloys under high-strain-rate deformation[J]. Int J Plast. 2016;85:203–29.
69. Lieou CKC, Bronkhorst CA. Thermomechanical conversion in metals: dislocation plasticity model evaluation of the Taylor-Quinney coefficient[J]. Acta Mater. 2021;202:170–80.
70. Cho HH, Hong ST, Roh JH, et al. Three-dimensional numerical and experimental investigation on friction stir welding processes of ferritic stainless steel[J]. Acta Mater. 2013;61(7):2649–61.
71. Meng X, Huang Y, Cao J, et al. Recent progress on control strategies for inherent issues in friction stir welding[J]. Prog Mater Sci. 2021;115: 100706.
72. Geng K, Li S, Yang YF, et al. 3D printing of Al matrix composites through in situ impregnation of carbon nanotubes on Al powder[J]. Carbon NY. 2020;162:465–74.
73. Xie Y, Meng X, Huang Y, et al. Deformation-driven metallurgy of graphene nanoplatelets reinforced aluminum composite for the balance between strength and ductility[J]. Compos Part B Eng. 2019;177: 107413.
74. Vogel T, Ma S, Liu Y, et al. Impact of alumina content and morphology on the mechanical properties of bulk nanolaminated Al_2O_3–Al composites[J]. Compos Commun. 2020;22: 100462.
75. Venkateswara Reddy K, Bheekya Naik R, Rao GR, et al. Microstructure and damping capacity of AA6061/graphite surface composites produced through friction stir processing[J]. Compos Commun. 2020;20: 100352.
76. Abraham SJ, Dinaharan I, Raja Selvam JD, et al. Microstructural characterization of vanadium particles reinforced AA6063 aluminum matrix composites via friction stir processing with improved tensile strength and appreciable ductility[J]. Compos Commun. 2019;12:54–8.
77. Han T, Li J, Zhao N, et al. Microstructure and properties of copper coated graphene nanoplates reinforced Al matrix composites developed by low temperature ball milling[J]. Carbon NY. 2020;159:311–23.
78. Shao P, Chen G, Ju B, et al. Effect of hot extrusion temperature on graphene nanoplatelets reinforced Al6061 composite fabricated by pressure infiltration method[J]. Carbon NY. 2020;162:455–64.
79. Dinaharan I, Kalaiselvan K, Murugan N. Influence of rice husk ash particles on microstructure and tensile behavior of AA6061 aluminum matrix composites produced using friction stir processing[J]. Compos Commun. 2017;3:42–6.
80. Li J, Meng X, Li Y, et al. Friction stir extrusion for fabricating Mg-RE alloys with high strength and ductility[J]. Mater Lett. 2021;289: 129414.

81. Zhang ZW, Liu ZY, Xiao BL, et al. High efficiency dispersal and strengthening of graphene reinforced aluminum alloy composites fabricated by powder metallurgy combined with friction stir processing[J]. Carbon NY. 2018;135:215–23.
82. Sakai T, Belyakov A, Kaibyshev R, et al. Dynamic and post-dynamic recrystallization under hot, cold and severe plastic deformation conditions[J]. Prog Mater Sci. 2014;60(1):130–207.
83. Yang M, Weng L, Zhu H, et al. Leaf-like carbon nanotube-graphene nanoribbon hybrid reinforcements for enhanced load transfer in copper matrix composites[J]. Scr Mater. 2017;138:17–21.
84. Wang H, Wang Y, Cao X, et al. Vibrational properties of graphene and graphene layers[J]. J Raman Spectrosc. 2009;40(12):1791–6.
85. Wu G, Yu Z, Jiang L, et al. A novel method for preparing graphene nanosheets/Al composites by accumulative extrusion-bonding process[J]. Carbon NY. 2019;152:932–45.
86. Komarasamy M, Wang T, Liu K, et al. Hierarchical multi-phase microstructural architecture for exceptional strength-ductility combination in a complex concentrated alloy via high-temperature severe plastic deformation[J]. Scr Mater. 2019;162:38–43.
87. Liu ZY, Xiao BL, Wang WG, et al. Singly dispersed carbon nanotube/aluminum composites fabricated by powder metallurgy combined with friction stir processing[J]. Carbon NY. 2012;50(5):1843–52.
88. Kocks UF, Mecking H. Physics and phenomenology of strain hardening: the FCC case[J]. Prog Mater Sci. 2003;48(3):171–3.
89. Yu P, Wu C, Shi L. Analysis and characterization of dynamic recrystallization and grain structure evolution in friction stir welding of aluminum plates[J]. Acta Mater. 2021;207: 116692.

Chapter 10
Anti-corrosion Aluminum Matrix Composites

10.1 Homogeneously Dispersed Graphene Nanoplatelets as Corrosion Inhibitors

10.1.1 Introduction

As a typical carbonaceous nanomaterial, graphene nanoplatelets (GNPs), sparked with their extraordinary strength and elastic modulus, have gained extensive attention for their application in next-generation aluminum matrix composites [1–3]. The load transfer effect induced by the two-dimensional morphology and large specific area of GNPs introduces a better strengthening efficiency than traditional nanoscale ceramic reinforcements [4–6]. However, the wider use of GNP-reinforced metal matrix composites is still restricted by their potential corrosion risk induced by the high electrochemical activity difference between aluminum and carbon [7, 8]. These composites tend to exhibit dissolution rates that are too high in complex environments to satisfy application requirements including durability and reliability. Latief et al. [9] stated that the presence of graphite nanoplatelets and an increase in their content raised the corrosion rate and reduced the polarization resistance of the aluminum matrix. Attempts to overcome the intrinsic shortcomings of improved corrosion resistance are consistently desired to expand the applications of GNP-reinforced aluminum matrix composites in high-value engineering structures, such as lightweight aerospace components and next-generation battery anodes [10–14].

The corrosion behavior of carbonaceous nanomaterial-reinforced aluminum matrix composites mainly depends on their microstructural features [15]. Under different fabrication and preparation techniques, they exhibit significant differences in microstructure and properties [16–19]. Severe plastic deformation processing techniques, such as high-pressure torsion [20], equal channel angular pressing [21], accumulative roll bonding [22], and friction stir welding/processing [23], are

Y. Huang et al., *Friction Stir Welding and Processing*, Materials Forming, Machining and Tribology, https://doi.org/10.1007/978-981-99-8688-0_10

usually used to enhance the strengthening efficiency of aluminum matrix composites. However, the corrosion behavior of composites processed by severe plastic deformation techniques remains unclear. The uniform microstructure formed by friction stir processing was found to comprise a homogeneous distribution of second phase particles embedded in the aluminum matrix with an ultrafine-grained structure, contributing to the improvement of corrosion resistance [24]. The homogenization of the friction stir processed microstructure with refined primary particles was also reported to change the semiconducting nature of the passive film formed on the aluminum alloy, resulting in enhanced corrosion resistance [25]. As such, the homogeneous microstructures and corresponding inhomogeneity-alleviated phase interfaces between reinforcements and aluminum matrices prepared by a severe plastic deformation strategy may contribute to the antagonism of corrosion resistance weakening due to the addition of GNPs into the composites.

Deformation-driven metallurgy (DDM), based on the basic principle of severe plastic deformation, was proposed as a high-efficiency solid-state preparation technique for bulk metallic composites via the one-step consolidation of the raw composite powders [26, 27]. This may provide homogenization and modification of the microstructure of GNP-reinforced aluminum matrix composites for better anti-corrosion performance. Our work aims to examine the effectiveness of DDM techniques on aluminum matrix composites in terms of corrosion behavior in chloride-containing environments. The microstructure factors, including interfacial transition, reinforcement redispersion, and grain refinement were evaluated. Electrochemical and intergranular corrosion tests were carried out. The effect of homogeneously dispersed GNPs as long-term corrosion inhibitors on the anti-corrosion behavior was discussed in detail.

10.1.2 Experimental Procedures

The raw pure aluminum powders (99.6% purity) were supplied as shown in Fig. 10.1a. The raw GNPs ((99% purity, 1–5 nm in thickness, 1–3 μm in diameter, Nanjing XFNANO Materials Tech Co., Ltd, China) were used to fabricate the aluminum matrix composites. The morphologies of the 2-propanol-dispersed GNPs are shown in Fig. 10.1b and c. The GNPs were large and thin flakes with average diameters on the order of micrometers. A higher content of GNPs would bring about an excessive lubrication effect to weaken the material flow, resulting in undesirable agglomeration of these reinforcements. 1.5 wt.% GNPs and a balanced weight fraction of spherical aluminum powders were selected and placed in a steel jar and processed by ball milling in a planetary ball milling machine at a rotational velocity of 200 r/min. A milling time of 10 h and a ball-to-powder ratio of 10:1 were chosen. The diameter of the ZrO_2 milling balls was 6 mm.

The milled powders were then cold pressed at approximately 200 MPa to obtain a pre-densified green compact with a thickness of 1.5 mm and a diameter of 16 mm. A processing tool rotating at a certain rotational velocity pressed the materials for 60 s.

Fig. 10.1 Morphologies of **a** spherical pure aluminum powders and GNPs in **b** SEM and **c** TEM

The densified aluminum matrix composites were obtained by frictional heat input and severe plastic deformation. The final diameter and thickness of the composite disc were 16 mm and 1.0 mm, respectively. Several bulk composites were prepared by adjusting the rotational velocity from 600 r/min to 1400 r/min with a step interval of 200 r/min. The detailed DDM procedure is illustrated in Fig. 10.2.

The GNP distribution on the milled aluminum powders was examined using field emission scanning electron microscopy (FE-SEM, HITACHI SU5000). The microstructures of the grain sizes were characterized by electron backscatter diffraction (EBSD) via a ZEISS SUPRA 55 FE-SEM instrument. The dispersion, microstructural evolution, and interfacial transition of the GNPs were observed by transmission electron microscopy (TEM, JEOL JEM-2100). Raman spectroscopic measurements were conducted using a Renishaw in Via-Reflex with an argon-ion 532 nm laser source to evaluate the disordering and defect density of graphene structures X-ray photoelectron spectroscopy (XPS, ThermoFisher ESCALAB 250Xi) was utilized to observe the chemical bonding evolution during the DDM process.

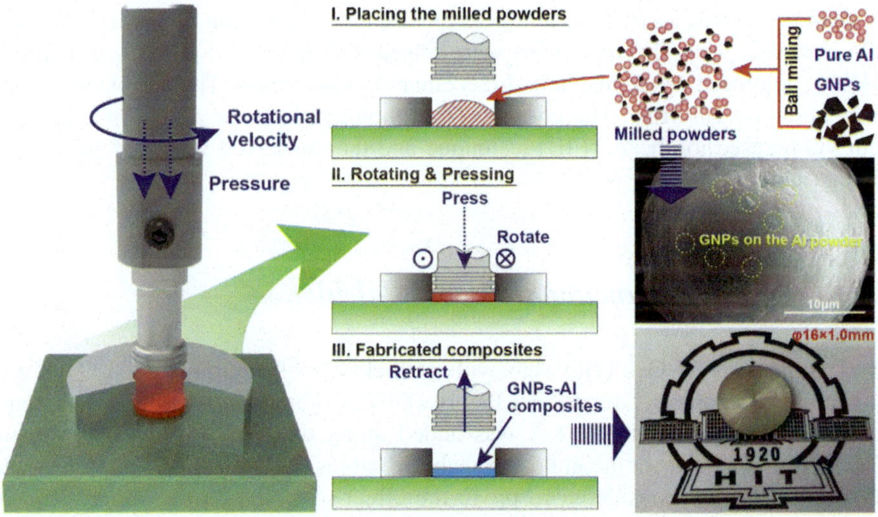

Fig. 10.2 Schematic of the DDM preparation route and the typical sample

Quadrant specimens were cut from the DDM samples and commercial pure aluminum sheets with a radius of 8 mm. Electrochemical tests were performed with a CHI 760E system for potentiodynamic polarization (PDP) and a PARSTAT 4000A system for electrochemical impedance spectroscopy (EIS) and Mott-Schottky analyses. The electrolyte solution was a 3.5 wt.% NaCl aqueous solution at room temperature. Electromagnetic stirring was used to maintain the homogeneity of the solution. A three-electrode system was used in the test. The DDM specimens were used as working electrodes, with Pt foil as the counter electrode and a saturated calomel electrode (SCE) as the reference electrode. The exposed area of the quadrant specimen was 0.50 cm^2. A one-hour stabilization time was applied before PDP tests. The data of each specimen was recorded by continuously scanning at a scanning rate of 50 mV/min for better stability and reliability. The PDP tests were repeated three times. EIS measurements were taken at the open circuit at frequencies ranging from 1 MHz to 0.001 Hz, with an applied sinusoidal perturbation of 5 mV in amplitude. The acquired data were analyzed by OriginLab and ZSimpWin software for Tafel extrapolation and curve fitting. Mott-Schottky analyses were performed by monitoring the frequency response at 1 kHz and 10 kHz between -1.5 V to 0.0 V versus SCE with a scanning interval of 25 mV.

Intergranular corrosion tests were conducted to evaluate the tendency of preferential corrosion behavior. The intergranular corrosion solution was prepared by weighing 57 g of NaCl dissolved in deionized water, adding 10 mL of H_2O_2, and diluting to 1 L according to GB/T 7998–2005 standards. After grinding and mechanical polishing, the specimens were soaked in 10 wt.% NaOH aqueous solution for 10 min, washed with deionized water, and then immersed in 30% HNO_3 aqueous solution to rinse the corrosion products. The pre-treated specimens were hung with an insulative wire and placed into the newly configured solution. The intergranular corrosion test was carried out at (35 $\pm$ 2) °C and the corrosion time was 6 h. After the corrosion tests, the specimens were rinsed with deionized water, concentrated HNO_3, deionized water, and anhydrous ethanol, respectively. The corrosion surface was observed via SEM. The corrosion morphology along the depth direction was cut via focused ion beam (FIB) technique and then observed via TEM (FEI Talos F200x).

10.1.3 Density Functional Theory Calculation

The simplest α-Al_2O_3(0001) slab with a thickness of approximately 11 Å was constructed for density functional theory (DFT) calculations. A sufficiently large vacuum of approximately 20 Å was added along the [0001] direction to avoid interaction between the neighboring slabs. C atoms, Cl atoms, Al vacancies, and O vacancies were considered to prove the effect of GNPs on the suppression of corrosion in NaCl solution. Norm-conserving pseudopotentials were implemented via QUANTUM ESPRESSO [28] for the DFT calculation. The system of Kohn–Sham equations was self-consistently solved on the plane wave basis to determine the electron density of the system. We performed integration in reciprocal space on a $5 \times 5 \times 1$ Monkhrost-Pack lattice of k-points [29]. The threshold for the maximum kinetic energy of plane waves included in the basis set was 830 eV. Calculations were performed with the PBE exchange–correlation function in the generalized gradient approximation (GGA). The simplest α-Al_2O_3(0001) slab, single-carbon-atom-doped α-Al_2O_3(0001)-1C slab, and dual-carbon-atom-doped α-Al_2O_3(0001)-2C slab were first relaxed until the energy, residual force and displacement were less than 5.0×10^{-6} eV/atom, 0.01 eV/Å, and 5.0×10^{-4} Å, respectively. Then the bottom two layers of atoms were fixed in the relaxed slabs. Vacancy and ingress of Cl atoms were added manually to calculate the featured energy fluctuation. The vacancy formation energy can be expressed as:

$$E_{vac} = \left[E_{vac@s} + \frac{1}{2}E_{O_2} or E_{Al}\right] - E_s \tag{10.1}$$

where $E_{vac@s}$ and E_{vac} are the total energy of the slab with/without manual vacancy, respectively. E_{O_2} and E_{Al} are the energy of O_2 molecule and Al atom, respectively.

The insertion energy of the ingress Cl atom into the corresponding vacancy can be calculated by:

$$E_{in} = E_{Cl@s} - \left[E_{vac@s} + \frac{1}{2}E_{Cl_2}\right] \tag{10.2}$$

where $E_{Cl@s}$ and $\frac{1}{2}E_{Cl_2}$ are the total energy of the slab with manual Cl atoms and the energy of the Cl_2 molecule, respectively. The work function is defined as the difference between the energy in a vacuum and the Fermi energy level of the upper surface.

10.1.4 Microstructural Characteristics

Severe plastic deformation was applied to the disc with a formation diameter of 16 mm and thickness of 1 mm, as shown in Fig. 10.2. Raman spectroscopy was implemented to reveal and evaluate the thermomechanical effect, as shown in Fig. 10.3a. The D-band (about 1350 cm^{-1}), G-band (about 1580 cm^{-1}) and Al_4C_3 region (about 860 cm^{-1}) correspond to the A_{1g} symmetrical breathing vibration, sp^2 pair stretching, and formation of the brittle intermetallic compound Al_4C_3, respectively. The damage degree of GNPs can be characterized by the intensity ratio of the D-band to the G-band (I_D/I_G) [30], which has a positive correlation with the brokenness/enfoldment and a negative correlation with the thinning of GNPs. The value of raw GNPs is 0.055 and increases by more than 20 times after DDM, implying that the GNPs have been significantly broken and dispersed into the aluminum matrices. This phenomenon produces many two-dimensional edges on GNPs to provide active reactional sites for diffusion and reaction between reinforcements and matrices in the DDM process.

Additionally, this value increases with the rotational velocity, indicating that the more severe plastic strain contributes to the further brokenness of the GNPs to bring about a more uniform distribution of the reinforcements, that is, the homogenization of the microstructure. The characteristic peak of Al_4C_3 appears when the rotational velocity reaches 1400 r/min. This intermetallic compound is an easily-hydrolyzed substance, and is often produced at the interface of GNPs and the aluminum matrix. The existence of Al_4C_3 can seriously damage the corrosion resistance of the composites, implying that the strategy of increasing the corrosion resistance by excessively enhancing severe plastic deformation on the composites to fully homogenize the microstructure is not feasible. Figure 10.3b shows the deconvolution results of the XPS C 1 s peak of the raw GNPs and composites at different rotational velocities. The ratio of the sp^2 to sp^3 peak gradually increases with the rotational velocity, indicating that the degree of fragmentation and redistribution of GNPs increases. In addition, the characteristic peak corresponding to the Al–O–C bond increases first and then decreases, while the Al–C bond gradually increases. This implies that the mechanical/chemical bonding state between GNPs and the matrix evolves gradually. Aluminum powders that have not been subjected to severe plastic deformation maintain oxide films on the surface, and directly mechanically interlock with the GNPs to form Al–O–C bonds. In addition, the surficial oxide film on the other aluminum matrix is broken under the plastic strain effect of severe plastic deformation. The aluminum matrix without oxide is exposed and reacts with the oxygen-containing groups on the surface of the graphene, which also form many Al–O–C bonds. With a further increase in the rotational velocity, aluminum can directly react with carbon due to more deformation energy to form Al–C bonds. These two bonds form simultaneously to constitute efficient bonding between the aluminum matrices and GNPs. However, it is worth noting that the Al–C bond also implies the potential formation of the hydrolysable intermetallic compound Al_4C_3. Thus, the Al–C bond needs to be controlled to enhance the corrosion resistance via microstructural homogenization during the DDM process.

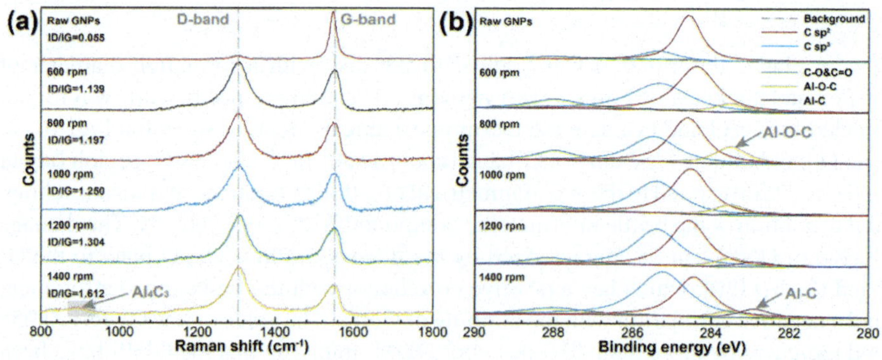

Fig. 10.3 Characterization of GNP evolution: **a** Raman spectra and **b** XPS results

EBSD and TEM observations were used to intuitively evaluate the microstructural characteristics of the composites. The average grain sizes at 600 r/min, 800 r/min, 1000 r/min, 1200 r/min, and 1400 r/min are 191.3 nm, 216.3 nm, 267.0 nm, 326.8 nm, and 501.6 nm, respectively. Higher plastic deformation contributes to the more uniform dispersion of GNPs but also leads to a gradual increase in the grain size due to higher heat input. Figure 10.4 shows the TEM images of the typical DDM composites. An important phenomenon is that the reinforcements, GNPs, are not only agglomerated intergranularly (Fig. 10.4a). There are plenty of GNPs distributed inside grains. For the conventional pressure infiltration [31] or hot sintering [32], GNPs with high van der Waals forces tend to gather at the grain boundaries, which significantly improves the energy heterogeneity adjacent to grain boundaries [33]. This phenomenon effectively avoids the adverse effect of second phases at the boundaries on corrosion behavior [34]. Additionally, partial GNPs dispersed at the grain boundaries exhibit a Zener-Smith pinning effect on the migration of the grain boundaries, as shown in Fig. 10.4b, implying that GNPs play an important role in dynamic recovery and recrystallization. These two-dimensional nanophases have a significant inhibitory effect on the grain boundary migration of the newly nucleated grains via recrystallization, thereby avoiding the growth and coarsening of these grains to improve the microstructural homogeneity.

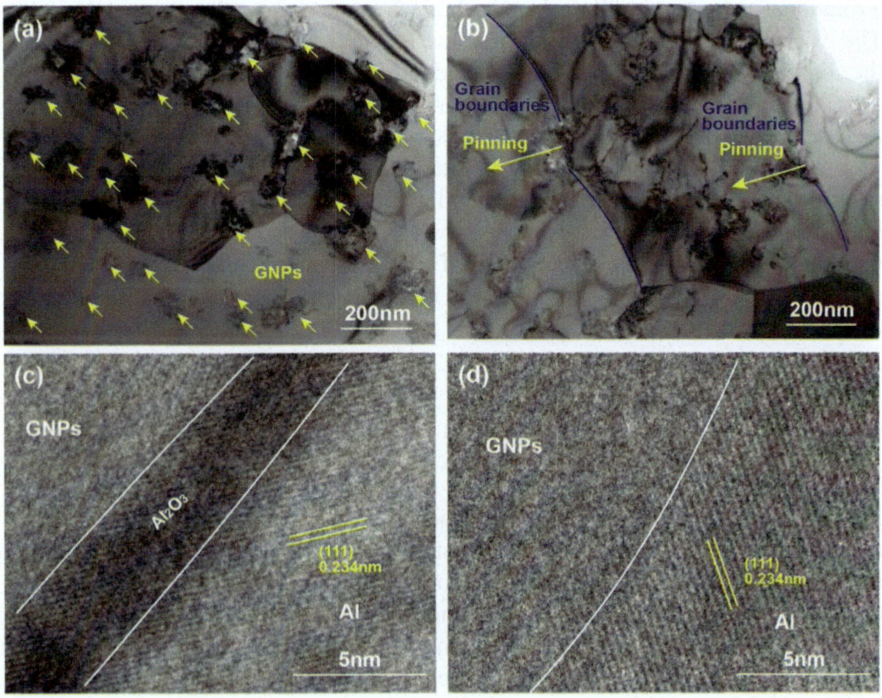

Fig. 10.4 TEM images of DDM composites: **a** GNP distribution, **b** grain boundaries and smith-zener pinning effect, **c** indirect phase interface, and **d** direct phase interface

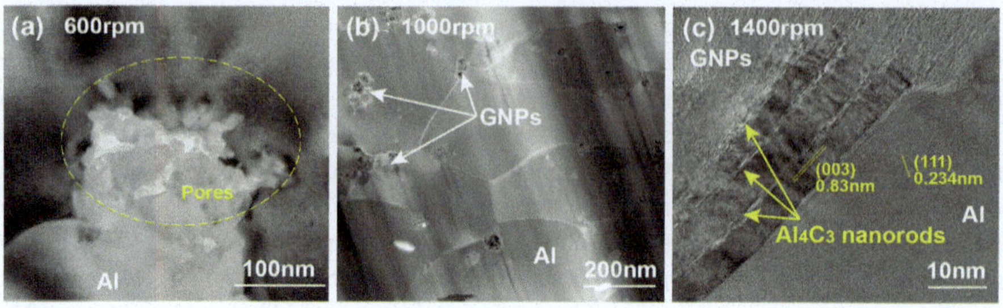

Fig. 10.5 TEM images of DDM composites: **a** porosity at 600 r/min, **b** ideal interface at 1000 r/min, and **c** undesirable Al_4C_3 nanorods at 1400 r/min

Figure 10.4c and d show two modes of the phase interface between GNPs and the aluminum matrix. The indirect GNPs-amorphous Al_2O_3-Al interface corresponds to the Al–O–C bonds [35], while the direct GNP-Al interface corresponds to Al–C bonds [36]. The phase interface changes from the indirect interface in dominant to the direct interface dominating. Obviously, in the case where the indirect interface dominates the interphase bonding, there are still many mechanical bonding zones between GNPs and the aluminum matrix [27]. This indicates that the degree of composite consolidation is still in its infancy. As shown in Fig. 10.5a, pores can be observed at a rotational velocity of 600 r/min, which is when the indirect interface is dominant. The existence of the porous microstructure becomes a hidden danger that damages the continuity of the surface oxide film and induces crevice corrosion [37]. Additionally, the direct interface becomes the dominant interface when the rotational velocity is higher. The Al_4C_3 rod-like phase is observed at the phase interface in Fig. 10.5c. The presence of such hydrolysable intermetallic compounds impairs the corrosion resistance of the composites. A multimodal phase interface such as Fig. 10.5b may be the microstructure with the best corrosion resistance. Therefore, it is necessary to evaluate the corrosion behavior of DDM composites with different processing parameters to reveal the comprehensive influence of microstructural evolution on corrosion performance.

10.1.5 Corrosion Behaviors

Figure 10.6 shows the PDP curves of all specimens after immersion in 3.5 wt.% NaCl aqueous solution for one hour to form a stabilized oxide film. The extracted data of all the curves are shown in Table 10.1. Compared with commercial pure aluminum, the corrosion potential of DDM pure aluminum is nobler, while its corrosion current/rate is also increased. According to the microstructural characterization, the severe plastic deformation process greatly refines the grain sizes, which helps to increase the homogeneity of the composites. Additionally, it introduces a higher density of dislocations and grain boundaries, contributing to an increase in the sensitivity of the intergranular interface in terms of the diffusion and penetration of oxygen and chloride ions. The dissolution of the aluminum matrix is thus accelerated, implying that grain refinement was likely to increase the overall reactivity on the surface and deteriorate the corrosion performance without the formation of a passive layer [38]. This shows that simply refining the grain microstructure is not applicable for enhancing the corrosion resistance. However, with the addition of GNPs, the corrosion resistance of the composites prepared by DDM is significantly improved compared to that of commercial pure aluminum and DDM pure aluminum. When the rotational velocity is 800–1200 r/min, the corrosion current density is significantly reduced. Scanning electrochemical microscopy testing was also conducted to evaluate the electrochemical homogeneity of composites at 1000 r/min by extracting the corrosion potential and current density from local Tafel measurements, as shown in Fig. 10.7. Twenty sampling points were selected from the center to the edge of the composite disc. The results showed that the distribution of anti-corrosion characteristics was basically uniform along the radius direction, indicating that the microstructures and reinforcement dispersion were homogeneous. Compared with the pure aluminum specimen prepared by DDM, the composites contain fully redispersed GNPs through severe plastic deformation. These nanophases improve the corrosion resistance from a phenomenological point of view. The interaction between the GNPs and the surface oxide film forms a dense protective layer to inhibit further corrosion. Similar to the surface adsorption strategy of carbon-based corrosion inhibitors, carbon dots added into the corrosion electrolyte can adsorb on the surface of the aluminum alloy and react with aluminum ions to form a reticular resistant film, thereby reducing the corrosion rate of the aluminum alloy [39]. This adsorption obeying the Langmuir adsorption isotherm could promote the formation of corrosion inhibition mechanism [40]. For the GNP-reinforced aluminum matrix composites prepared by DDM, a large number of GNPs gather on the surface of the composites due to the dissolution and consumption of the aluminum matrix as the corrosion process progresses. These GNPs react with the surface oxide film (and corrosion products) to form a continuous dense corrosion-resistant film. This film can effectively reduce the corrosion rate so that an obvious decrease in the corrosion current densities of the DDM composites in the Tafel curve is observed. It is worth noting that not all GNPs in the composites can act as corrosion inhibitors to reduce the corrosion rate. The addition of GNPs into aluminum reduced the corrosion resistance in hot press sintered composites [9]. These GNPs cannot be sufficiently broken and redispersed in the aluminum matrix, and tend to aggregate adjacent to the grain boundaries. This results in serious heterogeneity of the composites, thereby weakening the corrosion resistance. Through the DDM strategy, severe plastic deformation significantly promotes the uniform distribution of GNPs and avoids the tendency towards aggregation, thereby enhancing the corrosion resistance of the composite.

In addition, it is worth noting that the corrosion rates of the GNP-reinforced aluminum matrix composites at rotational velocities of 600 and 1400 r/min increase drastically. These pores formed in the 600 r/min specimen (Fig. 10.5a) lead to a connected penetration path for the corrosive electrolyte into the composites and thus induce detrimental crevice corrosion. A similar conclusion can also be drawn for

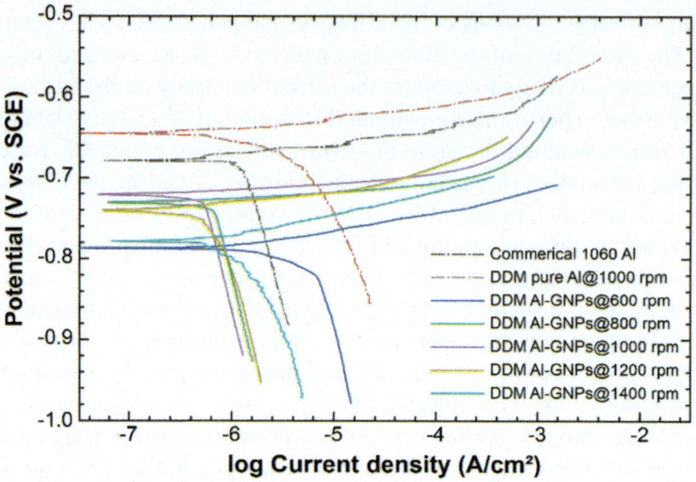

Fig. 10.6 PDP curves of DDM pure aluminum and composites with several preparation parameters, as well as commercial pure aluminum

Table 10.1 Electrochemical parameters via Tafel extrapolation

Sample	E_{corr} (V versus SCE)	I_{corr}/(μA·cm^{-2})	Corrosion rate/ (μm·year^{-1})
Commercial 1060 Al	−0.679 ± 0.013	0.895 ± 0.077	10.8 × 10^0
DDM pure Al@1000 r/min	−0.645 ± 0.009	2.813 ± 0.201	3.1 × 10^1
DDM Al-GNPs composites@600 r/min	−0.783 ± 0.029	12.259 ± 1.345	1.3 × 10^2
DDM Al-GNPs composites@800 r/min	−0.731 ± 0.010	0.702 ± 0.092	7.7 × 10^0
DDM Al-GNPs composites@1000 r/min	−0.721 ± 0.015	0.378 ± 0.051	4.1 × 10^0
DDM Al-GNPs composites@1200 r/min	−0.737 ± 0.008	0.429 ± 0.102	4.7 × 10^0
DDM Al-GNPs composites@1400 rpm	−0.777 ± 0.021	4.657 ± 0.651	5.1 × 10^1

the 1400 r/min specimen. Due to the excessively high heat input and strain rate, the hydrolysable intermetallic compound Al_4C_3 is produced at the interface between GNPs and aluminum matrices (Fig. 10.5c). Particles containing Al_4C_3 dissolve quickly in corrosive media and induce crevice corrosion into the phase interface. Since there is a significant intrinsic corrosion potential difference between GNPs and the aluminum matrix [41], local galvanic corrosion reactions are formed to impair the corrosion resistance. Appropriate parameters can both induce a serious reduction in the corrosion resistance of the composites.

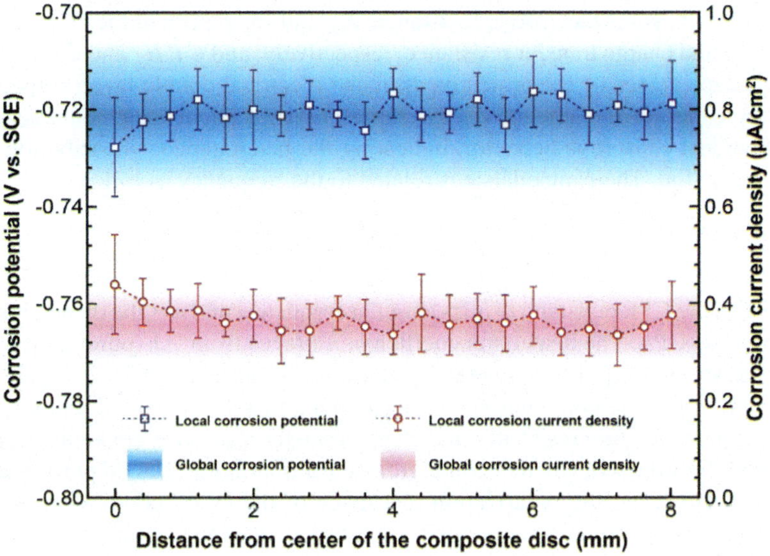

Fig. 10.7 Distribution of local corrosion characteristic by scanning electrochemical microscopy

To further analyze the corrosion inhibitor behavior of heterogeneously dispersed GNPs in the DDM composites, EIS tests were implemented to assess the impedance characteristics with different immersion times. Figures 10.8 and 10.9 show the Nyquist and Bode fitting results of all the specimens after 1 h, 48 h, and 120 h of exposure in 3.5 wt.% NaCl aqueous solution, respectively. These Nyquist plots show one or two capacitance loops, whose radii describe the protection and stability of the protective film (passivation layer). A larger radius indicates a weaker tendency of corrosion reaction kinetics. Due to the continuous corrosion and destruction of the protective film by chloride ions, the corrosion resistance of commercial pure aluminum declines continuously as the exposure time increases. In contrast, the initial Nyquist loop radius of the DDM pure aluminum is smaller than that of commercial pure aluminum, while the response characteristics of these two specimens tend to be similar after the immersion time reaches 120 h. This shows that although the high-density grain boundaries as the fast diffusion path in DDM pure aluminum has increased, the fine-grained microstructures are no longer responsible for the weakening of corrosion resistance when the chloride ions have severe damaged the surface oxide film. For the GNP-reinforced aluminum matrix composites prepared by DDM, the initial loop radius is also smaller. However, when the exposure time reaches 120 h, the actual impedance represented by the loop radius becomes significantly larger. This proves that the GNPs on the surface of the aluminum matrix composites can react with the surface oxide film to form a more protective passivation film. This passivation film shows better inhibition of chloride ion ingress, which improves the electrochemical stability of the composites in a chlorine-containing environment. A simplified equivalent circuit R(Q(R(QR))), as shown in Fig. 10.10, is applied to fit the

EIS results. R_S is the electrolyte resistance; R_f and R_{ct} represent the protective film resistance and charge transfer resistance, respectively; and CPE_f and CPE_{dl} represent the capacitance of the protective film and electrical double layer, respectively. These constant phase elements (CPEs) represent the capacitive elements associated with nonideal capacitive behavior, indicating the heterogeneous distribution of the surface reaction. Their impedance is related to the frequency as follows:

$$Z_{CPE} = \left[T(i\omega)^n\right]^{-1} \tag{10.3}$$

where T is the CPE constant, ω is the angular frequency and n is the dispersion coefficient of the CPE. The dispersion coefficient n ranges from -1 (pure inductor) to 1 (pure capacitor) [42]. The extracted results were listed in Table 10.2. Figure 10.11 shows the equivalent polarization resistance of all the results. Pure aluminum shows that both the oxide film resistance and charge transfer resistance gradually decreased over time, regardless of whether commercial pure aluminum or DDM pure aluminum was used. However, the polarization resistance of the DDM composites increases over time, which proves the corrosion inhibition of GNPs. This effect becomes more significant with the increasing exposure time, similar to a long-term corrosion

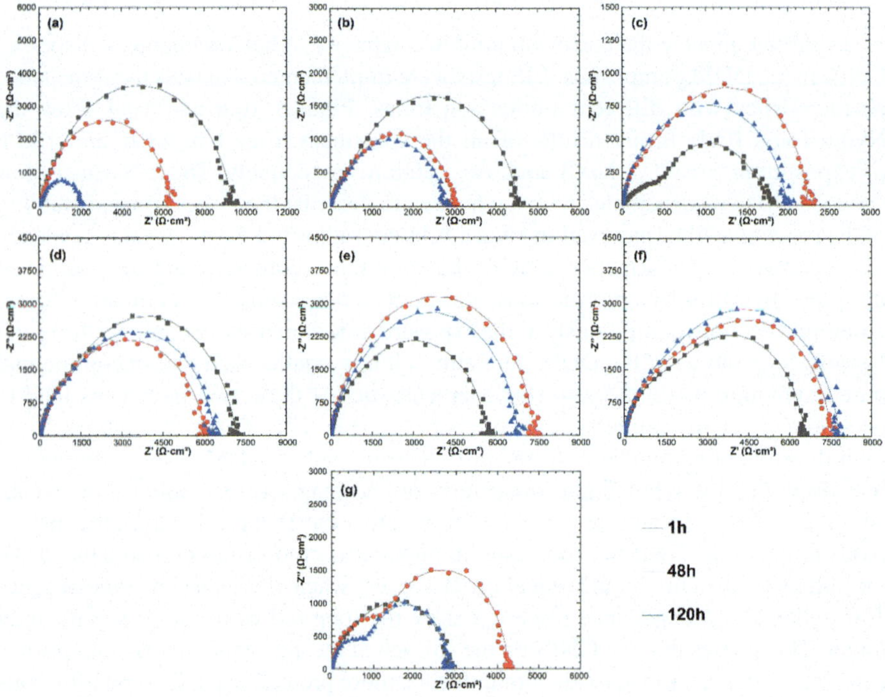

Fig. 10.8 Fitted Nyquist spectra: **a** commercial pure aluminum, **b** DDM pure aluminum, and **c–g** DDM composites at rotational velocities of 600 ~ 1400 r/min

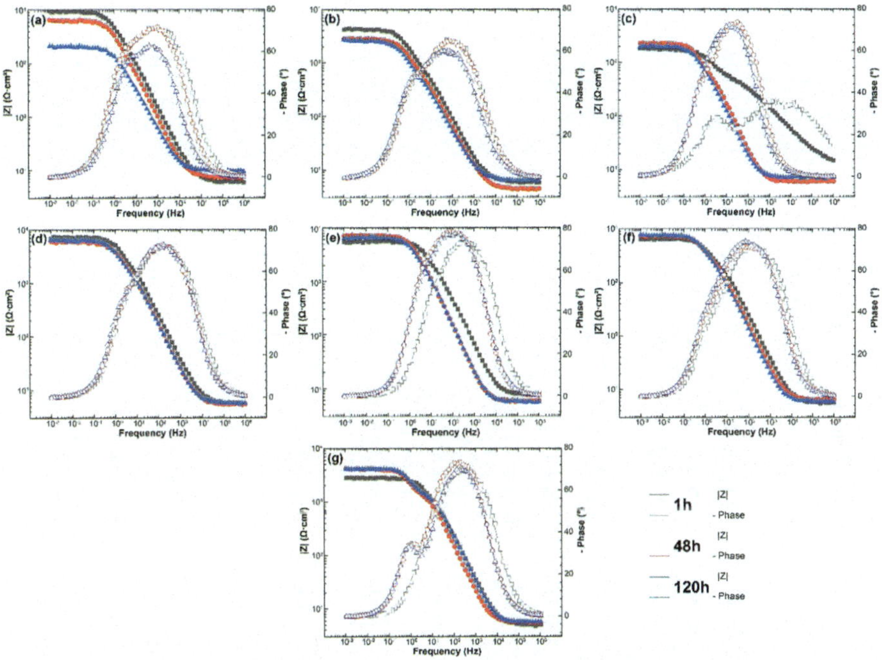

Fig. 10.9 Fitted Bode spectra: **a** commercial pure aluminum, **b** DDM pure aluminum, and **c**–**g** DDM composites at rotational velocities of 600–1400 r/min

Fig. 10.10 Equivalent circuit for modeling the EIS results

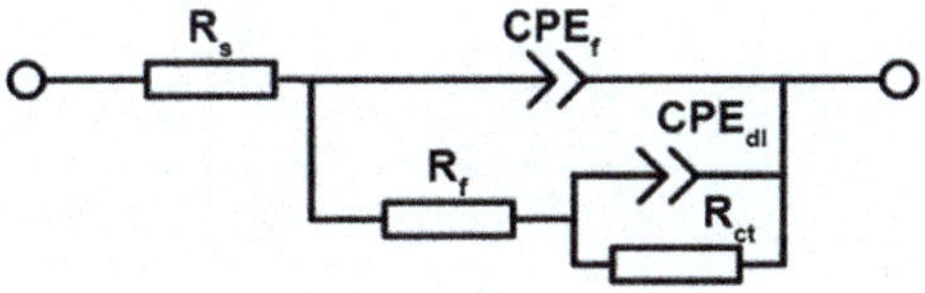

inhibitor. Additionally, the values T of CPE_f and CPE_{dl} in the DDM composites decrease compared to those in commercial pure aluminum, proving that the homogeneity of the composites is enhanced significantly via the DDM process. Moreover, the polarization resistance of the composites decreased sharply due to the presence of crevice corrosion with inappropriate processing parameters (600 r/min and 1400 r/min), proving that densification of the composites is the prerequisite for higher corrosion resistance brought by GNP corrosion inhibitors. In other words, if the parameters are controlled to avoid the occurrence of these two undesirable phenomena, the long-term corrosion inhibition effect of GNPs will significantly suppress the corrosion rate of the aluminum matrix composites produced by the DDM technique. This indicates that a large parameter window exists for our technique towards long-term anti-corrosion performance, which ensures the reproducibility of the control of the process and excludes the randomness of the selection of micro-characterization.

Table 10.2 Fitting results of the EIS spectra corresponding to the equivalent circuit

Sample	R_s/(Ω cm^2)	R_f/(kΩ cm^2)	CPE_f		R_{ct}/(kΩ cm^2)	CPE_{dl}	
Immersion time			T_f/(μΩ^{-1} cm^{-2}s^{-n})	n_f		T_{dl}/(μΩ^{-1} cm^{-2} s^{-n})	n_{dl}
Commercial 1060 Al							
1 h	6.00 ± 0.31	4.97 ± 0.29	210.8 ± 2.2	0.83	4.44 ± 0.36	18.2 ± 1.8	0.98
48 h	7.26 ± 0.44	2.45 ± 0.11	310.9 ± 2.6	0.86	3.96 ± 0.21	20.4 ± 1.6	0.96
120 h	5.78 ± 0.25	0.98 ± 0.09	810.7 ± 6.5	0.81	1.08 ± 0.10	55.2 ± 4.2	0.96
DDM pure Al@1000 r/min							
1 h	6.22 ± 0.27	1.58 ± 0.12	58.6 ± 3.8	0.80	2.97 ± 0.22	66.4 ± 3.3	0.92
48 h	4.61 ± 0.22	1.26 ± 0.16	87.5 ± 4.8	0.75	1.76 ± 0.18	93.7 ± 5.6	0.92
120 h	6.67 ± 0.29	1.28 ± 0.16	128.6 ± 7.2	0.77	1.53 ± 0.13	123.7 ± 3.8	0.99
DDM Al-GNPs composites@600 r/min							
1 h	6.82 ± 0.55	0.81 ± 0.21	97.6 ± 10.4	0.48	1.04 ± 0.35	83.3 ± 3.5	0.97
48 h	5.99 ± 0.60	1.00 ± 0.12	125.2 ± 6.5	0.91	1.24 ± 0.08	290.5 ± 12.1	1.00
120 h	7.18 ± 0.37	1.02 ± 0.10	184.7 ± 8.7	0.91	0.97 ± 0.12	334.7 ± 8.8	0.81
DDM Al-GNPs composites@800 r/min							
1 h	5.63 ± 0.26	3.77 ± 0.26	18.2 ± 1.2	0.84	3.42 ± 0.19	23.4 ± 1.9	0.99
48 h	5.52 ± 0.34	2.61 ± 0.21	23.1 ± 1.6	0.86	3.47 ± 0.23	38.9 ± 1.0	0.80
120 h	5.70 ± 0.41	2.87 ± 0.13	23.9 ± 1.3	0.85	3.52 ± 0.19	40.8 ± 2.2	0.92
DDM Al-GNPs composites@1000 r/min							
1 h	7.78 ± 0.28	3.27 ± 0.08	7.0 ± 0.9	0.88	2.37 ± 0.20	12.5 ± 0.7	0.92
48 h	5.95 ± 0.17	4.02 ± 0.25	17.1 ± 1.2	0.92	3.32 ± 0.12	14.9 ± 1.1	0.98
120 h	5.75 ± 0.38	3.65 ± 0.24	18.0 ± 1.0	0.91	3.12 ± 0.16	24.2 ± 0.9	1.00

(continued)

Table 10.2 (continued)

Sample	R_s/(Ω cm^2)	R_f/(kΩ cm^2)	CPE_f		R_{ct}/(kΩ cm^2)	CPE_{dl}	
Immersion time			T_f/(μΩ^{-1} cm^{-2}s^{-n})	n_f		T_{dl}/(μΩ^{-1} cm^{-2} s^{-n})	n_{dl}
DDM Al-GNPs composites@1200 r/min							
1 h	5.72 ± 0.39	3.55 ± 0.33	17.7 ± 0.8	0.84	3.01 ± 0.20	50.6 ± 2.2	0.98
48 h	6.57 ± 0.45	4.31 ± 0.13	24.9 ± 1.3	0.85	3.22 ± 0.21	80.1 ± 3.4	0.97
120 h	6.04 ± 0.26	4.33 ± 0.22	27.1 ± 1.1	0.87	3.48 ± 0.17	63.5 ± 3.9	0.94
DDM Al-GNPs composites@1400 r/min							
1 h	5.07 ± 0.66	1.71 ± 0.52	18.2 ± 2.1	0.84	1.13 ± 0.67	310.0 ± 6.2	0.92
48 h	5.86 ± 0.41	1.78 ± 0.17	21.7 ± 1.0	0.89	2.45 ± 0.12	1110.6 ± 2.3	1.00
120 h	5.90 ± 0.31	1.07 ± 0.09	26.1 ± 1.5	0.87	1.74 ± 0.11	197.7 ± 3.1	0.99

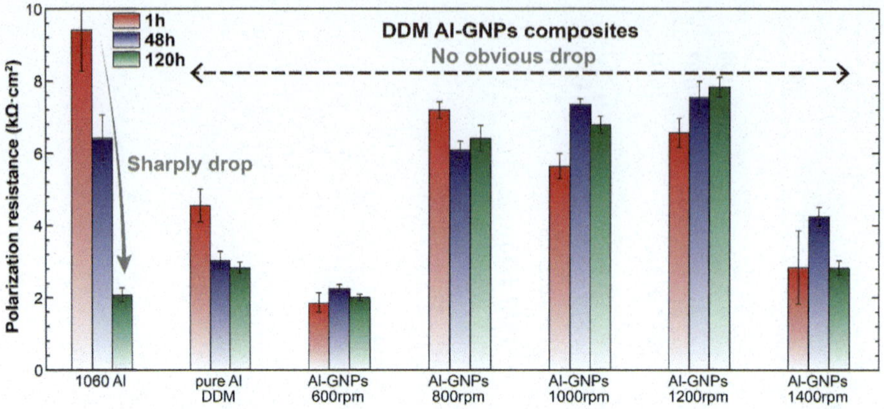

Fig. 10.11 Comparison of polarization resistance with different time in different specimens

The corrosion performance of aluminum mainly depends on the characteristics of its surface protective passivation film [43, 44]. Although films containing aluminum oxide and aluminum hydroxide are generally considered insulators due to the high energy gap between their valence and conduction bands [45], the thickness of the surface oxide films is small and the lattice defects are greatly different from those of ideal aluminum oxide, so they can also be regarded as semiconductors. Therefore, Mott-Schottky analyses at 1 kHz and 10 kHz were implemented in 3.5 wt.% NaCl aqueous solution to describe the semiconductor properties of the protective films. Figure 10.12 shows the Mott-Schottky characteristic curves and corresponding extracted data of all the specimens. All the specimens at both frequencies exhibit N-type semiconductor characteristics; that is, pure aluminum or DDM composites, interstitial defects, and oxygen vacancies show more dominance than cation vacancies [46, 47]. Based on the Mott-Schottky theory, the space charge capacitance of an N-type semiconductor can be expressed as follows:

$$\frac{1}{C_{sc}^2} = \frac{2}{\varepsilon_0 \varepsilon_r e N_D}\left(E - \varphi_{FB} - \frac{kT}{e}\right) \tag{10.4}$$

where C_{sc} is the space charge capacitance, ε_0 is the vacuum permittivity (8.854×10^{-14} F/cm), ε_r is the dielectric constant for the oxide film (10 for Al_2O_3 [24]), e is the electron charge (1.602×10^{-19} C), N_D is the donor density (cm^{-3}), E is the scanning potential, φ_{FB} is the flat band potential, k is Boltzmann's constant and T is the absolute temperature. Therein, the donor density N_D can be calculated from the slope $\left(\frac{2}{\varepsilon_0 \varepsilon_r e N_D}\right)$ of the linear correlation between $\frac{1}{C_{sc}^2}$ and E. The oxygen vacancies and interstitial defects in the protective film (determined by its density and diffusivity) are the major factors that control the growth and breakdown kinetics. The adsorption and penetration of corrosive electrolytes (such as chloride ions) in protective films is usually the incubation stage of corrosion and is obtained by oxygen vacancies

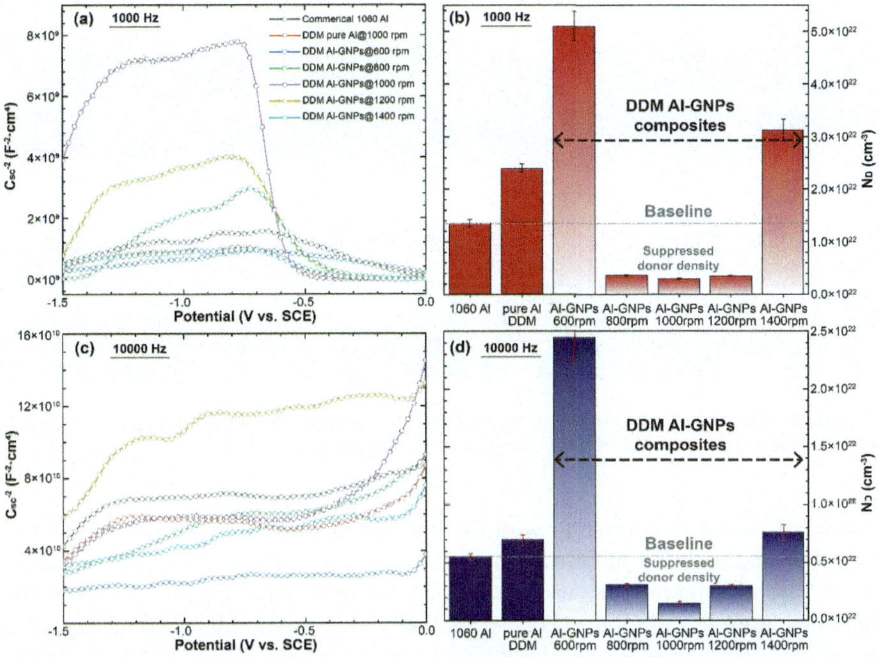

Fig. 10.12 Mott-Schottky plots and the variations of the donor density: **a** curves and **b** donor density at 1 kHz, **c** curves, and **d** donor density at 10 kHz

acting as electron donors. Therefore, lower corrosion resistance can be related to a higher donor density of the protective film. The DDM composites at appropriate rotational velocities show a significantly lower donor density than pure aluminum. The generation of oxygen vacancies in the protective film becomes more difficult when it contains GNPs and is doped by carbon atoms. In particular, due to the severe plastic deformation process induced by DDM, GNPs can be evenly embedded and distributed in the aluminum matrix, contributing to the homogeneity of the carbon-doped oxide film and resulting in suppressed local preferential corrosion caused by the heterogeneity of the surface protective film.

10.1.6 Long-Term Corrosion Inhibitor Evaluation

An intergranular corrosion test was carried out to prove the long-term corrosion inhibition mechanism of GNPs. Commercial pure aluminum, DDM pure aluminum, and GNP-reinforced aluminum matrix composites at a rotational velocity of 1000 r/min were tested. Figure 10.13 shows the corrosion morphology after the corrosion specimens are rinsed by intergranular corrosion. The density of corrosion pits on the surface of the DDM specimens is significantly higher than that of commercial pure

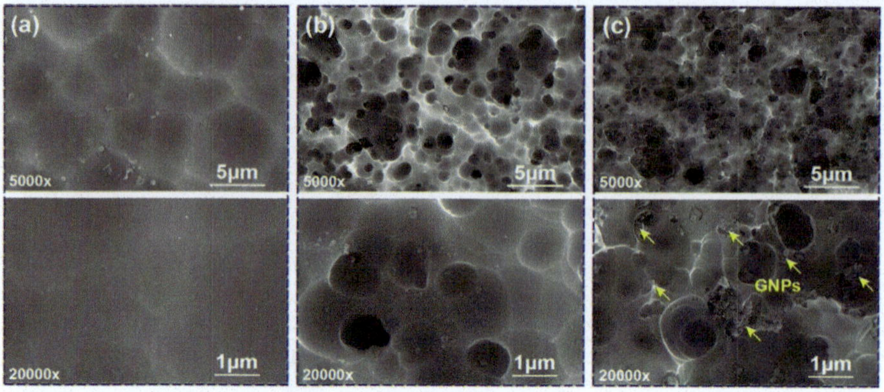

Fig. 10.13 Surficial morphologies of intergranular corrosion: **a** commercial pure aluminum, **b** DDM pure aluminum, and **c** DDM composites at 1000 r/min

aluminum, which is consistent with the finer-grained microstructures. Since more grain boundaries bring more rapid ion transport paths, the corrosion resistance of DDM pure aluminum is lower than that of commercial pure aluminum. However, due to the presence of GNPs (Fig. 10.13c), the composites enhance their resistance to chloride ion ingress formed by the interaction between GNPs and oxide protective film, which is reflected in the fact that the GNP-reinforced aluminum matrix composites with high homogeneity showed better corrosion resistance, that is, the GNPs contained therein exhibit the characteristics of long-term corrosion inhibitors.

Figure 10.14 shows the surface morphology of the intergranular corroded DDM composite cross-section prepared by the FIB technique. The oxide film is observed to be uniformly distributed on the surface of the composites to suppress further corrosion. The exposed and semi-exposed GNPs are embedded in the oxide film and directly bond well with the oxide film with interfacial defects. This allows GNPs to act as corrosion inhibitors to alleviate the ingress of chloride ions, thereby enhancing the corrosion resistance of the composites. As time progresses, more GNPs gather on the surface, which further enhances the difficulty of corrosion [48]. This time-dependent corrosion resistance enhancement presents a significant long-term mechanism, as shown in Fig. 10.8, proving the long-term corrosion inhibitor essence of GNPs. Figure 10.15 shows the local element distribution assessed via energy dispersive spectroscopy. The Pt-coating is added after intergranular corrosion as the boundary between the specimen and the FIB C-coating. GNPs and the oxide film are observed to form a protective film together against corrosion. The presence of carbon is also observed in the part of the protective film without GNPs, which proves that diffusion and chemical reaction did occur between GNPs and the oxide film. The carbon-doped oxide film may provide a better defense against the ingress of chloride ions. The phenomenon that chloride accumulates on the surface of the specimen but hardly enters the carbon-doped oxide film or even the aluminum matrix also proves this conclusion.

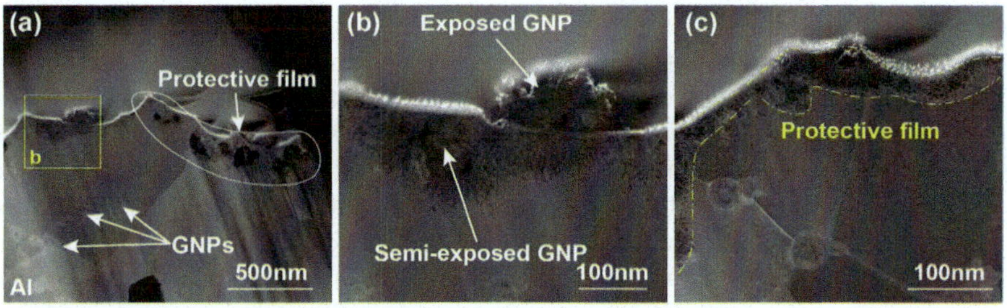

Fig. 10.14 Protective film and adjacent GNPs in DDM composites: **a** protective film and adjacent GNPs, **b** exposed and semi-exposed GNPs and **c** continuous protective film

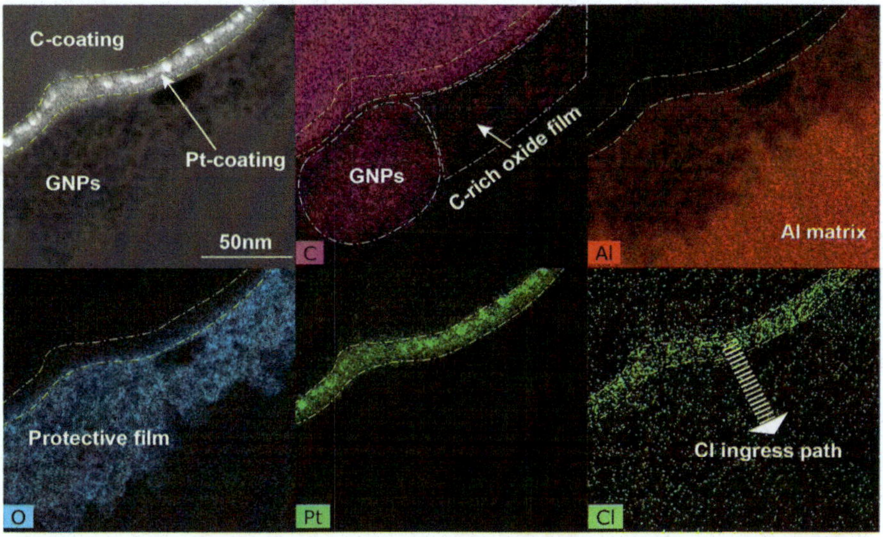

Fig. 10.15 Long-term corrosion inhibitor effect of GNPs observed by energy dispersive spectroscopy

DFT calculations were implemented to elucidate the long-term corrosion inhibitor behavior of GNPs from the aspects of carbon doping into aluminum oxide induced by diffusion and chemical reaction. To verify the reasonability of the calculation strategy, the bandgap of bulk α-Al_2O_3 is calculated to be 6.06 eV, in good agreement with other peer-reviewed literature [49]. The calculated bandgaps for the α-Al_2O_3(0001) slab, single-carbon-atom-doped α-Al_2O_3(0001)-1C slab, and dual-carbon-atom-doped α-Al_2O_3(0001)-2C slab (Fig. 10.16) are 4.77 eV, 4.26 eV, and 3.92 eV, respectively. Carbon doping led to a decrease in bandgaps. However, due to the presence of chloride ions and galvanic corrosion behavior, the actual protective film has more components than pure Al_2O_3. There are also chloride ions and vacancies generated for charge transfer in the protective film. The vacancy formation energy was also calculated. The formation energy of oxygen vacancies and aluminum vacancies is an endothermic process with a positive value. The formation energy of aluminum vacancies is approximately 30 ~ 60% higher than that of oxygen vacancies, indicating a more favorable formation of oxygen vacancies within the oxide film. This is in accordance with other work [50]. According to Mott-Schottky analysis, the semiconductor behavior of the protective film of pure aluminum and DDM materials is dominated by oxygen vacancies and interstitial defects. The result that oxygen vacancies are easier to produce is reasonable. As carbon is doped into the protective film, the formation energy of oxygen vacancies is significantly improved, implying that the donor density is reduced. Fewer carriers make the corrosion rate of the DDM composite drop significantly, thereby improving the corrosion resistance.

The insertion energy of Cl atoms into α-Al_2O_3 was also calculated. An assumption that Cl ingress occurs through vacancies was made. Due to the alternate distribution characteristics of aluminum and oxygen in α-Al_2O_3, the ingress of Cl atoms must pass through aluminum vacancies. Although the formation of oxygen vacancies is much easier than that of aluminum oxygen, the Cl atom was introduced into calculations for both types of vacancies. The insertion energy and energy barrier of ingress of Cl atoms from oxygen-site to aluminum-site are listed in Table 10.3. Cl ingress in aluminum vacancies is less favorable, but the ingress of Cl in oxygen vacancies has a negative value. This is attributed to the essence of positively charged aluminum atoms and negatively charged oxygen atoms. The strong electron affinity makes it easier to accommodate the ingress of the Cl atom into oxygen vacancies. In addition,

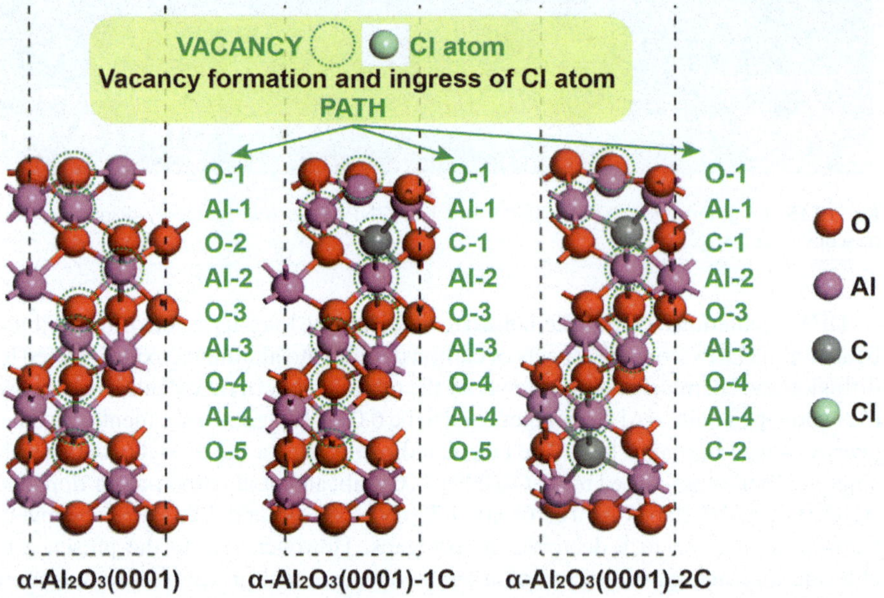

Fig. 10.16 Schematics of the α-$Al_2O_3(0001)$ slab, single-carbon-atom-doped α-$Al_2O_3(0001)$-1C slab, and dual-carbon-atom-doped α-$Al_2O_3(0001)$-2C slab

the presence of carbon doping induced by GNPs significantly increases the energy barrier, thereby inhibiting the ingress and diffusion of Cl atoms. GNPs, as a long-term corrosion inhibitor, avoid the rapid breakage and failure of the surface oxide film, so that the protective film can work for a longer time.

The surface Al–O distance with the insertion of Cl atoms at a certain site was also calculated. This distance characterizes the strength of specific bonds within α-Al_2O_3 due to the introduction of the Cl atom. The surface interatomic distance for optimized α-Al_2O_3(0001) is 1.70 Å, while it increases with the ingress of the Cl atom. With carbon doping, the increase in interatomic distance is significantly smaller than that of the simplest model. This proves that carbon doping strengthens the surface bond energy, thereby increasing the density of the protective film. Denser oxide films usually indicate higher corrosion resistance [51]. The existence and aggregation of

Table 10.3 Calculated results from DFT calculation

Position	O-1	Al-1	O-2 (C-1)	Al-2	O-2	Al-3	O-4	Al-4	O-5 (C-2)
Vacancy formation energy/eV	Formation of the vacancy								
α-Al_2O_3(0001)	8.11	12.57	8.85	13.96	10.02	12.33	8.93	13.56	8.88
α-Al_2O_3(0001)-1C	8.13	11.66	8.85	13.81	10.60	12.65	10.07	13.59	8.91
α-Al_2O_3(0001)-2C	8.19	11.64	10.39	13.94	10.67	13.01	10.45	12.89	8.91
Insertion energy/ eV	Insertion of Cl atom into the corresponding vacancy								
α-Al_2O_3(0001)	−2.38	−0.26	−1.51	1.37	−1.63	0.46	−1.76	0.50	−1.68
α-Al_2O_3(0001)-1C	−2.39	1.04	−1.51	2.37	−1.52	0.53	−1.74	0.53	−1.66
α-Al_2O_3(0001)-2C	−2.37	1.04	−1.34	2.43	−1.49	0.72	−1.37	2.69	−1.66
Energy barrier/eV	Ingress of Cl atom from O-site to Al-site								
α-Al_2O_3(0001)	2.12		2.88		2.09		2.26		
α-Al_2O_3(0001)-1C	3.43		3.88		2.05		2.27		
α-Al_2O_3(0001)-2C	3.41		3.77		2.21		4.06		
Surface interatomic distance/Å	Surface Al-O distance with insertion of Cl atom at the corresponding site (Perfect surface Al-O distance is 1.70 Å)								
α-Al_2O_3(0001)	1.78		1.79		1.77		1.76		1.74
α-Al_2O_3(0001)-1C	1.76		1.79		1.75		1.76		1.73
α-Al_2O_3(0001)-2C	1.76		1.75		1.73		1.75		1.73
Work function/eV	Work function with Cl insertion in the corresponding site (Perfect work function without insertion: α-Al_2O_3(0001) $\rightarrow$ 6.61 eV; α-Al_2O_3(0001)-1C $\rightarrow$ 5.37 eV; α-Al_2O_3(0001)-2C $\rightarrow$ 5.53 eV)								
α-Al_2O_3(0001)	4.79		4.42		4.01		3.89		3.72
α-Al_2O_3(0001)-1C	5.21		4.42		4.83		4.46		4.02
α-Al_2O_3(0001)-2C	5.22		4.75		4.88		4.81		4.02

GNPs enhance the corrosion resistance of GNP-reinforced aluminum matrix composites. The calculated work function also proves this phenomenon. This indicates that carbon doping within the protective film increases the resistance against electron transfer with Cl ingress, implying an improvement in the corrosion resistance of the oxide-covered aluminum matrix composites.

10.1.7 Summary

The effect of homogeneously dispersed GNPs as long-term corrosion inhibitors on the corrosion behavior of aluminum matrix composites was studied. DDM applied to prepare GNP-reinforced aluminum matrix composites can effectively obtain grain refinement and brokenness/redispersion of GNPs via severe plastic deformation, contributing to the microstructural homogeneity of the composites. Appropriate rotational velocity during the DDM process can suppress porosity and the formation of hydrolysable compounds to avoid potential crevice corrosion. The sound composites without the formation of Al_4C_3 were found to decrease the corrosion rate due to the formation of GNP-containing protective oxide film. Homogeneously dispersed GNPs induced by severe plastic deformation contributed to higher corrosion resistance and a more homogeneous protective oxide film. The long-term corrosion inhibition mechanism of GNPs versus exposure time was proven. The lower donor density showed better resistance of the composites to chloride ion attack and charge transfer compared to pure aluminum. GNPs and oxide films are observed to work together to form a protective film against corrosion via diffusion and chemical reactions. A higher vacancy formation energy, ingress energy barrier of Cl atoms, and work function for charge transfer were obtained with the doping of carbon atoms, proving the long-term corrosion inhibition mechanism of GNPs.

10.2 Heteroatom Modification Towards Enhanced Corrosion Resistance

10.2.1 Introduction

Sparked with its in-plane impermeability to nearly all molecules [52–54], graphene and its derivatives are emerging as anti-corrosion supplementary substances for metals. This barrier characteristic and chemical stability make it not surprising to shield the metals underneath it from unexpected reactions and corrosion [55]. However, when these nanomaterials are applied to prepare aluminum matrix composites in practical applications, especially for high specific-strength functions, fatal acceleration of the corrosion rate is often to be obtained instead of corrosion improvement [9]. The undesirable anti-corrosion weakening of graphene reinforced aluminum matrix composites can be attributed to three major inherent issues: ① Graphene can trigger corrosion-promotion activity (CPA) to enhance the localized galvanic corrosion due to the heterogeneous corrosion potential, such as graphene (0.2 V versus. saturated calomel electrode, SCE) and aluminum matrix (−1.1 V versus SCE) [56]. When coupling with a relatively active metal, the electrochemical stable and high electrically conductive graphene tends to behave like noble cathode, contributing to the formation of micro-galvanic corrosion that initiates localized corrosion and boosts the anodic dissolution [57, 58]; ② Small molecules such as O_2 and H_2O may infiltrate from the edge defects of graphene over the long-term application, which consequently accelerated the corrosion rate [59]; ③ Graphene is easy to agglomerate and exhibits poor compatibility due to its large specific surface area and strong van der Waals interaction, which severely reduces their dispersibility in the aluminum matrices and deteriorates the corrosion resistance performances [60].

In light of these, several tailoring strategies were proposed to suppress the CPA behavior towards highly applicable aluminum matrix composites. Heteroatom tailoring of graphene shows great potential for changing the electronic density to tailor its electronic and electrochemical properties [61]. It can reduce the electrical conductivity by increasing the electron scattering in graphene lattice, as well as modify the local electron density to inhibit the electrochemical activity of graphene. Fluorinated graphene is promising to enhance anti-corrosion performance for the insulating nature induced by F doping, as well as the molecule impermeability inherited from graphene [62]. An alternative to suppressing the corrosion rate is to design microstructures that are intrinsically electrochemically homogeneous (for example, through severe plastic deformation engineering [24, 25, 63]). However, the combination of these two approaches loses effectiveness since the composites reinforced by fluorinated graphene are susceptible to weak interfacial bonding and low load-bearing capacity due to the chemical inertness of this modified graphene. The desirable interfacial bonding cannot be simply obtained by the high-value plastic strain induced by severe plastic deformation. Developing cheap but effective microstructure solutions that enable both high corrosion resistance and mechanical performances thus still remain a fundamental challenge.

Here we propose a bottom-up strategy based on the principle of exploiting great interfacial bonding between fluorinated graphene nanoplatelets (F-GNPs) and aluminum matrices via Mg alloying and severe plastic deformation. A severe plastic deformation technique, deformation-driven metallurgy (DDM), was applied to obtain the homogeneous dispersion of F-GNPs to realized extraordinary corrosion resistance. Corrosion-suppression activity (CSA) induced by the synergistic strategy of active metal and heteroatomic modification was studied. This work aims to examine the effectiveness of synergistic modification and DDM techniques in terms of corrosion behavior in an aggressive chloride-containing environment. Microstructural factors, interfacial characteristics, reinforcement dispersion, and electrochemical response were evaluated and discussed in detail.

10.2.2 Experimental Procedures

Mg-doped F-GNP reinforced aluminum matrix composites were fabricated by a bottom-up preparation route consisting of sequential ball milling and DDM. Atomized aluminum powders with an average diameter of 10 μm and a purity of 99.6% were blended with the appropriate amount of atomized magnesium powders (10 μm in diameter) and F-GNPs (XF079, Nanjing XFNANO Materials Tech Co., Ltd, China) by a planetary ball milling equipment. The addition of F-GNPs is 1.5 wt.%. Yttria stabilized zirconia ball of 6 mm (70%) and 8 mm (30%) in diameter were used, and the ball powder ratio was 10:1. Polyethylene glycol 2000 with a weight fraction of 4% was added as process control agent. The mixed powder was ball milled under argon atmosphere at a rotational velocity of 200 r/min. Two kinds of ball milling sequences were designed as shown in Fig. 10.17. The as-milled composite powders were sintered at 400 °C in argon atmosphere to remove the process control agent.

The obtained powders were cold-pressed at about 200 MPa into a pre-densified compact with a weight of (0.82 ± 0.01) g. A DDM tool rotating at a specific rotational velocity of 1000 r/min pressed the compact for 60 s to introduce severe plastic deformation and corresponding heat input for the sintering of the final composite disc. The final diameter and thickness were 16 mm and 1.0 mm, respectively. For comparison, aluminum matrix composites reinforced by Mg-doped GNPs (XF022-1, Nanjing XFNANO Materials Tech Co., Ltd, China) were also fabricated by the same route.

SEM (HITACHI SU5000) equipped with backscatter electron (BSE) detector and EDS was utilized to show the dispersion and homogeneity of the microstructure and the morphologies of the corroded surfaces. The grain features of samples were characterized by the EBSD. The data of EBSD results were analyzed by AZtecHKL software. The phase composition and lattice distortion of the samples was analyzed

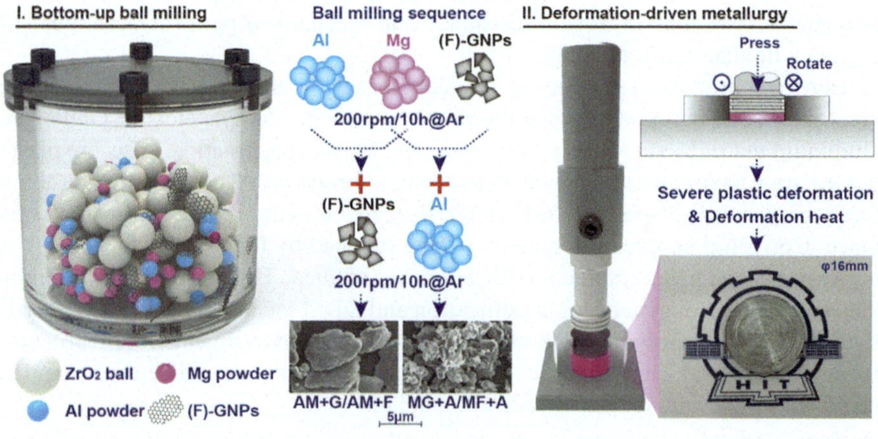

Fig. 10.17 Bottom-up preparation route of the aluminum matrix composites

based on the XRD (PANalytical X'PERT) data. TEM (FEI Talos F200x) was applied to examine the detailed microstructure among Al matrix, Mg coating, and F-GNPs with an accelerating voltage of 200 kV. Raman spectroscopy was conducted using Renishaw in Via-Reflex to evaluate the disordering and defect density of F-GNP structures XPS (ThermoFisher ESCALAB 250Xi) was utilized to obtain the chemical bonding evolution during the DDM process.

Specimens cut from Mg-coated F-GNP/GNP reinforced aluminum matrix composite disc were tested by electrochemical measurement with a quadrant shape and a diameter of 16 mm. A CHI 760E system for potentiodynamic polarization (PDP) and a PARSTAT 4000A system for electrochemical impedance spectroscopy (EIS) and Mott-Schottky analyses. A three-electrode system was used in the test and the applied electrolyte was a 3.5 wt.% NaCl aqueous solution at room temperature, as shown in Fig. 10.18. The cut specimens were used as working electrodes. A Pt foil and an SCE were used as the counter electrode and reference electrode. A stabilization time by immersing the working electrodes into the electrolyte for one hour was used before initial PDP and EIS tests. The PDP data of each specimen were recorded by continuously scanning at a scanning rate of 50 mV/min, and each test was repeated for three times. EIS measurements were conducted at the open circuit at frequencies ranging from 1 MHz to 0.01 Hz with a sinusoidal perturbation of 5 mV in amplitude. Mott-Schottky characteristics were acquired by monitoring the frequency-potential response at 1 kHz and 10 kHz between −1.5 V to 0.0 V versus SCE with a scanning interval of 25 mV.

The corrosion rate of these aluminum matrix composites was estimated from the continuous hydrogen evolution collections. The applied experimental procedure for hydrogen collection was shown in Fig. 10.19. The evolved hydrogen gas was collected for each quadrant specimen and measured by the volume of discharged deionized water. 30 days of immersion in 500 mL 3.5 wt.% NaCl aqueous solution at room temperature were selected to monitor the discharged volume. Each specimen was mounted by polymethyl methacrylate with an exposed area of 0.5 cm^2 and fine ground by 7000 grit SiC paper. No renewing or agitating of the static solution was conducted during the entire measurement. The immersion tests were repeated three times for better statistics.

10.2.3 Density Functional Theory Calculation

As shown in Fig. 10.20, a simple α-Al_2O_3 (0001) slab with a thickness of approximately 11 Å was constructed for DFT calculations. F-GNP@Al_2O_3 (0001) slab and F-GNP+Mg@Al_2O_3 (0001) slab were also established by replacing partial Al_2O_3 layer with F-GNP and Mg-alloyed F-GNP, respectively. A sufficient large vacuum of approximately 20 Å was added along the [0001] direction to avoid interaction between the neighboring slabs along Z-axis. Norm-conserving pseudopotentials were implemented via QUANTUM ESPRESSO [28] for the DFT calculation. The system of Kohn–Sham equations was self-consistently solved on the plane wave basis to

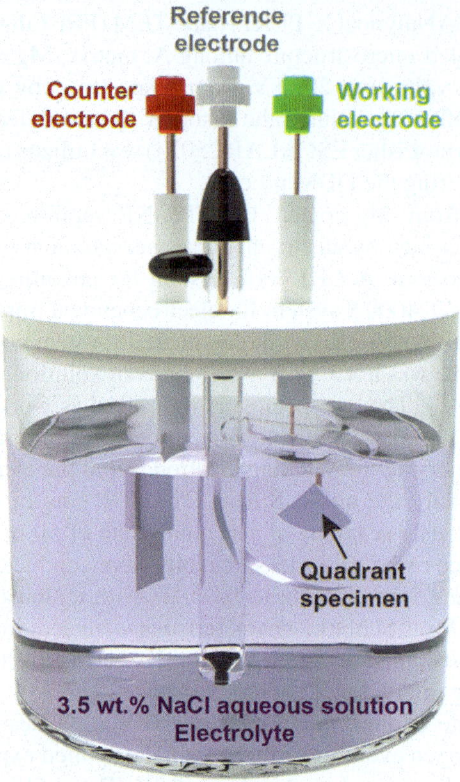

Fig. 10.18 Schematic of the three-electrode system

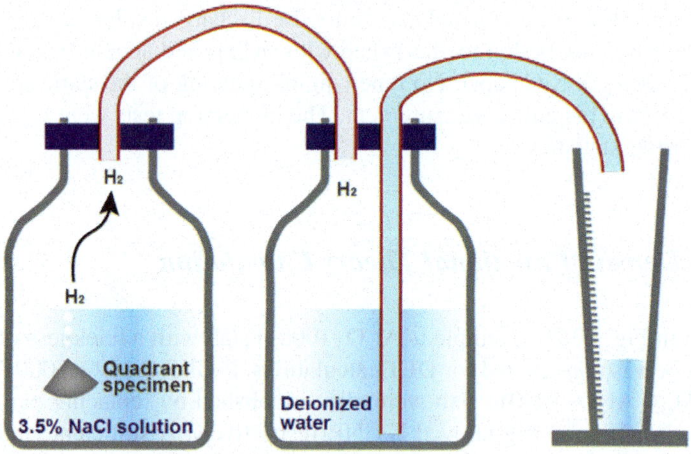

Fig. 10.19 Schematic of the hydrogen collection system

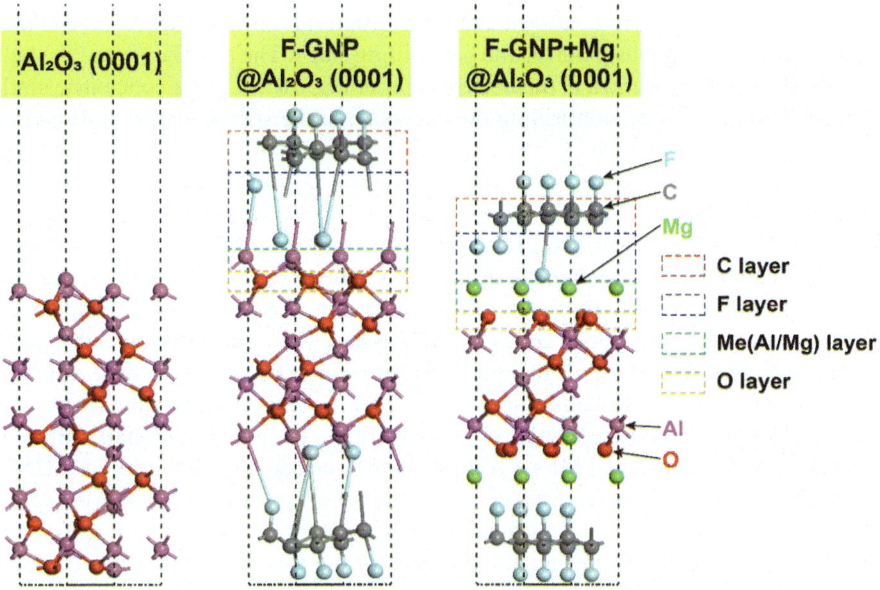

Fig. 10.20 Schematics of Al_2O_3 (0001) slab, F-GNP@Al_2O_3 (0001) slab, and F-GNP+Mg@Al_2O_3 (0001) slab

determine the electron density of the system. We performed integration in reciprocal space on a $5 \times 5 \times 1$ Monkhrost-Pack lattice of k-points [29]. The threshold for the maximum kinetic energy of plane waves included in the basis set was 830 eV. Calculations were performed with the PBE exchange–correlation function in the generalized gradient approximation (GGA). These slabs were firstly relaxed until the energy, residual force and displacement were less than 5.0×10^{-6} eV/atom, 0.01 eV/Å, and 5.0×10^{-4} Å, respectively. Work function, electron density difference, and local partial density of states were calculated to illustrate the corrosion suppression activity brought by Mg-alloyed F-GNPs.

10.2.4 Microstructural Factors

Four kinds of ball milling sequences were selected and denoted as AM+G, AM+F, MG+A, and MF+A. AM+G and AM+F refer to mix the Al powders and Mg powders firstly, and then mix these blended powders with graphene nanoplatelets (GNPs) and F-GNPs via ball milling, respectively. MG+A and MF+A refer to mix the Mg powders and GNPs and F-GNPs, respectively, and then mix these blended powders with Al powders. Since the (F)-GNPs added in the first step separate the Mg powders and Al powders physically and avoid the formation of viscous Al–Mg intermetallic compounds, the ball-milled powders of MG+A/MF+A are significantly smaller than those of AM+G/AM+F, indicating that these powders have lower viscosity and tend to be blended homogeneously during the ball milling stage. The morphology of the bulk sample after DDM is shown at the bottom right of Fig. 10.17. Clear concentric circle streamlines can be seen, confirming the existence of the severe plastic deformation process.

To access the influence of different bottom-up preparation routes on the microstructural factors of the (F)-GNPs and the matrices, we conducted Raman spectroscopy, as shown in Figure 10.21a–c. The ratio of D-band to G-band in the Raman spectrum characterizes the damage degree of graphene structures, which is denoted as I_D/I_G [30]. A higher I_D/I_G value refers to the more broken graphene, indicating that there are more active open edges to provide necessary conditions for subsequent metallurgical bonding [64]. Compared with the original (F)-GNPs, an increase in this value can be seen in all the ball milling routes. Due to the more significant lubricating effect of the fluorinated graphene [65], the crushing effect caused by ball milling is weakened in the AM+F and MF+A routes. On the one hand, the integrity of F-GNPs themselves is better preserved, resulting in a more efficient load transfer effect for mechanical performances and surficial absorbing integrity for corrosion resistance. On the other hand, it also means the reduction of the active reaction sites. It is also worth noting that the characteristic peak of MgF_2 is observed in Fig. 10.21b and c. Since MgF_2 has high chemical stability and bond energy [66], the existence of this substance implies the strong chemical bonding between the carbonaceous phases and the matrices. This phenomenon compensates for the adverse effects of fewer active sites during the ball milling process. In addition, the ball milling routes of preferential mixing of (F)-GNPs in the first step have a low I_D/I_G value, indicating that the physical isolation effect between Al and Mg brought by (F)-GNPs can reduce the overreaction of the ball milled powders. Figure 10.21c shows the evolution characteristics of the bulk DDM composites, in which all the I_D/I_G value have decreased correspondingly. Obviously, graphene cannot be re-spliced into a larger piece via severe plastic deformation [4]. The decrease in this value actually indicates that the decrease in the thickness and layers of the (F)-GNPs, and finally results in the increase of the specific surface area of graphene. This provides better prerequisites for the load transfer effect. In addition, the results of X-ray diffraction (XRD) tests (Fig. 10.21d) shows that only the α-Al phase can be seen in all the composite materials. Since the XRD has the concentration detection limit, homogeneously dispersed nanophase with low concentration cannot be detected in the bulk materials. This indicates that there are no large-sized Mg clusters or undesirable large-sized Al–Mg intermetallic compounds in the matrix. In other words, the presence of a large number of nano-magnesium particles in the matrix provides sufficient reaction sites for the solid-state alloying between the (F)-GNPs and the matrices.

X-ray photoelectron spectroscopy (XPS) was also applied to analyze the microstructural evolution and further illustrate the bonding behavior of the DDM composites. Figure 10.22 shows the deconvolution results of C 1 s peak. Compared with the GNP-reinforced composites, F-GNP-reinforced composites have a higher sp^3 to sp^2 ratio, which is in line with the bonding form of F-GNPs, that is, one C atom forms a bond with two in-plane C atoms and one out-of-plane F atom [67].

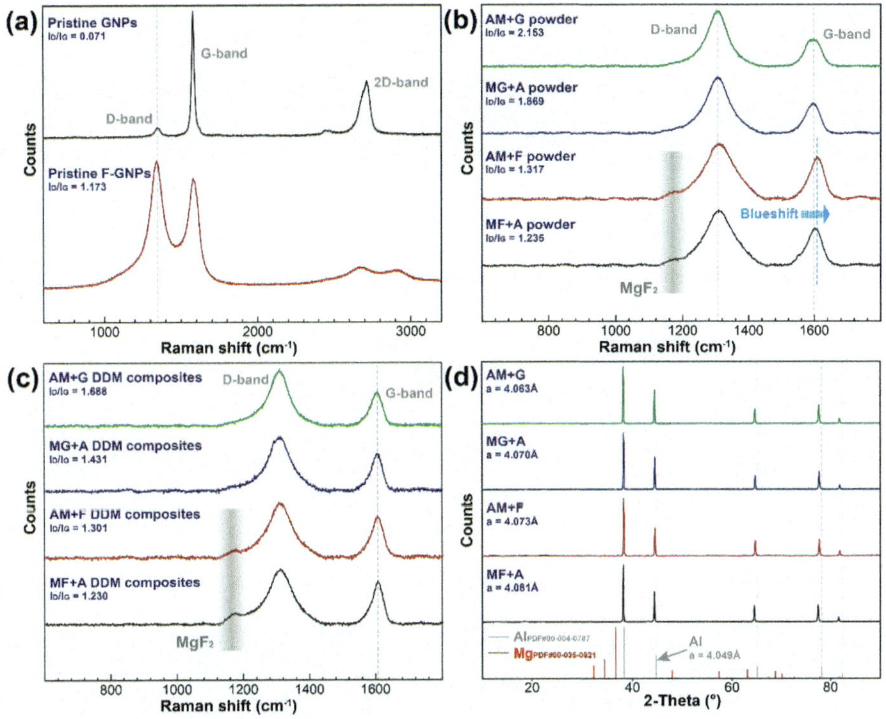

Fig. 10.21 Characterization of the microstructural evolution: **a** Raman spectra of the pristine reinforcements, **b** Raman spectra of the ball-milled powders, **c** Raman spectra of the DDM composites, and **d** XRD results of the DDM composites

Besides, it can be seen that there are basically no C-Al bonds in the F-GNPs DDM composites, which proves that the existence of F atoms isolates the direct reaction between Al matrix and C atoms, thereby avoiding the formation of the easily hydrolyzable compounds Al_4C_3. Since this substance is detrimental to the structural integrity of materials exposed to the aqueous environment [36], this phenomenon provides the possibility to improve corrosion resistance. Figure 10.23 depict the deconvolution results of Al 2p peak. Different ball milling routes do not seem to change the ratio distribution of the chemical bonds, but the binding energy of the composite added with the F element has been significantly reduced, indicating that the number of electrons lost by the Al atom during the metallurgical process is reduced. This implies that Al is less directly involved in the reaction with carbonaceous substance and partially reacts with Mg to form nanophases (Al in Al–Mg intermetallic compounds tends to gain electrons [68, 69]). This is consistent with our vision properly, which is to achieve indirect bonding between aluminum and (F)-GNPs through Mg alloys, thereby avoiding the CPA effect and improving the comprehensive corrosion resistance of the DDM composites.

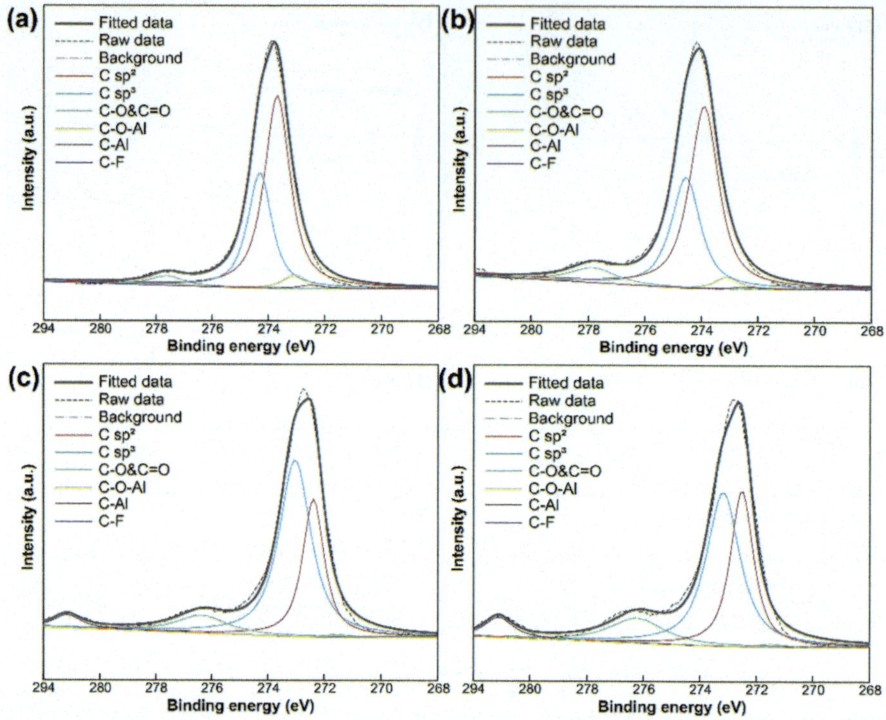

Fig. 10.22 XPS results C 1 s of the DDM composites: **a** AM+G, **b** MG+A, **c** AM+F, and **d** MF+A

Figure 10.24a depicts the grain size distribution of the composites prepared by four bottom-up routes. This result can be explained from the following two aspects: ① F-GNP has an excellent lubricating effect similar to polytetrafluoroethylene [70], which significantly promotes the flow intensity of the composites during the severe plastic deformation process. More obvious dynamic recovery and recrystallization are triggered by the enhanced lattice distortion and defect energy [23]; ② by adding (F)-GNPs in the first step of the ball milling process, the excessive reaction of Al and Mg can be avoided, thereby reducing the dynamic viscosity of the thermo-plasticized composites during DDM process. This makes the material more prone to plastic deformation and further improves the grain refinement effect. From the above two explanations, it can be found that the DDM composites obtained via the MF+A route has achieved the finest microstructures with an average grain diameter of (1.14 ± 0.45) μm. We further observed the grain characteristics of the MF+A DDM composites, as shown in Fig. 10.24b. The equiaxed grains occupy the primary component and the residual stress remaining in the microstructures is low to show a limited geometric necessary dislocation density. This implies that the internal defect energy after DDM process is low, thus avoiding the CPA effect caused by the energy release of the internal stress.

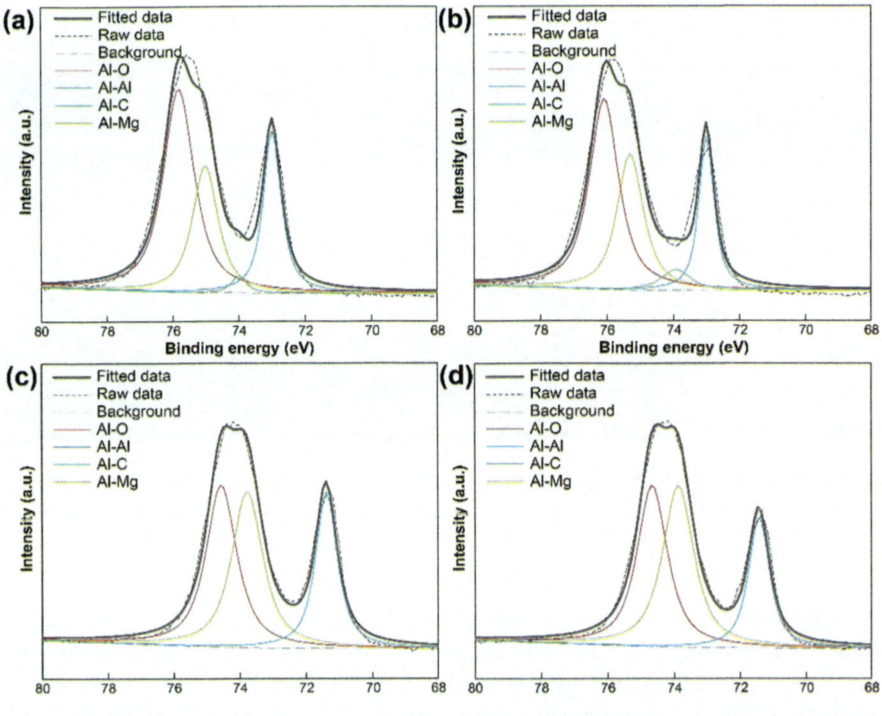

Fig. 10.23 XPS results Al 2p of the DDM composites: **a** AM+G, **b** MG+A, **c** AM+F, and **d** MF+A

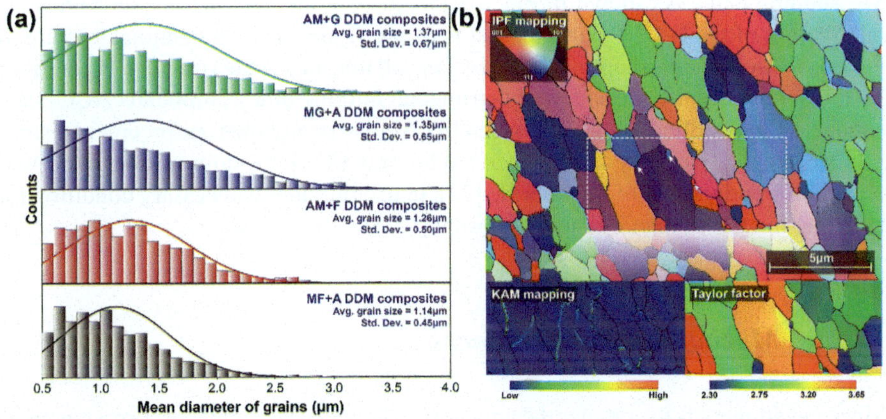

Fig. 10.24 Microstructural characterization images of the DDM composites via EBSD: **a** distribution of the grain sizes and **b** grain morphologies of the MF+A DDM composites

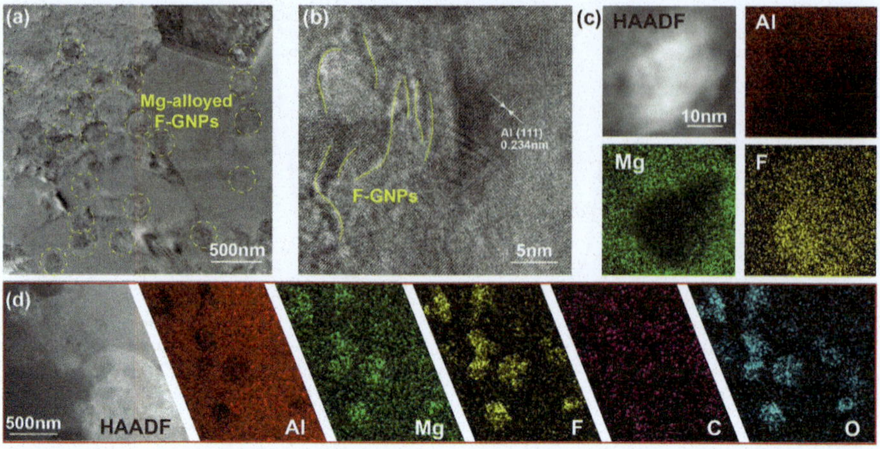

Fig. 10.25 Microstructural characterization images of the DDM composites via TEM: **a** distribution of the Mg-alloyed F-GNPs, **b** local phase interface, **c** core–shell structures formed by Mg alloying, and F-GNPs, and **d** EDS characterization of the Mg-alloyed F-GNPs

Figure 10.25a shows the transmission electron microscopy (TEM) morphologies of these DDM composites. F-GNPs are homogeneously dispersed in the matrix, and they have a good lattice matching with the Al matrix (Fig. 10.25b). From the energy spectrum results, it can be seen that the distribution area of Mg and F-GNPs basically overlaps, indicating that a self-assembled nanoscale structure is formed between Mg and graphene. As shown in Fig. 10.25c, the Mg element mainly surrounds the distribution of the F-GNP particle, that is to say, it presents a transition structure of the F-GNP core to Mg shell to Al matrix. We call this nanoscale structure Mg-alloyed F-GNPs. It avoids the formation of detrimental intermetallic compounds Al_4C_3, and enhances the bonding strength between the Al matrix and the reinforcements, since F-GNPs are relatively chemically inert to aluminum [71]. In addition, due to the poor electronic conductivity of the fluoride [72], it also provides a necessary condition for reducing the CPA effect on the composites.

10.2.5 Mechanical Performances

Since our strategy is to conquer the antagonism between strength and corrosion resistance in graphene-reinforced aluminum matrix composites, we tested the mechanical performances of the DDM composites with four different preparation routes. Figure 10.26a shows the mechanical properties of the composites with specific Mg addition (5.5 wt.%). The ball milling sequences show great influences on the mechanical response. The strength and ductility of the composites obtained by the ball milling route of adding (F)-GNPs firstly are significantly higher than the other two strategies. The tensile strength and elongation of the materials obtained via

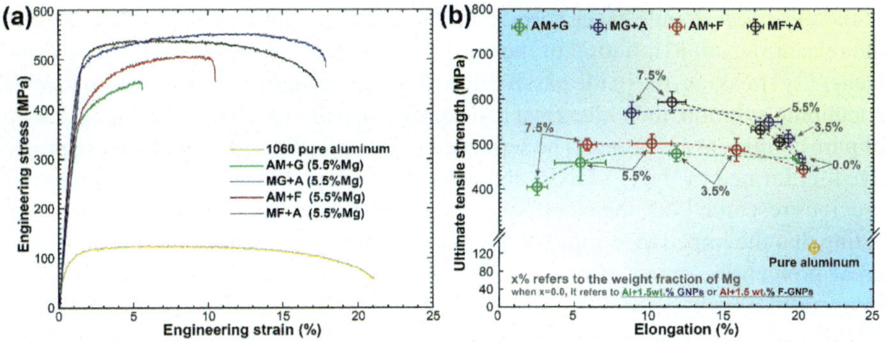

Fig. 10.26 Mechanical performances of DDM composites: **a** engineering stress–strain curves of composites alloyed with 5.5% Mg and **b** comparison among pure aluminum and DDM composites alloyed with different routes and additions

MG+A and MF+A reached (549 ± 32) MPa/(17.9 ± 1.8%) and (532 ± 39) MPa/(17.3 ± 1.2%), respectively, indicating that GNPs and F-GNPs play a very close strengthening effect. This shows that although F-GNPs are relatively chemically inert, they still have a sound metallurgical bond with the Al matrix with the help of the Mg alloying strategy. With the load transfer effect brought by the superior specific surface area of the two-dimensional carbonaceous nanomaterials [73], the precipitation/solid solution strengthening produced by the Mg element [74], and the Hall–Petch strengthening [75] brought by the severe plastic deformation, the DDM composites we prepared has obtained the extraordinary mechanical performances. We further studied the comprehensive mechanical responses with different Mg addition. One can see that the tensile strength performance with all the compositions is several times that of pure aluminum, which confirms the good mechanical properties. According to our previous research [26], the comprehensive mechanical performance has reached the highest known level in related carbonaceous nanomaterial-reinforced aluminum matrix composites. In the two preferred preparation routes, the mechanical properties of the composites basically increase gradually with the increase in the Mg addition. Regrettably, the elongation has shown a downward trend. Since ductility is also the primary component of the comprehensive mechanical properties [76], we need to strictly consider the balance of strength and ductility. As such, a 5.5% Mg addition can achieve the relatively ideal mechanical responses.

10.2.6 Electrochemical Corrosion Behaviors

Figure 10.27a shows the EIS response of the MF+A DDM composite versus exposure time to the 3.5 wt.% NaCl aqueous solution. A capacitive loop can be seen with similar characteristics in each curve, which is caused by the overlying of two electrochemical processes. The first is caused by the charge transfer of Al to Al^{3+}

at the electrical double layer formed at the corrosion surface, while the second is the relaxation and exfoliation of the corrosion products at the exposed surface [34]. Generally speaking, the oxide passive film of pure aluminum is porous in the chlorine-containing solution, their electrical resistance is usually low. The two response loops of pure aluminum are prone to be separated [77]. In contrast, due to the dense protective film formed by the (F)-GNPs and heteroatom-doped oxide film according to our previous research [78], the electrochemical responses cannot be distinguished, indicating that the impedance value of the doped oxide film is relatively high and shows great protection against corrosion. Besides, the larger diameter of the capacitive loop means a bigger electrochemical impedance. The corrosion resistance of MF+A DDM composites gradually increases and decreases over time. It is worth noting that even when the exposure time reaches 1440 h, that is, 2 months later, the diameter of the capacitive loop still does not show a sharp weakening trend, which shows that our composites are capable of showing good anti-corrosion performance under long-term service conditions. We further conducted EIS tests on various preparation routes, as well as pure aluminum, and fitted them according to the equivalent circuit in Fig. 10.27a. Figure 10.27b shows the polarization resistance change extracted versus exposure time. The corrosion resistance of pure aluminum decreases drastically over time. Its initial value is close to or higher than the initial value of the DDM composites. But after 2 months of immersion, its polarization resistance was only less than 1/8 of the initial value. This shows that the oxide layer on the surface of pure aluminum can no longer maintain corrosion resistance due to the ingress of chloride ions. In contrast, the corrosion resistance of DDM composites obtained via any preparation route has only slightly decreased. Especially for the preferred preparation route MF+A, its 2-month final state polarization resistance is about 89% of the initial state, which can be considered that there is no significant drop compared to the initial state.

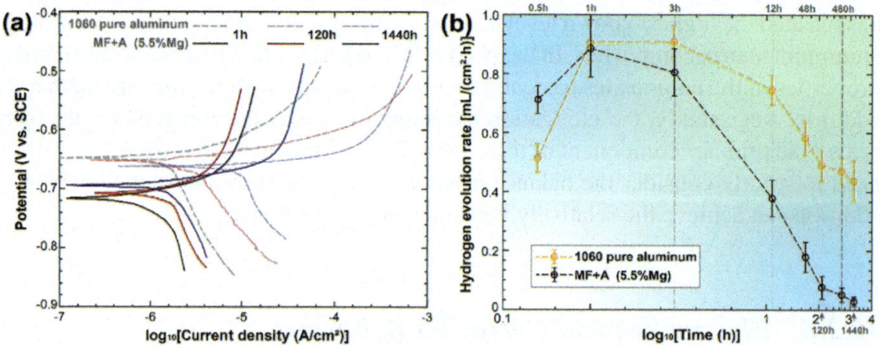

Fig. 10.27 Electrochemical responses of DDM composites: **a** EIS Nyquist spectra with different exposure time and **b** polarization resistance comparison among pure aluminum and DDM composites with different routes and exposure time

Figure 10.28a shows the results of PDP of the MF+A DDM composites versus exposure time to the 3.5 wt.% NaCl aqueous solution. Although the corrosion potentials of the composites are lower than those of the pure aluminum (this is because the equilibrium potential of Mg is lower than that of Al [79]), the corrosion current densities are significantly lower than that of pure aluminum. The Tafel curves only show a small change over time, indicating that the corrosion rate is not sensitive to time. Figure 10.28b shows the hydrogen evolution rate of the composite materials and pure aluminum in the 3.5 wt.% NaCl aqueous solution. During the first hour of exposure, both of them can be seen that the oxide film naturally formed on the surface gradually broke and the hydrogen evolution process is accelerated. With the further increase of time, the corrosion products begin to accumulate on the exposed surface. The diffusion distance required for the chloride-containing corrosion medium to react with the surface of the composites gradually increases, which is manifested as the gradual decrease in the rate of hydrogen evolution. It can be seen that in addition to the initial hydrogen evolution rate of the DDM composite is slightly higher than that of pure aluminum, while the subsequent evolution rate is significantly lower than that of pure aluminum. Especially when the time reaches 2 months. The hydrogen evolution rate is even less than 5% of pure aluminum, indicating that Mg-alloyed F-GNPs have a significant CSA effect. It can be presumed that the initial corrosion of the aluminum matrix contributes to the formation of a dense chemical bond between the exposed Mg-alloyed F-GNPs and the surface oxide film. Due to the chemical inertness of F-GNP and its in-plane impermeability to nearly all molecules, the diffusion distance of protons and aluminum ions in the corrosion interface is greatly increased, thereby reducing its corrosion rate.

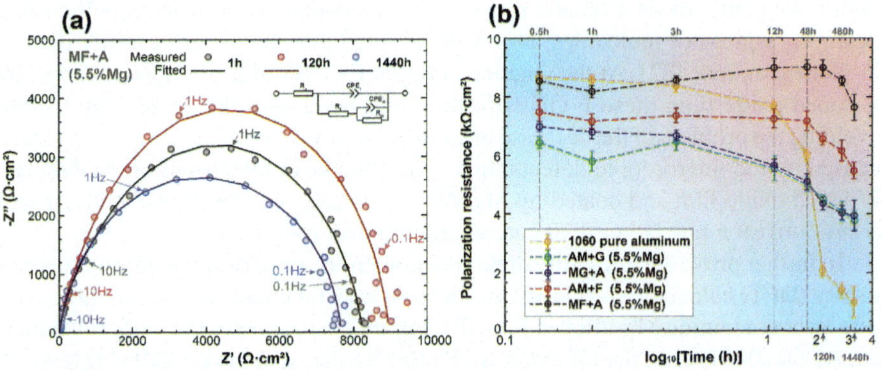

Fig. 10.28 Electrochemical responses of DDM composites: **a** Tafel curves extracted from PDP tests with different exposure time and **b** hydrogen evolution rate versus exposure time of pure aluminum and MF+A DDM composites

10.2.7 Corrosion Suppression Activity

Electrochemical results and hydrogen evolution rate measurements seem to prove that the corrosion resistance of the DDM composites is higher than that of pure aluminum. However, according to Sun et al. [80], undesirable anti-corrosion weakening is easily triggered due to the CPA effect, high-density graphene defects, and agglomeration. The problem of agglomeration can be well solved by the severe plastic deformation process [81]. Our bottom-up DDM process can realize homogeneous distribution of (F)-GNPs in the matrix, and even further reduce the thickness of graphene through interlayer shear to improve the service performance of graphene. Nevertheless, how the corrosion acceleration caused by the CPA effect and high-density defects can be avoided or even achieved the CSA effect is still puzzling. We therefore performed elemental distribution analysis and characterization of the electrochemical corroded surface, as shown in Fig. 10.29. Taking the Pt coating as the boundary, the corroded surface is shown in the right side. It can be seen that there is a continuous protective layer composed of dense F-GNPs and an oxide film doped with F elements on the surface. There is also two layers of Mg as an alloying element on both sides of the protective layer. This dense protective layer plays an effective in corrosion protection. Due to the inertness of fluoride, this protective layer is theoretically unable to produce a strong metallurgical bond with the matrix. However, the presence of Mg makes up for this serious issue. Its good double-sided metallurgical compatibility provides a strong interfacial bonding force and avoids the exfoliation of the protective layer. This continuous and dense film avoids the direct three-phase interface among the matrix, graphene, and the corrosion medium to suppress the CPA effect. In addition, due to the honeycomb structure, the F-GNPs form a dense delocalized cloud to block the gap within the center of its aromatic rings, which contributes to a strong repelling field again the ingress of molecules. It is well accepted that the impermeability through in-plane graphene [82]. At the junction between F-GNPs, due to the existence of the F-doped oxide film, these F-GNPs form a tight bond with the oxide film, thereby avoiding the problem of the decrease in the diffusion distance of the corrosive medium caused by the microscopic defects. In a word, this layer composed of F-GNPs and F-doped oxide film and coated by Mg alloying finally realizes the CSA effect with a low corrosion rate better than pure aluminum over long-term service.

To further prove the reliability of this phenomenon, we performed density function theory (DFT) calculation to clarify the electron transfer characteristics of the protective layer containing Mg-alloyed F-GNPs. Figure 10.30a shows the work function of the Al_2O_3 (0001) slab, the F-GNP@Al_2O_3 (0001) slab, and the F-GNP+Mg@Al_2O_3 (0001) slab, respectively. The existence of F-GNP significantly improves the work function of the composites, indicating that the difficulty of electrons removed from the matrix is increased [29]. Moreover, although F-GNP without Mg alloying seems to have a higher work function, we can see from the electron density difference image (Fig. 10.30b) that there is a larger vacuum area between the surficial F-GNP and the matrix. This proves that an effective chemical bond cannot be formed directly between F-GNP and the oxide film. Due to the addition of Mg with higher

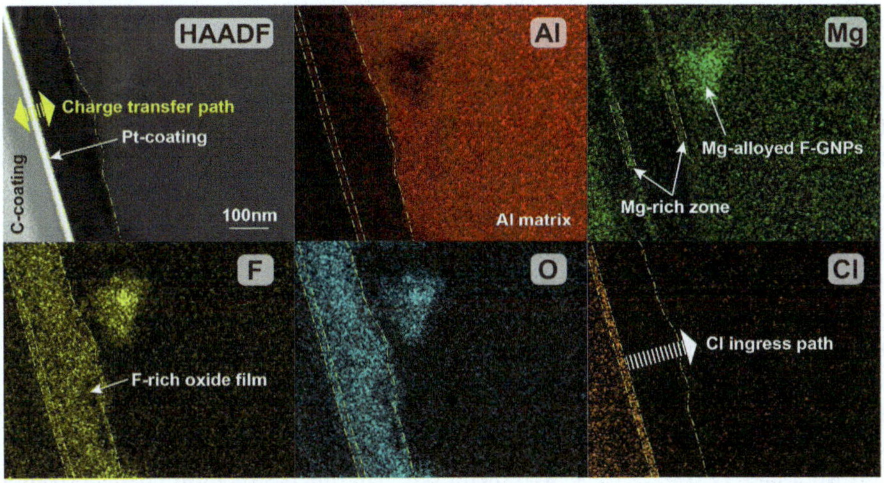

Fig. 10.29 Elemental distribution of the corroded surface

electronegativity, the vacuum slab area between F-GNP and the substrate is eliminated, which proves that the sound bonding of the interface. Combined with a higher work function and a dense and homogeneous protective layer, we finally obtained breakthrough results of DDM composites with excellent corrosion performance. To our best knowledge, this is the first time to obtain the carbonaceous nanomaterial-reinforced aluminum matrix composite with both high mechanical performances and corrosion resistance. As such, this economical bottom-up strategy of microstructural modification broadens the horizon for anti-corrosion engineering in aluminum matrix composites.

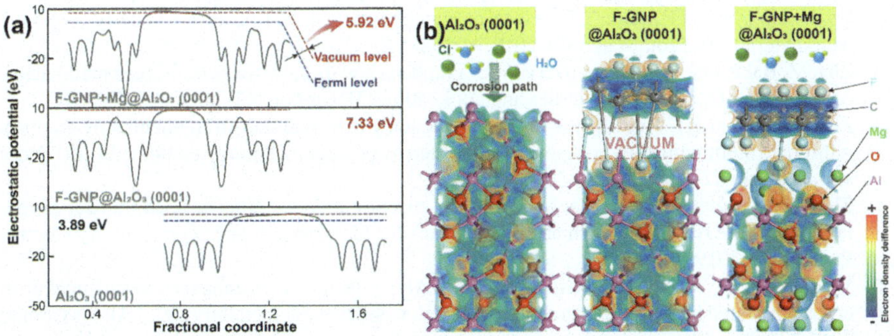

Fig. 10.30 Theoretical calculation of the corrosion suppression activity: **a** work functions and **b** electron density differences of three slabs

10.2.8 Summary

We designed a novel bottom-up strategy to achieve charming carbonaceous nanomaterial-reinforced aluminum matrix composites with a better balance between corrosion resistance and mechanical performances. This strategy is to conquer the corrosion promotion activity caused by heterogeneous corrosion potential, graphene defects, and agglomeration. Mg-alloyed fluorinated graphene nanoplatelets were introduced via severe plastic deformation and designed preparation routes. Homogeneously dispersed Mg-alloyed fluorinated graphene nanoplatelets and ultrafine-grained microstructures contributed to the formation of dense and strong protective film again chloride-containing aqueous environment. Corrosion suppression activity was discovered in the designed composites. The anti-corrosion polarization resistance remained 89% of the initial state after 2-month immersion in chloride containing environment, while the ultra-tensile strength and elongation of (532 ± 39) MPa and 17.3% ± 1.2% were achieved. The microstructural characterization and density function theory calculation confirmed the existence and rationality of the dense anti-corrosion layer composed of fluorinated graphene nanoplatelets and F-doped oxide film and coated by Mg alloying. Moreover, this work is significant for conquering the antagonism between mechanical performances and corrosion resistance, which is indeed critical for the design of carbonaceous nanomaterial-reinforced aluminum matrix composites to realize some desired performance for practical applications.

References

1. Wu G, Yu Z, Jiang L, et al. A novel method for preparing graphene nanosheets/Al composites by accumulative extrusion-bonding process[J]. Carbon NY. 2019;152:932–45.
2. Ovid'ko IA, Valiev RZ, Zhu YT. Review on superior strength and enhanced ductility of metallic nanomaterials[J]. Prog Mater Sci. 2018;94:462–540.
3. Zhao Z, Bai P, Du W, et al. An overview of graphene and its derivatives reinforced metal matrix composites: preparation, properties and applications[J]. Carbon NY. 2020;170:302–26.
4. Zhang ZW, Liu ZY, Xiao BL, et al. High efficiency dispersal and strengthening of graphene reinforced aluminum alloy composites fabricated by powder metallurgy combined with friction stir processing[J]. Carbon NY. 2018;135:215–23.
5. Zhou W, Mikulova P, Fan Y, et al. Interfacial reaction induced efficient load transfer in few-layer graphene reinforced Al matrix composites for high-performance conductor[J]. Compos Part B Eng. 2019;167:93–9.
6. Wu Y, Zhan K, Yang Z, et al. Graphene oxide/Al composites with enhanced mechanical properties fabricated by simple electrostatic interaction and powder metallurgy[J]. J Alloys Compd. 2019;775:233–40.
7. Wang J, Guo L, Lin W, et al. The effects of graphene content on the corrosion resistance, and electrical, thermal and mechanical properties of graphene/copper composites[J]. New Carbon Mater. 2019;34(2):161–9.
8. Rashad M, Pan F, Asif M, et al. Corrosion behavior of magnesium-graphene composites in sodium chloride solutions[J]. J Magnes Alloy. 2017;5(3):271–6.

9. Latief FH, Sherif ESM, Almajid AA, et al. Fabrication of exfoliated graphite nanoplatelets-reinforced aluminum composites and evaluating their mechanical properties and corrosion behavior[J]. J Anal Appl Pyrolysis. 2011;92(2):485–92.
10. Liu ZY, Xiao BL, Wang WG, et al. Tensile strength and electrical conductivity of carbon nanotube reinforced aluminum matrix composites fabricated by powder metallurgy combined with friction stir processing[J]. J Mater Sci Technol. 2014;30(7):649–55.
11. Gao C, Wang P, Wang Z, et al. The disordering-enhanced performances of the Al-MOF/graphene composite anodes for lithium-ion batteries[J]. Nano Energy. 2019;65: 104032.
12. Papageorgiou DG, Kinloch IA, Young RJ. Mechanical properties of graphene and graphene-based nanocomposites[J]. Prog Mater Sci. 2017;90:75–127.
13. Zhang Y, Liu S, Ji Y, et al. Emerging nonaqueous aluminum-ion batteries: challenges, status, and perspectives[J]. Adv Mater. 2018;30(38):1706310.
14. Li X, Deng S, Banis MN, et al. Suppressing corrosion of aluminum foils via highly conductive graphene-like carbon coating in high-performance lithium-based batteries[J]. ACS Appl Mater Interfaces. 2019;11(36):32826–32.
15. Sharma A, Gupta G, Paul J. A comprehensive review on the dispersion and survivability issues of carbon nanotubes in Al/CNT nanocomposites fabricated via friction stir processing[J]. Carbon Lett. 2021;31(3):339–70.
16. Liu ZY, Wang LH, Zan YN, et al. Enhancing strengthening efficiency of graphene nano-sheets in aluminum matrix composite by improving interface bonding[J]. Compos Part B Eng. 2020;199: 108268.
17. Turan ME. Investigation of mechanical properties of carbonaceous (MWCNT, GNPs and C60) reinforced hot-extruded aluminum matrix composites[J]. J Alloys Compd. 2019;788:352–60.
18. Shao P, Chen G, Ju B, et al. Effect of hot extrusion temperature on graphene nanoplatelets reinforced Al6061 composite fabricated by pressure infiltration method[J]. Carbon NY. 2020;162:455–64.
19. Yang Z, Gu H, Sha G, et al. TC4/Ag metal matrix nanocomposites modified by friction stir processing: surface characterization, antibacterial property, and cytotoxicity in vitro[J]. ACS Appl Mater Interfaces. 2018;10(48):41155–66.
20. Huang Y, Bazarnik P, Wan D, et al. The fabrication of graphene-reinforced Al-based nanocomposites using high-pressure torsion[J]. Acta Mater. 2019;164:499–511.
21. Brodova IG, Petrova AN, Shirinkina IG, et al. Mechanical properties of submicrocrystalline aluminium matrix composites reinforced by in situ graphene through severe plastic deformation processes[J]. J Alloys Compd. 2021;859: 158387.
22. Tiwari JK, Mandal A, Rudra A, et al. Evaluation of mechanical and thermal properties of bilayer graphene reinforced aluminum matrix composite produced by hot accumulative roll bonding[J]. J Alloys Compd. 2019;801:49–59.
23. Meng X, Huang Y, Cao J, et al. Recent progress on control strategies for inherent issues in friction stir welding[J]. Prog Mater Sci. 2021;115: 100706.
24. Rafieazad M, Mohammadi M, Gerlich A, et al. Enhancing the corrosion properties of additively manufactured AlSi10Mg using friction stir processing[J]. Corros Sci. 2021;178: 109073.
25. Rao AG, Katkar VA, Gunasekaran G, et al. Effect of multipass friction stir processing on corrosion resistance of hypereutectic Al–30Si alloy[J]. Corros Sci. 2014;83:198–208.
26. Xie Y, Meng X, Huang Y, et al. Deformation-driven metallurgy of graphene nanoplatelets reinforced aluminum composite for the balance between strength and ductility[J]. Compos Part B Eng. 2019;177: 107413.
27. Xie Y, Huang Y, Wang F, et al. Deformation-driven metallurgy of SiC nanoparticle reinforced aluminum matrix nanocomposites[J]. J Alloys Compd. 2020;823: 153741.
28. Giannozzi P, Baroni S, Bonini N et al. Quantum Espresso: a modular and open-source software project for quantum simulations of materials[J]. J Phys Condens Matter. 2009;21(39).
29. Liu M, Jin Y, Leygraf C, et al. A DFT-study of Cl ingress into α-Al_2O_3 (0001) and Al(111) and its possible influence on localized corrosion of Al[J]. J Electrochem Soc. 2019;166(11):C3124–30.
30. Wang H, Wang Y, Cao X, et al. Vibrational properties of graphene and graphene layers[J]. J Raman Spectrosc. 2009;40(12):1791–6.

31. Shao P, Yang W, Zhang Q, et al. Microstructure and tensile properties of 5083 Al matrix composites reinforced with graphene oxide and graphene nanoplates prepared by pressure infiltration method[J]. Compos Part A Appl Sci Manuf. 2018;109(2):151–62.
32. Jiang Y, Xu R, Tan Z, et al. Interface-induced strain hardening of graphene nanosheet/aluminum composites[J]. Carbon NY. 2019;146:17–27.
33. Chak V, Chattopadhyay H, Dora TL. A review on fabrication methods, reinforcements and mechanical properties of aluminum matrix composites[J]. J Manuf Process. 2020;56:1059–74.
34. Liu Q, Ma Q, Chen G, et al. Enhanced corrosion resistance of AZ91 magnesium alloy through refinement and homogenization of surface microstructure by friction stir processing[J]. Corros Sci. 2018;138:284–96.
35. Ju B, Yang W, Shao P, et al. Effect of interfacial microstructure on the mechanical properties of GNPs/Al composites[J]. Carbon NY. 2020;162:346–55.
36. So KP, Kushima A, Park JG, et al. Intragranular dispersion of carbon nanotubes comprehensively improves aluminum alloys[J]. Adv Sci. 2018;5(7):1800115.
37. Ni Y, Zhang F, Njoku DI, et al. Corrosion mechanism of CuAl–NiC abradable seal coating system—the influence of porosity, multiphase, and multilayer structure on the corrosion failure[J]. J Mater Sci Technol. 2021;88:258–69.
38. Birbilis N, Ralston KD, Virtanen S, et al. Grain character influences on corrosion of ECAPed pure magnesium[J]. Corros Eng Sci Technol. 2010;45(3):224–30.
39. Cen H, Zhang X, Zhao L, et al. Carbon dots as effective corrosion inhibitor for 5052 aluminium alloy in 0.1 M HCl solution[J]. Corros Sci. 2019;161:108197.
40. Cui M, Ren S, Xue Q, et al. Carbon dots as new eco-friendly and effective corrosion inhibitor[J]. J Alloys Compd. 2017;726:680–92.
41. Yang L, Wan Y, Qin Z, et al. Fabrication and corrosion resistance of a graphene-tin oxide composite film on aluminium alloy 6061[J]. Corros Sci. 2018;130:85–94.
42. Yu M, Dong H, Shi H, et al. Effects of graphene oxide-filled sol-gel sealing on the corrosion resistance and paint adhesion of anodized aluminum[J]. Appl Surf Sci. 2019;479:105–13.
43. Wang L, Liang J, Li H, et al. Quantitative study of the corrosion evolution and stress corrosion cracking of high strength aluminum alloys in solution and thin electrolyte layer containing Cl-[J]. Corros Sci. 2021;178: 109076.
44. Vijh AK. The corrosion potentials of some common metals in oxygenated solutions in relation to semi-conductivity of the corrosion films[J]. Corros Sci. 1972;12(2):105–11.
45. Lee C-K, Cho E, Lee H-S, et al. Comparative study of electronic structures and dielectric properties of alumina polymorphs by first-principles methods[J]. Phys Rev B. 2007;76(24):245110.
46. Mott NF. The theory of crystal rectifiers[J]. Proc R Soc London Ser A Math Phys Sci. 1939;171(944):27–38.
47. Schottky W. Zur halbleitertheorie der sperrschicht-und spitzengleichrichter[J]. Zeitschrift für Phys. 1939;113(5–6):367–414.
48. Zhou W, Yamaguchi T, Kikuchi K, et al. Effectively enhanced load transfer by interfacial reactions in multi-walled carbon nanotube reinforced Al matrix composites[J]. Acta Mater. 2017;125:369–76.
49. Santos RCR, Longhinotti E, Freire VN, et al. Elucidating the high-k insulator α-Al_2O_3 direct/indirect energy band gap type through density functional theory computations[J]. Chem Phys Lett. 2015;637:172–6.
50. Carrasco J, Gomes JRB, Illas F. Theoretical study of bulk and surface oxygen and aluminum vacancies in α-Al_2O_3[J]. Phys Rev B. 2004;69(6): 064116.
51. Li J, Wei H, Zhao K, et al. Effect of anodizing temperature and organic acid addition on the structure and corrosion resistance of anodic aluminum oxide films[J]. Thin Solid Films. 2020;713: 138359.
52. Bunch JS, Verbridge SS, Alden JS, et al. Impermeable atomic membranes from graphene sheets[J]. Nano Lett. 2008;8(8):2458–62.
53. Böhm S. Graphene against corrosion[J]. Nat Nanotechnol. 2014;9(10):741–2.
54. Berry V. Impermeability of graphene and its applications[J]. Carbon NY. 2013;62:1–10.

55. Cui C, Lim ATO, Huang J. A cautionary note on graphene anti-corrosion coatings[J]. Nat Nanotechnol. 2017;12(9):834–5.
56. Sun W, Wang L, Wu T, et al. Synthesis of low-electrical-conductivity graphene/pernigraniline composites and their application in corrosion protection[J]. Carbon NY. 2014;79:605–14.
57. Wlasny I, Dabrowski P, Rogala M, et al. Impact of electrolyte intercalation on the corrosion of graphene-coated copper[J]. Corros Sci. 2015;92:69–75.
58. Lee J, Berman D. Inhibitor or promoter: insights on the corrosion evolution in a graphene protected surface[J]. Carbon NY. 2018;126:225–31.
59. Schriver M, Regan W, Gannett WJ, et al. Graphene as a long-term metal oxidation barrier: worse than nothing[J]. ACS Nano. 2013;7(7):5763–8.
60. Yang H, Li F, Shan C, et al. Covalent functionalization of chemically converted graphene sheets via silane and its reinforcement[J]. J Mater Chem. 2009;19(26):4632.
61. Kumar R, Sahoo S, Joanni E, et al. Heteroatom doped graphene engineering for energy storage and conversion[J]. Mater Today. 2020;39:47–65.
62. Yang Z, Sun W, Wang L, et al. Liquid-phase exfoliated fluorographene as a two dimensional coating filler for enhanced corrosion protection performance[J]. Corros Sci. 2016;103:312–8.
63. Sinhmar S, Dwivedi DK. A study on corrosion behavior of friction stir welded and tungsten inert gas welded AA2014 aluminium alloy[J]. Corros Sci. 2018;133(February):25–35.
64. Liu ZY, Xiao BL, Wang WG, et al. Developing high-performance aluminum matrix composites with directionally aligned carbon nanotubes by combining friction stir processing and subsequent rolling[J]. Carbon NY. 2013;62:35–42.
65. Jeon I, Park GH, Wang P, et al. Dynamic fluid-like graphene with ultralow frictional molecular bearing[J]. Adv Mater. 2019;31(43):1903195.
66. Casper P, Hünig R, Gomard G, et al. Optoelectrical improvement of ultra-thin Cu(In, Ga)Se2 solar cells through microstructured MgF2 and A_12_O3 back contact passivation layer[J]. Phys Status Solidi-Rapid Res Lett. 2016;10(5):376–80.
67. Feng W, Long P, Feng Y, et al. Two-dimensional fluorinated graphene: Synthesis, structures, properties and applications[J]. Adv Sci. 2016;3(7):1500413.
68. Bauer E, Kaldarar H, Lackner R, et al. Superconductivity in the complex metallic alloy β-Al_3Mg_2[J]. Phys Rev B. 2007;76(1): 014528.
69. Zhuang H, Chen M, Carter EA. Elastic and thermodynamic properties of complex Mg–Al intermetallic compounds via orbital-free density functional theory[J]. Phys Rev Appl. 2016;5(6): 064021.
70. Zhou S, Li W, Zhao W, et al. Tribological behaviors of polyimide composite coatings containing carbon nanotubes and fluorinated graphene with hybrid phase or blend phase[J]. Prog Org Coat. 2020;147: 105800.
71. Şahin H, Topsakal M, Ciraci S. Structures of fluorinated graphene and their signatures[J]. Phys Rev B. 2011;83(11): 115432.
72. Wei W, Jacob T. Electronic and optical properties of fluorinated graphene: a many-body perturbation theory study[J]. Phys Rev B. 2013;87(11): 115431.
73. Chen B, Shen J, Ye X, et al. Length effect of carbon nanotubes on the strengthening mechanisms in metal matrix composites[J]. Acta Mater. 2017;140:317–25.
74. Kleiven D, Ødegård OL, Laasonen K, et al. Atomistic simulations of early stage clusters in Al–Mg alloys[J]. Acta Mater. 2019;166(7491):484–92.
75. Shanmugasundaram T, Heilmaier M, Murty BS, et al. On the Hall-Petch relationship in a nanostructured Al–Cu alloy[J]. Mater Sci Eng A. 2010;527(29–30):7821–5.
76. Zhang Y, Li X. Bioinspired, graphene/Al_2O_3 doubly reinforced aluminum composites with high strength and toughness[J]. Nano Lett. 2017;17(11):6907–15.
77. Mansouri K, Ibrik K, Bensalah N, et al. Anodic dissolution of pure aluminum during electrocoagulation process: influence of supporting electrolyte, initial pH, and current density[J]. Ind Eng Chem Res. 2011;50(23):13362–72.
78. Xie Y, Meng X, Mao D, et al. Homogeneously dispersed graphene nanoplatelets as long-term corrosion inhibitors for aluminum matrix composites[J]. ACS Appl Mater Interfaces 2021;acsami.1c07148.

79. Grimm M, Lohmüller A, Singer RF, et al. Influence of the microstructure on the corrosion behaviour of cast Mg–Al alloys[J]. Corros Sci. 2019;155:195–208.
80. Sun W, Yang Y, Yang Z, et al. Review on the corrosion-promotion activity of graphene and its inhibition[J]. J Mater Sci Technol. 2021;91:278–306.
81. Xie Y, Meng X, Li Y, et al. Insight into ultra-refined grains of aluminum matrix composites via deformation-driven metallurgy[J]. Compos Commun. 2021;26: 100776.
82. Wu Y, Zhu X, Zhao W, et al. Corrosion mechanism of graphene coating with different defect levels[J]. J Alloys Compd. 2019;777:135–44.

Chapter 11
SiC Reinforced Aluminum Matrix Composites via Deformation-Driven Metallurgy

11.1 Effect of the SiC Particle Size on the Strength-Ductility Synergy of the Reinforced Aluminum Matrix Composites

11.1.1 Introduction

Silicon carbide particles reinforced aluminum matrix composites (SiC/AMCs) are highly regarded for low density, excellent wear resistance, and high specific strength, which have great application prospects in the field of aerospace, automobile, and military industries [1–4]. However, the SiC/AMCs are usually limited by a great loss of ductility in parallel with the enhancement of strength.

It is a complicated issue to overcome the strength-ductility trade-off, which is determined by the size, dispersion site, and uniformity of particles [5–7]. Particles dispersed in different sites, such as intragranularly-dispersed or grain boundary segregated, served as different roles in various strengthening mechanisms. Controlling the reinforcement particles intragranularly-dispersed was a typical method to balance the strength and ductility, which could enhance the strength of AMCs homogeneously under the effect of Orowan strengthening [8–12]. However, the Orowan strengthening would be weakened with the coarsening of reinforcement particles in micro-sized particulate-reinforced metal matrix composites, as particles with different sizes have different capabilities to hinder the movement dislocations [13, 14]. The relationship between particle size and load-bearing characteristics was illustrated by several theoretical models [15, 16] that the strength could be enhanced better with the smaller reinforcement particles. The premise of these theories was also emphasized that the particles should be dispersed uniformly as the fatal ductility loss was usually attributed to the nonuniform dispersion of reinforced particles [17]. Remarkable attention should be paid to the fact that the solutions for the nonuniform dispersion are different in various preparation methods [18–21]. For the liquid-phase methods,

Y. Huang et al., *Friction Stir Welding and Processing*, Materials Forming, Machining and Tribology, https://doi.org/10.1007/978-981-99-8688-0_11

the reinforcement particles were usually pre-oxidized [22] or graphene-encapsulated [23] to counteract the localized cluster at the liquid–solid interface during the solidification process as small-sized particles were tended to cluster with each other under the effect of the high specific areas. As the particles could not be redispersed in the solid-phase method, the dispersion of particles was usually promoted by the pretreatment of appropriate ball-milling [24]. A comprehensive understanding of the influence of particle size and its dispersion characteristics was needed, which had significance in balancing the strength and ductility of SiC/AMCs.

A novel solid-phase method with the features of low heat-input and severe plastic deformation was proposed, namely deformation-driven metallurgy, which could promote the redispersion of particles during the preparation process. AMCs reinforced with micro/nano-SiC particles were investigated. The dispersion characteristics of particles and strengthening-efficiency mechanism were evaluated. An effective strategy to promote strength-ductility synergy was provided through the optimization of SiC particles in deformation-driven metallurgy of SiC/AMCs.

11.1.2 Experimental Procedures

Spherical 1060 pure aluminum powders (5 μm in average diameter, purity > 99.6%) were reinforced with 10 wt.% SiC particles of two different size levels. Nano-SiC particles (10 nm in average diameter, purity > 99.0%) and micro-SiC particles (2 μm in average diameter, purity > 99.0%) were chosen. The blended powders of SiC/Al were mixed by a planetary ball milling machine (QM-3SP2, Nanjing University Instrument Plant). The rotational velocity, milling time, and ball-to-powder ratio were 200 r/min, 10 h, and 5:1, respectively. The diameter of the ZrO_2 milling balls was 6 mm. As shown in Fig. 11.1, both nano-SiC particles and micro-SiC particles were mixed with Al powders uniformly, especially the easy-to-aggregate nano-SiC particles. The surface of Al powders was covered homogeneously by nano-SiC particles (Fig. 11.1a), which provided the precondition for obtaining sound nano-SiC reinforced AMCs (SiC_{np}/AMCs).

The DDM process was applied with a processing time of 30 s and a rotational velocity of 1200 r/min. The fabricated composites with 16 mm in diameter and 1 mm in height were obtained. The specific process was showed in Fig. 11.2.

The morphology of the raw powders and the microstructures of the bulk composites were characterized by the scanning electron microscope (SEM, Zeiss MERLIN Compact) equipped with attachments for energy dispersive spectroscopy (EDS). The nature of the grains was characterized by electron backscattered diffraction (EBSD) with a step size of 0.1 μm. A high-resolution transmission electron microscope (HR-TEM, FEI Talos F200x) at a working voltage of 200 kV was applied to characterize the morphologies of grains and reinforcements. The density of each composite was determined in accordance with the Archimedean principle.

An Instron 5569 tester was applied to evaluate the tensile properties of the Al-SiC composites with a tensile speed of 0.5 mm/min. HXD-1000TM microhardness

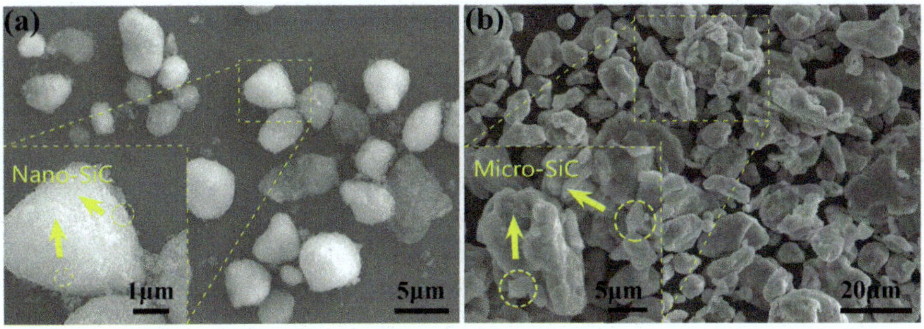

Fig. 11.1 SEM images of ball-milled **a** Al/nano-SiC powders and **b** Al/micro-SiC powders

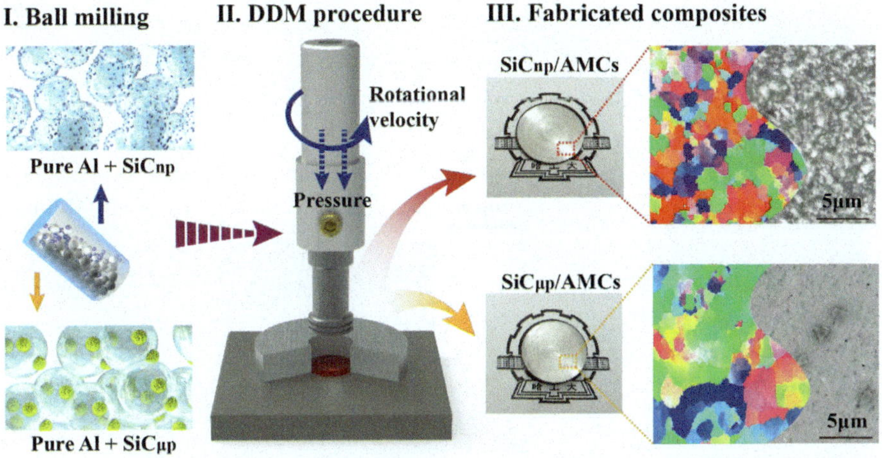

Fig. 11.2 Schematic of DDM procedure and the typical composites

measurement was used to measure Vicker's microhardness. The experimental force was 100 gf, and the dwelling time was 10 s. The nano-indentation was measured using an Agilent G200 nanoindenter with an applied force of 10 mN.

11.1.3 Microstructural Integrity

Figure 11.3 shows the dispersion condition of nano-SiC and micro-SiC particles after DDM. The average relative density of the SiC_{np}/AMCs and $SiC_{\mu p}$/AMCs were (98.0 $\pm$ 0.2%) and (92.4 $\pm$ 0.1%), respectively. The density of SiC_{np}/AMCs was significantly higher than that of the $SiC_{\mu p}$/AMCs as the DDM characterized by severe plastic deformation is a solid-state preparation method. The dispersion of nano-SiC particles was promoted (Fig. 11.3a) with the non-existence of a solid–liquid interface. With the pressure and rotation of the DDM tool (Fig. 11.2), the particles were redispersed during the fabrication process. The formation of nano-particle clusters was further alleviated, which was beneficial to the increase of the density in SiC_{np}/AMCs. Although the dispersion of micro-SiC particles was promoted, the particles were broken during the DDM, as shown in Fig. 11.3b. Micro-holes were formed among micro-SiC particles, which made the density lower.

Figure 11.3c depicts the dispersion of nano-SiC particles in the SiC_{np}/AMCs. The two kinds of particles with different size-level were observed, which were formed

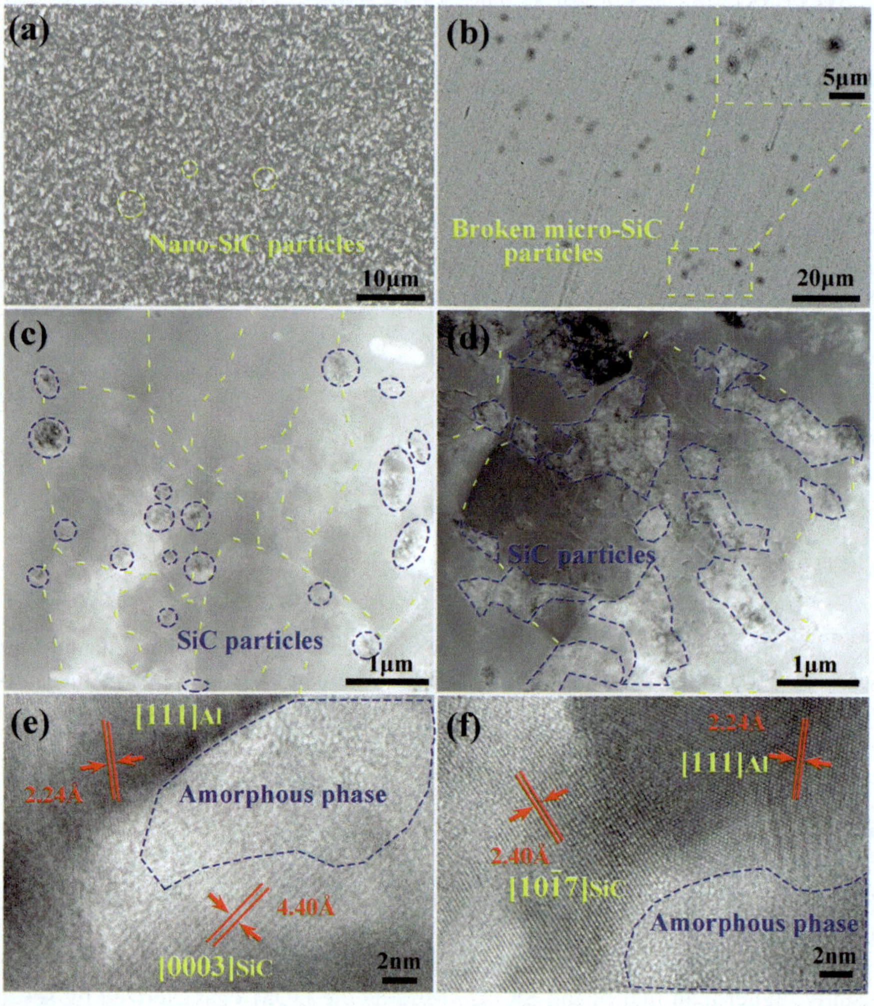

Fig. 11.3 The typical dispersion of **a/c** nano-SiC particles and **b/d** micro-SiC particles in Al matrix; the interface of **e** SiC_{np}/Al and **f** $SiC_{\mu p}$/Al

by dozens of slightly-clustered raw SiC powders. The size of large particles was controlled to 200 nm approximately, while the smaller particles with a size of about tens of nanometers were observed. In $SiC_{\mu p}$/AMCs, micro-SiC particles were broken during the DDM process (Fig. 11.3b). In addition to the large particles with a size of about 1.5 μm, some smaller particles of different sizes were dispersed around, whose diameter ranged from nano-level to micro-level (Fig. 11.3d).

The amorphous oxide was formed on the surface of the SiC particles both in SiC_{np}/AMCs and $SiC_{\mu p}$/AMCs due to the severe plastic deformation of DDM as circled by blue dot lines (Fig. 11.3e, f). The oxide layer enhanced the wetting behavior between the reinforcements and the matrices, which could restrain the formation of the Al_4C_3 brittle intermetallic compounds.

11.1.4 Dynamic Recrystallization Process

With severe plastic deformation, SiC_{np}/AMCs and $SiC_{\mu p}$/AMCs both undergo dynamic recrystallization (DRX) during the DDM process. The grain sizes were decreased, but the refinement extent was different in the SiC_{np}/AMCs and $SiC_{\mu p}$/AMCs, as shown in Fig. 11.4. The average grain size of the SiC_{np}/AMCs was much smaller, about 0.5 μm lower than that of the $SiC_{\mu p}$/AMCs. The grains refinement in the $SiC_{\mu p}$/AMCs were nonuniform as quite a few grains were over the size of 4 μm compared with that of SiC_{np}/AMCs.

The specific grain refinement process of SiC_{np}/AMCs is expressed in Fig. 11.5. Figure 11.5a shows the grain undergoing DRX with about three times the average grain size (Fig. 11.4c), and Fig. 11.5b shows the equiaxed grains finally obtained after DRX. In Fig. 11.5a, multiple sub-grains were formed, and the sub-grain boundaries were marked with red dotted lines, which was consistent with the EBSD data that low-angle grain boundaries (LAGBs) with a considerable proportion were measured (Fig. 11.4e). Inside the equiaxed grains, with sufficient DRX, the typical stable grain boundary with an angle of 120° was observed (indicated by the yellow arrow in Fig. 11.5b).

Dislocations with high density were formed in the Al grains under the effect of deformation stress (Fig. 11.5c). As a material with high stacking fault energies, the grain refinement of Al induced by deformation is mainly caused by dislocation accumulation and rearrangement [25, 26]. With the further interaction of dislocations, sub-grains with LAGBs appeared (Fig. 11.5a). As shown in Fig. 11.5d, the migration of dislocations inside the Al grain was impeded by SiC particles. The dislocation interaction was transformed into the mutual action between SiC and dislocations. Compared with the DRX process through dislocations interaction, the hindrance effect of SiC particles activated the accumulation of dislocations at the early stage, even with the low dislocation density (Fig. 11.5d). As pointed by red dotted lines in Fig. 11.5b, nano-SiC particles pinned at the grain boundaries inhibited the grain growth, as the plastic deformation can also result in grain coarsening. The final grain size was achieved by the dynamic balance between the grain refinement and growth

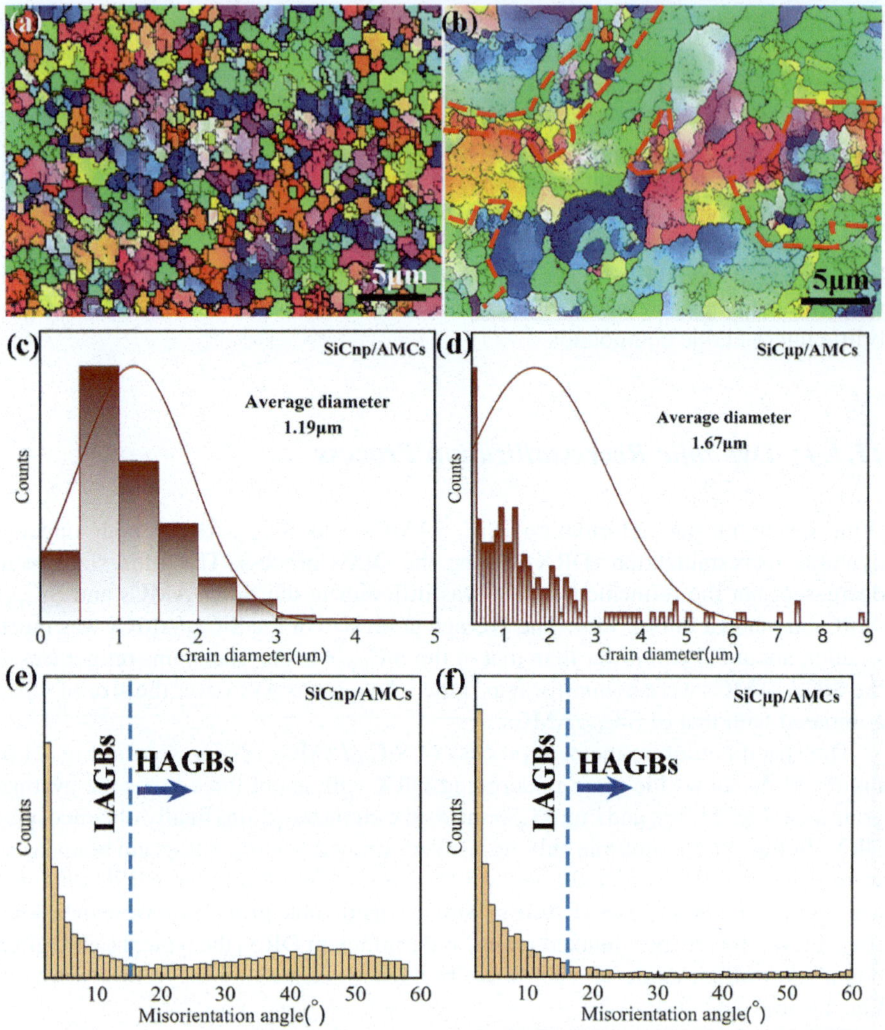

Fig. 11.4 Typical EBSD images of **a** SiC_{np}/AMCs and **b** $SiC_{\mu p}$/AMCs; Grain diameter and misorientation angle statistics dispersion of **c/e** SiC_{np}/AMCs and **d/f** $SiC_{\mu p}$/AMCs

process [27, 28]. With the pinning effect of nano-SiC particles, the stability of the ultra-fine-grained microstructure was increased, and the refinement of AMCs was further improved, which was known as the Zener pinning mechanism [29].

In $SiC_{\mu p}$/AMCs, SiC particles with sizes of about 1.5 μm (Fig. 11.3) were beneficial to the refinement of Al grains as the SiC particles could function as the nucleation sites. According to several researches [30], a specific critical size of the particles as nucleation sites should be larger than 0.1 μm, with optimum particle size ranging from 0.5 to 2.0 μm, known as particle stimulated nucleation mechanism.

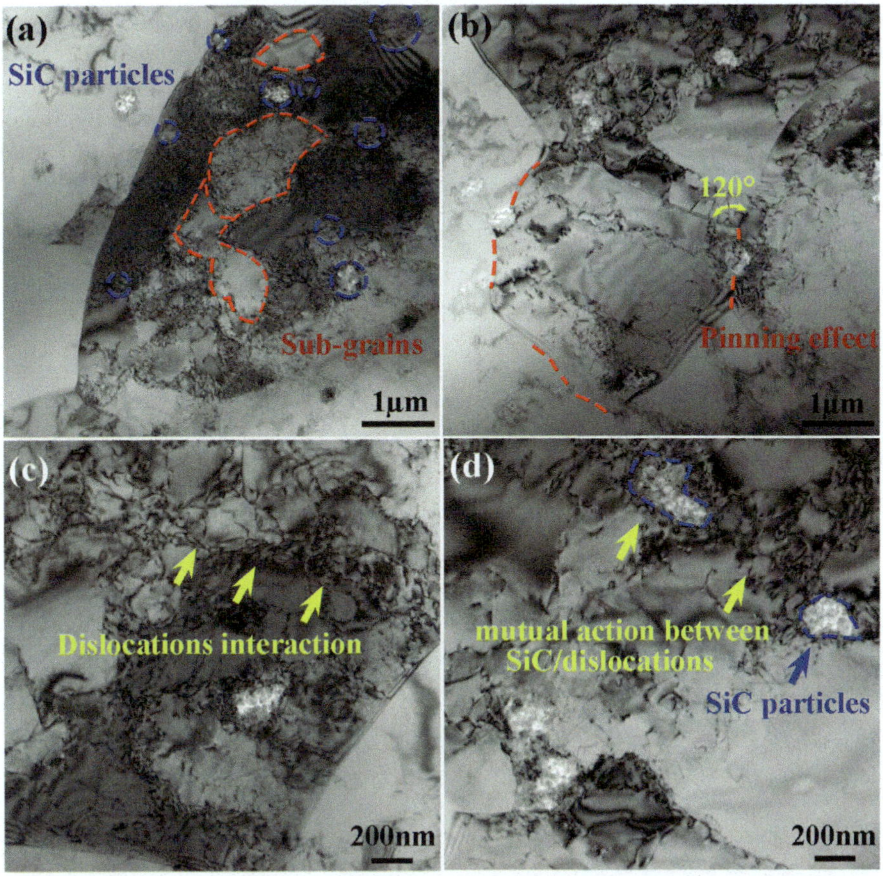

Fig. 11.5 TEM images of SiCnp/AMCs: **a** sub-grains and internal dispersed SiC particles in grains **b** the equiaxed recrystallized grains and Zener pinning effect **c** dislocation–dislocation interaction **d** SiC-dislocation mutual action

The dislocations were produced around the SiC particles with the plastic deformation of the Al matrix during the DDM process. The dislocation densities around these deformed zones were high (Fig. 11.6a), which were preferable sites for recrystallization nucleation. The nucleation occurs via the rapid sub-boundary migration around non-deformable SiC particles and eventually contributes to the creation of high-angle grain boundaries (HAGBs). As shown in Fig. 11.6d, the deformed zones in the $SiC_{\mu p}$/AMCs were localized dispersed. Only the DRX process of these deformed grains was promoted, which was consistent with the localized refined microstructure pointed out by red dotted lines in Fig. 11.4b. Thus, a typical multimodal grain microstructure with micro-sized and ultrafine grains was obtained. As characterized by low heat input, the trigger of the particle stimulated nucleation mechanism was further guaranteed during the DDM process, as the dislocations could climb around the reinforcement particles at high temperatures [31, 32].

In $SiC_{\mu p}$/AMCs, combined with the particle stimulated nucleation mechanism, nano-scale SiC particles (i.e., broken micro-SiC particles pointed by blue dotted circles in Fig. 11.6a and b around micro-SiC particles had also promoted grain

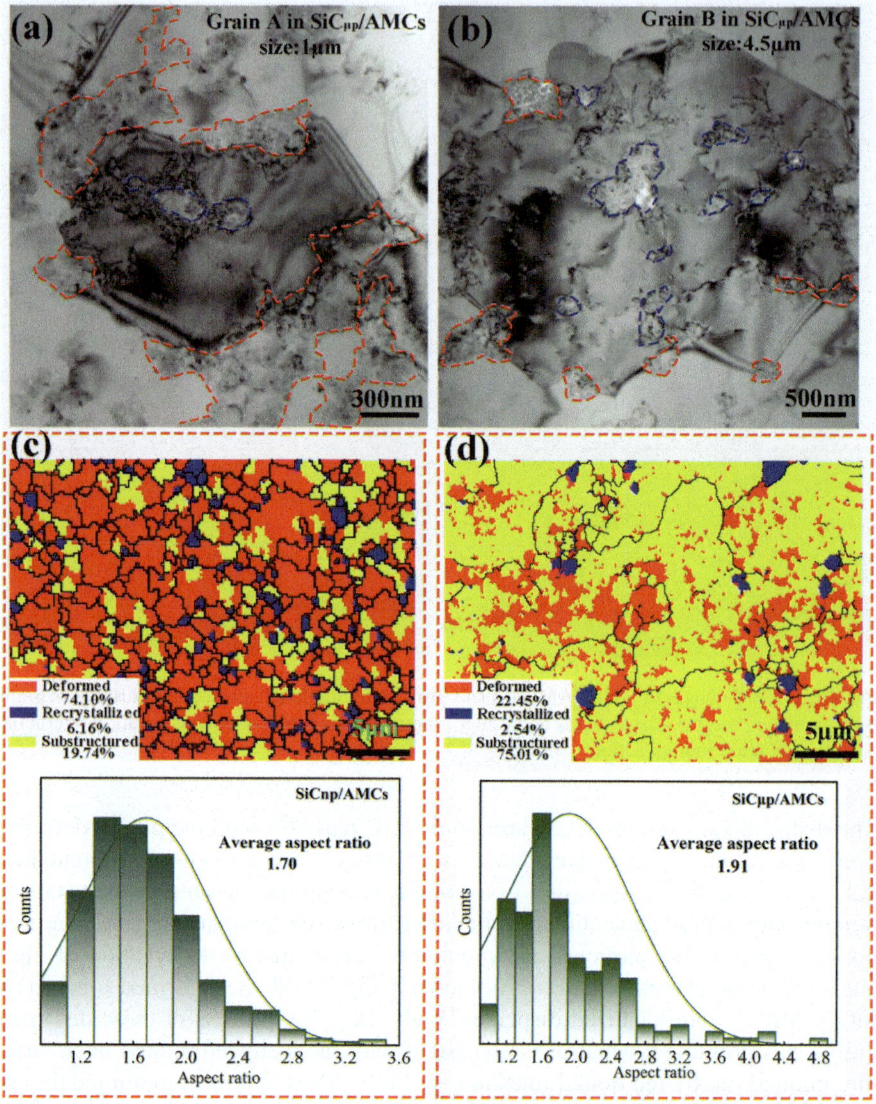

Fig. 11.6 TEM images of $SiC_{\mu p}$/AMCs: **a** the equiaxed recrystallized grain **b** the grain with insufficient DRX; the fraction of recrystallized/deformed grains and aspect ratio statistics dispersion of **c** SiC_{np}/AMCs and **d** $SiC_{\mu p}$/AMCs

refinement. The mechanism was the same as that in SiC_{np}/AMCs, as the DRX was promoted by intragranularly-dispersed SiC particles, and the stability of the fine recrystallized grains was increased under the effect of the Zener pinning mechanism (Fig. 11.5). Under the effect of these broken micro-SiC particles, the DRX process in $SiC_{\mu p}$/AMCs was facilitated, and the typical multimodal grain microstructure (Fig. 11.4) was further intensified. The different grain refinement content between SiC_{np}/AMCs and $SiC_{\mu p}$/AMCs was caused by the different SiC dispersion modes. The nano-SiC particles in SiC_{np}/AMCs were dispersed homogeneously (Fig. 11.3a). The grains were all ultra-fine and uniform in size (Fig. 11.4a). However, the amount of SiC particles in $SiC_{\mu p}$/AMCs were not as much as that of SiC_{np}/AMCs, and the SiC particles were localized dispersed (Fig. 11.3b). Figure 11.6a and b show two grains (grain A and B) with different sizes in $SiC_{\mu p}$/AMCs. The number of SiC particles around grain B was less than that of grain A (Fig. 11.6a, b), and even the size of SiC particles around grain B was also much smaller. The refinement of grain B was restrained, as its size was about 4.5 times grain A. The DRX degree of $SiC_{\mu p}$/AMCs was far less than that of SiC_{np}/AMCs, which makes the average aspect ratio of $SiC_{\mu p}$/AMCs larger (Fig. 11.6c, d), even quite a few grains with an aspect ratio > 3.6 were found (Fig. 11.6d). Moreover, due to the insufficient DRX, the LAGBs proportion of $SiC_{\mu p}$/AMCs was also higher than that of the SiC_{np}/AMCs (Fig. 11.4e, f).

11.1.5 Strengthening Mechanism

As shown in Table 11.1, the microhardness, elastic modulus, and nanohardness of SiC_{np}/AMCs were improved significantly, reaching 300, 150, and 223% of the $SiC_{\mu p}$/AMCs, respectively. These differences were attributed to the different grain sizes of the two AMCs. The grain refinement-induced strengthening can be described in terms of the empirical Hall–Petch relation [33, 34]. According to formula (11.1), with the smaller grain size, the hardness of SiC_{np}/AMCs was enhanced better.

$$H = H_0 + K_H d^{-1/2} \tag{11.1}$$

where H is the hardness, H_0 and K_H are material constants, and d is the grain size.

Figure 11.7 depicts the dispersion mode of the SiC_{np}/AMCs and $SiC_{\mu p}$/AMCs. With more uniform dispersion of the nano-SiC particles in the matrix (Fig. 11.7a), the hardness of the SiC_{np}/AMCs improved obviously as a large amount of dispersed

Table 11.1 The mechanical performance of the SiC/Al composites

Samples	Microhardness (HV)	Elastic modulus (GPa)	Nanohardness (GPa)
SiC_{np}/AMCs	119.5 ± 3.1	121.1 ± 2.3	1.79 ± 0.05
$SiC_{\mu p}$/AMCs	39.7 ± 1.6	80.6 ± 1.7	0.80 ± 0.07

nano-SiC particles bear the load simultaneously compared with that of $SiC_{\mu p}$/AMCs (Fig. 11.7a, b). In SiC_{np}/AMCs, the particles were evenly dispersed under the indenter after the measurement, while the micro-SiC particles have peeled off from the matrix. Micro-cracks sprouted at the edge of the interface and gradually expanded into the matrix. This phenomenon illustrated that the nano-SiC particles were better combined with the matrix.

As shown in Fig. 11.8, the engineering stress of SiC_{np}/AMCs and $SiC_{\mu p}$/AMCs both increased compared with pure Al. The ultimate tensile strength of SiC_{np}/AMCs enhanced to 435 MPa, reaching 292% of the pure Al, while that of the $SiC_{\mu p}$/AMCs increased to 200 MPa, reaching only 135% of the pure Al. Following the Hall–Petch relationship [35], plastic deformation becomes more difficult at smaller grains with a higher grain boundaries density, which hinders dislocation motion.

Uniformly dispersed SiC particles can be clearly observed in the dimples (Fig. 11.9a, b). As can be seen in the magnified figure (Fig. 11.9c), the nano-SiC particles in the dimples still maintain a sound bonding with the Al matrix, while obvious cracks were observed at the interface of the micro-SiC/matrix (Fig. 11.9d)

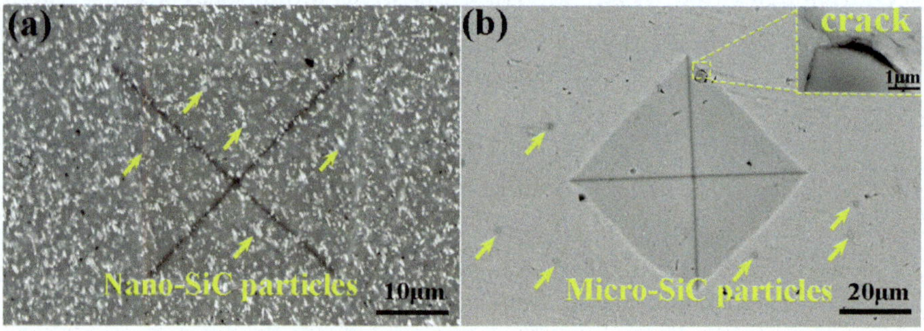

Fig. 11.7 SEM images of **a** SiCμp/AMCs and **b** SiCnp/AMCs after micro-hardness test

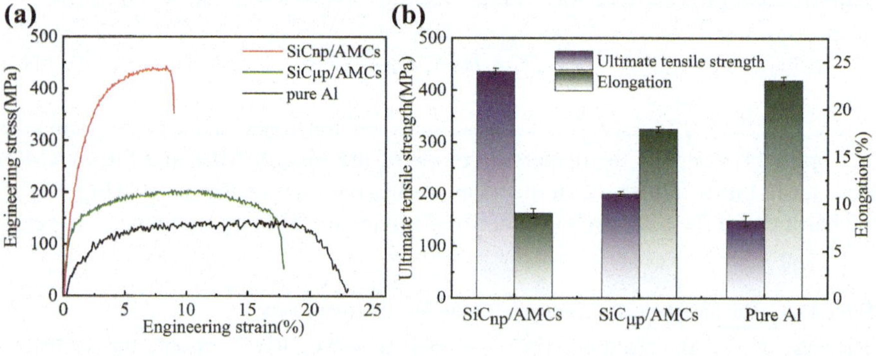

Fig. 11.8 Mechanical performances of the DDMed samples: **a** typical engineering stress–strain curves and **b** ultimate tensile strength/elongation of SiC_{np}/AMCs, $SiC_{\mu p}$/AMCs, and pure Al

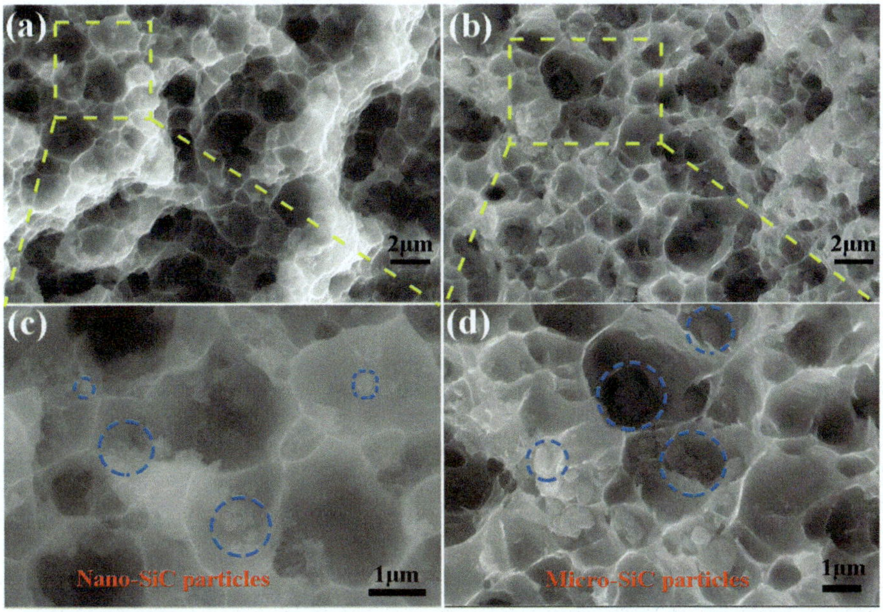

Fig. 11.9 Typical fractography surfaces of **a/c** SiC_{np}/AMCs and **b/d** $SiC_{\mu p}$/AMCs

as the micro-pores were more likely to nucleate at the particles/matrix interface. Some micro-SiC particles had even completely peeled off from the matrix and were scattered in the dimples (Fig. 11.9d). The decrease in the mechanical properties may also be affected by such a weak strength of the micro-SiC/matrix interface.

The dislocation movement was inhibited by SiC particles both in SiC_{np}/AMCs and $SiC_{\mu p}$/AMCs, as shown in Fig. 11.10a and b. The strain field caused by dislocations around the SiC/Al interface was observed (Fig. 11.10c, d), as can be figured out through the strong contrast (i.e., the grains appear dark), which indicated that the dislocation movement was hindered to some extent exactly. The SiC particles in $SiC_{\mu p}$/AMCs with the size of about 500 nm (Fig. 11.10b) hindered the dislocation movement more effectively compared with that of SiC_{np}/AMCs, which has small SiC particles with the size of about 200 nm (Fig. 11.10a). With more dislocations hindered by SiC particles in $SiC_{\mu p}$/AMCs, the dislocation density around the SiC particles tended to be much higher with the interaction of accumulated dislocations, which confirmed that SiC particles with different sizes have a different hinder effect on dislocations.

In the SiC_{np}/AMCs, the two kinds of particles with different size-level were observed (Fig. 11.11). The large particles with a size of about 200 nm were pointed out by yellow arrows, while blue arrows highlighted the smaller particles with a size of about tens of nanometers. As illustrated in Fig. 11.11a, the dislocations (white lines) were distributed homogeneously. With the block of small particles, the dislocation lines were stacked and bent around the particles until they were enclosed completely.

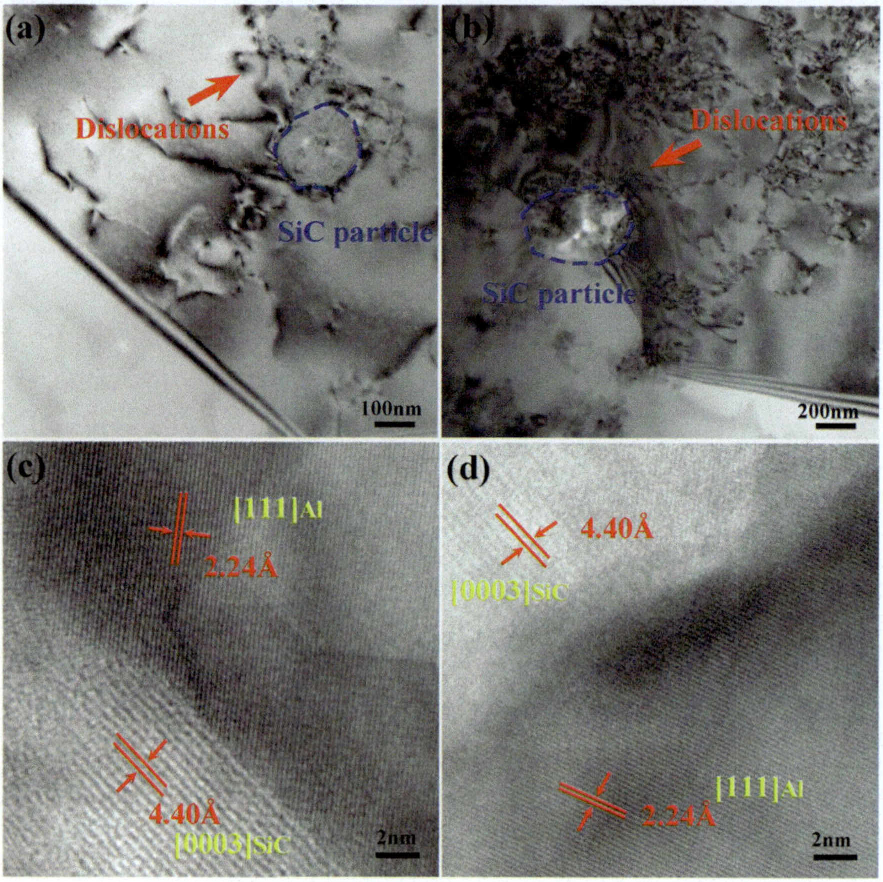

Fig. 11.10 The dislocation/SiC_p interaction of **a** SiC_{np}/AMCs and **b** $SiC_{\mu p}$/AMCs; the SiC/Al interface of **c** SiC_{np}/AMCs and **d** $SiC_{\mu p}$/AMCs

The classical "bow-out" process of dislocation lines (Fig. 11.11c) was observed as the feature of the Orowan strengthening. Under the resistance of closely spaced nano-SiC particles, the dislocation density of the SiC_{np}/AMCs was elevated entirely with the continued passing of dislocations. The dislocation density in each region of the grain was high, regardless of whether there were SiC particles around. The large SiC particles in SiC_{np}/AMCs were served as the pinning obstacles (Fig. 11.11c), which impeded the further movement of dislocations. Hence, the strength of SiC_{np}/AMCs was enhanced homogeneously under the coupling effect of the Orowan mechanism and pinning effect. In $SiC_{\mu p}$/AMCs, the Orowan strengthening mechanism was not the main factor due to the coarsening of SiC particles, though the dislocation "bow-out" process occasionally occurs (red line in Fig. 11.11d). As shown in Fig. 11.11b, the size of micro-SiC particles reached about 0.5 ~ 1 μm. Instead of the passing-by mode, the dislocations were stopped and gathered around coarse SiC particles. Thus, compared with the SiC_{np}/AMCs, the strength of the $SiC_{\mu p}$/AMCs was enhanced heterogeneously. As shown in Fig. 11.11d, the dislocation pile-ups induced by the interaction of dislocations were observed around the micro-SiC particles. Localized reinforcement and the high dislocation density around the SiC/Al interface made it a preferred site for crack initiation, which weakened the strength of the SiC/Al interface.

The kernel averaged misorientation (KAM) of $SiC_{\mu p}$/AMCs, and SiC_{np}/AMCs were also tested, shown in Fig. 11.11e and f, which measured the local grain misorientation as a representation of dislocation density dispersion in different grains. As a general rule, the KAM value is high (> 1°) in deformed grains due to the high density of dislocations, while the value is low (< 1°) in recrystallized grains [36, 37] conversely. Some recrystallized grains were found in $SiC_{\mu p}$/AMCs and SiC_{np}/AMCs (shown in Fig. 11.11e, f). Except for the recrystallized equiaxed grains, the dislocation density of each grain in the SiC_{np}/AMCs was almost at the same level, as nearly the whole grains were in green with the KAM values ranging from 1 to 2, while the KAM values of $SiC_{\mu p}$/AMCs were uneven. Excluding a small amount of recrystallized equiaxed grains, the KAM values in the $SiC_{\mu p}$/AMCs were various as the grains in blue and green were separated locally. Even within a single grain, the values of KAM were diverse. In brief, the dislocation bypass mode enhanced the dislocation density all over the SiC_{np}/AMCs homogeneously. On the contrary, the dislocations were blocked by large-size SiC particles and made the dislocation density localized raised in the $SiC_{\mu p}$/AMCs.

As shown in Fig. 11.8, the ductility of the $SiC_{\mu p}$/AMCs was kept at 18%. Although the strengthening-efficiency of the $SiC_{\mu p}$/AMCs was inferior to that of the SiC_{np}/AMCs, it maintained a great elongation. The entire fracture of the two AMCs showed a typical ductile feature (Fig. 11.9). The dimples in the SiC_{np}/AMCs were small and shallow, while the dimples were large and deep in the $SiC_{\mu p}$/AMCs, which confirmed that the $SiC_{\mu p}$/AMCs had undergone sufficient plastic deformation. Such great ductility of the $SiC_{\mu p}$/AMCs was attributed to the multimodal grain microstructure, which consists of micro-sized grains mixed with ultrafine grains (shown in Fig. 11.4). In the $SiC_{\mu p}$/AMCs, the coarse grains had plenty of space for accommodating newly generated dislocations. The slip of dislocations inside the coarse grains was easy as the coarse grains had a larger free pathway, which the dislocation density was low (colored by blue in Fig. 11.11f). The dislocation storage efficiency inside the ultrafine grains was low, and it is prone to plastic instability according to the Considere' criterion [38, 39]. With the tensile strain, the dislocation density in ultrafine grains will be saturation quickly, and consequent stress concentration would be relaxed by transferring local loads to the soft coarse grains. The geometrically necessary dislocations will build up to accommodate the deformation incompatibility near grain boundaries between soft and hard grains, and the local complex 3D stress states induced by the strain gradients would promote the stress-relaxing process. Further, the slip of the dislocations was less hindered as the SiC particles around the coarse Al grains were less and small in the $SiC_{\mu p}$/AMCs (Fig. 11.6b). The ductility of the $SiC_{\mu p}$/AMCs with multimodal grain microstructure was exactly promoted, which attributed to the broken micro-SiC particles with sizes ranging

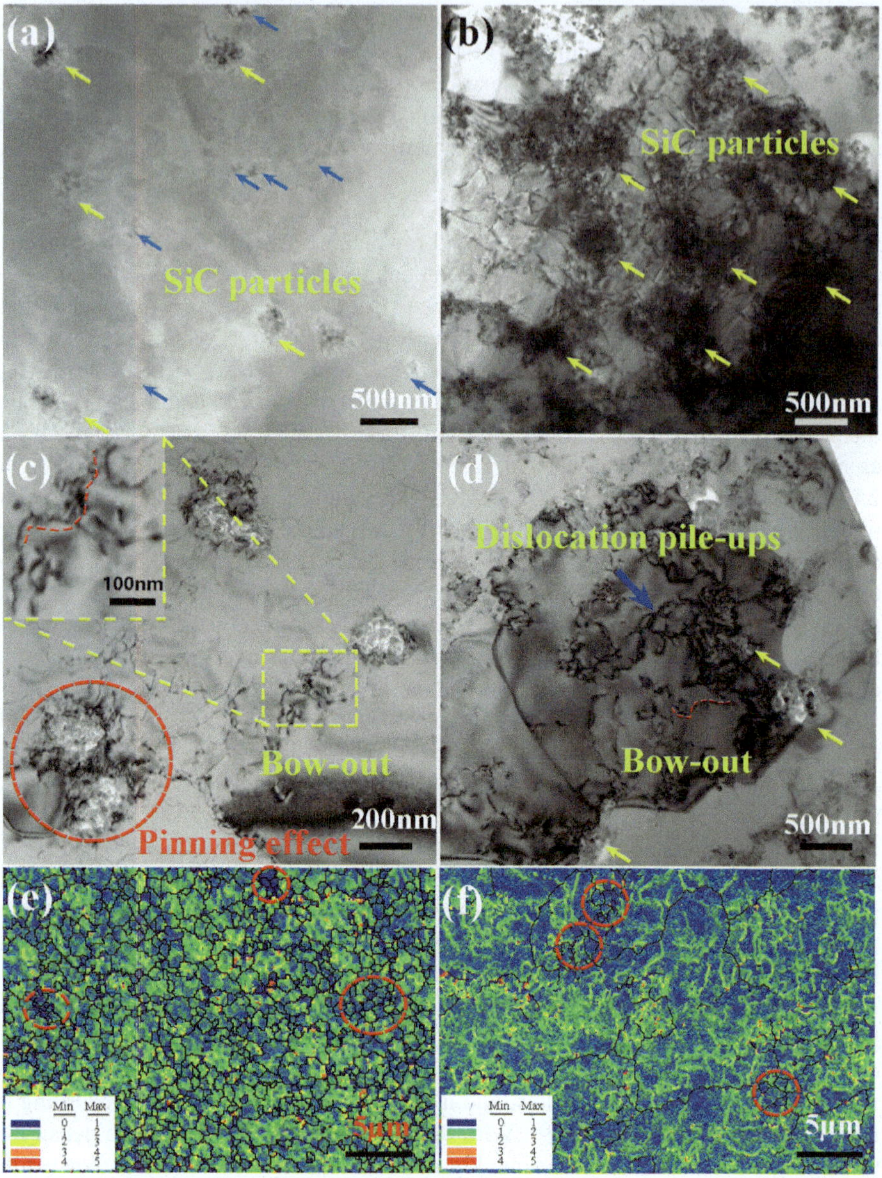

Fig. 11.11 The dispersion of SiC particles and its hinder effect on dislocations in **a/c** SiC_{np}/AMCs and **b/d** $SiC_{\mu p}$/AMCs; Kernel average misorientation dispersion of the **e** SiC_{np}/AMCs and **f** $SiC_{\mu p}$/AMCs

from nano-level to micro-level (Fig. 11.3b, d) during the DDM process. However, the broken micro-SiC particles also made the estimation of the average diameters of the reinforcements harder, which enlarges the error range of the quantitative analysis [40–42]. Thus, in the present work, the quantitative analysis of the strength mechanism was not conducted further.

To further expound the balance strategy in the strength and ductility via optimizing the SiC particles, a schematic was proposed in Fig. 11.12. The grains in SiC_{np}/AMCs were ultra-fine and uniform in size. The intragranularly-dispersed SiC particles promoted the DRX process, and the stability of such ultra-fine-grained particles microstructure was increased under the effect of the Zener pinning mechanism. A typical multimodal grain microstructure of $SiC_{\mu p}$/AMCs was obtained with micro-sized grains and ultrafine grains as the micro-SiC particles promoted the localized grain refinement and were intensified by the broken micro-SiC particles during the DDM process. The great strength could be obtained in SiC_{np}/AMCs under the coupling effect of the grain boundaries strengthening and the Orowan strengthening, while the $SiC_{\mu p}$/AMCs showed a great ductility as the coarse grains could bear the stress-relax from the ultrafine grains and undergone the sufficient plastic deformation. According to the needs of practical applications, the specific microstructure of SiC/AMCs could be designed to balance the strength and ductility via optimizing the SiC particles: matrix strength is improved by selecting a certain amount of nano-SiC particles, while an excellent ductility could be obtained by adding the appropriate amount of micro-SiC particles.

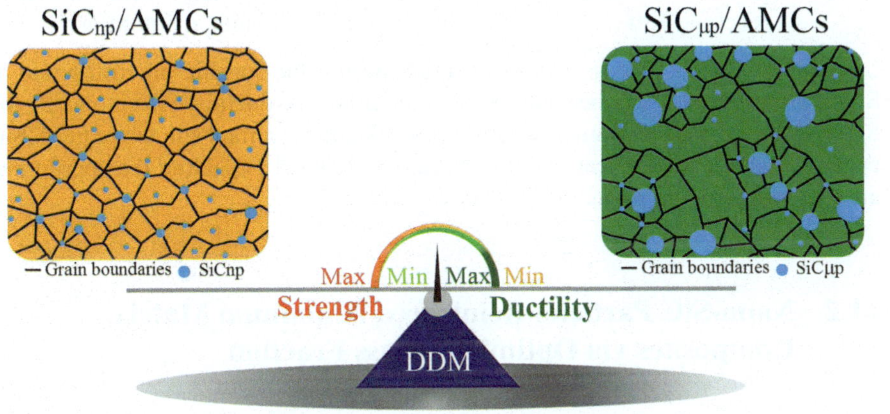

Fig. 11.12 Strategy in the strength-ductility balance via optimizing SiC particles

11.1.6 Summary

The strength-ductility balance strategy and theoretical basis were proposed for fabricating SiC-reinforced aluminum matrix composites via deformation-driven metallurgy. The dispersed characteristic of particles and strengthening-efficiency mechanisms in the SiC_{np} or $SiC_{\mu p}$ reinforced AMCs were evaluated:

(1) The grain refinement of SiC_{np}/AMCs was promoted with the intragranularly-dispersed nano-SiC particles, as the dislocation interaction was transformed into the mutual action between SiC particles and dislocations. The stability of the ultra-fine-grained microstructure was improved under the effect of the Zener pinning mechanism. The great strength could be obtained in SiC_{np}/AMCs under the coupling effect of the grain boundaries strengthening and the Orowan strengthening.

(2) A typical multimodal grain microstructure of $SiC_{\mu p}$/AMCs was obtained with micro-sized grains and ultrafine grains as the localized grain refinement was promoted by the particle stimulated nucleation mechanism and was intensified by the broken micro-SiC particles. The $SiC_{\mu p}$/AMCs showed great ductility as the coarse grains could bear the stress-relax from the ultrafine grains and undergone sufficient plastic deformation.

(3) The ultimate tensile strength of the SiC_{np}/AMCs was enhanced homogeneously, reaching 435 MPa, and kept the ductility at 9% under the combined effect of the Orowan strengthening and the grain boundaries strengthening. With a typical multimodal grain microstructure, the ductility of the $SiC_{\mu p}$/AMCs was kept well, reaching 18%, and kept the ultimate tensile strength at 200 MPa.

According to the needs of practical applications, the specific microstructure of SiC/AMCs could be designed to balance the strength and ductility via optimizing the SiC particles: matrix strength is improved by selecting a certain amount of nano-SiC particles, while an excellent ductility could be obtained by adding the appropriate amount of micro-SiC particles.

11.2 Nano-SiC Particles Reinforced Aluminum Matrix Composites via Optimized Mass Fraction

11.2.1 Introduction

The distribution of SiC particles in the Al matrix has a significant effect on the properties of the composites, as the agglomeration cannot transfer shear and tensile stresses [43]. The agglomeration of the particles contributes to the unacceptable brittle nature of the fabricated composites. Additionally, the severity of agglomeration becomes more serious as the scale of the reinforcements decreases, which is the case when using nanoparticles as reinforcements. The best characteristics of composites can be achieved when the reinforcement is homogeneously distributed in the matrix, as confirmed by experimental and theoretical studies [44, 45]. According to the Orowan mechanism, the reinforcement efficiency of the SiC particles was based on the average inter-particle spacing, which is influenced by the mass fraction and the SiC particles size as shown in the following formula:

$$L = r\left[\left(\frac{2\pi}{3f}\right)^{0.5} - 2\left(\frac{2\pi}{3}\right)^{0.5}\right] \tag{11.2}$$

where f is the volume fraction of the particles and r is the mean radius of the particles, and L is the average inter-particle spacing.

With the pressure and rotation of the DDM tool, the particles were redispersed during the fabrication process, as discussed in Table 11.1. Although the agglomeration of nanoparticles was further alleviated, the slight SiC clusters still existed. The sizes of some clusters were 200 nm approximately, which was about 20 times than the raw nano powders. Thus, in this section, the mass fraction of the nano-SiC particles in the composites was adjusted to modify the distribution of the particles. The microstructure evolution of the composites and the DDM mechanism was further explored. The strengthening mechanism was quantitatively analyzed.

11.2.2 Experimental Procedures

Spherical 1060 pure aluminum powders (5 μm in average diameter, purity ≥ 99.6%, supplied by Southwest Aluminum Co. Ltd.) and SiC_{np} (10 nm in average diameter, purity ≥ 99.0%, supplied by Haoxi Research Nanomaterials Inc.) were chosen as raw materials due to its simplicity and basicity for the investigation of AMCs, as shown in Fig. 11.13. 5 wt.% SiC_{np} with balanced aluminum powders were blended by a planetary ball milling machine (QM-3SP2, Nanjing University Instrument Plant) without a process control agent. The rotational velocity, milling time, and ball-to-powder ratio were respectively 250 r/min, 5 h, and 10:1. The mean diameter of the ZrO_2 milling balls was 6 mm. Process control agent (0.4 wt.% stearic acids) was used during ball milling to avoid the coating of milling balls and change the milling kinetics. After ball milling, the blended powders were firstly placed in an aluminum mold and cold pre-compacted at 25 MPa to obtain the preform with the shape of 16 mm in diameter and 2 mm in height. Then, the DDM process was applied to the preform with a certain rotational velocity and processing time.

Figure 11.14 schematically shows the DDM process used in this work. To further improve the density of the composites, additional compaction was added during the DDM process under the compaction ratio of 2:1. When the rotation was applied, the disc with an AA6061-T6 mold was tightly clamped to the bottom die. The rotational velocity was selected as the variation for 400, 600, and 800 r/min. The processing

Fig. 11.13 Raw materials: **a** aluminum powders, **b** and **c** SiC nanoparticles with different resolutions

time of 30 s was constant. Through the above process, the features of ultrafine-grained microstructures and uniformly dispersed SiC_{np} can be obtained, which was cut from the mold via wire electrical discharge machining. Besides, DDMed sample made of pure aluminum at the rotational velocity of 400 r/min was also conducted for contrast.

Scanning electron microscope (SEM, Zeiss MERLIN Compact) at a working voltage of 20 kV was used to observe the morphology of the raw powders, the microstructures, and the fractography of the bulk composites with different rotational velocities. Electron backscattered diffraction (EBSD) was performed with an

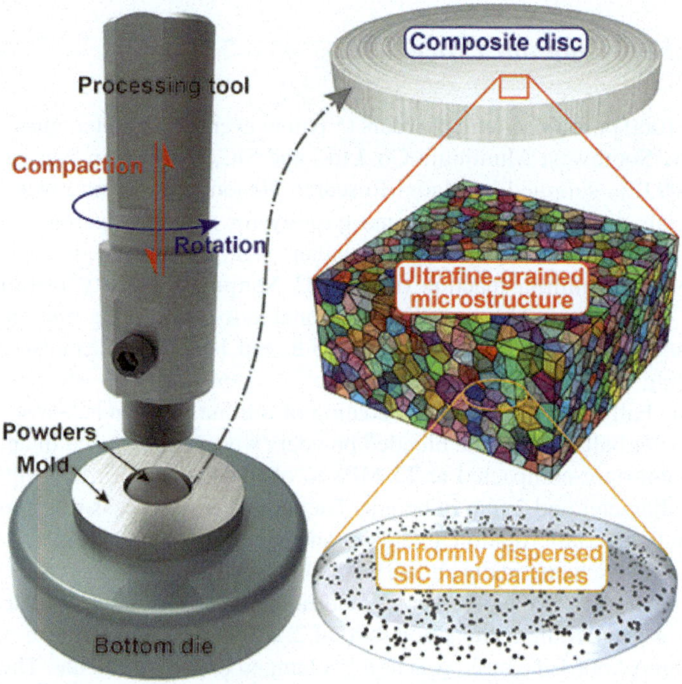

Fig. 11.14 Schematic illustration of DDM method

HKL EBSD detector. The specimen for SEM microstructure and EBSD was polished by a JEOL IB-09020CP cross-section polisher. The step size of EBSD was 0.1 μm. The specimen for TEM was ion-milled by a GATAN695 Precision ion polishing system. A high-resolution transmission electron microscope (HR-TEM, FEI Talos F200X) at a working voltage of 200 kV was applied to characterize the morphologies of grains, reinforcements, and phase interfaces of the composites. The phase composition was analyzed by X-ray diffractometer (XRD, PANalytical Empyrean). The relative porosity was determined from Archimedes method at room temperature.

To evaluate the bending and tensile properties of the Al-SiC_{np} composites at room temperature, an Instron 5569 tester at a work mode of 1 kN was applied. The bending specimens were measured at a constant crosshead speed of 0.5 mm/min, while the tensile specimens were clamped at both ends and stretched along the length direction at a constant crosshead speed of 1 mm/min. The standard and geometry of the tensile samples were given in our previous research [46], while the dimensions of bending tests were 14 mm × 5 mm × 1 mm. The nano indentation was measured using a nano indenter (Agilent G200). Specimens were mounted on an aluminum stub and indentation was carried out with an applied force of 200 mN. The elastic modulus and hardness measurements were done with respect to indentation curves.

11.2.3 Microstructures

Figure 11.15 shows the microstructures of the fabricated composites at different rotational velocities. The three specimens are generally well-formed, but some pores still can be seen. The porosity becomes lower as the rotational velocity increases. The porosities at the rotational velocity of 400, 500, and 600 r/min are 0.32 ± 0.09, 0.22 ± 0.10, and 0.17 ± 0.04%, which are significantly lower than the porosity obtained in other literatures [47–49]. During the DDM process, the revolution numbers per unit time of the processing tools are proportional to the rotational velocity, which contribute to the increase of cumulative strain and heat input generated in the composites. This makes the milled powders flow easier and enhances the wettability between the SiC_{np} and aluminum matrix [50]. These powders are more likely to produce relative motion under the synergistic effect of plastic strain and thermal diffusion, thereby promoting the rapid formation of sintering necks [51]. Subsequently, the oxide films on the surface of the aluminum powders are broken and the bare aluminum surfaces are exposed under the external shear strain. The aluminum powders thus encounter with the SiC_{np}. Diffusion and metallurgical bonding form at the interface of these two powders due to the heat input and defect energies of DDM process, resulting in the dense sintering of the composites. The rotational velocity of 600 r/min is 50% higher than that of 400 r/min, which leads to about 50% improvement of the cumulative strain and corresponding increase of the heat input. This leads to a more intense plastic flow, which further reduces the porosity of the composites. In addition, the agglomeration of SiC_{np} has a detrimental effect on the formation of pores. Due to the large difference in the coefficient of thermal expansion between SiC_{np} and

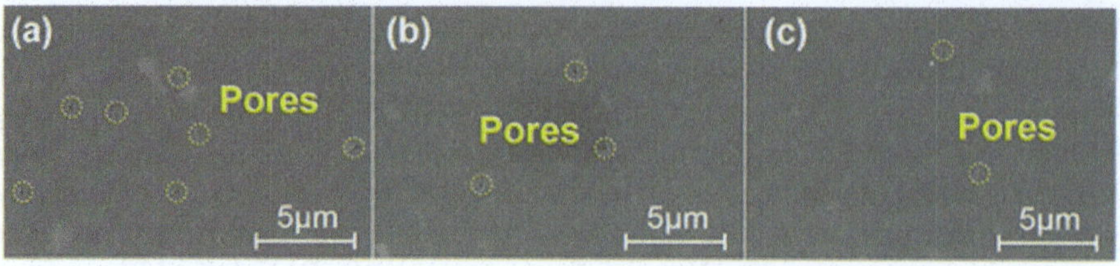

Fig. 11.15 Metallography microstructures of Al–SiC nanocomposites at different rotational velocities: **a** 400 r/min, **b** 500 r/min, and **c** 600 r/min

aluminum powders, the agglomerated SiC_{np} leads to the nucleation of pores [47]. Due to the mechanical agitation caused by severe plastic deformation of DDM process, a blending effect like the ball milling process is applied to the composites, thereby avoiding the agglomeration of the SiC_{np}, which provides favorable conditions for the decrease of the porosities.

The porosities, grain sizes of matrices, and distributions of reinforcements together affect the mechanical performances of the composites. Although the relatively high rotational velocity reduces the porosity, one should note that the lower heat input at lower rotational velocity contributes to the significant refinement of the microstructures. Figure 11.16 shows the grain morphologies at different rotational velocities. The grain sizes increase with the increasing rotational velocity. Since the DDM process brings about severe plastic deformation and low heat input, the composites undergo dynamic recovery and recrystallization [46]. The grain boundary migration of newly formed recrystallized grains is mainly affected by the following two parts: ① Thermal activation. According to the migration rate of grain boundaries V [52]:

$$V = M\exp\left(\frac{Q}{R_g T}\right)(P_{dis} + P_c - P_z) \tag{11.3}$$

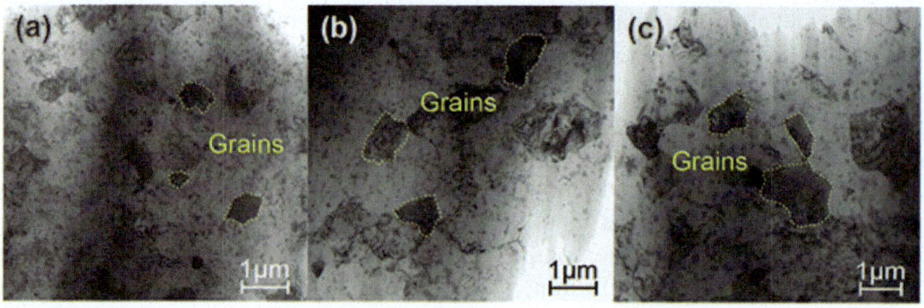

Fig. 11.16 Sizes and morphologies of grains at different rotational velocities: **a** 400 r/min, **b** 500 r/min, and **c** 600 r/min

where M is a constant, Q is the activation energy, R_g is the gas constant, T is the temperature, P_{dis} is the driven force due to the dislocations and point defects, P_c is the geometrical driving force, and P_z is the Zener pinning force. Since the quasi-steady temperature increases with the increasing heat input, the migration rate of grain boundaries is significantly enhanced, resulting in the coarsening of the grains; ② Pinning effect of SiC_{np}. Large amounts of dispersed nanoparticles significantly improve the Zener pinning force, which makes the pinning force more easily balanced with the dislocation driving force and the curvature force. In this investigation, due to the effect of large plastic strain, the distribution of SiC_{np} can be considered to be relatively uniform [53]. As such, the grain size is directly related to the heat input. Higher rotational velocity results in high heat generation, which finally coarsens the microstructures.

Figure 11.17 shows the grain morphology at the minimum rotational velocity of 400 r/min. One can see that the entire composite is composed of fine recrystallized microstructures with slight orientation. Besides, a gradient of grain size can be seen in the composites, as shown in Fig. 11.17c. This is due to the non-uniform plastic deformation caused by local turbulence including local shear strain vortices and double-swirl patterns [54]. Although this turbulent or non-linear flow increases the heterogeneity of the composites to a certain extent, it also improves the mass transfer and blending on the macroscopic scale. This fully explains the uniform distribution of SiC_{np} on the spatial scale of the aluminum matrix.

Figure 11.18 is the detailed statistics of the EBSD map in Fig. 11.17a. The average grain size and aspect ratio of the composites are 1.19 μm and 1.57, respectively. This is consistent with the results obtained above, indicating that the DDM method can effectively achieve ultrafine-grained structures. Additionally, partial grains with a relatively high aspect ratio prove that the final state of the dynamic recrystallization is insufficient during the process. The insufficient dynamic recrystallization can be

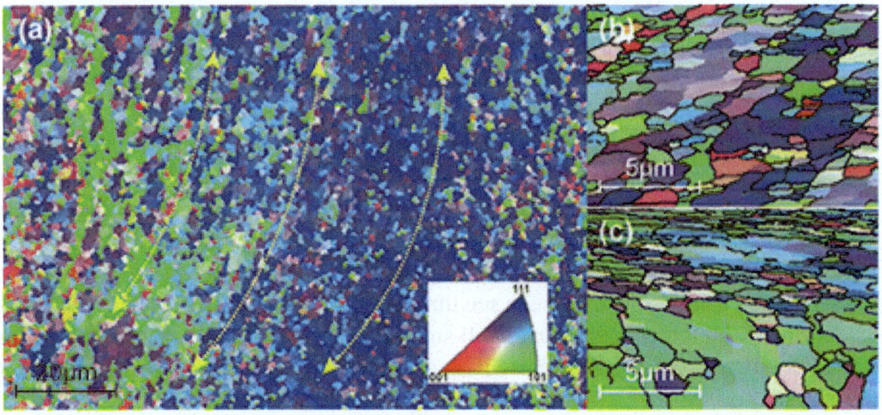

Fig. 11.17 Typical ultrafine-grained microstructures at the rotational velocity of 400 r/min: **a** Overall morphology, **b** equiaxed grains, and **c** gradient distribution of grain sizes

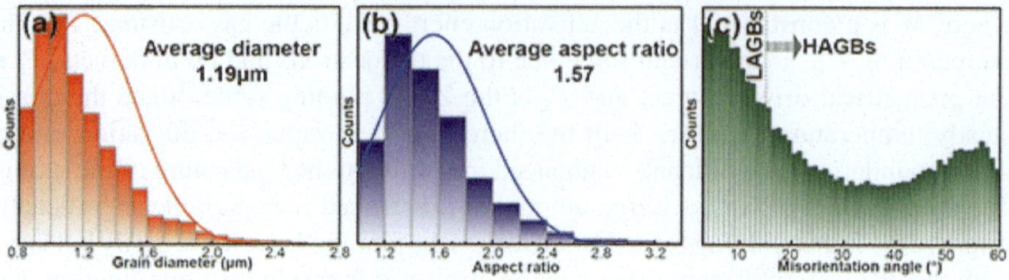

Fig. 11.18 Statistics distribution extracted from EBSD analysis: **a** Grain diameter, **b** aspect ratio, and **c** misorientation angle

attributed to the extremely low heat input, implying that the processing temperature only slightly exceeds the lower temperature limit required for dynamic recrystallization of the aluminum matrix. To explain this phenomenon, the self-heating temperature conditions during the DDM process can be decoupled into the following two parts: (a) the frictional heat generation at the interface of processing tools and the powders; (b) the deformation heat generation due to the extrusion and shear deformation of the powders. According to the Sellars-Tegart constitutive equation [55] or Johnson–Cook constitutive equation [56], the flow plastic stress drops with the increasing local temperature. High flow stress at relatively low temperature leads to the sharp increase of the temperature, that is, the self-heating condition for sintering the composites is obtained in just several seconds. Subsequently, the self-heating behavior dependent on the flow stress strikes a dynamic balance between heat generation and dissipation, which maintains an extremely low temperature for dynamic recrystallization. This temperature condition also demonstrates that the migration of grain boundaries is inhibited on the other hand, thereby refining the microstructure of the composites. Furthermore, the composites have a high proportion of large-angle grain boundaries, as shown in Fig. 11.18c.

Figure 11.19 shows the results of XRD analysis at different rotational velocities. Only the presence of aluminum matrix and SiC_{np} can be seen in all results, and no formation of other intermetallic compounds is observed. The microscopic energy dispersive spectroscope (EDS) results under TEM observation are shown in Fig. 11.20. There are two kinds of second phase particles inside the composites: SiC_{np} and iron nanoparticles. The SiC_{np} is derived from the artificial addition in the experimental section, while iron nanoparticles are derived from the detachment of processing tool during the DDM process. These iron nanoparticles also provide some assistance for the mechanical strengthening of the composites. Besides, as shown in Fig. 11.21a and b, one can see that partial surfaces of SiC_{np} are coated with oxide films, which forms the core–shell structure of SiC_{np}-oxide. This phenomenon avoids the direct contacts between SiC_{np} and aluminum matrix, and suppresses the formation of brittle Al_4C_3 intermetallic compounds. As mentioned above, this brittle phase severely impairs the plasticity and corrosion resistance of the composites. In other words, the formation of these core–shell structures effectively circumvents the decrease of elongation.

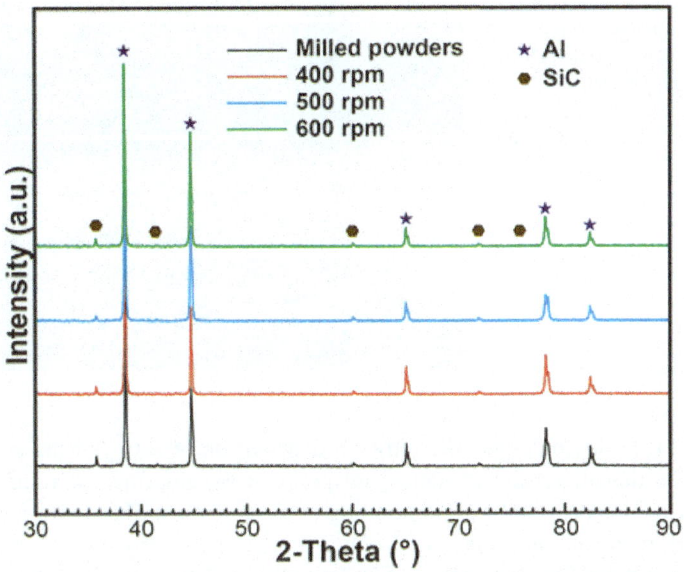

Fig. 11.19 XRD patterns of milled powders and composites at different rotational velocities

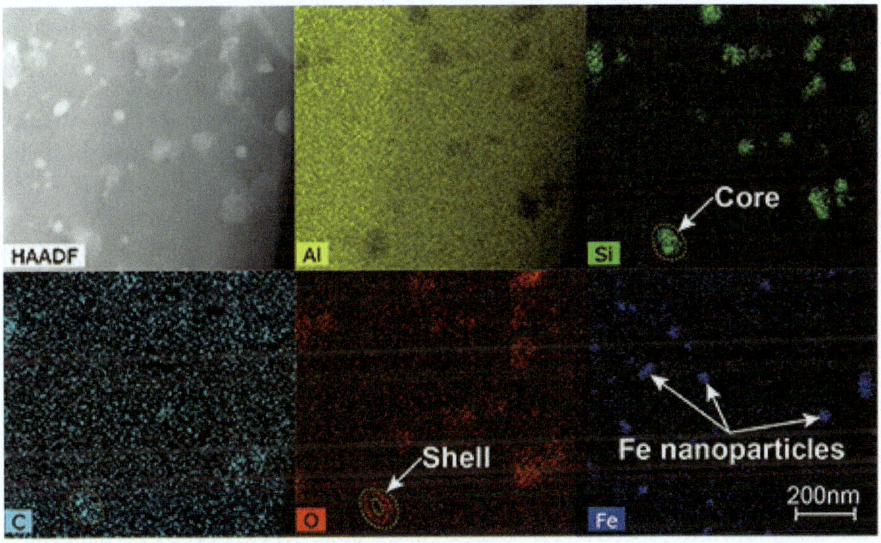

Fig. 11.20 Elemental mappings for typical composite at the rotational velocity of 400 r/min

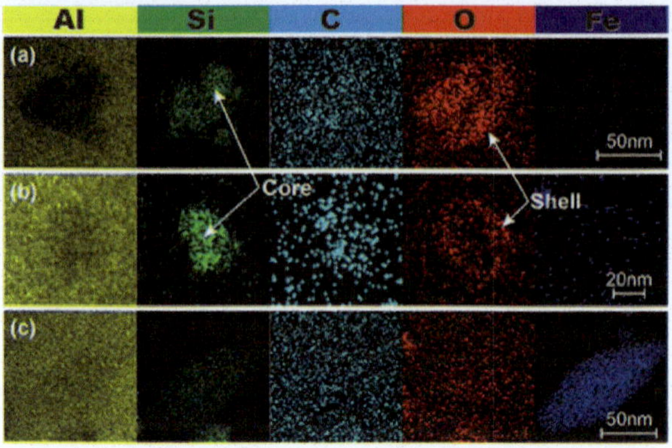

Fig. 11.21 Three elemental mappings for nanoparticles: **a** and **b** SiC nanoparticles, and **c** Fe nanoparticles

Figure 11.22 demonstrates the uniform distribution of SiC_{np} in the composites. One can see that there are no nanosized pores at the interface of reinforcements and matrix, which proves the good bonding of these phases. The uniform dispersion of the reinforcements and their good bonding with the matrix ensures the Orowan strengthening behavior of the SiC_{np} [57]. In addition, there are two modes of typical contact between SiC_{np} and matrix. The indirect contact mode is shown in Fig. 11.23a. The amorphous phase, also called oxide above, is physical separation to avoid the direct contact between the two phases, thereby inhibiting the formation of harmful intermetallic compounds. This local amorphous oxide layer was formed due to the severe plastic deformation [58–61], which is direct evidence of DDM on the composites. Figure 11.23b shows the direct contact mode. The SiC_{np} and aluminum matrix are bonded to each other by a semi-coherent interface, which realizes the better reinforcing effect of the composites. As shown in Fig. 11.22a, the rod-like Al_4C_3 phase is not observed in both two contact modes, implying that the DDM process can effectively inhibit the damage of the composites by this brittle intermetallic compound.

11.2.4 Mechanical Properties

The representative elastic modulus and hardness results with respect to nano indentation depth are shown in Fig. 11.24. The elastic modulus and hardness of the reference pure aluminum are respectively 69.3 ± 0.4 and 0.65 ± 0.11 GPa [62]. As expected, a significant increase in the hardness and modulus values was obtained. The best results are achieved at the rotational velocity of 400 r/min, which is 102 and 151% higher than the reference in terms of elastic modulus and hardness. This can be attributed to the ultrafine-grained microstructures and uniformly dispersed nanoparticles. One should note that this enhancement of the composites is achieved with 5 wt.% addition of SiC_{np}, indicating that the ultrafine grains play a key role in the composites.

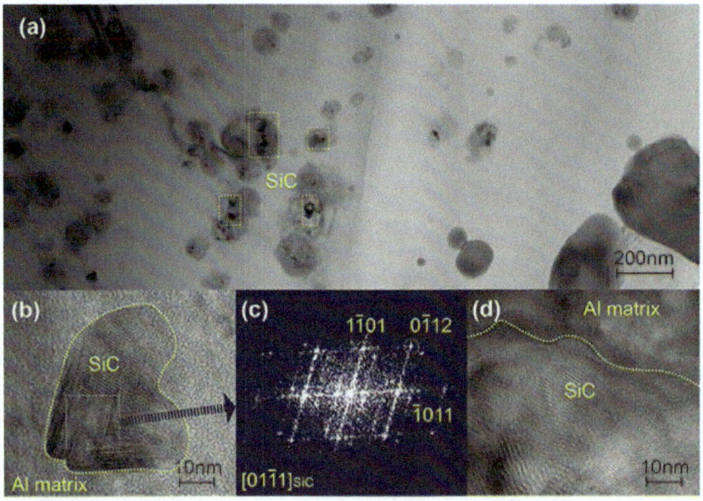

Fig. 11.22 Uniformly distribution of SiC nanoparticles in DDMed composites: **a** Uniformly distribution, **b** single SiC nanoparticles, **c** corresponding SAED (Selected area electron diffractio pattern), and **d** interface between aluminum matrix and nanoparticles

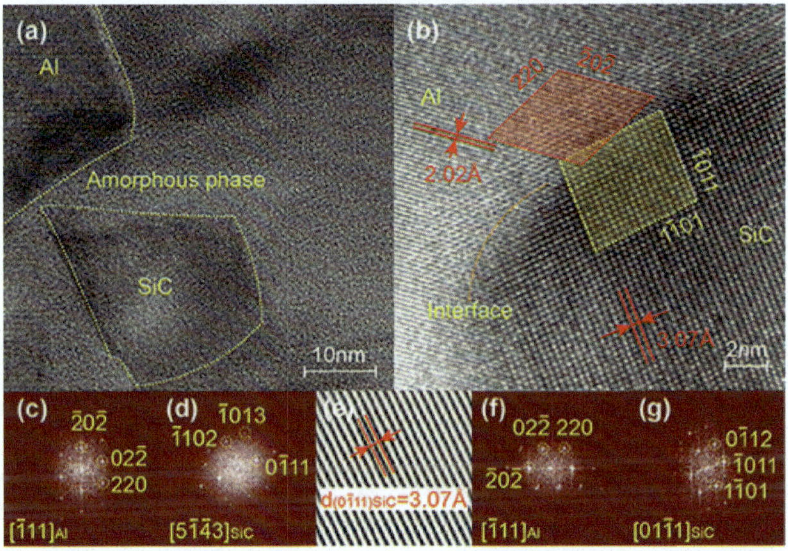

Fig. 11.23 Interfaces between aluminum matrix and SiC nanoparticles: **a** Indirect interface with amorphous aluminum intermediate layer, **b** direct interface. **c** Al SAED pattern, **d** SiC SAED pattern and **e** SiC interplanar distance corresponding to Fig. 11.23a. **f** Al SAED pattern and **g** SiC SAED pattern corresponding to Fig. 11.23b

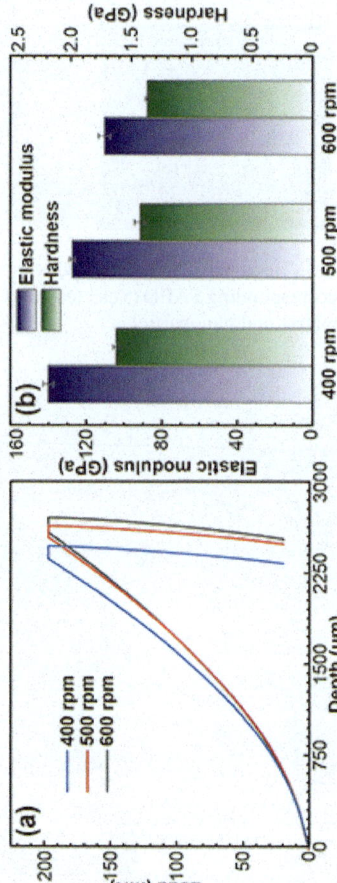

Fig. 11.24 Nanoindentation results of the composites: **a** load-depth curves, and **b** variation of elastic modulus and hardness

Table 11.2 Mechanical performance of the composites

Sample	Rotational velocity/ ($r{\cdot}min^{-1}$)	Elastic modulus/GPa	Hardness/ GPa	Flexural strength/MPa	Ultimate tensile strength/ MPa	Elongation/ %
1	400	140.1 ± 2.9	1.63 ± 0.04	288.1 ± 12.1	361	14.2
2	500	127.4 ± 1.7	1.43 ± 0.06	297.5 ± 22.6	329	14.8
3	600	110.4 ± 3.4	1.37 ± 0.03	255.3 ± 13.7	280	16.1

Table 11.2 shows the mechanical performance of the composites at different rotational velocities. The flexural strength of the composites at the rotational velocities of 400, 500, and 600 r/min is respectively 288.1 ± 12.1, 297.5 ± 22.6, and 255.3 ± 13.7 MPa. The best results occur at the rotational velocity of 500 r/min instead of 400 r/min, because the flexural properties of the composites are more sensitive to the porosity. As mentioned above, the porosity at low rotational velocity is relatively high, which offsets the effect of ultrafine grains induced by extremely low heat input to a certain extent. Therefore, the best bending performance does not occur at the lowest rotational velocity. The ultimate tensile strength is consistent with the nano indentation results and is substantially proportional to the hardness. When the rotational velocity is 400 r/min, the ultimate tensile strength and elongation reach 361 MPa and 14.2%, respectively, which is the highest strength that can be obtained with 5 wt.% addition of SiC_{np} to our best knowledge. In contrast, the ultimate tensile strength of pure aluminum without silicon carbide prepared by DDM at the rotational velocity of 400 r/min is 147 MPa and 23.7%.

Figure 11.25 is the typical tensile fracture morphology of the DDMed composites. The entire fracture shows a ductile feature, and extremely small dimples can be observed, which prove that the composites has good plasticity and toughness. In addition, the SiC_{np} is uniformly distributed inside the aluminum matrix, but not only locates at the nucleation site of the dimples, indicating that the plasticity of the composites is not substantially adversely affected by the added nanoparticles.

11.2.5 Principle of DDM

Since the DDM method was firstly proposed to fabricate nanocomposites directly from powders, there is no generally accepted theory yet. However, due to the nature of severe plastic deformation, the principle of DDM can refer to in-situ FSP or HPT (high pressure torsion). Based on the assumption of simple shear plastic deformation with localized strain proportional to the distance from the normal axis [63], the DDMed samples receive cumulative strain associated with rotational velocity and dwelling time. Furthermore, several investigations show deviations from the laminar flow model. In the friction stir welding or FSP of aluminum matrix, there is evidence for the formation of significant local turbulence [64]. There is also recent simulation

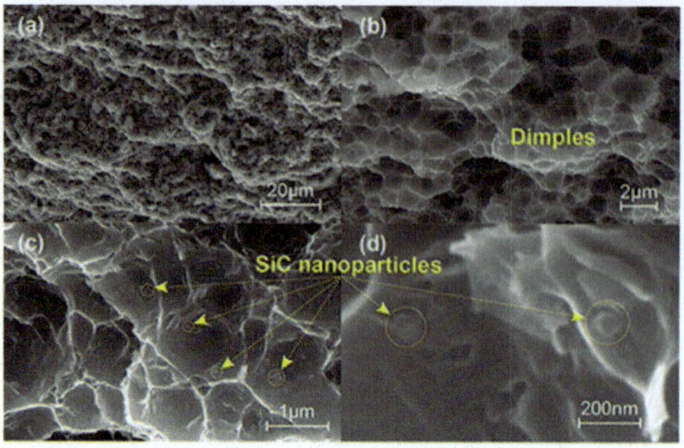

Fig. 11.25 Fractography surfaces of typical composites with different resolution: **a** 2000 ×, **b** 6000 ×, **c** 20,000 ×, and **d** 100,000 ×

demonstrating that turbulent or non-laminar flow contributes to intense mass transfer and mixing of the deformed materials [65]. This explains the improved uniformity of SiC_{np} distribution in the pure aluminum matrix after heavy shear deformation of DDM. Based on these experimental observations and simulations, it is reasonable to anticipate that the plastic deformation applied by DDM can fragment the agglomerated SiC_{np} in the aluminum matrix and redistribute the nanoparticles of iron and SiC homogenously through the advent of turbulent flow. Figure 11.22 confirms the uniform distribution of nanoparticles in the composites, implying that DDM may provide an effective manufacturing route for obtaining composites with ultrafine grains and well dispersed nanoparticles through severe plastic deformation.

11.2.6 Strengthening Behaviors

As expected [66], the large refinement in grain size comes together with the increase of strength and hardness: from 119 MPa for the ultimate tensile strength of reference matrix to 361 MPa for that of the sample at the rotational velocity of 400 r/min. Nevertheless, this evolution of mechanical performance is likely not only induced by grain refinement, since other strengthening obstacles to moving dislocations generate with DDM process: dislocation density, number density of reinforcements, and porosity [67].

Reinforcements contribute to one of the most important strengthening of composites because the redistribution leads to a progressive increase of the actual number density. Assuming that SiC_{np} does not undergo mass loss or form detrimental intermetallic compounds throughout the DDM process, the total remaining SiC_{np} leads to the volume fraction of reinforcement 4.3%. Besides, the actually observed volume fractions of iron nanoparticles with their intermetallic compounds at the rotational velocity of 400, 500, and 600 r/min from TEM observation are 1.7, 1.3, and 0.7%, respectively. The average diameters of the SiC_{np} and iron nanoparticles are respectively 10 and 12 nm. The corresponding Orowan strengthening contribution can be estimated by this equation [40]:

$$\Delta\sigma_{Orowan} = \sum M \frac{0.4Gb}{\pi\sqrt{1-\nu}} \frac{\ln\left(\frac{d}{b}\right)}{\lambda} \tag{11.4}$$

where $M = 3.06$ is the Taylor factor [68], $G = 25.4$ GPa is the shear modulus of aluminum matrix, $b = 0.286$ nm is the magnitude of the Burgers vector of aluminum matrix, $\nu = 0.33$ is the Poisson's ratio of aluminum matrix, d is the average diameters of the reinforcements, and λ is the average inter-reinforcement spacing which can be calculated from the volume fraction.

The contribution of grain boundaries can be estimated through the Hall–Petch relationship [41]:

$$\Delta\sigma_{GB} = \left(\sigma_0 + \frac{k}{\sqrt{d}}\right) - \left(\sigma_0 + \frac{k}{\sqrt{d_0}}\right) = \frac{k}{\sqrt{d}} - \frac{k}{\sqrt{d_0}} \approx \frac{k}{\sqrt{d}} \tag{11.5}$$

where σ_0 is the friction stress, $k = 0.07$–0.09 MPa m$^{0.5}$ is a constant [69], $d_0 \approx 300$–400 μm is the reference coarse grain size of pure aluminum, and d is the average grain size of the composites estimated from EBSD mapping. Since the value of $\frac{k}{\sqrt{d_0}}$ is much smaller than that of $\frac{k}{\sqrt{d}}$, it can be ignored to simplify the calculation.

The contribution of the dislocation density can be estimated via Bailey-Hirsh law [42]:

$$\Delta\sigma_{dis} = \alpha MGb\rho^{\frac{1}{2}} \tag{11.6}$$

where $\alpha = 0.2$ is the strength factor for dislocation strengthening of aluminum [70], and ρ is the dislocation density which is extracted by Williamson-Hall method [71].

Table 11.3 shows the components of each strengthening mechanism at different rotational velocities and their comparison with the experimental results. Grain boundary, volume fraction (mainly from the detachment of the steel processing tools), and dislocation density increase with the decreasing of heat input, which is consistent with the experimental results. This proves that the severe plastic deformation introduced by DDM can improve the mechanical properties of the composites from these triple aspects. In addition, one should note that the actual change of the ultimate tensile strength at the rotational velocities of 400 and 500 r/min is close to the estimated upper limit, while that at 600 r/min is close to the lower limit. It is presumed that this is due to the composition of the grain boundaries or the twin structures [72]. However, since there is no clear evidence to illustrate it, this hypothesis remains to be confirmed in further investigation.

Table 11.3 Calculation of the strengthening mechanism

	400 r/min	500 r/min	600 r/min
λ	121.3	126.2	138.7
$\Delta\sigma_{Orowan}$	104	100	91
$d_{grainsize}$	1.19 μm	1.87 μm	2.78 μm
$\Delta\sigma_{GB}$	64–83	51–66	42–54
ρ	1.53–20.1 × 10^{14}	0.85–12.1 × 10^{14}	0.43–0.73 × 10^{14}
$\Delta\sigma_{dis}$	55–63	41–49	29–38
$\Delta\sigma_{total}$	223–250	192–215	162–183
$\Delta\sigma_{measured}$	242	210	161

11.2.7 Summary

(1) SiC_{np}-Al nanocomposites with 5 wt.% SiC addition were successfully fabricated via DDM at different rotational velocities. Significant microstructural refinement was achieved in the composites with the average grain size of 1.19 μm.

(2) Agglomerated SiC_{np} was fragmented and redistributed into the aluminum matrix during the DDM process, and tended to become more uniformly dispersed under the effect of turbulent flow. Detached iron nanoparticles from the processing tool were observed and further improved the Orowan strengthening effect of the composites.

(3) Two modes of interfaces between SiC_{np} and aluminum matrix were found. The indirect interface provides effective physical isolation and avoids the formation of detrimental intermetallic compounds. The direct interface enhanced the strengthening effect of the nanoparticles.

(4) The DDMed composites have improved modulus, hardness by 102 and 151%. The synergistic effects of Orowan strengthening, grain boundary strengthening, and dislocation strengthening effectively improved the bending strength and ultimate tensile strength of the composites.

References

1. Rodrigo P, Poza P, Utrilla V, et al. Effect of reinforcement geometry on precipitation kinetics of powder metallurgy AA2009/SiC composites. J Alloys Compd. 2009;479(1–2):451–6.
2. Safi M, Hassanzadeh-Aghdam MK, Mahmoodi MJ. Effects of nano-sized ceramic particles on the coefficients of thermal expansion of short SiC fiber-aluminum hybrid composites. J Alloys Compd. 2019;803:554–64.
3. Bembalge OB, Panigrahi SK. Development and strengthening mechanisms of bulk ultrafine grained AA6063/SiC composite sheets with varying reinforcement size ranging from nano to micro domain. J Alloys Compd. 2018;766:355–72.

4. Knowles AJ, Jiang X, Galano M, et al. Microstructure and mechanical properties of 6061 Al alloy based composites with SiC nanoparticles. J Alloys Compd. 2015;615(S1):S401–5.
5. Chen J, Bao C, Ma Y, et al. Distribution control of AlN particles in Mg–Al/AlN composites. J Alloys Compd. 2017;695:162–70.
6. Inoue R, Nakano J, Nakamura T, et al. Mechanical and thermoelectric properties of intragranular SiC-nanoparticle/Mg_2Si composites. J Alloys Compd. 2019;775:657–66.
7. Xie Y, Meng X, Mao D, et al. Homogeneously dispersed graphene nanoplatelets as long-term corrosion inhibitors for aluminum matrix composites. ACS Appl Mater Interfaces. 2021;13(27):32161–74.
8. Guo H, Zhang Z, Zhang Y, et al. Improving the mechanical properties of B4C/Al composites by solid-state interfacial reaction. J Alloys Compd. 2020;829:154521.
9. Xiong B, Liu K, Yan Q, et al. Microstructure and mechanical properties of graphene nanoplatelets reinforced Al matrix composites fabricated by spark plasma sintering. J Alloys Compd. 2020;837:155495.
10. Xie Y, Meng X, Wang F, et al. Insight on corrosion behavior of friction stir welded AA2219/AA2195 joints in astronautical engineering. Corros Sci. 2021;192:109800.
11. Dong Z, Ma Z, Yu L, et al. Achieving high strength and ductility in ODS-W alloy by employing oxide@W core-shell nanopowder as precursor. Nat Commun. 2021;12(1):297.
12. Xie Y, Meng X, Li Y, et al. Insight into ultra-refined grains of aluminum matrix composites via deformation-driven metallurgy. Compos Commun. 2021;26:100776.
13. Zhang Z, Chen D. Consideration of Orowan strengthening effect in particulate-reinforced metal matrix nanocomposites: a model for predicting their yield strength. Scr Mater. 2006;54(7):1321–6.
14. Li J, Li Y, Wang F, et al. Friction stir processing of high-entropy alloy reinforced aluminum matrix composites for mechanical properties enhancement. Mater Sci Eng A. 2020;792:139755.
15. Tan M, Zhang X. Powder metal matrix composites: selection and processing. Mater Sci Eng A. 1998;244(1):80–5.
16. Etemadi R, Wang B, Pillai KM, et al. Pressure infiltration processes to synthesize metal matrix composites: a review of metal matrix composites, the technology and process simulation. Mater Manuf Process. 2018;33(12):1261–90.
17. Naher S, Brabazon D, Looney L. Computational and experimental analysis of particulate distribution during Al-SiC MMC fabrication. Compos A Appl Sci Manuf. 2007;38(3):719–29.
18. Hamilton RW, Zhu Z, Dashwood RJ, et al. Direct semi-solid forming of a powder SiC–A1 PMMC: flow analysis. Compos A Appl Sci Manuf. 2003;34(4):333–9.
19. Ozden S, Ekici R, Nair F. Investigation of impact behaviour of aluminium based SiC particle reinforced metal-matrix composites. Compos A Appl Sci Manuf. 2007;38(2):484–94.
20. Tzamtzis S, Barekar NS, Hari Babu N, et al. Processing of advanced Al/SiC particulate metal matrix composites under intensive shearing: a novel Rheo-process. Compos A Appl Sci Manuf. 2009;40(2):144–51.
21. Barmouz M, Givi MKB. Fabrication of in situ Cu/SiC composites using multi-pass friction stir processing: evaluation of microstructural, porosity, mechanical and electrical behavior. Compos A Appl Sci Manuf. 2011;42(10):1445–53.
22. Luo Z. Crystallography of SiC/MgAlO/Al interfaces in a pre-oxidized SiC reinforced SiC/Al composite. Acta Mater. 2006;54(1):47–58.
23. Fadavi Boostani A, Tahamtan S, Jiang ZY, et al. Enhanced tensile properties of aluminium matrix composites reinforced with graphene encapsulated SiC nanoparticles. Compos A Appl Sci Manuf. 2015;68:155–63.
24. Bastwros M, Kim G-Y, Zhu C, et al. Effect of ball milling on graphene reinforced Al6061 composite fabricated by semi-solid sintering. Compos B Eng. 2014;60:111–8.
25. Sauvage X, Wilde G, Divinski SV, et al. Grain boundaries in ultrafine grained materials processed by severe plastic deformation and related phenomena. Mater Sci Eng A. 2012;540:1–12.
26. Meng X, Huang Y, Cao J, et al. Recent progress on control strategies for inherent issues in friction stir welding. Prog Mater Sci. 2021;115:100706.

27. Liao XZ, Kilmametov AR, Valiev RZ, et al. High-pressure torsion-induced grain growth in electrodeposited nanocrystalline Ni. Appl Phys Lett. 2006;88(2):021909.
28. Guo X, Guo Q, Nie J, et al. Particle size effect on the interfacial properties of SiC particle-reinforced Al–Cu–Mg composites. Mater Sci Eng A. 2018;711:643–9.
29. Li Z, Wang J, Huang H. Influences of grain/particle interfacial energies on second-phase particle pinning grain coarsening of polycrystalline. J Alloys Compd. 2020;818:152848.
30. Huo W, Hou L, Cui H, et al. Fine-grained AA 7075 processed by different thermo-mechanical processings. Mater Sci Eng A. 2014;618:244–53.
31. Humphreys FJ, Kalu PN. Dislocation-particle interactions during high temperature deformation of two-phase aluminium alloys. Acta Metall. 1987;35(12):2815–29.
32. Ramesh CS, Keshavamurthy R, Koppad PG, et al. Role of particle stimulated nucleation in recrystallization of hot extruded Al 6061/SiCp composites. Trans Nonferrous Met Soc China. 2013;23(1):53–8.
33. Gupta R, Chaudhari GP, Daniel BSS. Strengthening mechanisms in ultrasonically processed aluminium matrix composite with in-situ Al_3Ti by salt addition. Compos B Eng. 2018;140:27–34.
34. Bazarnik P, Huang Y, Lewandowska M, et al. Enhanced grain refinement and microhardness by hybrid processing using hydrostatic extrusion and high-pressure torsion. Mater Sci Eng A. 2018;712:513–20.
35. Hu J, Shi YN, Sauvage X, et al. Grain boundary stability governs hardening and softening in extremely fine nanograined metals. Science. 2017;355(6331):1292–6.
36. Saraf L. Kernel average misorientation confidence index correlation from FIB sliced Ni–Fe–Cr alloy surface. Microsc Microanal. 2011;17(S2):424–5.
37. Li H, Hsu E, Szpunar J, et al. Deformation mechanism and texture and microstructure evolution during high-speed rolling of AZ31B Mg sheets. J Mater Sci. 2008;43(22):7148–56.
38. Wu SH, Xue H, Yang C, et al. Hierarchical structure in Al–Cu alloys to promote strength/ductility synergy. Scr Mater. 2021;202:113996.
39. Hart EW. Theory of the tensile test. Acta Metall. 1967;15(2):351–5.
40. Wang X, Guan RG, Misra RDK, et al. The mechanistic contribution of nanosized Al_3Fe phase on the mechanical properties of Al–Fe alloy. Mater Sci Eng A. 2018;724:452–60.
41. Chen B, Kondoh K, Li JS, et al. Extraordinary reinforcing effect of carbon nanotubes in aluminium matrix composites assisted by in-situ alumina nanoparticles. Compos B Eng. 2020;183:107691.
42. Ma K, Wen H, Hu T, et al. Mechanical behavior and strengthening mechanisms in ultrafine grain precipitation-strengthened aluminum alloy. Acta Mater. 2014;62:141–55.
43. Slipenyuk A, Kuprin V, Milman Y, et al. Properties of P/M processed particle reinforced metal matrix composites specified by reinforcement concentration and matrix-to-reinforcement particle size ratio. Acta Mater. 2006;54(1):157–66.
44. Boselli J, Pitcher PD, Gregson PJ, et al. Numerical modelling of particle distribution effects on fatigue in Al–SiCp composites. Mater Sci Eng A. 2001;300(1–2):113–24.
45. Slipenyuk A, Kuprin V, Milman Y, et al. The effect of matrix to reinforcement particle size ratio (PSR) on the microstructure and mechanical properties of a P/M processed AlCuMn/SiCp MMC. Mater Sci Eng A. 2004;381(1–2):165–70.
46. Xie Y, Meng X, Huang Y, et al. Deformation-driven metallurgy of graphene nanoplatelets reinforced aluminum composite for the balance between strength and ductility. Compos B Eng. 2019;177:107413.
47. Deb S, Panigrahi SK, Weiss M. Development of bulk ultrafine grained Al–SiC nano composite sheets by a SPD based hybrid process: experimental and theoretical studies. Mater Sci Eng A. 2018;738:323–34.
48. Wang D, Zheng Z, Lv J, et al. Multimodal particle distribution in 3D–SiC/Al–Si–Mg interpenetrating composite fabricated by pressureless infiltration. Ceram Int. 2018;44(16):19851–8.
49. Nayak KC, Date PP. Hot deformation flow behavior of powder metallurgy based Al–SiC and Al–Al_2O_3 composite in a single step and two-step uni-axial compression. Mater Charact. 2019;151:563–81.

50. Jain VKS, Yazar KU, Muthukumaran S. Development and characterization of Al5083-CNTs/SiC composites via friction stir processing. J Alloys Compd. 2019;798:82–92.
51. Rahimian M, Parvin N, Ehsani N. The effect of production parameters on microstructure and wear resistance of powder metallurgy Al–Al_2O_3 composite. Mater Des. 2011;32(2):1031–8.
52. Huang Y, Xie Y, Meng X, et al. Atypical grain coarsening of friction stir welded AA6082-T6: characterization and modeling. Mater Sci Eng A. 2019;740–741:211–7.
53. Sharma A, Narsimhachary D, Sharma VM, et al. Surface modification of Al6061–SiC surface composite through impregnation of graphene, graphite and carbon nanotubes via FSP: a tribological study. Surf Coat Technol. 2019;368:175–91.
54. Cao Y, Wang YB, Figueiredo RB, et al. Three-dimensional shear-strain patterns induced by high-pressure torsion and their impact on hardness evolution. Acta Mater. 2011;59(10):3903–14.
55. Su H, Wu CS, Pittner A, et al. Thermal energy generation and distribution in friction stir welding of aluminum alloys. Energy. 2014;77:720–31.
56. Kuykendall K, Nelson T, Sorensen C. On the selection of constitutive laws used in modeling friction stir welding. Int J Mach Tools Manuf. 2013;74:74–85.
57. Huang Y, Li J, Wan L, et al. Strengthening and toughening mechanisms of CNTs/Mg–6Zn composites via friction stir processing. Mater Sci Eng A. 2018;732:205–11.
58. Roy D, Mitra R, Ojo OA, et al. Evaluation of mechanical properties of partially amorphous and nanocrystalline $Al_{50}Ti_{40}Si_{10}$ composites prepared by mechanical alloying and hot isostatic pressing. Mater Sci Eng A. 2012;555:21–7.
59. Roy D, Mitra R, Chudoba T, et al. Structure and mechanical properties of $Al_{65}Cu_{20}Ti_{15}$-based amorphous/nanocrystalline alloys prepared by high-pressure sintering. Mater Sci Eng A. 2008;497(1–2):93–100.
60. Roy D, Mitra R, Ojo OA, et al. Microstructural evolution and mechanical properties of nanointermetallic phase dispersed $Al_{65}Cu_{20}Ti_{15}$ amorphous matrix composite synthesized by mechanical alloying and hot isostatic pressing. Metall Mater Trans A. 2011;42(8):2498–508.
61. Roy D, Chakravarty D, Mitra R, et al. Effect of sintering on microstructure and mechanical properties of nano-TiO_2 dispersed $Al_{65}Cu_{20}Ti_{15}$ amorphous/nanocrystalline matrix composite. J Alloys Compd. 2008;460(1–2):320–5.
62. Wagih A, Fathy A. Experimental investigation and FE simulation of nano-indentation on Al–Al_2O_3 nanocomposites. Adv Powder Technol. 2016;27(2):403–10.
63. Edalati K, Toh S, Iwaoka H, et al. Microstructural characteristics of tungsten-base nanocomposites produced from micropowders by high-pressure torsion. Acta Mater. 2012;60(9):3885–93.
64. Huang Y, Xie Y, Meng X, et al. Joint formation mechanism of high depth-to-width ratio friction stir welding. J Mater Sci Technol. 2019;35(7):1261–9.
65. Li Z, Yue Y, Ji S, et al. Optimal design of thread geometry and its performance in friction stir spot welding. Mater Des. 2016;94:368–76.
66. Chen B, Shen J, Ye X, et al. Length effect of carbon nanotubes on the strengthening mechanisms in metal matrix composites. Acta Mater. 2017;140:317–25.
67. Duchaussoy A, Sauvage X, Edalati K, et al. Structure and mechanical behavior of ultrafine-grained aluminum-iron alloy stabilized by nanoscaled intermetallic particles. Acta Mater. 2019;167:89–102.
68. Vachhani SJ, Doherty RD, Kalidindi SR. Studies of grain boundary regions in deformed polycrystalline aluminum using spherical nanoindentation. Int J Plast. 2016;81:87–101.
69. Shanmugasundaram T, Heilmaier M, Murty BS, et al. On the Hall-Petch relationship in a nanostructured Al–Cu alloy. Mater Sci Eng A. 2010;527(29–30):7821–5.
70. Huskins EL, Cao B, Ramesh KT. Strengthening mechanisms in an Al–Mg alloy. Mater Sci Eng A. 2010;527(6):1292–8.
71. Koizumi T, Kuroda M. Grain size effects in aluminum processed by severe plastic deformation. Mater Sci Eng A. 2018;710:300–8.
72. Lu K. Stabilizing nanostructures in metals using grain and twin boundary architectures. Nat Rev Mater. 2016;1(5):16019.

GPSR Compliance

The European Union's (EU) General Product Safety Regulation (GPSR) is a set of rules that requires consumer products to be safe and our obligations to ensure this.

If you have any concerns about our products, you can contact us on ProductSafety@springernature.com

In case Publisher is established outside the EU, the EU authorized representative is:

Springer Nature Customer Service Center GmbH
Europaplatz 3
69115 Heidelberg, Germany

The manufacturer's authorised representative in the EU is Springer Nature Customer Service Centre GmbH, Europaplatz 3, 69115 Heidelberg, Germany. If you have any concerns regarding our products, please contact ProductSafety@springernature.com

Printed and bound by CPI Group (UK) Ltd, Croydon, CR0 4YY
01/10/2026
02225866-0017